Lecture Notes in Operations Research

Lecture Notes in Operations Research is an interdisciplinary book series which provides a platform for the cutting-edge research and developments in both operations research and operations management field. The purview of this series is global, encompassing all nations and areas of the world.

It comprises for instance, mathematical optimization, mathematical modeling, statistical analysis, queueing theory and other stochastic-process models, Markov decision processes, econometric methods, data envelopment analysis, decision analysis, supply chain management, transportation logistics, process design, operations strategy, facilities planning, production planning and inventory control.

LNOR publishes edited conference proceedings, contributed volumes that present firsthand information on the latest research results and pioneering innovations as well as new perspectives on classical fields. The target audience of LNOR consists of students, researchers as well as industry professionals.

Oliver Grothe · Stefan Nickel · Steffen Rebennack ·
Oliver Stein
Editors

Operations Research Proceedings 2022

Selected Papers of the Annual International Conference of the German Operations Research Society (GOR), Karlsruhe, Germany, September 6–9, 2022

Springer

Editors
Oliver Grothe
Institute for Operations Research
Karlsruhe Institute of Technology
Karlsruhe, Germany

Stefan Nickel
Institute for Operations Research
Karlsruhe Institute of Technology
Karlsruhe, Germany

Steffen Rebennack
Institute for Operations Research
Karlsruhe Institute of Technology
Karlsruhe, Germany

Oliver Stein
Institute for Operations Research
Karlsruhe Institute of Technology
Karlsruhe, Germany

ISSN 2731-040X ISSN 2731-0418 (electronic)
Lecture Notes in Operations Research
ISBN 978-3-031-24906-8 ISBN 978-3-031-24907-5 (eBook)
https://doi.org/10.1007/978-3-031-24907-5

This Springer imprint is published by the registered company Springer Nature Switzerland AG
The registered company address is: Gewerbestrasse 11, 6330 Cham, Switzerland

Organization

OrganizingCommittee

Steffen Rebennack (Chair)
Oliver Stein (Co-chair)
Wolf Fichtner
Kai Furmans
Oliver Grothe
Veit Hagenmeyer
Stefan Nickel
Frank Schultmann

Program Committee

Stefan Nickel, KIT (Chair)
Christina Büsing, RWTH Aachen University
Alf Kimms, University of Duisburg-Essen
Gisela Lanza, KIT
Peter Letmathe, RWTH Aachen University
Steffen Rebennack, KIT
Norbert Trautmann, University of Bern
Guido Voigt, Universität Hamburg

Stream Chairs

Valentin Bertsch
Michael Breitner

Dirk Briskorn
Stephan Bütikofer
Wolf Fichtner
Andreas Fink
Jutta Geldermann
Jochen Gönsch
Ralph Grothmann
Martin Grunow
Florian Jaehn
Patrick Jochem
Robert Klein
Natalia Kliewer
Max Klimm
Max Krüger
Ulf Lorenz
Elmar Lukas
Teresa Melo
Maximilian Moll
Dominik Möst
Sven Müller
Britta Peis
Peter Pelz
Marc Pfetsch
Boris Reuter
Lorena Reyes-Rubiano
Franz Rothlauf
Stefan Ruzika
Katja Schimmelpfeng
Marie Schmidt
Hanno Schülldorf
Jens Schulz
Andrea Seidl
Thomas Setzer
Claudius Steinhardt
Philipp Trotter
Guido Voigt
Hans-Jörg von Mettenheim
Stefan Wagner
Gerhard-Wilhelm Weber
Hans-Georg Zimmermann

Sponsors

Additive
Deutsche Bahn
EnBW
FICO
GAMS
Gurobi
IBM
INFORM
LocalSolver
Log-Hub
OPTANO
PTV
Quantagonia
Springer

Partners

DFG
EURO
GOR
IAI
IFL
IOR
IPP
KIT
MathSEE

Preface

The annual scientific conference OR 2022 of the German Operations Research Society (GOR) was held at the Karlsruhe Institute of Technology on 6–9 September 2022. The Operations Research Institute (IOR) hosted the conference together with the Institute for Industrial Production (IIP), the Institute for Automation and Applied Informatics (IAI) and the Institute for Material Handling and Logistics (IFL).

After two years of COVID-19, the OR 2022 was held as an in-person conference again; the conference in 2020 was cancelled and in 2021 it was held as an online conference hosted at the University of Bern. The organization committee faced uncertainty while preparing for the event; until about 6 weeks before the conference, we were preparing for both an online format and an in-person conference.

Over 600 scientists from about 30 different countries have attended the OR 2022 in Karlsruhe. Three plenary and nine semi-plenaries covered theoretical aspects of Operations Research, applications and real-world practices. In addition, more than 400 contributed presentations were held over three days in up to 21 parallel sessions.

This conference proceedings contains only papers presented at the OR 2022. All accepted papers have been reviewed by two domain experts, led by the corresponding stream chairs. All papers went through at least one round of revisions. In total, seventy-nine papers have been accepted for this proceedings. The papers are sorted by the corresponding streams as presented at the conference. This also includes the program committee stream which hosted the GOR master thesis award as well as the GOR PhD thesis award.

We thank all participants for their contribution to OR 2022. In addition, we are grateful for our sponsors (Gurobi, Additive, FICO, GAMS, LocalSolver, Optano, Quantagonia, PTV Group, Springer, IBM, DB, Log-hub, EnBW, INFORMS), our internal partner (MathSEE) and external partners (DFG and EURO), the stream chairs as well as the GOR. We also thank the program committee and all the many helpers from KIT at the conference.

Karlsruhe, Germany
November 2022

Oliver Grothe
Stefan Nickel
Steffen Rebennack
Oliver Stein

Contents

Part I
Awards of GOR

Chapter 1
A Two-Stage Stochastic Optimisation Model for Urban Same-Day Delivery with Micro-hubs

Charlotte Ackva

Abstract To compete with the rapid growth in e-commerce, many local shops provide a delivery service to their customers. To increase consolidation opportunities, shops start cooperating in local delivery by using shared vehicles and micro-hubs for joint transportation of parcels. Stores deposit their orders at close-by micro-hubs for further delivery by the shared vehicles, which conduct consistent routes between the micro-hubs. As long as it is in line with their schedule, the vehicles collect the parcels and drop them off close to customers' locations. Hence, it is very important to find effective schedules which is particularly challenging since order placements vary from day to day. We propose a two-stage stochastic program. In the first stage, the vehicle schedules are determined. In the second stage, the realised orders are routed. The goal is to maximise the expected amount of fulfilled parcel orders with the shared vehicles. We solve the problem with the Progressive Hedging algorithm. We consider the optimal solution without consistency constraints and a practically-inspired heuristic solution as benchmarks. We find that Progressive Hedging behaves rather poorly on random data, but performs particularly well on highly structured demand patterns.

Keywords Micro-hubs · Same-day delivery · Routing consistency · Two-stage stochastic programming · Progressive hedging

Motivation

In view of the increasing volume of e-commerce, many local shops have started offering a delivery service to their customers. They are facing many challenges in the delivery process: low transportation volumes, frequent stops on busy roads, and access restrictions in inner city areas turn conventional delivery by motorised vans to be inefficient. At the same time, customer expectations regarding the reliability and speed of their delivery continue to increase. As a response to this, many local shops are

C. Ackva (✉)
Otto von Guericke Universität Magdeburg,Magdeburg, Germany
e-mail: charlotte.ackva@ovgu.de

O. Grothe et al. (eds.), *Operations Research Proceedings 2022*, Lecture Notes in Operations Research, https://doi.org/10.1007/978-3-031-24907-5_1

re-thinking their delivery concepts, integrate transshipment facilities into the supply chain, and start collaborating for more efficient delivery [2]. Both the integration of transshipment facilities, called micro-hubs, and the joint delivery of parcels to the same region increase consolidation opportunities in the delivery process. Local shops bring their orders to a near-by micro-hub for further transportation. The shared vehicles conduct consistent tours on a daily basis to pick up the parcels and drop them off at micro-hubs close to the parcels' destinations. Each day, the vehicles visit the micro-hubs in the same order and at the same time. Such consistent tours are desirable by different stakeholders: They give planning security to storekeepers, provide operational stability for drivers, and allow reliable and punctual deliveries which increases customers' trust and satisfaction. However, finding such a consistent tour is a difficult task given the daily variability in order placements. If orders cannot be fulfilled, they have to be outsourced or served the next day, which is expensive or may lead to dissatisfaction, respectively. Hence, we are looking for consistent tours maximising the expected number of fulfilled parcel orders per day. A survey on different concepts of consistent routing is given by [4].

The problem is stated as a two-stage stochastic program. It represents a special version of a 2-echelon vehicle routing problem, see Sluijk et al. [8] for a review, and is formulated as a team-orienteering problem [9]. The first stage concerns the long-term tactical planning of a consistent daily schedule between the micro-hubs before the actual demand is revealed. The second stage addresses the daily operational routing of realised parcel orders. The problem is solved with the Progressive Hedging (PH) algorithm [7], an approach particularly suited for two-stage stochastic programs. It considers a set of possible future demand scenarios to come up with a solution that can be expected to perform well under any realised demand setting. In the following, we provide a problem description of the two-stage model, and state the PH algorithm. We analyse the convergence behaviour of the algorithm, and show that PH performs well for highly structured demand, but rather poorly for random demand.

Problem Description

We assume one depot to be located at the outskirts of the city and a set of micro-hubs to be placed at fixed locations in the city. Micro-hubs may either be located in selected stores, or close to customer's locations or shopping areas. We consider a finite fleet of vehicles each with a given, finite capacity and velocity. Vehicles start and end their service at the depot. While operating, vehicles are allowed to wait at micro-hubs in order to include later parcels. Also, vehicles are allowed to perform pickup and delivery on the same route, and simultaneously at one micro-hub. Moreover, we consider a limited service time horizon within which the service is operated. We consider a duration of stay at each micro-hub a vehicle visits on its itinerary to load and unload parcels.

Parcel orders are placed on a daily basis. We hence consider a set of several daily scenarios, each associated with a certain probability of occurrence. We assume that each parcel order has a homogeneous volume, and consists of a pickup location (store) with a release time, and a delivery location (customer). Further, we assume that each storekeeper brings a parcel to her closest micro-hub as soon as the order is placed. Similarly, when a parcel reaches its final micro-hub, the customer picks it up there directly. To represent this in our model, pickup and delivery locations are mapped to their closest micro-hubs such that each parcel has a pickup micro-hub and a delivery micro-hub. The release time of a parcel indicates the earliest time it can be picked up. We set a delivery time promise of four hours for all orders, restricting the delivery time at the parcel's destination micro-hub to be not later than 4 h after the parcel's release time. We do not have to serve all orders placed, but aim to pickup and deliver as many parcels as possible. Each parcel that is picked up must also be delivered on time.

Decisions are made in two stages. The first stage concerns the long-term planning of vehicle routes. Although demand varies from day to day, vehicles are to follow a fixed, consistent schedule. The second stage concerns the operational daily planning of which parcels to serve and how to transport them given the vehicles' schedule from the first stage. The objective of the model is to maximise the expected number of delivered parcels per day. A mathematical formulation of the described problem can be found in [1].

The Progressive Hedging Algorithm

The PH approach is a scenario-based decomposition technique and especially suitable for two-stage programs. It is based on the optimal solution under each scenario, which allows us to capture the fact that demand is still unknown at time of planning while incorporating a-priori knowledge about different possible demand scenarios. It was introduced by Rockafellar and Wets in 1991 [7]. The greatest challenge in the algorithm lies in the question of how to consolidate the different scenario-dependent solutions into an overall decision policy. Since the vehicles' schedules are determined before demand is revealed, the schedules must be independent of the resulting scenario. This property is called *non-anticipativity* – we have no foresight into the future. In the following, we specify in more detail how such a solution is generated.

The complete algorithm for our two-stage program is stated in Algorithm 1. It assumes a set of different scenarios S with associated probabilities of occurrence p_s, objective functions and feasible sets to be known for any $s \in S$. It further requires a penalty parameter $\rho > 0$, and a termination threshold $\epsilon > 0$. To initialise the algorithm, each scenario subproblem is solved to optimality, yielding scenario-dependent optimal first stage solutions $x_s^{(0)}$ for any $s \in S$.

In order to be meaningful, the final first stage solution should satisfy two properties. First, it should be non-anticipative, i.e. not depend on the realised demand scenario. Second, it should be feasible for any scenario $s \in S$. Non-anticipativity can be ensured by forcing the first stage decision variables x_s to have the same value for every scenario $s \in S$. This common value is chosen as the weighted average sum of the first stage solutions of the scenario subproblems of the previous PH iteration, i.e.:

$$x_s = \sum_{s' \in S} p_{s'} x_{s'}^{(k-1)} =: \overline{x}^{(k)} \quad \forall s \in S. \tag{1.1}$$

To derive such a non-anticipative solution, Eq. (1.1) is included to the objectives of the subproblems in a Lagrangian sense through multipliers $w_s^{(k)}$:

$$w_s^{(k)} := w^{(k-1)} + \rho \left\| x_s^{(k)} - \overline{x}^{(k)} \right\|_2 \quad \forall s \in S, \tag{1.2}$$

where $w_s^{(0)}, s \in S$, can be chosen arbitrarily [7]. To lead to feasibility of $\overline{x}^{(k)}$, a second penalty term is added to the objective function of the subproblems:

$$\frac{\rho}{2} \left\| x - \overline{x}^{(k)} \right\|_2^2. \tag{1.3}$$

Altogether, the PH algorithm iteratively solves a modified version of each scenario subproblem with the two penalty terms Eqs. (1.1) and (1.3) added. To converge to a solution that is both non-anticipative and feasible, the Lagrangian weights $w_s^{(k)}$ are adjusted over time according to Eq. (1.2). This way, the algorithm generates an improving sequence of policies $\overline{x}^{(k)}$, terminating when non-anticipativity is reached up to some threshold ϵ. The resulting decision policy is the non-anticipative solution which can be expected to perform best for any demand realisation, and thus provides a consistent vehicle routing without knowing the future demand.

Algorithm 1: Progressive Hedging Algorithm

Input: A set of possible scenarios S with a probability of occurrence p_s, an objective function and a feasible set for any $s \in S$; a penalty parameter $\rho > 0$, and a threshold $\epsilon > 0$.

Initialisation:

- $\forall s \in S$: find $x_s^{(0)}$ as an optimal first stage decision of subproblem s.
- $\forall s \in S : w_s^{(0)} := 0$.
- $k := 1$.

1. Determine current best first stage solution: $\overline{x}^{(k)} := \sum_{s \in S} p_s x_s^{(k-1)}$.
2. $\forall s \in S$: find $x_s^{(k)}$ as an optimal first stage decision of *modified* subproblem s, where two penalty terms are added for:
   a. non-anticipativity: $w_s^{(k-1)} \|x\|_2$,
   b. non-feasibility: $\frac{\rho}{2} \left\| x - \overline{x}^{(k)} \right\|_2^2$.
3. Update weights: $\forall s \in S : w_s^{(k)} := w_s^{(k-1)} + \rho \left\| x_s^{(k)} - \overline{x}^{(k)} \right\|_2$. Set $k := k + 1$.

Repeat step 1 to step 3 until $\sum_{s \in S} p_s \left\| x_s^{(k)} - \overline{x}^{(k)} \right\|_2 < \epsilon$.

Computational Study

In a computational study we investigate whether the PH algorithm can successfully be applied to the team orienteering problem stated in Section "Problem Description". We analyse the convergence behaviour of the algorithm for different parameter settings, and evaluate the solution quality for different demand distributions.

Convergence Behaviour of PH

In their paper, [7] prove that the algorithm converges with a linear convergence rate if the problem is convex. Since convergence is not guaranteed for integer problems, we analyse the behaviour of the algorithm on our problem for different choices of the penalty parameter ρ, the main influencing factor of the algorithm. According to Fan et al. [3], Listes and Dekker [5], and Mulvey and Vladimirou [6], there is no unique best choice of ρ, rather it has to be determined empirically. Testing different values for ρ, we find a trade-off between convergence and objective values. We observe that the algorithm does not converge if ρ is chosen too small. In this case, little importance is given to the non-anticipativity constraints in the modified subproblems such that no common average solution is found. In contrary, if ρ is chosen too large, the algorithm converges within a few iterations, but to solutions of very poor quality. Here, too much weight is given to the non-anticipativity constraints in the first iterations, leading to a local optimum too quickly. The best results are obtained with intermediate values. For such values, PH requires a few more iterations, but leads to solutions with better objective values.

The PH algorithm is also driven by the number of scenarios used. While a higher number of scenarios better represents the variability in demand, they also increase computation time significantly since a modified version of each scenario-dependent problem is to be solved in each PH iteration. Hence, a trade-off between satisfying solution quality and runtime has to be found. In our experiments, we find that using 20 scenarios leads to the best objective values in a computation time of 91.30 min on average. Using more scenarios does not improve the solution quality. Fewer scenarios (5, 10, or 15) in the algorithm lead to slightly worse objective values (1.68 to 4.35% worse), while decreasing the computation time tremendously (from 28.10 to 81.67%).

Service Rates on Different Demand Patterns

To assess the method's performance, we evaluate the solutions obtained via PH for a practically inspired, fixed benchmark, and for a non-consistent upper bound where a new schedule is determined on a daily basis. We consider scenarios with

Table 1.1 Average relative gaps of PH and fixed solution to non-consistent upper bound

Demand pattern	Gap PH solution (%)	Gap fixed solution (%)
Highly structured	13.38	43.27
Semi-structured	32.60	43.01
Random	36.48	34.87

highly structured, semi-structured, and random demand. For each demand pattern, we compute the average relative gap of the objective values obtained by PH and the fixed solution to the non-consistent upper bound, respectively, displayed in Table 1.1.

Progressive Hedging is designed to uncover common trends among scenarios. It hence performs better if some common demand structure is available. On highly structured demand, PH derives solutions that increase the average service rates by more than 50% compared to the fixed solution. The average relative gap to the upper bound is notably lower for the PH than for the fixed solution. On semi-structured demand, PH still performs better than the fixed benchmark, although showing less improvement than on highly structured demand. Increasing the variability in demand, as on the random instances, the fixed solution even outperforms PH. In this case, demand fluctuates too much from day to day, making it hard to find an "average" schedule performing well for any demand realisation. Since the problem is formulated in a time-expanded network, there are many arcs not chosen in the solution. When determining the average solution in the PH algorithm, they tend towards zero, i.e. favour no movement.

Conclusion and Future Work

In this paper, we have set up a framework to find consistent schedules for shared vehicles between micro-hubs with the PH algorithm. We found that intermediate values for the penalty parameter and the number of scenarios have to be found to balance solution quality and computational time. The solutions produced by PH are closer to the upper bound the more structure is shared among the scenarios. Future work may convert the problem to a dynamic environment where parcel orders arrive over time. In this, the consistent tours between micro-hubs may be combined with dynamic couriers. This allows to incorporate the transportation of parcels from shops to micro-hubs, and from micro-hubs to customers.

References

1. Ackva, C. & Ulmer, M. (2022). Consistent routing for local same-day delivery via micro-hubs. *FEMM Working Paper No. 7/2022*. Available at https://www.fww.ovgu.de/fww_media/femm/femm_2022/2022_07.pdf
2. Burns, T., Davis, A., Harris, T. & Kuzmanovic, A. (2022). Beyond the distribution center. https://www.mckinsey.com/industries/retail/our-insights/beyond-the-distribution-center, 2022. Accessed 24 June, 2022.
3. Fan, Y., & Liu, C. (2010). Solving stochastic transportation network protection problems using the progressive hedging-based method. *Networks and Spatial Economics., 10*, 193–208.
4. Kovacs, A., Golden, B., Hartl, R., & Parragh, S. (2014). Vehicle routing problems in which consistency considerations are important: A survey. *Networks., 64*, 192–213.
5. Listes, O., & Dekker, R. (2005). A scenario aggregation-based approach for determining a robust airline fleet composition for dynamic capacity allocation. *Transportation Science., 39*(3), 367–382.
6. Mulvey, J., & Vladimirou, H. (1991). Applying the progressive hedging algorithm to stochastic generalized networks. *Annals of Operations Research., 31*, 399–424.
7. Rockafellar, R., & Wets, R. (1991). Scenarios and policy aggregation in optimization under uncertainty. *Mathematics of Operations Research., 16*(1), 119–147.
8. Sluijk, N., Florio, A., Kinable, J., & Dellaert, N., & Van Woensel, T. (2022). Two-echelon vehicle routing problems: A literature review. *European Journal of Operational Research*.
9. Song, Y., Ulmer, M., Thomas, B., & Wallace, S. (2020). Building trust in home services-stochastic team-orienteering with consistency constraints. *Transportation Science., 54*(3), 823–838.

Chapter 2
Computational Linear Bilevel Optimization

Thomas Kleinert

Abstract In this article, we summarize a subset of the findings of the cumulative dissertation "Algorithms for Mixed-Integer Bilevel Problems with Convex Followers"; see [4]. First, we present a result that renders the application of the well-known and widely used big-M reformulation of linear bilevel problems infeasible for many practical applications. Second, we present valid inequalities and demonstrate that an SOS1-based approach is a competitive alternative to the error-prone big-M method in case both approaches are equipped with these valid inequalities. Third, we introduce a penalty alternating direction method, which computes (close-to-)optimal feasible points in extremely short computation times and outperforms a state-of-the-art local method.

Keywords Bilevel optimization · Computational optimization · Mixed-integer programming

Introduction to Linear Bilevel Problems

Hierarchical decision making processes naturally appear in an enormous amount of applications from areas including energy markets, revenue management, machine learning, critical infrastructure defense and many more. Such processes are formalized by bilevel optimization problems, i.e., optimization problems for which a subset of variables is constrained to be an optimal solution of a follower optimization problem. The "easiest" bilevel optimization problems are linear bilevel problems, i.e., problems of the form

T. Kleinert (✉)
Friedrich-Alexander Universität Erlangen-Nürnberg, Discrete Optimization, Cauerstr.11, 91058 Erlangen, Germany
e-mail: thomas.kleinert@fau.de

O. Grothe et al. (eds.), *Operations Research Proceedings 2022*, Lecture Notes in Operations Research, https://doi.org/10.1007/978-3-031-24907-5_2

$$\begin{aligned}\min_{x\in\mathbb{R}^n, y\in\mathbb{R}^m} \quad & c^\top x + d^\top y \\ \text{s.t.} \quad & Ax + By \geq a, \\ & y \in \arg\min_{\bar{y}\in\mathbb{R}^m}\{e^\top \bar{y} \colon Cx + D\bar{y} \geq b\},\end{aligned} \tag{2.1}$$

with $c \in \mathbb{R}^n$, $d, e \in \mathbb{R}^m$, $A \in \mathbb{R}^{k\times n}$, $B \in \mathbb{R}^{k\times m}$, $C \in \mathbb{R}^{\ell\times n}$, $D \in \mathbb{R}^{\ell\times m}$, $a \in \mathbb{R}^k$, and $b \in \mathbb{R}^\ell$. Even linear bilevel problems are intrinsically nonconvex and strongly NP-hard: see [3]. Note that in case of ambiguity in the follower problem, we assume the optimistic solution, in which the follower chooses the optimal solution y most in favor of the leader's objective.

In practice, almost all global solution approaches for Problem (2.1) rely on a single-level reformulation that replaces the follower problem by its necessary and sufficient KKT conditions. This yields the following mathematical problem with complementarity constraints (MPCC)

$$\min_{x,y,\lambda} \quad c^\top x + d^\top y \tag{2.2a}$$

$$\text{s.t.} \quad (x, y) \in \Omega := \{(x, y) \colon Ax + By \geq a, Cx + Dy \geq b\}, \tag{2.2b}$$

$$\lambda \in \Omega_D := \{\lambda \geq 0 \colon D^\top\lambda = e\}, \tag{2.2c}$$

$$\lambda^\top(Cx + Dy - b) \leq 0, \tag{2.2d}$$

in which the set of inequalities (2.2c) denotes the KKT nonnegativity and stationarity conditions and (2.2d) is the KKT complementarity condition of the follower problem.

Branch-and-Bound Methods for Linear Bilevel Problems

Due to the combinatorial nature of the KKT complementarity constraint (2.2d), it is reasonable to tackle Problem (2.2) in a branch-and-bound fashion.

The Big-M Approach: Not Your Everyday Friend

The KKT complementarity constraint (2.2d) can be reformulated in a mixed-integer linear way by introducing additional binary variables $z_i \in \{0, 1\}$, $i = 1, \ldots, \ell$, sufficiently large big-M constants M_P and M_D, and the following conditions:

$$(Cx + Dx - b)_i \leq M_P z_i, \quad \lambda_i \leq M_D(1 - z_i), \quad i = 1, \ldots, \ell. \tag{2.3}$$

This reformulation goes back to [2] and is by far the most-used approach to solve linear bilevel problems in practice. The reason is obvious: The single-level problem (2.1) with the complementarity conditions (2.2d) replaced by (2.3) can be easily implemented and the resulting model can be solved without further ado by ordinary mixed-integer solvers.

However, this approach is problematic. On the one hand, large values M_P and M_D may cause numerical instabilities that, in the worst case, result in infeasible "solutions" for Problem (2.2). On the other hand, choosing these constants too small may cut off feasible solutions and global optimality may be lost. In fact, we show that verifying that given values M_P and M_D do not cut off bilevel-feasible points is as hard as solving the original bilevel problem (2.1); see Theorem 3 in the dissertation [4]. This result is also published in [6] and implies that the big-M approach should not be used in practice unless big-M constants can be derived from problem-specific knowledge.

The SOS1 Approach: A Lame Duck?

Instead of reformulating Problem (2.2) to a mixed-integer problem at the cost of additional binary variables and big-M constants, one can also branch on the complementarity constraints (2.2d); see [4] for more details. In Fortuny-Amat and McCarl [2], it is proposed to express the KKT complementarity condition (2.2d) as SOS1 conditions:

$$s_i = (Cx + Dy - b)_i, \quad \{s_i, \lambda_i\} \text{ is SOS1}, \quad i = 1, \ldots, \ell. \tag{2.4}$$

This turns the MPCC (2.2) into a linear problem with SOS1 constraints. Such problem types can be handled by modern mixed-integer solvers in a way that guarantees the correctness of the obtained solutions—in contrast to the big-M approach stated above. From a practical point of view, the SOS1-based reformulation is not more difficult to model and implement than the mixed-integer-based reformulation. Still, the latter is by far more popular in practice. One reason for this might be that it is regarded as computationally more effective. This is supported by an extensive and detailed computational study that we conducted in [4]; see also the next subsection.

The Game Changer: Valid Inequalities Based on Strong Duality

The dissertation [4] also contributes the following valid inequality for Problem (2.1), which is derived from the strong duality condition of the lower-level problem:

$$\lambda^\top C^- + e^\top y - \lambda^\top b \leq 0. \tag{2.5}$$

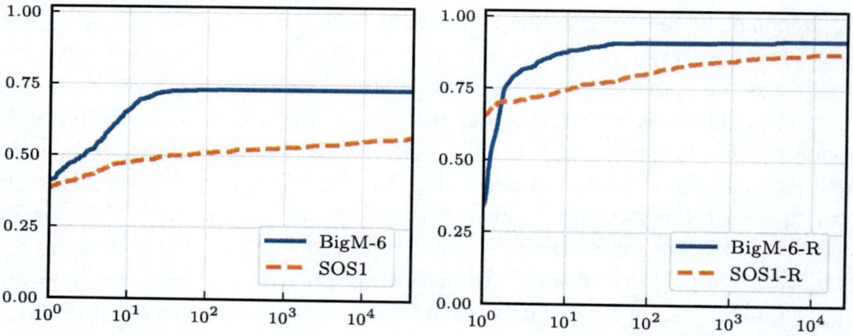

Fig. 2.1 Performance profile of running times of BigM-6 and SOS1 (left) and of BigM-6-R and SOS1-R (right)

Here, C^- denotes the vector of lower bounds $C_i^- \leq C_{i\cdot}x$. Such bounds exist, if the joint feasible set Ω is bounded. Note that the derivation of the valid inequality along with a detailed computational evaluation on the effectiveness within the SOS1 approach is published in [5]. There, we also generalize the valid inequality (2.5) to a family of inequalities by applying McCormick envelopes. In the dissertation [4], we further show that applying inequality (2.5) at the root node of each of the two branch-and-bound approaches yields surprising results, which are summarized in the performance profiles of running times in Fig. 2.1. For details on the test set, implementation, and hardware, we refer to [4]. Figure 2.1 (left) reveals that the plain big-M formulation (with $M_P = M_D = 10^6$) indeed outperforms the plain SOS1 formulation and significantly "solves" more instances—at least if the big-M constants would be provably correct. The story changes if we equip both formulations with the derived inequality at the root node. In this case, the SOS1 approach is the faster approach for 65% of the instances and solves almost as many instances as the big-M approach (see Fig. 2.1 (right)). We follow that whenever valid and tight values for M_P and M_D are available, one may use the big-M-based approach. In all other cases, the SOS1-based approach constitutes an easy-to-use alternative that is very competitive in terms of computational performance if it is equipped with the valid inequality (2.5)—without requiring troublesome big-M values.

A Penalty Alternating Direction Method

When facing NP-hard problems, it is quite common to develop heuristics that quickly deliver feasible points. In the cumulative dissertation [4], we propose a heuristic based on a penalty alternating direction method (PADM). The main idea is to reformulate Problem (2.1) to a single-level problem by replacing the follower problem by primal and dual feasibility as well as a single nonconvex inequality enforcing strong duality:

$$\min_{x,y,\lambda} \quad c^\top x + d^\top y \tag{2.6a}$$

$$\text{s.t.} \quad (x, y) \in \Omega, \quad \lambda \in \Omega_D, \tag{2.6b}$$

$$e^\top y \le (b - Cx)^\top \lambda. \tag{2.6c}$$

Penalizing the single nonconvex constraint (2.6c) in the objective function with a penalty parameter $\rho > 0$ yields the following problem with a nonconvex objective function over linear constraints:

$$\begin{aligned} \min_{x,y,\lambda} \quad & h(x, y, \lambda) := c^\top x + d^\top y + \rho \left(e^\top y - (b - Cx)^\top \lambda\right) \\ \text{s.t.} \quad & (x, y) \in \Omega, \quad \lambda \in \Omega_D. \end{aligned} \tag{2.7}$$

Note that the constraint set decomposes into a (x, y)-block and a λ-block. In addition, $h(x, y, \lambda)$ is linear, if either one of the variable blocks is fixed. This motivates the PADM described in Algorithm 1.

Please see [4] for a formal convergence result. The main takeaway is that Algorithm 1 provides a stationary point of the single-level reformulation (2.7), if it terminates. We note that the derivation, convergence analysis and a computational evaluation are also published in [7]. In the dissertation [4], we provide an additional computational analysis, in which we compare Algorithm 1 to an established local regularization method published in [1] as well as to the global SOS1 approach stated above; see Fig. 2.2 for the results and [4] for details on the setup.

Algorithm 1 A Penalty Alternating Direction Method for Problem (2.7).

1: Choose initial values $\lambda^{0,0} \in \Omega_D$ and a penalty parameter $\rho^0 > 0$.
2: **for** $j = 0, 1, \ldots$ **do**
3: Set $i = 0$.
4: **while** $(x^{j,i}, y^{j,i}, \lambda^{j,i})$ is not a partial minimum of Problem (2.7) with $\rho = \rho^j$ **do**
5: Solve Problem (2.7) with fixed $\lambda = \lambda^{j,i}$ and $\rho = \rho^j$ to obtain $(x^{j,i+1}, y^{j,i+1})$.
6: Solve Problem (2.7) with fixed $(x, y) = (x^{j,i+1}, y^{j,i+1})$ and $\rho = \rho^j$ to obtain $\lambda^{j,i+1}$.
7: Set $i \leftarrow i + 1$.
8: **end while**
9: Update the penalty parameter $\rho^{j+1} > \rho^j$.
10: **end for**

Figure 2.2 (left) is a performance profile of running times and shows that the PADM is by far the fastest method for every tested instance. Figure 2.2 (right) is an ECDF plot of relative gaps. We observe that PADM clearly outperforms the local method REG and that it solves 50% of the instances indeed to global optimality. Taking into account both the very fast running times and the good quality of the computed solutions—particularly also in comparison to the global approach SOS1-R—PADM cannot only be used as a primal heuristic within global approaches for linear bilevel problems but also as a standalone method. In addition, Algorithm 1 is extendable to a much broader class of bilevel problems. In [7], Algorithm 1 is

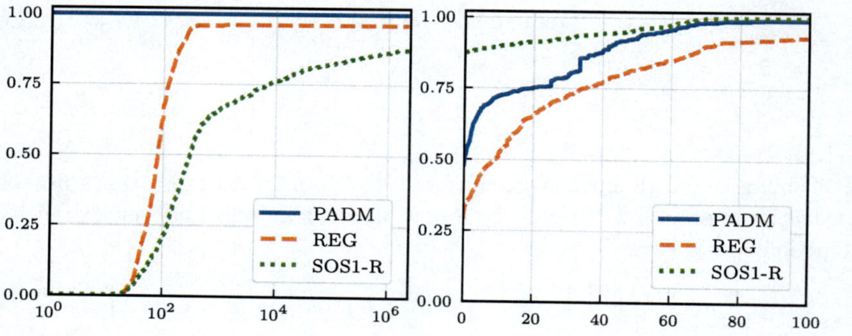

Fig. 2.2 Left: Performance profile of running times of PADM, REG, and SOS1-R. Right: ECDF plot of relative gaps of PADM, REG, and SOS1-R

successfully applied to convex MIQP-QP bilevel problems. Going further, even the class of MINLP-QP bilevel problems could be tackled with Algorithm 1 in case the involved MINLPs can be solved effectively.

Conclusion

In this article, we summarized some contributions of the dissertation [4]. We introduced linear bilevel problems and discussed what might go wrong when applying the well-established and often used big-M approach. Further, we derived a valid inequality applicable for the big-M approach as well as the less popular SOS1 approach. The presented computational results show that applying these inequalities renders the SOS1 approach a very competitive alternative to the error-prone big-M approach. Finally, we introduced a primal heuristic applicable for a broad class of bilevel problems, which is shown to be extremely fast and to provide close-to-optimal solutions.

Overall, the presented techniques enhance the solution of linear bilevel problems significantly. Still, there are many directions for future research. For example, the derivation of additional valid inequalities or cutting planes could further strengthen the SOS1 approach. In addition, bilevel-specific presolving techniques may help to tackle even larger and more challenging instances.

References

1. Dempe, S. (2019). Computing locally optimal solutions of the bilevel optimization problem using the KKT approach. In M. Khachay, Y. Kochetov, & P. Pardalos (Eds.), *Mathematical Optimization Theory and Operations Research* (pp. 147-157). Springer,2019. https://doi.org/10.1007/978-3-030-22629-911
2. Fortuny-Amat, J., & McCarl, B. (1981). A representation and economic interpretation of a two-level programming problem. *The Journal of the Operational Research Society 32*(9), pp. 783-792. https://doi.org/10.1057/jors.1981.156
3. Hansen, P., Jaumard, B., & Savard, G. (1992). New branch-and-bound rules for linear bilevel programming. *SIAM Journal on Scientific and Statistical Computing 13*(5), 1194-1217. https://doi.org/10.1137/0913069
4. Kleinert, T. (2021). *Algorithms for mixed-integer bilevel problems with convex followers*. PhD thesis. Friedrich-Alexander-Universität Erlangen-Nürnberg (FAU), pp. 1 -272. url: https://opus4.kobv.de/opus4-fau/files/16225/dissertationkleinert published digital.pdf (visited on 08/15/2022).
5. Kleinert, T., Labbé, M., Plein, F., & Schmidt, M. (2021). Closing the gap in linear bilevel optimization: a new valid primal-dual inequality. *Optimization Letters, 15*, 1027–1040. https://doi.org/10.1007/s11590-020-01660-6
6. Kleinert, T., Labbé, M., Plein, F., Schmidt, M. (2020). Technical note-there's no free lunch: On the hardness of choosing a correct big-M in cilevel optimization. *Operations Research 68*(6), pp. 1716-1721. https://doi.org/10.1287/opre.2019.1944
7. Kleinert, T., & Schmidt, M. (2021). Computing feasible points of bilevel problems with a penalty alternating direction method. *INFORMS Journal on Computing 33*(1) , 198-215. https://doi.org/10.1287/ijoc.2019.0945

Chapter 3
Faster Algorithms for Steiner Tree and Related Problems: From Theory to Practice

Daniel Rehfeldt

Abstract The Steiner tree problem in graphs (SPG) is a classic $\mathcal{NP}$-hard problem. Many applications can be modeled as SPG or closely related problems. This article describes a state-of-the-art solver for finding optimal solutions to classic Steiner tree and 14 related problem classes. It was developed as part of the author's doctoral dissertation.

Keywords Combinatorial optimization · Steiner tree problems · Exact methods

Introduction

Given an undirected connected graph $G = (V, E)$, edge costs $c : E \rightarrow \mathbb{Q}_{\geq 0}$ and a set $T \subseteq V$ of *terminals*, the *Steiner tree problem in graphs* (SPG) is to find a tree $S \subseteq G$ with $T \subseteq V(S)$ such that $c(E(S))$ is minimized. The $\mathcal{NP}$-hard SPG is one of the most studied problems in combinatorial optimization, and many applications can be modeled as SPG or closely related problems [13]. The SPGhas seen numerous theoretical advances in the last 10 years, bringing forth significant improvements in approximability, see e.g. [1, 6], and complexity, see e.g. [11, 14, 25]. However, the state of the art in (practical) exact SPG solving, set in the joint PhD theses of Polzin and Vahdati Daneshmand [2, 16], has remained largely unchallenged for almost 20 years. While the DIMACS Challenge 2014 [3] and the PACE Challenge 2018 [15], both dedicated to Steiner tree problems, brought renewed interest into the solution of SPGs, even the best new solvers fall far short of reaching the state of the art.

This article provides an overview of the algorithmic base and the implementation of a new SPG solver that finally manages to outperform the long-reigning state of

D. Rehfeldt (✉)
Zuse Institute Berlin, Takustr. 7, 14195 Berlin, Germany
e-mail: rehfeldt@zib.de

O. Grothe et al. (eds.), *Operations Research Proceedings 2022*, Lecture Notes in Operations Research, https://doi.org/10.1007/978-3-031-24907-5_3

the art. The new Steiner tree solver is named SCIP- JACK[1], is freely available for academic purposes and comes with source code. In contrast, the previous state-of-the-art SPG solver by Polzin and Vahdati Daneshmand is not publicly available. The new solver was developed as part of the author's doctoral thesis [17]. For more theoretical underpinnings of this solver, we refer the reader to the articles [18–21] or the thesis [17]. For a parallelization of SCIP- JACK see [22].

The directed equivalent of the SPG is called *Steiner arborescence problem* (*SAP*), see e.g. [8]. Given a directed graph $D = (V, A)$, costs $c : A \to \mathbb{Q}_{\geq 0}$, a set $T \subseteq V$ of terminals, and a root $r \in T$, the SAP is to find a directed tree $S \subseteq D$ such that: First, for all $t \in T$ the tree S contains exactly one directed path from r to t. Second, $c(A(S))$ is minimized. A classic integer programming (IP) formulation for SAP is given by [26] Associate with each arc $a \in A$ a binary variable $y(a)$, indicating whether a is contained in the Steiner arborescence ($y(a) = 1$) or not ($y(a) = 0$).

Formulation 1 *Directed Cut Formulation* $(DCut)$

$$\begin{aligned} &min \;\; c^T y && (3.1)\\ y(\delta^-(W)) &\geq 1 \qquad for\ all\ W \subset V,\ r \notin W,\ W \cap T \neq \emptyset, && (3.2)\\ y(a) &\in \{0, 1\} \quad for\ all\ a \in A. && (3.3) \end{aligned}$$

Constraints 3.2 ensure that for each terminal t, there is a directed path from r to t in the solution. Any SPG can be readily transformed to an equivalent SAP by replacing each edge by two anti-parallel arcs of the same weight and choosing an arbitrary terminal as the root. We will refer to the $DCut$ formulation applied to this transformed SAP as $BDCut$. In practice, the $BDCut$ formulation provides very tight LP bounds, although its theoretical strength remains an important open question, see e.g. [1], only an (almost trivial) upper bound of 2 is known.

Finding Minimum Steiner Trees by Branch-and-Cut

The three main algorithmic classes of our branch-and-cut framework for solving SPG are:

- Reduction techniques,
- Heuristics (primal and dual), and
- IP formulation and cutting planes.

Notably, the three algorithmic classes are deeply intertwined. For example, reduction methods are used within two primal heuristics, while the quality of the primal bound obtained by the heuristics determines the effectiveness of other reduction methods. Additionally, reduced problems usually show a smaller integrality gap for

[1] https://scipjack.zib.de/.

the IP formulation, and require less time for solving the LP-relaxation. In turn, the reduced-costs from the LP-relaxation can be used for further reductions.

In the following, we briefly list the main components of the branch-and-cut framework.

Presolving For presolving, the reduction methods described in the thesis are executed iteratively within a loop. This loop is reiterated as long as a predefined percentage of edges has been eliminated during the previous round. For computing reduced-costs during presolving (which can be used for powerful reduction methods [16]), we employ a dual-ascent algorithm [26].

Domain propagation During branch-and-bound we use the reduced-costs from the LP-relaxations for fixing arcs of the $BDCut$ formulation to 0. To this end, we use a path-based criterion. Additionally, whenever a predefined percentage of all arcs has been newly fixed during the branch-and-bound procedure, further reduction techniques are applied, and the achieved reductions are re-translated into arc fixing.

Decomposition It is well-known that the biconnected components of the graph underlying an SPGinstance can be solved separately. Given the super-linear run time of most algorithms that we employ, such a decomposition can lead to significant speed-ups. While SPG instances usually do not have articulation points in their original form, this property sometimes changes after the application of reduction techniques. Therefore, we use decomposition into biconnected components both during presolving and during branch-and-cut. Additionally, we also use (more sophisticated) techniques to exploit triconnected components.

Dynamic programming There have been notable efforts during the last years to make the classic SPG dynamic programming algorithms by Dreyfus and Wagner [4] and Erickson et al. [5] competitive in practice, see e.g. [7, 9] for prominent examples. Our dynamic programming implementation combines the node-separator concept from [9] with several reduction techniques. Unfortunately, however, our implementation is only competitive with branch-and-cut for instances with less than 20 terminals—and these instances are usually solved quickly by either approach. Still, in the context of decomposition techniques, where we often obtain many small sub-problems, the dynamic programming algorithm has turned out to be useful.

Primal heuristics We try to retain the best solution found during presolving, to provide it as an initial primal solution. During branch-and-bound, we periodically employ a classic construction heuristic [23] and a new recombination heuristic. We use the solution to the current LP-relaxation to guide the construction heuristic—an idea already utilized by other authors, e.g. [12]. Additionally, we use local heuristics from [24] to improve high-quality primal solutions.

Separation After presolving, SCIP- JACK runs the dual-ascent heuristic to select a set of constraints from the $BDCut$ formulation to be included in the initial LP. Additionally, we use all $0-1$ constraints. We separate the (remaining) constraints (3.2) of the $BDCut$ formulation by using a newly implemented maximum-flow algorithm with warm-start capabilities. We also (trivially) separate other classes of cuts, described in [12], to strengthen the LP-relaxation.

Branching Classic variable branching for the $BDCut$ formulation often leads to a badly balanced branch-and-bound tree, since the inclusion of an arc has a far larger impact than its exclusion. Thus, a well-known strategy is to branch on vertices instead, see e.g. [8]: A selected vertex is made a terminal in one branch-and-bound child node, and is removed in its sibling. Such a change is reflected in the IP formulation by adding one additional constraint. We note, however, that branching is rarely required, due to the various powerful algorithms that we apply before.

Computational Results

Notably, SCIP- JACK significantly outperforms the long-reigning state-of-the-art solver by Polzin and Vahdati Daneshmand, see [19]. Here, we provide results on the instances from the 3rd *Parameterized Algorithms and Computational Experiments Challenge*[2]. The challenge was dedicated to fixed-parameter tractable SPGs and attracted 75 participants from 16 countries. A previous version of SCIP-JACK reached first (Track B), second (Track A), and third (Track C) place in the three tracks of the challenge. The current version of SCIP- JACK is much faster, and would significantly outperform all competitors in each track. Figure 3.1 provides computational results on the exact tracks of the 3rd PACE Challenge. We use Gurobi 9.5 (*Commercial*), the best other solver from the PACE Challenge (*SPDP*), and SCIP- JACK with SoPlex (SCIP- JACK/*spx*) and Gurobi 9.5 (SCIP- JACK/*grb*) as LP-solver. A timelimit of one hour was set. Average times are given as arithmetic mean (as in the PACE Challenge) with time-outs counted as one hour. We note that the gap between SCIP- JACK and the leading commercial MIP solvers is usually much larger than shown in Fig. 3.1. However, the PACE Challenge benchmark set contains many small instances (with less than 1000 edges) that can be considered trivial for modern Steiner tree solvers and can also be solved quickly by state-of-the-art MIP solvers.

From Classic Steiner Tree to Related Problems

A central property of SCIP- JACK is its ability to not only solve the SPG, but also 14 related problems. Also for these problems, including the prize-collecting Steiner tree and the maximum-weight connected subgraph problem, SCIP- JACK significantly

[2] https://pacechallenge.org/2018/.

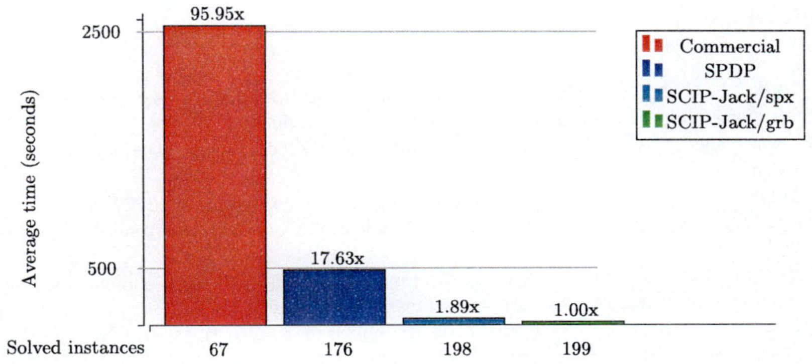

Fig. 3.1 Computational results on the 200 benchmark instances of tracks A and B of the PACE challenge 2018

Table 3.1 Computational comparison of the solvers *GeoSteiner* (*GeoS*) and SCIP- JACK (*S.-J.*)

Test-set	#	# Solved		Mean time (sh. geo. mean)			Maximum time		
		GeoSt.	S.-J.	GeoS. [s]	S.-J. [s]	Speedup	GeoS. [s]	S.-J. [s]	Speedup
R25	15	15	15	238.5	43.2	**5.5**	6250.6	54.6	**114.4**
R50	15	8	**15**	54844.5	128.2	**427.8**	TL	196.5	**3077.8**
R100	15	3	**15**	260311.5	477.9	**544.7**	TL	729.7	**828.8**

TL signifies the time limit (of one week)
Bold signifies a superior performance of the software

outperforms specialized solvers. Particularly notable results are achieved for the Euclidean Steiner tree problem: Table 3.1 shows results on large-scale test-sets from the 11th DIMACS Challenge with 25,000, 50,000, and 100,000 points in the plane. We use a time-limit of one week and compare SCIP- JACK and the leading geometric Steiner tree solver GeoSteiner 5.1 [10]. Notably, we solve 19 of the instances for the first time to optimality.

Conclusion

This article has described several central components of the leading Steiner tree solver SCIP- JACK. Notably, SCIP- JACK is the faster solver for all 15 problem classes it can handle. Besides being freely available for academic use, SCIP- JACK has been employed in several industry projects. The arguably most successful employment is the planning of fiber-optic networks by Open Grid Europe in Germany. The company reports that the SCIP- JACK has been employed for hundreds of planning problems and estimates the monetary savings to be several hundred thousand euros per project.

References

1. Byrka, J., Grandoni, F., Rothvoss, T., & Sanità, L. (2013). Steiner tree approximation via iterative randomized rounding. *The Journal of the ACM, 60*(1), 6.
2. Daneshmand, S. V. (2004). *Algorithmic approaches to the Steiner problem in networks*. PhD thesis, Universität Mannheim.
3. Dimacs challenge 2014. https://dimacs11.zib.de/. Accessed October 18, 2022.
4. Dreyfus, S. E., & Wagner, R. A. (1971). The Steiner problem in graphs. *Networks, 1*(3), 195–207.
5. Erickson, Ranel E., Monma, Clyde L., & Veinott, Arthur F. (1987). Send-and-split method for minimum-concave-cost network flows. *Mathematics of Operations Research, 12*(4), 634–664.
6. Goemans, M. X., Olver, N., Rothvoß, T., & Zenklusen, R. (2012). Matroids and integrality gaps for hypergraphic steiner tree relaxations. In *Proceedings of the Forty-Fourth Annual ACM Symposium on Theory of Computing*, STOC '12, pp. 1161-1176, New York, NY, USA, 2012. Association for Computing Machinery.
7. Hougardy, Stefan, Silvanus, Jannik, & Vygen, Jens. (2017). Dijkstra meets Steiner: A fast exact goal-oriented Steiner tree algorithm. *Mathematical Programming Computation, 9*(2), 135–202.
8. Hwang, F. K., Richards, D. S., & Winter, P. (1992). *The Steiner Tree Problem*. Annals of Discrete Mathematics: Elsevier Science.
9. Iwata, Yoichi, & Shigemura, Takuto. (2019). Separator-based pruned dynamic programming for Steiner tree. *In Proceedings of the AAAI Conference on Artificial Intelligence, 33*, 1520–1527.
10. Juhl, D., Warme, D. M., Winter, P., & Zachariasen, M. (2018). The GeoSteiner software package for computing Steiner trees in the plane: An updated computational study. *Mathematical Programming Computation, 10*(4):487–532.
11. Kisfaludi-Bak, S., Nederlof, J., & van Leeuwen, E. J. (2020). Nearly ETH-tight algorithms for planar Steiner tree with terminals on few faces. *ACM Transactions on Algorithms (TALG), 16*(3), 1–30.
12. Koch, Thorsten, & Martin, Alexander. (1998). Solving Steiner tree problems in graphs to optimality. *Networks, 32*, 207–232.
13. Ljubić, Ivana. (2021). Solving steiner trees: Recent advances, challenges, and perspectives. *Networks, 77*(2), 177–204.
14. Nederlof, J. (2009). Fast polynomial-space algorithms using Möbius inversion: Improving on Steiner tree and related problems. In *International Colloquium on Automata, Languages, and Programming*, pp. 713–725. Springer.
15. Pace challenge 2018. https://pacechallenge.org/2018/. Accessed October 18, 2022.
16. Polzin, T. (2003). *Algorithms for the Steiner problem in networks*. PhD thesis, Saarland University.
17. Rehfeldt, D. (2021). *Faster algorithms for Steiner tree and related problems: From theory to practice*. PhD thesis, Technische Universität Berlin.
18. Rehfeldt, D., & Koch, T. (2019). Combining NP-Hard Reduction Techniques and Strong Heuristics in an Exact Algorithm for the Maximum-Weight Connected Subgraph Problem. *SIAM Journal on Optimization, 29*(1), 369–398.
19. Rehfeldt, D., & Koch, T. (2021). Implications, conflicts, and reductions for steiner trees. *Mathematical Programming*, pp. 1–64.
20. Rehfeldt, D., & Koch, T. (2022). On the exact solution of prize-collecting steiner tree problems. *INFORMS Journal on Computing, 34*(2), 872–889.
21. Rehfeldt, D., Franz, H., & Koch, T. (2022). Optimal connected subgraphs: Integer programming formulations and polyhedra. *Networks, 80*(3), 314–332.
22. Shinano, Y., Rehfeldt, D., & Koch, T. (2019). Building optimal steiner trees on supercomputers by using up to 43,000 cores. In *Integration of Constraint Programming, Artificial Intelligence, and Operations Research. CPAIOR 2019, 11494*, pp. 529 – 539.

23. Takahashi, H., & Matsuyama, A. (1980). An approximate solution for the Steiner problem in graphs. *Mathematica Japonicae, 24*, 573–577.
24. Uchoa, E., & Werneck, R. F. F. (2010). Fast Local Search for Steiner Trees in Graphs. In G. E. Blelloch & D. Halperin (Eds). editors, *ALENEX*, pp. 1–10. SIAM Journal of Applied Mathematics.
25. Vygen, Jens. (2011). Faster algorithm for optimum Steiner trees. *Information Processing Letters, 111*(21), 1075–1079.
26. Wong, R. T. (1984). A dual ascent approach for Steiner tree problems on a directed graph. *Mathematical Programming, 28*, 271–287.

Chapter 4
Prescriptive Analytics for Data-Driven Capacity Management

Pascal M. Notz

Abstract Prescriptive analytics approaches integrate machine learning prediction and optimization to directly derive decisions for planning problems from historical observations of demand and a large set of features (co-variates). This paper summarizes the key results of the author's dissertation and presents two prescriptive analytics approaches, kernelized empirical risk minimization and weighted sample average approximation, to solve complex capacity planning problems. It demonstrates the applicability of both approaches to a real-world two-stage capacity planning problem and evaluates their performance relative to traditional parametric approaches that first estimate a demand distribution and then solve a stochastic optimization problem, and a traditional non-parametric approach (sample average approximation). The results of numerical analyses demonstrate that the new prescriptive analytics approaches can lead to substantial performance improvements of up to 58% compared to traditional approaches.

Keywords Prescriptive analytics · Machine learning · Data-driven operations management · Capacity planning

Introduction

Digitization and artificial intelligence are impacting many areas in business and society, and are leading to a newly emerging field of research within Operations Research: Prescriptive analytics (see, e.g., [5]). Prescriptive analytics approaches combine machine learning methods with optimization techniques to solve planning problems under uncertainty. Based on a large amount of historical observations of an uncertain quantity (e.g., demand) and corresponding features (covariates), they *learn* a prescription function that maps from a (future) feature vector directly to a decision, e.g., a capacity level for a specific week in the future. Therefore,

P. M. Notz (✉)
Chair of Logistics and Quantitative Methods, Julius-Maximilians-Universität Würzburg, Sanderring 2, 97070 Würzburg, Germany
e-mail: pascal.notz@uni-wuerzburg.de

O. Grothe et al. (eds.), *Operations Research Proceedings 2022*, Lecture Notes in Operations Research, https://doi.org/10.1007/978-3-031-24907-5_4

prescriptive approaches contrast traditional two-step approaches that first estimate the (unknown) distribution of demand, and then solve a stochastic optimization problem to arrive at a capacity decision. While first contributions to this new field of research studied approaches to solving comparatively simple planning problems in the area of inventory management (e.g., [1, 2]), this paper focuses on solving capacity planning problems with a more complex structure. In particular, it proposes two prescriptive analytics approaches—kernelized empirical risk minimization (kERM) and weighted sample average approximation (wSAA)—to solve a two-stage stochastic problem with recourse, requiring multiple interdependent decisions depending on matrix-valued observations of demand.

The capacity planning problem studied in this paper is an adaptation of the two-stage planning problem introduced in [4] and based on our work with a logistics service provider in Germany that collects, sorts, and delivers mail (letters, parcels), newspapers, and advertising material. Every day, the company receives mail items that need to be sorted manually (service line 3), semi-automatically (service line 2), or on a fully automated line (service line 1)—with each service line requiring a certain staffing level and service line 1 staff capacity being most expensive, followed by service line 2 capacity. While the staffing levels need to be planned in advance for each week (stage 1 of the planning problem), the company has the option of upgrading after demand has realized on each day (stage 2 of the planning problem): staff of service line 1 can also operate service lines 2 and 3, and staff of service line 2 can also operate service line 3. All mail items need to be sorted on the day of arrival, therefore the company uses costly overtime as required. A formal statement of the problem can be found in [6].

The remainder of the paper is structured as follows: The Section "Prescriptive Analytics Approaches" introduces the approaches kERM and wSAA, and the Section "Numerical Evaluation Based on a Real-World Problem" presents an evaluation of both approaches using historical observations of demand from our case company and realistic cost parameters. The paper concludes that prescriptive approaches can lead to performance improvements of up to 58% compared to traditional approaches.

Prescriptive Analytics Approaches

Prescriptive analytics approaches *learn* a function $\vec{q}(\vec{x})$ that directly maps from feature vectors $\vec{x}$ to decisions $\vec{q}$, based on a data set of historical observations of demand and feature vectors. Assume historical observations of demand $\boldsymbol{d}^n \in \mathcal{D}$ and feature vectors $\vec{x}^n \in \mathcal{X} \subseteq \mathbb{R}^p$, which constitute a data set $S_N = \{(\boldsymbol{d}^1, \vec{x}^1), ..., (\boldsymbol{d}^N, \vec{x}^N)\}$. The features that form the vector $\vec{x}^n$ describe, for example, seasonality (day, month, week), weather conditions, or other quantities that are predictive of the demand values (for further details see [6]). In the following, we present two prescriptive analytics approaches: kERM and wSAA.

Kernelized Empirical Risk Minimization

One approach to learning a prescription function $\vec{q}(\vec{x})$ is to minimize the true risk $R(\vec{q}(\cdot))$, defined as expected loss over the joint distribution of features and demand (as random vectors) $\vec{X} \times \boldsymbol{D}$:

$$\min_{\vec{q}(\cdot)\in\mathcal{F}} R(\vec{q}(\cdot)) := \min_{\vec{q}(\cdot)\in\mathcal{F}} \mathbb{E}_{\vec{X}\times \boldsymbol{D}}\left[L\left(\vec{q}(\vec{X}), \boldsymbol{D}\right)\right] \tag{4.1}$$

with $L(\vec{q}, \boldsymbol{D})$ being the loss function of the planning problem and $\mathcal{F}$ being a function space. $\vec{X} \times \boldsymbol{D}$ describes the underlying joint distribution of features and demand, from which our data set S_N is drawn. Because the distribution of $\vec{X} \times \boldsymbol{D}$ is unknown, and the decision maker only has a data set S_N available, we minimize the *empirical* risk to determine the prescription function $\vec{q}(\cdot)$ (see, e.g., [1, 2, 6]):

$$\min_{\vec{q}(\cdot)\in\mathcal{F}} R_N(\vec{q}(\cdot)) := \min_{\vec{q}(\cdot)\in\mathcal{F}} \frac{1}{N}\sum_{n=1}^{N} L(\vec{q}(\vec{x}^n), \boldsymbol{d}^n), \tag{4.2}$$

using the data set S_N.

We solve (4.2) using a reproducing kernel Hilbert space with kernel function $K(\vec{x}, \vec{x})$ as (nonlinear) function space $\mathcal{F}$ and determine the prescription function of the kERM approach (as in [6]):

$$\vec{q}^{\,\text{kERM}}(\vec{x}) = \sum_{n=1}^{N} \vec{u}^n K\left(\vec{x}^n, \vec{x}\right) - \vec{b}, \tag{4.3}$$

with $\vec{u}^n$ defined as $u_j^n = \frac{1}{2\lambda_j}\left(\sum_{t=1}^{T}(\beta_j^{tn}) + \epsilon_j^n - f_j\right)$, and β_j^{tn}, ϵ_j^n being the optimal solution to the dual problem:

$$\begin{aligned}
\max_{\{\alpha_i^{tn}\},\{\beta_j^{tn}\},\{\epsilon_j^n\}} \quad & L_{\text{d}} := -\sum_{j=1}^{I}\lambda_j \sum_{p,q=1}^{N}\left(u_j^p u_j^q K(\vec{x}^p, \vec{x}^q)\right) + \sum_{n=1}^{N}\sum_{i=1}^{I}\sum_{t=1}^{T}(c_i - \alpha_i^{tn})d_i^{tn} \\
\text{s.t. } & \alpha_i^{tn},\ \beta_j^{tn},\ \epsilon_j^n \geq 0\ \forall i, j, n, t \\
& \alpha_i^{tn} + \beta_j^{tn} \geq a_{ij}\ \forall i \geq j,\ \forall n, t \\
& \sum_{n=1}^{N} u_j^n = 0\ \forall j.
\end{aligned} \tag{4.4}$$

We use a random forest-based kernel function (see, e.g., [3, 6, 7]), because this has lead to superior results in our numerical experiments:

$$K^{\mathrm{RF}}(\vec{x}_1, \vec{x}_2) := \frac{1}{L} \sum_{l=1}^{L} \frac{\mathbb{1}[\mathcal{R}^l(\vec{x}_1) = \mathcal{R}^l(\vec{x}_2)]}{\sum_{j=1}^{N} \mathbb{1}[\mathcal{R}^l(\vec{x}_1) = \mathcal{R}^l(\vec{x}_j)]}, \tag{4.5}$$

where $\mathcal{R}^l(\vec{x})$ is the terminal node of tree l within the forest. We refer the reader to [6] for a more detailed description of the kERM approach.

Weighted Sample Average Approximation

The second approach defines the prescription function $\vec{q}(\vec{x})$ point-wise by solving

$$\min_{\vec{q} \in \mathcal{Q}} \mathbb{E}_{\boldsymbol{D}} \left[L\left(\vec{q}, \boldsymbol{D}\right) \middle| \vec{X} = \vec{x} \right] \tag{4.6}$$

for each new feature vector $\vec{x}$, assuming a conditional probability distribution of $\boldsymbol{D}$, given $\vec{X} = \vec{x}$. The conditional distribution, however, is unknown, but the decision maker has a data set S_N available. Based on local learning techniques, [2] proposes the wSAA approach that aims at "optimizing the decision [$\vec{q}$] against a reweighting of the data" ([2], p. 1030):

$$\vec{q}^{\,\mathrm{wSAA}}(\vec{x}) = \arg\min_{\vec{q} \in \mathcal{Q}} \sum_{n=1}^{N} w_n(\vec{x}) L(\vec{q}, \boldsymbol{d}^n), \tag{4.7}$$

with a weight function $w_n(\vec{x}) \in [0, 1]$ that describes the similarity between the feature vectors $\vec{x}^n$ and $\vec{x}$.

The performance of a wSAA approach is determined by the weight function, and a number of weight functions are introduced in [2], based on: k-nearest-neighbor regression, kernel regression, local linear regression, regression trees, and random forests. In our numerical experiments, we use a random forest-based weight function, which is similar to the random forest kernel. We refer the reader to [6] for a more detailed description of the wSAA approach.

Numerical Evaluation Based on a Real-World Problem

This section presents the numerical evaluation of the prescriptive approaches using the real-world capacity planning problem of our case company. We use realistic cost parameters that specify the planning problem and historical demand data of our case company to compare the performance of kERM and wSAA with that of traditional approaches.

Data Set and Feature Engineering

The data set S_N required for applying the prescriptive approaches consists of demand and feature data. The historical demand data was provided by the case company as daily demand for each of the three sorting lines for a total of $N = 209$ weeks between 2014 and 2017. For constructing the corresponding feature vectors $\vec{x}^n \in \mathbb{R}^{162}$, we determined 162 features including date-based features (e.g., year number, quarter, week number, and so on), lagged demand features (e.g., demand for the same service line and week one year ago), and features encoding information on public holidays. More details on the features used in our experiments can be found in [6].

Evaluation Procedure

To evaluate the performance of the prescriptive and benchmark approaches, we split the data set S_N into a training data set (of $N = 157$ weeks) and a test data set (of 52 weeks). For both prescriptive approaches—kERM and wSAA—we determine the prescription function using a random forest kernel or weight function, and derive capacity decisions for the test period. As benchmark approaches, we use the non-parametric sample average approximation (SAA) approach and two traditional parametric approaches that we call *sequential estimation and optimization* (SEO) approaches: Support Vector Regression (SVR, a machine learning-based approach), and ARIMA (a time series-based approach). Both SEO approaches first predict a demand distribution and then solve a stochastic optimization problem using Monte Carlo sampling, to determine the capacity decision.

The performance of each approach is evaluated in terms of absolute gap to optimal profit (achieved profits based on the approach's capacity decisions in comparison to the maximum achievable profit):

$$\Delta_{\Pi,\text{abs}} = \Pi^*(\boldsymbol{d}) - \Pi(\vec{q}, \boldsymbol{d}) \tag{4.8}$$

with the ex-post optimal profit $\Pi^*(\boldsymbol{d})$.

Results and Discussion

Figure 4.1 shows the absolute gap to optimal profit for all approaches for the test period. We observe that the prescriptive approaches (kERM, wSAA) lead to the best performance, while the traditional SEO approaches lead to the lowest performance. wSAA, which performs best in this particular setting, shows a gap to optimality that is 58% lower compared to ARIMA-SEO, the traditional time series-based approach.

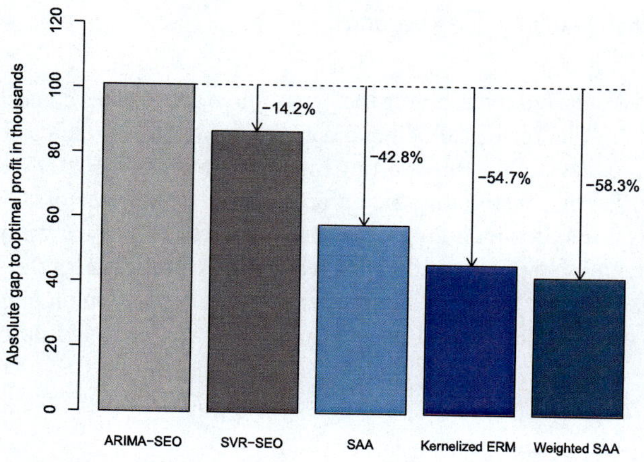

Fig. 4.1 Absolute gap to optimal profit, as shown in [6], p. 1769

The largest performance improvement occurs between the traditional two-step approaches (SVR-SEO, ARIMA-SEO) and SAA, which is the most basic non-parametric approach that does not explicitly estimate the underlying demand distribution. To explain this observation, we assume—in analogy to a newsvendor setting—that each prescription consists of a capacity to serve the mean demand and a safety buffer. Because in our particular case the cost parameters induce a high "optimal" service level (for service line 1), the importance of the prescribed safety capacity is high compared to that of the mean demand. Therefore, the results suggest that the differences in performance are rooted primarily in how the approaches explicitly (SVR-SEO, ARIMA-SEO) or implicitly (SAA, kERM, wSAA) account for demand uncertainty (see [6] for further analyses and details that substantiate this conjecture).

Conclusion

This paper summarizes the key results of the author's dissertation and presents two prescriptive analytics approaches to solve the real-world capacity planning problem of a logistics service provider. We evaluate the performance of both approaches (kERM, wSAA) in comparison with three traditional approaches using historical observations of demand and corresponding features, and observe that the prescriptive approaches lead to a lower gap to optimal profit than the traditional contenders. wSAA leads to the best performance in this particular setting, with a performance improvement of up to 58% compared to the traditional approaches.

One limitation of our work is related to the problem of generalization beyond a specific data set, as commonly observed in the field of machine learning: the (relative) performance of the approaches may be different when using other data sets. In addition, because the prescriptive approaches depend on the problem-specific loss function, their performance may also be sensitive to the specific choice of loss function. This should be addressed by future research.

References

1. Ban, Gah-Yi, & Rudin, Cynthia. (2019). The Big Data Newsvendor: Practical Insights from Machine Learning. *Operations Research, 67*(1), 90–108.
2. Bertsimas, Dimitris, & Kallus, Nathan. (2020). From Predictive to Prescriptive Analytics. *Management Science, 66*(3), 1025–1044.
3. Breiman, Leo. (2000). Some infinity theory for predictor ensembles. Technical Report 577, University of California, Berkeley, CA.
4. Netessine, Serguei, Dobson, Gregory, & Shumsky, Robert A. (2002). Flexible service capacity: Optimal investment and the impact of demand correlation. *Operations Research, 50*(2), 375–388.
5. Notz, Pascal M. (2021). *Prescriptive Analytics for Data-driven Capacity Management. Dissertation*, Würzburg.
6. Notz, Pascal M., & Pibernik, Richard. (2021). Prescriptive Analytics for Flexible Capacity Management. *Management Science, 68*(3), 1756–1775.
7. Scornet, Erwan. (2016). Random forests and kernel methods. *IEEE Transaction Information Theory, 62*(3), 1485–1500.

Chapter 5
Resident Scheduling in Teaching Hospitals

Sebastian Kraul

Abstract After graduation, physicians receive further training in a medical domain like anesthesiology. There are 57 medical specialties in Germany in total. The high cost pressures of hospitals and the changing view of the medical profession regarding the work-life balance have led to recruitment problems and low employee satisfaction in many places. A promising approach to counter this problem is objective and structured training planning. This research project mainly deals with medical residents' strategic and tactical-operative training scheduling. In addition to relieving the medical staff currently responsible for the planning process, this research project increases the predictability of structured training. This allows hospitals to increase the quality of their training and, consequently, their attractiveness to other hospitals. In addition, supervisors from different departments can better assess residents' knowledge and thus keep the level of service, which is particularly important in hospitals, permanently high even when changing residents. From the residents' point of view, a well-structured training schedule enables a high degree of information. Therefore, residents are no longer surprised by a short-term change of department and have a direct insight into their training progress. A real-world case study evaluates the mathematical formulations and the solution approaches.

Keywords OR in health services · Mixed integer programming · Real-world application · Stochastic optimization

Motivation

After graduation, physicians receive further training in a medical domain like anesthesiology or urology. There are 57 medical specialties in Germany in total, in which further training as a specialist is possible. Physicians are called residents in this special phase of training. In Germany, there are about 60,000 residents in specialist

S. Kraul (✉)
Department of Operations Analytics, Vrije Universiteit Amsterdam, Amsterdam, Netherlands
e-mail: s.kraul@vu.nl

O. Grothe et al. (eds.), *Operations Research Proceedings 2022*, Lecture Notes in Operations Research, https://doi.org/10.1007/978-3-031-24907-5_5

training [1]. A characteristic of the German model—representative for many non-English speaking countries in Europe—is the training and further education accompanying the work. Entirely in keeping with the "dual training" concept, the residents are deployed in regular hospital operations and are an essential part of the value chain. In contrast to the American or British rotation model, mainly considered in the current literature, the German specialist training lasts longer. It is tied to specific interventions (tasks), i.e., the possibility of training and further education depends on the hospital's treatment spectrum. However, since the occurrence of interventions relevant to the resident program cannot be planned with certainty in advance and therefore is not taken into account in personnel planning, this form of specialist training is subject to uncertainty concerning training duration [2]. The basic formulation of the resident scheduling problem can be described as follows. A function $f{:}I \times J \times T \rightarrow \{0, 1\}$ is sought, which fulfills:

$$\sum_{i \in I} f(i, j, t) = D(j, t) \qquad \forall j \in J, t \in T \tag{5.1}$$

$$\sum_{t \in T} f(i, j, t) = M(i, j) \qquad \forall i \in I, \ j \in J \tag{5.2}$$

$$\sum_{j \in J} f(i, j, t) = 1 \qquad \forall i \in I, t \in T \tag{5.3}$$

Constraints (5.1) ensure that a teaching service $D(j, t)$ must be fulfilled for all rotations $j \in J$ and periods $t \in T$—in personnel planning problems this is often seen as a demand. Rotations can then be described as departments. Constraints (5.2) consider the training requirement $M(i, j)$ that must be met over the entire time horizon. This means that each resident $i \in I$ must fulfill each rotation to a specified extent. Constraints (5.3) ensure that each resident should be assigned to exactly one rotation in each period.

Personnel planning that is not appropriate to the resident training can have two economic consequences, both of which are based on the fact that every resident training provider (hospital) is obliged to guarantee the residents' training in the time allocated. Suppose the resident can prove that the training program must be extended for reasons the provider is responsible. In that case, the hospital must pay a specialist's salary despite the lack of specialist status. In this context, an initial judgment was made in 2015 by the State Labor Court of Baden-Württemberg. It was stated here that an objective training plan must already be available at the start of employment; otherwise, a time limit is not justified [3]. Even more important is the fact that departments that are repeatedly unable to offer the contents of the resident program within the specified time are threatened with restrictions on the authorization of further training by the medical associations. The departments concerned would then have to hire more expensive specialists instead of the more cost-effective residents if available on the market [4].

The main contributions are the following. First, a strategic problem for determining the maximum number of residents is presented, which is able to consider an uncertain set of interventions via a robustness concept. Second, a possibility is described to deal with the uncertainty between annual planning and daily planning by introducing priorities in annual planning. Third, a model is presented to measure the training program's impact on service quality in the departments.

The remainder of this work is as follows. Next, the methodologies will be presented. Afterward, details of the results are discussed. Eventually, it will close with a conclusion.

Methodology

The interventions have an elementary role in the concept of task-related resident scheduling and are an essential innovation to the existing literature in resident scheduling [5]. Since the distribution of the interventions over the planning horizon of 5 years is difficult if at all, to determine, robustness concepts offer an excellent possibility to consider the uncertainty. Historical data over 26 months makes it possible to determine the upper and lower limits of executable interventions per department quarterly. These limits allow the usage of the price of robustness [6] for the model formulation. One of the advantages of this concept is that the mapped uncertainty does not change the model, i.e., if the model is linear, it can also be formulated linear using the price of robustness. This is important because resident scheduling problems are NP-hard [7]. The same applies to this formulation with interventions, which can be seen as a task allocation and sequencing problem with upper and lower bound temporal constraints. Bertsimas and Sim [8] have proven that this type of problem is NP-hard and, therefore, not tractable. Since the problem's block structure suggests using a decomposition based on a Dantzig-Wolfe reformulation, the compact model is decomposed by residents. The result is a master problem that handles each department's capacity and a subproblem that generates feasible training schedules for one generic resident. The decomposition's unique feature is that the robustness concept is completely included in the subproblem, i.e., each solution of the subproblem is robust according to the defined parameters. A sophisticated column generation heuristic that detects near-optimal solutions (on avg. $< 5\%$ optimality gap) within minutes is developed to solve the decomposition (see Fig. 5.1). In particular, a new pattern generation approach for cyclic problems is presented that further decreases solution times significantly. For each subproblem with negative reduced cost, multiple columns are generated by shifting the start period of the generic resident.

A second aspect of uncertainty is considered in the tactical problem of annual resident scheduling by taking absences into account. Due to absences, a resident might be assigned to a department deviating from the training schedule. One approach to take these types of changes into account in the training schedule is using backup schedules, i.e., a resident is assigned to more than one department. Setting different schedules

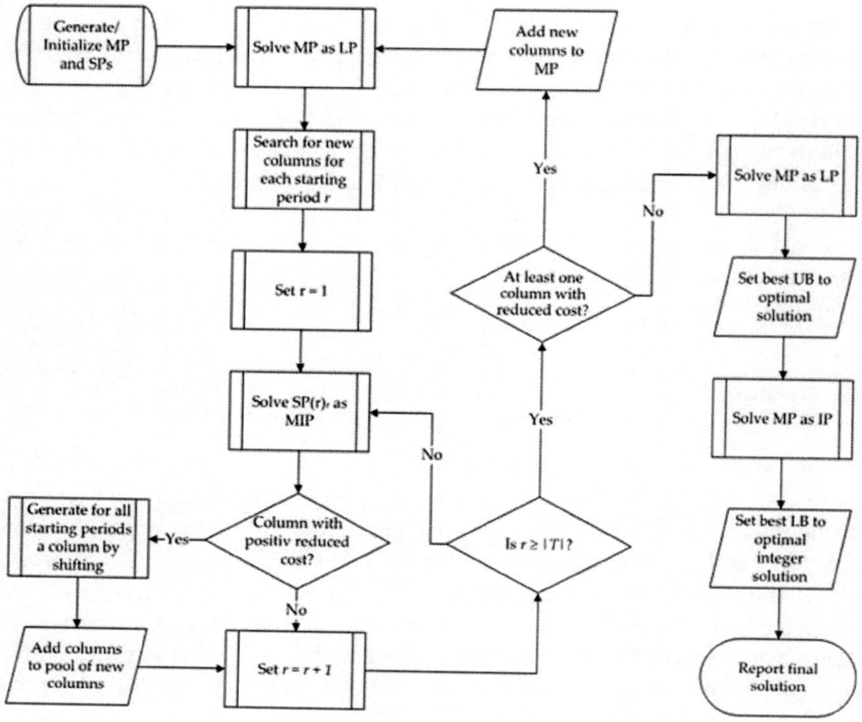

Fig. 5.1 Illustrative flow chart of the column generation heuristic [5]

using priorities gives residents additional information about possible activity fields and ensures that they are not surprised by short-term changes. This kind of training schedule design is not yet considered in the current literature. Therefore, one of the main contributions is the novel mathematical formulation of the annual resident scheduling problem with priorities. The problem is set up as a two-stage stochastic program to map the previously identified absences of residents in the model. Since the deterministic resident scheduling problem is already NP-hard, the stochastic version causes several problems at once. On the one hand, the model cannot be solved with commercial solvers in a suitable time. On the other hand, it fails already because of a finite bound, i.e., the solver cannot determine a linear programming (LP) bound for realistic problem sizes within one hour. Therefore, an analytical bound is derived for the problem. The formulation is decomposed between the first and second stage and integrated into an iterative algorithm to solve the problem.

The violation of continuity is considered in a quadratic formulation of the annual resident scheduling problem [9]. The quadratic formulation is necessary since more changeovers have a stronger effect on the service level. Since non-linear problems are usually more challenging to solve, a linearization of the model is presented. Good, feasible solutions cannot be found in the original quadratic or the linearized formulation within 24 h with standard solvers. For this reason, a metaheuristic approach is used to solve the problem. The genetic algorithm has the advantage that several annual schedules can be made available for selection to the hospital management.

Results

The robustness concept and the integration in a column generation heuristic are tested in a real-world case study of a German university hospital with more than 1,200 beds. The case study shows that the hospital takes a very liberal approach to uncertainty. The hospital trains 84 residents. This value corresponds to approximately a 5% occurrence of the lower limit of interventions per department. Consequently, the hospital is not able to finish the education of all residents on time. The algorithm's performance is analyzed by using 60 instances divided into three different sizes. No significant difference in both the optimality gap and the solution time was found. A comparison of the pattern generation approach showed that the classical column generation took on average 6.1 times longer to solve than the newly developed cyclic approach (see Fig. 5.2). While an increase of the robustness level does not differ according to the problem size, the effect of a change of the robustness level is quite high across all problem sizes, with differences of up to 90%.

The developed algorithm and the analytical bound for the annual resident scheduling problem with absences are analyzed in an experimental study based on the same data set as before. The bounds analysis shows that the analytical bound is superior to the LP bound by more than 8% on average. In addition, it is shown that superiority increases with a rising number of selectable priorities. Besides the analytical bound, an approximation of the upper bound based on the solution algorithm is developed. A parameter is estimated based on numerical tests, which ensures that the approximate bound underestimates the upper bound by a maximum of 12%. However, this does not occur during the entire experimental study. In a further study, the performance of the solution algorithm is analyzed. For this purpose, the algorithm is tested using a cold and warm start as well as a batching approach to reduce the number of scenarios considered in parallel. This study shows that a warm start with the solution of the deterministic problem and in combination with a batching scheme

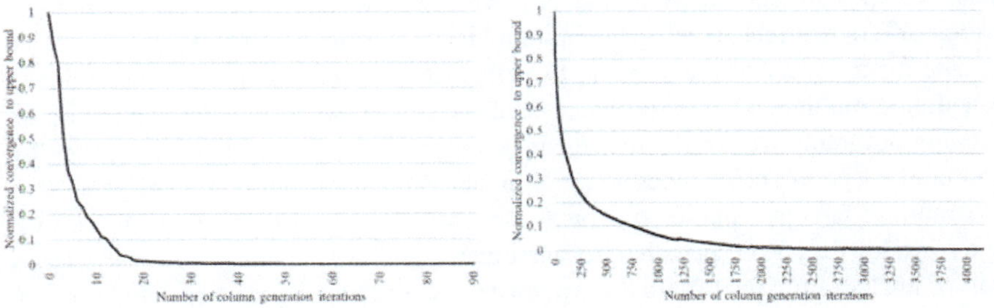

Fig. 5.2 Convergence cyclic pattern generation (left), convergence standard generation (right) [5]

is advantageous both in terms of solution quality and solution time. The results show that a training schedule with two priorities already eliminates almost all unexpected assignments, i.e., from initially over 10% to only 1%. Finally, the study shows that a training schedule with two priorities is already superior to traditional planning with only one priority.

The genetic algorithm is evaluated and compared with standard software solutions analyzing a German training hospital's real-world situation from 2016. When comparing the real-world case with the genetic algorithm's solutions, it becomes clear that the hospital has not exhausted all potentials concerning the objectives investigated. It would have been possible to improve continuity by 25% or improve fair training progress by up to 30%. The analysis of both models and the comparison of the genetic algorithm solutions with those of the standard solver shows that the standard solver solutions are always in the same solution region regardless of the selected objective function weights. This is because the solver often does not get beyond the heuristically constructed start solution within 24 h. It has been shown that the solver has problems finding and improving solutions concerning fair training progress. Even in small instances with only 20 residents, a standard solver could not solve the problem optimally. These results again underline the genetic algorithm's strength compared to the solver for the problem under investigation. Nevertheless, the solver results show that there are possibilities for small instances to sharpen the MIP formulation further to get a good solution. In general, the study shows that the costs for continuity of care and fair training progress are very high in the respective extreme, i.e., improving continuity of care by a small amount has a high impact on the fairness level if the schedule has a good fairness level and vice versa. However, it is also possible to determine a medium range in which the relative improvement of fair training progress is associated with low costs of continuity.

Conclusion

The resident scheduling problem has a unique structure due to the simultaneous demands of departments and residents. According to specialization, hospital, and country, the individual characteristics provide for a large number of variations of the basic model. This work deals with a German case of the resident scheduling problem, which differs from the existing literature by its task-related structure.

The overall goal of the work is to get a deeper understanding of the resident scheduling problem. Current problems such as non-compliance with the duration of training programs are considered. Besides, generalizable interconnections, such as continuity of care or differences between annual and daily planning, are analyzed. The Operations Management literature on resident scheduling can be divided into staffing and rostering decisions, as with other personnel planning problems. However, a large part of the literature deals with the American resident training system, which shows notable differences to the training system investigated in this work. The first contribution combines a column generation method with an existing robustness concept based on surgical data. An advantage of this approach is that decision-makers can define a suitable level of robustness independently. Interventions that are difficult to predict in this time horizon have a fundamental impact on the duration of the training program and the possible number of residents a hospital can employ. Uncertain events, such as absences, are another contribution. In addition, the loss of information due to operational processes' aggregation on a tactical level is considered here. It is shown that the use of priority assignments positively affects adherence to the tactical schedule on an operational level. Planning processes often influence other areas, as well. Thus, the final study shows the influence of departmental changes within the training program and how to deal with them.

To summarize, the socio-economic change on the one hand and medical progress on the other will increasingly present hospitals with new challenges. Residents will play a decisive role in this, representing a more cost-effective alternative to specialists. Therefore, it will become increasingly important to deploy residents effectively and efficiently.

Acknowledgements This research project is funded by the Deutsche Forschungsgemeinschaft (DFG, German Research Foundation) and Grant no. 405488489.

References

1. German Medical Association. (2019). Ärztestatistik 2019.
2. Miani, C., Hinrichs, S., Pitchforth, E. et al. (2015). *Best practice: Medizinische Aus- und Weiterbildung aus internationaler Perspektive.* RAND Corporation.
3. LAG, Baden-Württemberg. (2015). Voraussetzung für eine Befristung eines Arbeitsvertrags mit einem Arzt in Weiterbildung ist nach § 1 Abs. 1 ÄArbVtrG, dass die Beschäftigung des Arztes seiner zeitlich und inhaltlich strukturierten Weiterbildung dient. Dies bedeutet, dass der Arbeitgeber bei Abschluss des befristeten Arbeitsvertrags zu diesem Zweck eine Weiterbildungsplanung erstellen muss, die zeitlich und inhaltlich auf die konkrete Weiterbildung zugeschnitten ist. Die Planung muss nicht Inhalt der (schriftlichen) Befristungsabrede sein; sie muss aber objektiv vorliegen. openJur 2015, 19265.
4. Knichwitz, G., & Wenning, M. (2009). Gehen Deutschland die Anästhesisten aus. *Anästh Intensivmed, 50*, 276–282.
5. Kraul, S., Fügener, A., Brunner, J. O., et al. (2019). A robust framework for task-related resident scheduling. *European Journal of Operational Research, 276*, 656–675. https://doi.org/10.1016/j.ejor.2019.01.034
6. Bertsimas, D., & Sim, M. (2004). The price of robustness. *Operations Research, 52*, 35–53. https://doi.org/10.1287/opre.1030.0065
7. Guo, J., Morrison, D. R., Jacobson, S. H., et al. (2014). Complexity results for the basic residency scheduling problem. *Journal of Scheduling, 17*, 211–223. https://doi.org/10.1007/s10951-013-0362-9
8. Bertsimas, D., Weismantel, R. (2005). *Optimization over integers.* Dynamic Ideas.
9. Kraul, S. (2020). Annual scheduling for anesthesiology medicine residents in task-related programs with a focus on continuity of care. *Flexible Services and Manufacturing Journal, 32*, 181–212. https://doi.org/10.1007/s10696-019-09365-4

Chapter 6
Solving Customer Order Scheduling Problems with an Iterated Greedy Algorithm

Julius Hoffmann

Abstract In this paper, three configurations of the customer order scheduling problem are presented. In contrast to classical scheduling problems, the customer order scheduling problem considers the scheduling of jobs that belong to customer orders and each order is only completed when each job of the order has finished. The studied configurations are the minimization of the sum of order completion times and the minimization of the earliness-tardiness in a machine environment where each order places one job on each machine. Furthermore, the minimization of the sum of order completion times in a flow shop environment is investigated. This paper states properties of the three problem configurations and describes developed solution methods that performed well in a computational experiment.

Keywords Customer order scheduling · Metaheuristics · Optimization · Iterated greedy algorithm · Manufacturing

Introduction

In classical scheduling problems, jobs are scheduled on machines for the purpose of minimizing or maximizing a job related objective. However, in many real life situations, multiple jobs belong to a customer order. To save transportation costs and execution time, it is reasonable to ship all jobs of an order at once [1]. Various studies address this issue by investigating the so-called customer order scheduling problem (COSP), see [1, 2] for examples. In the following, the COSP is defined as a problem in which several customers order multiple products. An order is completed when all jobs of the order are finished. The corresponding objective function of the problem is related to the completion times of the orders.

J. Hoffmann (✉)
Technische Universität Dresden, 01062 Dresden, Germany
e-mail: julius.hoffmann1@tu-dresden.de

O. Grothe et al. (eds.), *Operations Research Proceedings 2022*, Lecture Notes in Operations Research, https://doi.org/10.1007/978-3-031-24907-5_6

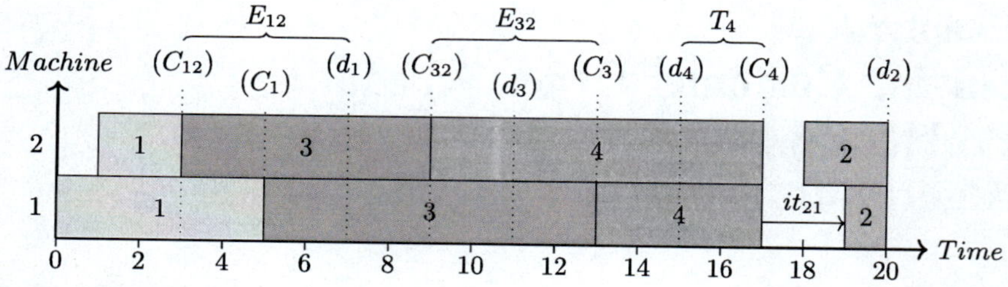

Fig. 6.1 Example Gantt chart of the COSP in a dedicated machine environment

Recently, the COSP has been applied in the context of 3D printing [3]. Nevertheless, the concept of the COSP is not limited to the manufacturing context and can be transferred to other disciplines, e.g., parallel computing. Here, several tasks are processed independently on multiple processors and are gathered afterwards to finish the main task [4]. Various configurations of the COSP are studied in the literature, e.g., minimizing the sum of weighted completion times of the orders with multiple identical machines [5], or minimizing the sum of tardinesses of the orders in an environment with so-called dedicated machines [6].

The objective of this paper is to present three problem configurations of the COSP, with corresponding properties and solution methods of the configurations.

General Problem Description

To introduce the notation, the COSP is explained by an example which can be found in Fig. 6.1. In this example, the dedicated machine environment is considered where each customer places exactly one job on each machine. The number inside the scheduled job represents the order of the job.

In the given example, the number of orders n is four, the number of jobs per order o is two, and the number of machines m is two. In this paper, the number of a job in an order corresponds to the machine number if the dedicated machine environment is considered. Each job j of an order i has a production time t_{ij} and a completion time C_{ij}. An order is completed when all of its jobs are finished. Consequently, the completion time of an order is calculated by $C_i = \max_{1 \leq j \leq o}\{C_{ij}\}$. This property can be seen by C_1, which equals C_{11}.

Furthermore, each order has a due date d_i if the problem configuration requires it. By the due date, the tardiness of an order is calculated by $T_i = \max\{0, C_i - d_i\}$, see d_4 and T_4 as example. Furthermore, the earliness of a job is defined as $E_{ij} = \max\{d_i, C_i\} - C_{ij}$. The definition of the earliness here is based on the idea that each finished job must wait until shipping, which occurs when all jobs of an order are completed or the due date is reached, whichever is later. The earliness E_{12} is given as an example for the case when the completion time is lower than the due date of the order, E_{32} is shown for the opposite case. If idle times are considered, the decision maker can insert idle times it_{ij} in front of jobs, e.g., it_{21} in front of job 1 of order 2.

Minimizing the Total Completion Time in a Dedicated Machine Environment

In this problem configuration it is aimed to minimize $\sum_{i=1}^{n} C_i$ in a dedicated machine environment. This configuration has already been studied in the literature, see [1, 2] for examples. The following lemma holds for this configuration [2]:

Lemma 1 *There exists an optimal schedule for each problem instance in which all orders are processed in the same sequence on each machine.*

Consequently, a solution Π of this problem configuration is represented as a permutation schedule of orders, e.g., $\Pi = \{1, 3, 4, 2\}$. Idle times and preemptions are not considered for this configuration. The iterated greedy algorithm (IGA) was used for solving this problem configuration because it is a simple and effective metaheuristic and has successfully solved permutation flow-shop scheduling problems [7]. The IGA consists of four or five functions, depending on the use of the local search function, and a termination criterion, each of which can be designed differently [7]. The pseudocode of the IGA can be found in Alg. 1.

Algorithm 1 General procedure of the IGA (based on [7])

procedure IGA
  $\Pi \leftarrow Initialization()$
  $\Pi \leftarrow LocalSearch(\Pi)$
  $\Pi^* \leftarrow \Pi$
  while Termination Condition not met **do**
    $\pi_d, \pi_{n-d} \leftarrow Destruction(\Pi)$
    $\Pi' \leftarrow Construction(\pi_d, \pi_{n-d})$
    $\Pi' \leftarrow LocalSearch(\Pi')$
    $\Pi^*, \Pi \leftarrow Acceptance(\Pi^*, \Pi, \Pi')$
  end while
  return $\Pi^*, F(\Pi^*)$ ▷ Best found solution Π^* and belonging value $F(\Pi^*)$
end procedure

Four different IGAs were developed which differ in their initialization and construction. Two IGAs use a modified *NEH heuristic* [8], the other IGAs the *ECT heuristic* [2] for the initialization. The ideas for the local search, destruction size and strategy, and acceptance of the four IGAs were based on the descriptions of *multiple insertion*, *variable number based on VND*, *random selection*, and *simulated-annealing-based acceptance with a solution-based temperature* in [7]. For the construction function, all IGAs use the idea of the *greedy reinsertion* [7]. However, the construction functions differ by their reinsertion sequence method. To obtain the reinsertion sequence, the idea of the *greedy selection* is used by two IGAs, while a *heuristic selection* according to the priority rule used during the modified NEH heuristic is performed by the other IGAs (see [7] for descriptions).

Two important adjustments have been made to the mentioned functions. Both use the destruction size d, which represents the number of iterations no better solution was found, but at the same time is limited to the range $[d_{\min}, d_{\max}]$. The local search function uses d by removing and subsequently reinserting single orders $\left\lfloor z * \frac{d}{d_{\min}} \right\rfloor$ times and the acceptance function by setting the probability of accepting a worse solution to $q \leq e^{-y * \frac{d_{\min}}{d} * \frac{F(\Pi') - F(\Pi)}{F(\Pi)}}$, where q is randomly drawn each iteration from $\mathcal{U}_{[0,1]}$ and $F(\cdot)$ is the objective function value of a solution. The parameters y and z are determined experimentally and $d_{\min} = 1$ and $d_{\max} = \frac{n}{2}$ are adopted from the literature [9]. Consequently, the local search is intensified and it is more likely to accept a worse solution in order to escape local optima when no better solution was found during recent iterations.

The proposed IGAs were compared with two IGAs from the literature [10] and a modeled MILP solved by Gurobi (version 9.0.3) in a computational experiment according to the criterion of the best found solution after given run times. In this experiment, the best performing method was the IGA which uses the modified NEH heuristic as initialization function and the priority rule of the modified NEH heuristic for determining the construction sequence.

Minimizing the Total Completion Time in a Flow Shop Environment

This section addresses the minimization of the sum of order completion times in a flow-shop environment with m stages ($m \geq 2$) and one machine per stage. To the best of our knowledge, this particular problem configuration has not yet been studied. An example with two orders, three jobs per order, and two machines for this environment can be found in Fig. 6.2. The string inside the scheduled job (i, j) represents the job j of order i.

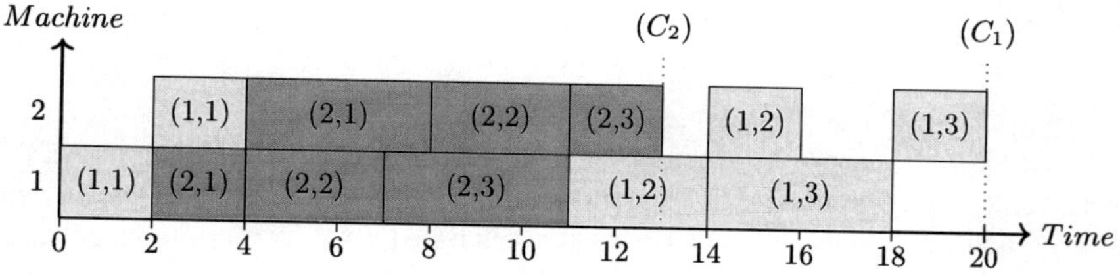

Fig. 6.2 Example Gantt chart of the COSP in a flow-shop environment

It can be shown by a counter example that there are instances of this problem configuration where it is not possible to schedule all jobs of an order one after another to obtain an optimal solution. Therefore, a solution of this configuration here is a permutation of all jobs of all orders which are processed in this sequence on each machine, e.g., $\Pi = \{(1, 1), (2, 1), (2, 2), (2, 3), (1, 2), (1, 3)\}$ in Fig. 6.2.

Four algorithms were developed for this problem configuration. The algorithms were derived from the structure and the functions of the best performing IGA in Section "Minimizing the Total Completion Time in a Dedicated Machine Environment" and initialize a first solution with the same heuristic, which is based on the ideas of the NEH heuristic. The algorithm *IGA-String* has a lot of similarities with the mentioned IGA from Section "Minimizing the Total Completion Time in a Dedicated Machine Environment". However, instead of moving orders, the jobs are moved in the permutation solution which has a size of $n * o$.

In contrast, the other algorithms, called *IGA-Matrix* algorithms, first modify the permutation of the jobs of each order separately and subsequently the sequence of the orders. There exists an additional acceptance function in *IGA-Matrix1* and *IGA-Matrix2* which is directly executed after the job permutations of the single orders are determined. This function defines the job permutations of the single orders for the next iteration before the final solution of the iteration is generated. Because the jobs of an order do not have to be processed one after another, *IGA-Matrix2* and *IGA-Matrix3* perform an additional local search before the (last) acceptance step in each iteration. This local search function swaps adjacent jobs of the schedule and intends to check if a better solution is generated when not all jobs of an order are processed one after another.

The four algorithms were compared with each other in a computational experiment. Again, the comparison criterion was the best found solution after given run times. The best performing solution method was *IGA-Matrix3*.

Minimizing the Earliness-Tardiness in a Dedicated Machine Environment

The problem addressed here is to minimize $\sum_{i=1}^{n}\sum_{j=1}^{m}\left(E_{ij} + T_i\right)$ in a dedicated machine environment. To the best of our knowledge, this problem configuration has not yet been investigated. It can be shown by counterexamples that it may be necessary to insert idle times, and that for some problem instances it is not possible to schedule all orders in the same sequence on each machine, to obtain an optimal solution.

Consequently, idle times are considered here, and each machine can have its own schedule. It can be shown that inserting or increasing an idle time before the first job of a job-group g, where the jobs are not separated by an idle time from each other (idle times between the jobs are zero) but separated from the following jobs by an idle time larger than zero or because they are the last scheduled jobs, decreases the objective value if the following equation holds:

$$y_g < \frac{n_g}{2 * m}. \tag{6.1}$$

In Eq. (6.1), n_g is the number of jobs in g, and y_g the number of jobs in g which have an earliness of zero. An example for the described group g is the group of jobs from the orders 1, 3, and 4 on machine 2 in Fig. 6.1.

Six IGAs were developed for this configuration. They follow the structure of the best performing IGA from Section "Minimizing the Total Completion Time in a Dedicated Machine Environment". Additionally, each IGA has a refinement function that is executed after the local search. Five of the refinement functions start with setting an idle time before the first job on each machine. The value of this offset is either the minimum earliness on the machine (used by *IGA2*, *IGA5*) or a value determined according to the property that is related to Eq. (6.1) (*IGA1*, *IGA3*, *IGA6*). Subsequently, all IGAs except *IGA1* and *IGA4* use a method that swaps adjacent jobs on single machines if it results in a better solution. In a last step, *IGA4*, *IGA5*, and *IGA6* check a possible idle time insertion before each job on each machine according to the property that is related to Eq. (6.1).

In a computational experiment, the six IGAs were compared with each other and a MIP solved by Gurobi according to the criterion of the best found solution after given run times. The IGAs outperformed the Gurobi solver in this experiment. Furthermore, *IGA6* was the best performing solution method.

Conclusion

This paper presents three problem configurations of the COSP. Properties are stated for these configurations and, based on these properties, algorithms are described. The first studied problem configuration was successfully solved with IGAs. Consequently, the structure of the best performing IGA for the first configuration was used for developing algorithms for the other COSP configurations.

Future research could examine further COSP configurations with algorithms that are based on the proposed IGAs. Additionally, the performance of the presented algorithms could be compared with the performance of other solution methods, e.g., AI-based techniques.

References

1. Framinan, J. M., & Perez-Gonzalez, P. (2017). New approximate algorithms for the customer order scheduling problem with total completion time objective. *Computers and Operations Research, 78*, 181–192. https://doi.org/10.1016/j.cor.2016.09.010
2. Leung, J. Y. -T., Li, H., & Pinedo, M. (2005) Order scheduling in an environment with dedicated resources in parallel. *Journal of Scheduling, 8*(5), 355–386. https://doi.org/10.1007/s10951-005-2860-x
3. Zipfel, B., Neufeld, J. S., & Buscher, U. (2021). Customer order scheduling in an additive manufacturing environment. In A. Dolgui, A. Bernard, D. Lemoine, G. von Cieminski & Romero, D. (Eds.), *Advances in production management systems. Artificial intelligence for sustainable and resilient production systems. APMS 2021. IFIP advances in information and communication technology* (Vol. 633, pp. 101–109). Springer, Cham. https://doi.org/10.1007/978-3-030-85910-7_11
4. Sung, C. S., & Yoon, S. H. (1998). Minimizing total weighted completion time at a preassembly stage composed of two feeding machines. *International Journal of Production Economics, 54*(3), 247–255. https://doi.org/10.1016/S0925-5273(97)00151-5
5. Leung, J. Y. T., Lee, C., Ng, C., & Young, G. (2008). Preemptive multiprocessor order scheduling to minimize total weighted flowtime. *European Journal of Operational Research, 190*(1), 40–51. https://doi.org/10.1016/j.ejor.2007.05.052
6. Framinan, J. M., & Perez-Gonzalez, P. (2018). Order scheduling with tardiness objective: Improved approximate solutions. *European Journal of Operational Research, 266*(3), 840–850. https://doi.org/10.1016/j.ejor.2017.10.064
7. Zhao, Z., Zhou, M., & Liu, S. (2022). Iterated greedy algorithms for flow-shop scheduling problems: A tutorial. *IEEE Transactions on Automation Science and Engineering, 19*(3), 1941–1959. https://doi.org/10.1109/TASE.2021.3062994
8. Nawaz, M., Enscore, E. E., & Ham, I. (1983). A heuristic algorithm for the m-machine, n-job flow-shop sequencing problem. *Omega, 11*(1), 91–95. https://doi.org/10.1016/0305-0483(83)90088-9
9. Framinan, J. M., & Leisten, R. (2008). Total tardiness minimization in permutation flow shops: A simple approach based on a variable greedy algorithm. *International Journal of Production Research, 46*(22), 6479–6498. https://doi.org/10.1080/00207540701418960
10. Wu, C. C., Yang, T. H., Zhang, X., Kang, C. C., Chung, I. H., & Lin, W. C. (2019). Using heuristic and iterative greedy algorithms for the total weighted completion time order scheduling with release times. *Swarm and Evolutionary Computation, 44*, 913–926. https://doi.org/10.1016/j.swevo.2018.10.003

Chapter 7
The Stochastic Bilevel Selection Problem

Jannik Irmai

Abstract We consider a bilevel continuous knapsack problem where the leader controls the capacity of the knapsack, while the follower chooses a feasible packing maximizing his own profit. The leader's aim is to optimize a linear objective function in the follower's solution, but with respect to item values that can be different from the follower's item values. We address a stochastic version of this problem where the follower's profits are uncertain and only a probability distribution is known. This problem is #P-hard for the case of independently and uniformly distributed follower profits. In this paper, efficient algorithms are developed for the special case where all items have unit weight, as is the case in the bilevel selection problem. Generalizing these results to the case of arbitrary weights leads to pseudo-polynomial time algorithms for the bilevel continuous knapsack problem.

Keywords Bilevel optimization · Stochastic optimization · Complexity

Introduction

The classical 0–1 knapsack problem and variants thereof are much studied optimization problems. The interpretation as a knapsack packing problem dates back to Dantzig [5]: Given a set of items, each associated with a weight and a value, and a capacity of the knapsack, the objective is to find a selection of the items such that the total weight of the selected items does not exceed the capacity, and the total value of the selected items is maximal. If all items have the same weight, i.e., unit weight, an optimal solution is easily obtained by selecting the most valuable items until the capacity is reached. This special case of the knapsack problem is called the selection problem.

In the past decades, variants of the knapsack problem involving uncertainty and/or multiple levels have received increasing attention, see, for example, [4] for bilevel variants or [2, 7] for bilevel variants with uncertainty. In this paper we consider

J. Irmai (✉)
TU Dresden,Dresden, Germany
e-mail: jannik.irmai@tu-dresden.de

O. Grothe et al. (eds.), *Operations Research Proceedings 2022*, Lecture Notes in Operations Research, https://doi.org/10.1007/978-3-031-24907-5_7

the stochastic bilevel continuous knapsack problem with an uncertain follower's objective (SKP), which is proposed in [1].

Definition 1 *The stochastic bilevel continuous knapsack problem is defined as follows*

$$\begin{aligned} \max\ & \mathbb{E}_{\mathbf{c}}[d^\top x^{\mathbf{c}}] \\ \text{s.t.}\ & b^- \leq b \leq b^+ \\ & x^{\mathbf{c}} \in \operatorname{argmax}\ \mathbf{c}^\top x \\ & \qquad \text{s.t.}\quad a^\top x \leq b \\ & \qquad\qquad\quad x \in [0,1]^n, \end{aligned} \tag{SKP}$$

where $n \in \mathbb{N}$ is the number of items. The leader controls the knapsack's capacity $b \in \mathbb{R}$, which is limited by the lower and upper bounds $b^-, b^+ \in \mathbb{R}_{\geq 0}$. The follower's variables are $x \in [0, 1]^n$, the fractions determining how much of each item is packed. The items have weights $a \in \mathbb{Z}^n_{>0}$ and different profits for the leader and follower. The leader's profits are $d \in \mathbb{R}^n$ and the follower's profits are the n dimensional random variable $\mathbf{c}$ that only takes positive values. The vector $x^{\mathbf{c}} \in [0, 1]^n$ denotes an optimal solution of the follower's problem for follower profits $\mathbf{c}$ and can also be understood as a random variable. To emphasize the dependency on the capacity b, we may also write $x^{\mathbf{c}}(b)$.

The possibly negative leader profits d allow an interpretation where the leader pays, i.e., has negative profit, for providing capacity to the follower, for details see [1]. For deterministic follower profits c, the bilevel continuous knapsack problem can be solved efficiently, see, for example, [6]. In [1] it is shown that the SKP is already #P-hard for componentwise independently binary uniformly or componentwise independently continuous interval uniformly distributed follower profits. In the following, for the sake of brevity, we just write *componentwise* instead of componentwise independent.

In this paper, we present pseudo-polynomial time algorithms for the SKP for both componentwise finitely and componentwise continuous interval uniformly distributed follower profits. Both algorithms have a run time that is polynomial in the number of items and linear in the sum of the weights of the items, meaning they are efficient if the weights of the items are bounded by a constant. The main results from this paper have already been published in [3].

Problem Analysis

For certain item values $c \in \mathbb{R}^n$, the follower's problem can easily be solved by sorting the items according to their relative profits $\frac{c_i}{a_i}$ and packing the most valuable items until the capacity is reached. As a result, the follower's solution $x^c(b)$ is a piecewise linear monotonically increasing function.

We denote the expected value of the follower's solution with $\hat{x}(b) := \mathbb{E}_{\mathbf{c}}[x^{\mathbf{c}}(b)]$ and the leader's objective function with $f(b) := \mathbb{E}_{\mathbf{c}}[d^\top x^{\mathbf{c}}] = d^\top \hat{x}(b)$. Since there are only finitely many possible orderings of the items, the expected value $\hat{x}$ is, as a sum of finitely many piecewise linear functions, also a piecewise linear function in b and so is f. Therefore, the SKP can be solved by evaluating the objective function f at all vertices between the linear segments. Furthermore, it is already #P-hard to evaluate the objective function for a given capacity.

Theorem 1 *Evaluating the expected value $\hat{x}(b)$ for a given b for independently binary uniformly or independently continuous interval uniformly distributed follower profits is #P-hard.*

Since the SKP is #P-hard (for componentwise finite and componentwise continuous uniform follower profits) we cannot expect to find a polynomial time algorithm for the problem. However, the hardness is proven by a reduction from #Knapsack, which is only weakly #P-hard and leaves the possibility of a pseudo-polynomial time evaluation of the objective function. This motivates us to initially study the special case where all items have unit weight.

Definition 2 *The stochastic bilevel selection problem is the following optimization problem*

$$\begin{aligned} \max\ & \mathbb{E}_{\mathbf{c}}[d^\top x^{\mathbf{c}}] \\ s.t.\ & b \in \{0, \dots, n\} \\ & x^{\mathbf{c}} \in \mathit{argmax} \quad \mathbf{c}^\top x \\ & \qquad s.t. \qquad \textstyle\sum_{i=1}^n x_i \leq b \\ & \qquad\qquad\qquad x \in \{0, 1\}^n. \end{aligned} \tag{SSP}$$

where d and $\mathbf{c}$ are the same parameters as for the SKP.

The objective function of the leader in the stochastic bilevel selection problem is, as for the SKP, piecewise linear with at most $n + 1$ vertices in $\{0, \dots, n\}$. Therefore, an algorithm to evaluate $\hat{x}(b)$ in polynomial time for all $b = 0, \dots, n$ would enable evaluating $f(b)$ in polynomial time for all $b = 0, \dots, n$, which, in turn, would enable solving the SSP in polynomial time. For the SSP, the expected value $\hat{x}_i(b)$ can be understood as the probability that the follower selects item i given the capacity b for $i, b \in \{1, \dots, n\}$.

Remark 1 The SKP can be reduced to the SSP by splitting each item i of weight $a_i \in \mathbb{Z}_n$ into a_i many individual items of weight 1 with leader profit $\frac{d_i}{a_i}$ and follower profit $\frac{c_i}{a_i}$. The derived SSP instance has $\sum_{i=1}^n a_i$ many items, and the described reduction is only a pseudo-polynomial time reduction. However, if the follower profits $\mathbf{c}$ of the SKP instance are distributed independently, the follower profits of the derived SSP instance are no longer independent (the follower profits of two items of the derived instance that were obtained by splitting an item of the original instance are correlated).

The Selection Probability

As discussed above, the SSP can be solved efficiently if the selection probabilities $\hat{x}_i(b)$ can be computed efficiently for $i, b \in \{1, \dots, n\}$. Instead of computing these selection probabilities directly, it turns out to be easier to first compute the values $\hat{y}_{ib} := \hat{x}_i(b) - \hat{x}_i(b-1)$ for $i, b \in \{1, \dots, n\}$. For an item i and a capacity b, the value $\hat{y}_{ib}$ is the probability that item i has the b largest follower profit.

In the following, we present a method to efficiently compute the *position probabilities* $\hat{y}$ by means of dynamic programming. To formalize this dynamic approach, we introduce the following notation.

Let $N := \{1, \dots, n\}$ denote the set of items. For $i \in N$, $I \subseteq N \setminus \{i\}$, $b \in \{0, \dots, |I|\}$, we define

$$g(b, I, x) = g(b, I, x; i) := \mathbb{P}\big(\big|\{j \in I \mid \mathbf{c}_j > x\}\big| = b \mid \mathbf{c}_i = x\big) ,$$

the probability that exactly b of the items in I have a larger follower profit than x when item i has follower profit x.

With this notation we can compute for $i, b \in N$

$$\hat{y}_{ib} = \sum_{x \in \text{supp}(\mathbf{c}_i)} \mathbb{P}(\mathbf{c}_i = x) \cdot g(b-1, N \setminus \{i\}, x; i) \tag{1}$$

for discretely distributed $\mathbf{c}_i$ and

$$\hat{y}_{ib} = \int_{x \in \text{supp}(\mathbf{c}_i)} p_i(x) \cdot g(b-1, N \setminus \{i\}, x; i)\, dx \tag{2}$$

for continuously distributed $\mathbf{c}_i$ where $\text{supp}(\mathbf{c}_i)$ is the support of $\mathbf{c}_i$ and $p_i(x)$ is the probability density function of $\mathbf{c}_i$.

In the following we assume that the follower profits are distributed componentwise, i.e., for $i, j \in N$, $i \neq j$ the random variables $\mathbf{c}_i$ and $\mathbf{c}_j$ are independent. Then the function g can be computed, independently of i, with the following recursion:

Proposition 1 *For componentwise distributed follower profits, $x \in \mathbb{R}$, $I \subseteq N$, $I \neq \emptyset$, $b = 0, \dots, |I|$ and $k \in I$, the following holds:*

$$g(b, I, x) = \mathbb{P}(\mathbf{c}_k > x) \cdot g(b-1, I \setminus \{k\}, x) + (1 - \mathbb{P}(\mathbf{c}_k > x)) \cdot g(b, I \setminus \{k\}, x) .$$

Further, we have $g(0, \emptyset, x) = 1$.

In other words, the probability that exactly b items in I have a larger follower profit than x equals the probability that item k has a larger follower profit than x and $b-1$ items in $I \setminus \{k\}$ have a larger follower profit than x or that item k has a smaller follower profit than x and b items in $I \setminus \{k\}$ have a larger follower profit than x.

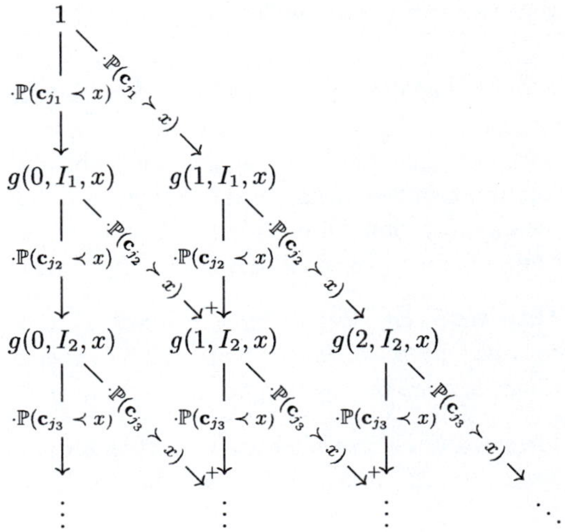

Fig. 7.1 Dynamic computation of $g(b, N \setminus \{i\}, x)$ according to Proposition 2.

Using the recursive formula according to the visualization in 7.1, the auxiliary values $g(b, N \setminus \{i\}, x)$ from (1) and (2) can be computed efficiently:

Proposition 2 *For $x \in \mathbb{R}$, $i \in N$ and given $\mathbb{P}(\mathbf{c}_j > x)$ for $j \in N \setminus \{i\}$, one can compute $g(b, N \setminus \{i\}, x)$ for $b = 0, \dots, n-1$ in time $\mathcal{O}(n^2)$.*

For componentwise finitely distributed follower profits, the position probabilities $\hat{y}$ can be computed efficiently with (1). From this, the selection probabilities $\hat{x}$ can be computed and with those the objective function can be evaluated. Together, we obtain a polynomial time algorithm for solving the SSP:

Theorem 2 *For componentwise finitely distributed follower profits, the SSP can be solved in time $\mathcal{O}(m^2n^2 + mn^3)$, where $m = \max_{i \in N} |supp(\mathbf{c}_i)|$ is the maximum number of different values that the follower profits take for each item.*

For componentwise follower profits distributed uniformly on continuous intervals, the auxiliary function g is a piecewise polynomial function in x and with (2), the SSP can be solved efficiently:

Theorem 3 *For componentwise continuous interval uniformly distributed follower profits, the SSP can be solved in time $\mathcal{O}(n^5)$.*

The complexity of the SSP clearly depends on the type of the distribution that describes the follower profits. For a more complex distribution that is not componentwise, even the SSP where all items have unit weight is hard to solve.

Theorem 4 *1. Computing $\hat{y}$ with follower profits that are uniformly distributed on a polytope, given by an outer description, is #P-hard.*

2. Computing $\hat{y}$ for follower profits that are uniformly distributed on the vertices of a box that satisfy a linear inequality is #P-hard.

Pseudo-polynomial Time Algorithms

As described in Remark 1, the more general SKP can be reduced to the special case of the SSP by splitting each item into multiple items of unit weight. By adapting the dynamic computation of the position probabilities to handle the correlation of the follower profits of the derived instance, we obtain algorithms for solving the SKP in pseudo-polynomial time. More precisely, the runtime is linear in $A = \sum_{i=1}^{n} a_i$, the sum of the weights of all items.

Theorem 5 *For componentwise finitely distributed follower profits, the SKP can be solved in time $\mathcal{O}(m^2n^2 + mn^2A)$, where m is the maximum number of different values that the follower profits take for each item.*

Theorem 6 *For componentwise continuous interval uniformly distributed follower profits, the SKP can be solved in time $\mathcal{O}(n^4A)$.*

Conclusion

This paper investigates the #P-hard SKP from an algorithmic perspective. For both componentwise finitely and componentwise continuous interval uniformly distributed follower profits, pseudo-polynomial time algorithms are developed. It is discussed how the SKP relates to the the SSP by splitting items into multiple items of unit weight.

The algorithm for continuous uniform follower profits can easily be extended for continuous distributions with polynomial density functions. This enables the possibility of approximating arbitrary density functions with polynomials to obtain approximation algorithms for the SSP and the SKP. It remains to be investigated how the approximation error carries forward in the problem's solution.

An additive approximation scheme for the SKP that approximates any componentwise distributed follower profits with a componentwise finitely uniformly distributed random variable is presented in [3].

References

1. Buchheim, C., Henke, D. (2020). The bilevel continuous knapsack problem with uncertain follower's objective. *Technical Report*. https://arxiv.org/abs/1903.02810v2
2. Buchheim, C., Henke, D., & Hommelsheim, F. (2021). On the complexity of robust bilevel optimization with uncertain follower's objective. *Operations Research Letters, 49*(5), 703–707.
3. Buchheim, C., Henke, D., & Irmai, J. (2022). The stochastic bilevel continuous knapsack problem with uncertain follower's objective. *Journal of Optimization Theory and Applications, 194*, 521–542.
4. Caprara, A., Carvalho, M., Lodi, A., & Woeginger, G. J. (2014). A study on the computational complexity of the bilevel knapsack problem. *SIAM Journal on Optimization, 24*(2), 823–838.

5. Dantzig, G. B. (1957). Discrete-variable extremum problems. *Operations Research, 5*(2), 266–288.
6. Dempe, S., Kalashnikov, V., Pérez-Valdés, G. A., & Kalashnykova, N. (2015). *Bilevel programming problems*. Berlin: Energy Systems. Springer.
7. Özaltın, O. Y., Prokopyev, O. A., & Schaefer, A. J. (2010). The bilevel knapsack problem with stochastic right-hand sides. *Operations Research Letters, 38*(4), 328–333.

Part II
Analytics and Learning

Chapter 8
A Combined Measure Based on Diversification and Accuracy Gains for Forecast Selection in Forecast Combination

Felix Schulz, Thomas Setzer, and Nathalie Balla

Abstract Recent innovations in the field of forecast combination include integrated methods for forecast selection, weighting and regularization. The methods proposed in related articles first label whether or not forecasters should remain in the selection using information criteria from statistical learning theory. Depending on the selection status, the optimal weights of all forecasters in the sample are then used as baseline to shrink the weights either toward zero or the mean, with the degree of regularization determining the final selection of forecasters. In this paper, we propose a new information criterion reflecting the importance of diversification and accuracy gains in the selection of forecasters for integrated methods. In an iterative procedure motivated by forward feature selection, each forecaster is selected sequentially, while at each step the increase in accuracy and diversification due to the addition of a forecaster to the previous selection is measured. To quantify the increase in diversity, the multiple correlation coefficient is used, which captures the correlation between the previously selected forecasters and a candidate, where the lower the correlation between the candidate and the selection, the higher the gain in diversity for the combination. For the accuracy increase, the accuracy achieved by optimal weight combinations with the previously selected forecasters is compared with the accuracy after adding a candidate. A hyperparameter further enables the trade-off between accuracy and diversification gains in the criterion. Simulation-based studies show scenarios in which our presented information criterion achieves advantages in out-of-sample prediction accuracy over previous criteria for selection by accounting for accuracy and diversification gains.

Keywords Forecast algorithms · Accuracy–Diversity · Forecast combination

F. Schulz (✉) · T. Setzer · N. Balla
Katholische Universität Eichstätt-Ingolstadt,Ingolstadt, Germany
e-mail: Felix.Schulz@ku.de
URL: https://www.ku.de/wfi/wi

O. Grothe et al. (eds.), *Operations Research Proceedings 2022*, Lecture Notes in Operations Research, https://doi.org/10.1007/978-3-031-24907-5_8

Introduction

Accuracy and diversity play crucial roles in forecasting. Accuracy, as the main goal in forecasting, refers to the actual predictive performance of a forecasting model and is usually measured by an error measure like the Mean Squared Error (MSE) [7]. Diversity, in turn, describes different expertise of forecasters and promotes learning from multiple perspectives. The latter is of special interest in the area of forecast combination, where forecasts from multiple forecast models are linearly combined into a single forecast by weighting each model. After Lichtendahl and Winkler [2], a key factor for efficient forecast combination is diversity among forecasts as diversity means low redundancy and more versatile knowledge. Atiya [1] similarly emphasizes that for successful forecast combination, the models must be either diverse or comparable in forecast ability.

In this paper, we combine the importance of accuracy and diversity for forecast combination in an information criterion called $Comb$, usable for selecting forecasts in methods such as the Linear Hybrid Shrinkage (*LHS*) proposed in [3]. *LHS* combines forecast selection, weighting, and regularization by shrinking optimal weights of forecasters to either zero or the mean, depending on the selection status of a forecast given by an information criterion. $Comb$ trade-offs accuracy and diversity gains by including a forecaster in the selection, where the diversity score can be controlled by a hyperparameter. In an experimental setup, we benchmark $Comb$ against previously presented information criteria and show scenarios with advantages for our presented criterion.

The paper is structured as follows. After the introduction in Sect. "Introduction", Sect. "Forecast Combination" introduces forecast combination techniques. Section "Measure Based on Diversification and Accuracy Gains" proposes the new information criterion, Sect. "Experimental Design" the experimental design, and Sect. "Experimental Evaluation" the experimental evaluation. The paper ends with a short conclusion in Sect. "Conclusion".

Forecast Combination

In a combination of forecasts, the combined forecast f_c is calculated as $f_c = \sum_i w_i f_i$, where f_i defines the forecast from forecast model i with $i = 1, \ldots, k$ and w_i the individual forecast model weight. w_i is usually constrained by $\sum_{i=1}^{k} w_i = 1$ with $w \in \mathbb{R}^k$. Weights w_i can be estimated using the inverse of the covariance matrix $\hat{\Sigma}_e^{-1}$ of the prediction errors e associated with the models on past data. With ι as a column vector of k 1s, Optimal Weights (OW) are defined in (8.1) [6].

$$\hat{w}^{OW} = \frac{\hat{\Sigma}_e^{-1}\iota}{\iota'\hat{\Sigma}_e^{-1}\iota} \tag{8.1}$$

Although the *OW* approach minimizes the inherent error in the sample (bias), *OW* are estimated on available data, which may lead to high variance in other data, e.g., in the case of sparse data and high uncertainty. Thus, a popular alternative approach is to use Equal Weights (*EW*), $w^{EW} = \frac{\iota}{k}$, which ignores available data (high bias) but reduces the sample-based error (low variance).

Researchers have tested and demonstrated the benefits of combining the *EW* and *OW* weighting approaches into a formula via a regularization hyperparameter λ tuned by cross-validation within a range between 0 and 1 to trade-off between bias and variance improvements. The so-called Linear Shrinkage (*LS*) approach can be found in (8.2) [5].

$$\hat{w}^{LS} = \lambda w^{EW} + (1 - \lambda)\, \hat{w}^{OW} \tag{8.2}$$

Newer methods in forecast combination unite the idea of forecast weighting and regularization with forecast selection [3, 4]. Using information criteria from statistical learning theory, each forecaster i is assigned a selection status, which is stored in vector v and describes whether a forecaster should remain (1) or not remain (0) in the selection. The forecasters are selected sequentially which, given k forecasters, results in k different forms of v. Depending on the selection status, the *OW* of all forecasters in the sample are then used as baseline to shrink the weights either toward zero or the mean, with the degree of regularization determining the final selection of forecasters. The so-called Linear Hybrid Shrinkage (*LHS*) approach can be found in (8.3).

$$\hat{w}^{LHS} = \lambda w^{sel} + (1 - \lambda)\, \hat{w}^{OW} \tag{8.3}$$

λ again ranges from 0 to 1, emphasizing the reduction of bias at $\lambda = 0$ and variance at $\lambda = 1$. w^{sel} sets a new hyperparameter given by $\frac{v}{\iota' v}$, containing either the values 0 or $\frac{1}{k'}$, with k' denoting the final number of forecasters selected. The main advantage of *LHS* is that poor-performing forecasts are removed from the selection, while *LS* weights each forecast equally in case of maximum shrinkage. To define the shape of v (and thus w^{sel}), three selection criteria have been proposed based on forecaster performance, forecaster importance, and forward feature selection [3, 4]. While the first is simply ranking and selecting the forecaster based on their in-sample performance, the second defines the importance of a forecaster in a cross-validation process by randomly shuffling the validation data and evaluating through resulting MSE changes the forecaster with the highest contribution for the combination. In forward feature selection, a forecaster is incrementally included in the selection based on the largest accuracy gain cross-validated by forming *OW* combinations on the data.

In the following chapter, we outline a new information criterion that explicitly reflects the importance of diversity in the selection of forecasters and can be seen as an extension of the forward feature selection approach proposed in [4].

Measure Based on Diversification and Accuracy Gains

We propose a new iteration-based information criterion that selects one forecaster at a time based on diversity and accuracy gains until all forecasters are selected. Let s be the set of selected forecasters and c the set of candidate forecasters. All forecasters in s get the value of 1 and all forecasters in c the value of 0 in v, where the number of iterations t depend on the number of forecasters, so $t = 1, ..., k$. In the first iteration, all available forecasters are in c, whereby the forecaster with the best forecast accuracy in-sample is selected and added to s — forming the first shape of v. In the next $k - 1$ iterations, each forecaster in c represents a candidate j for s, which is selected based on the diversity and accuracy gain of including j in s. To quantify the increase in diversity, the square root of the multiple correlation coefficient $\sqrt{R^2_{j,s}}$ is calculated in-sample after linear regressing the errors of candidate forecaster j by the selected forecaster(s) in s. The diversity between j and s, $\theta_{j,s}$, is then calculated as shown in (8.4).

$$\theta_{j,s} = 1 - \sqrt{R^2_{j,s}} \tag{8.4}$$

As $R^2_{j,s}$ is defined between 0 and 1, $\theta_{j,s}$ takes a value of 1 if there is no correlation between j and s and 0 if the correlation is 1. For the accuracy increase, *OW* combinations are formed between the previously selected forecaster(s) in s and j and evaluated by calculating the MSE by cross-validation. The $MSE_{s \cup j,t}$, after including j in s, is being compared with the MSE achieved without j in the previous iteration $MSE_{s,t-1}$. The resulting accuracy value $\kappa_{j,s}$ is calculated in the following way, where $\kappa_{j,s}$ leads to a value greater (less) than 1 if the addition of j to s results in a decrease (increase) in MSE:

$$\kappa_{j,s} = \frac{MSE_{s,t-1}}{MSE_{s \cup j,t}} \tag{8.5}$$

Finally, we multiply the diversity score $\theta_{j,s}$ and accuracy score $\kappa_{j,s}$ to form our new criterion *Comb* as in (8.6) which is being calculated in every iteration t for all j's in c whereby the higher the diversity or the higher the reduction in MSE, the higher the overall *Comb* score.

$$Comb_{j,t}(\alpha) = \theta^{\alpha}_{j,s} \times \kappa_{j,s} \tag{8.6}$$

The candidate forecaster j for which the criterion is maximized is added to s and deleted from c, defining a new shape of v. This process is repeated until c is $\emptyset$ and all forecasters are selected. A further feature from our combined measure *Comb* is hyperparameter α which defines the importance of diversity over accuracy in the selection process. When α is equal to 0, the diversity component has the value 1 and diversity is only implicitly considered in the formation of the *OW* combinations. In this case, the criterion leads to the same results as in forward feature selection and the candidate that leads to a higher MSE reduction is selected. When α is increased, more diverse candidates can compensate for MSE deficits measured after the formation of *OW* combinations against candidates that would result in greater κ. For α, values of 0, 0.5, 1 and 2 are tested in experiments as described in the following section.

Experimental Design

To test our new information criterion *Comb*, we set up an experimental design by synthetically generating error data for $k = 8$ forecasters following a multivariate normal distribution. The variance of forecaster 1, σ_1^2, is thereby fixed to 1, while for forecaster k, σ_k^2 takes one of the values 2.0, 3.5 or 5.0. For forecaster i, $1 < i < k$, σ_i^2 increases linearly from σ_1^2 to σ_k^2 as $\sigma_i^2 = \sigma_1^2 + \frac{(\sigma_k^2 - \sigma_1^2)(i-1)}{k-1}$. The correlations between forecasters ρ are drawn from a uniform probability distribution in intervals of [0.3,0.6] and [0.6,0.9]. The ρ values are then sorted in ascending order and stored in the correlation matrix such that worse-performing forecasters have lower ρ values than better-performing forecasters, i.e. $\rho_{k-1,k}$ is assigned the smallest ρ value, $\rho_{k-2,k}$ the second smallest, etc. A test set is given of 5,000 data points, while for training sets of $n = 30$ and $n = 70$ are available. Given the simulation parameters, every treatment combination is repeated 50 times. As benchmark, we use the methods *OW*, *EW*, *LS* and *LHS* as described in Sects. "Forecast Combination" and "Measure Based on Diversification and Accuracy Gains". For the *LHS* approach, we consider selection by performance (*LHS P*), importance (*LHS VI*) as well as diversity and accuracy gains using the *Comb* criterion (*LHS C*($\alpha = 0$), *LHS C(0.5)*, *LHS C(1)*, *LHS C(2)*).

Experimental Evaluation

To evaluate the performance of the methods per treatment combination of the simulation parameters k, σ_k^2, ρ, and n, the MSE on the test set is calculated using the models' learned weights and averaged over the 50 repetitions. We further compute the standard error (SE) per method and treatment combination. The results are shown in Table 8.1, where the model that resulted in the lowest average MSE in the test data for a given treatment combination is highlighted in **bold** and the SEs are given in parentheses after the MSE values. To assess the reliability of the results, a left-tailed Wilcoxon signed rank test for a significance level of 0.05 is calculated pairwise between the best and the remaining methods.

In low correlation ranges of $\rho \in [0.3, 0.6]$, for both training data sizes and a low variance of $\sigma_k^2 = 2.0$, *EW* achieves the lowest MSE, while only for $n = 30$ the result is significant, whereas for $n = 70$ the result is not significant compared to *LS* with a p-value of .304. For $\sigma_k^2 = 3.5$ and $n = 30$, *LHS P* reaches the statistically significant

Table 8.1 MSE (SE) on test data per method and treatment combination

	$n = 30$			$n = 70$		
	$\sigma_k^2 = 2.0$	$\sigma_k^2 = 3.5$	$\sigma_k^2 = 5.0$	$\sigma_k^2 = 2.0$	$\sigma_k^2 = 3.5$	$\sigma_k^2 = 5.0$
$\rho \in [0.3,0.6]$						
OW	0.982 (0.031)	1.159 (0.037)	1.250 (0.044)	0.828 (0.011)	0.986 (0.014)	1.038 (0.013)
EW	**0.762** (0.004)	1.106 (0.006)	1.441 (0.007)	**0.762** (0.004)	1.106 (0.006)	1.441 (0.007)
LS	0.779 (0.011)	1.046 (0.017)	1.158 (0.024)	0.767 (0.005)	**0.967** (0.01)	1.034 (0.012)
LHS P	0.841 (0.018)	**1.001** (0.015)	1.084 (0.025)	0.800 (0.007)	0.977 (0.01)	1.024 (0.012)
LHS VI	0.810 (0.016)	1.040 (0.023)	1.114 (0.031)	0.788 (0.007)	0.979 (0.012)	1.022 (0.013)
LHS C(0)	0.880 (0.019)	1.075 (0.029)	1.093 (0.032)	0.837 (0.009)	0.986 (0.013)	1.015 (0.011)
LHS C(0.5)	0.876 (0.019)	1.111 (0.027)	1.107 (0.036)	0.827 (0.012)	0.994 (0.014)	0.991 (0.009)
LHS C(1)	0.877 (0.018)	1.117 (0.027)	1.086 (0.034)	0.811 (0.012)	0.990 (0.015)	0.992 (0.011)
LHS C(2)	0.869 (0.018)	1.118 (0.027)	**1.083** (0.033)	0.806 (0.012)	**0.967** (0.014)	**0.986** (0.009)
$\rho \in [0.6,0.9]$						
OW	1.269 (0.043)	1.019 (0.04)	0.823 (0.029)	1.088 (0.016)	0.869 (0.018)	0.704 (0.018)
EW	1.150 (0.005)	1.671 (0.007)	2.179 (0.009)	1.150 (0.005)	1.671 (0.007)	2.180 (0.009)
LS	1.114 (0.018)	1.010 (0.038)	0.839 (0.04)	1.055 (0.011)	0.869 (0.019)	0.700 (0.017)
LHS P	**1.064** (0.018)	0.956 (0.03)	0.786 (0.022)	**1.027** (0.01)	0.857 (0.017)	**0.696** (0.016)
LHS VI	1.077 (0.02)	0.961 (0.037)	0.786 (0.022)	1.035 (0.012)	0.856 (0.017)	**0.696** (0.016)
LHS C(0)	1.070 (0.022)	0.961 (0.037)	**0.785** (0.022)	1.038 (0.012)	**0.855** (0.017)	**0.696** (0.016)
LHS C(0.5)	1.082 (0.022)	**0.945** (0.03)	**0.785** (0.022)	1.044 (0.015)	**0.855** (0.017)	**0.696** (0.016)
LHS C(1)	1.078 (0.022)	**0.945** (0.03)	**0.785** (0.022)	1.044 (0.015)	**0.855** (0.017)	**0.696** (0.016)
LHS C(2)	1.080 (0.023)	**0.945** (0.03)	**0.785** (0.022)	1.042 (0.015)	**0.855** (0.017)	**0.696** (0.016)

Models with the lowest average MSE values for a treatment combination are marked in *bold* best value with an MSE of 1.001, while, for $n = 70$, this is the case for *LS* and *LHS C(2)*. The best results of *Comb* are obtained for variances of $\sigma_k^2 = 5.0$. Here, for $n = 30$, the MSE of *LHS C(2)* is not significant compared with *LHS P*, *LHS VI*, and *LHS C(1)*, and, for $n = 70$, compared with *LHS C(0.5)* and *LHS C(1)*. Despite the non-significant results, the last observation indicates that when the amount of data is sufficient and the variance differences between forecasters are high, the deliberate selection of different forecasters is beneficial. In high correlation ranges of $\rho \in [0.6, 0.9]$, the integrated *LHS* methods outperform the classical *OW*, *EW*, and *LS* approaches with significant better MSE values across all treatment combinations. Thereby, the improvement is greatest when only a small amount of training data is available and uncertainty is high. For example, for $n = 30$ and $\sigma_k^2 = 5.0$, the best classical method *OW* achieves an MSE of 0.823, while all *LHS C*s achieve a 4.6% lower MSE of 0.785 in this treatment combination. Surprisingly, the test for statistical significance of the best *LHS* model(s) compared with the other *LHS* models shows that, over all variances for $n = 30$, no selection criterion yields significantly better results than any other selection criterion. The same is true for $n = 70$, with the exception of the MSE obtained by *LHS P* in $\sigma_k^2 = 2.0$ — the only significant result in high correlation ranges. Overall, the results for $\rho \in [0.6, 0.9]$ suggest that in areas of high correlation, the *LHS* selection criteria can be used largely interchangeably, and thus the selection of forecasters is independent of explicit diversity gains.

Conclusion and Future Work

Our proposed information criterion $Comb$ accounts for diversification and accuracy gains for forecast selection in integrated methods. The initial results are encouraging and open the direction of future research, like the review of more treatment combinations and the careful tuning of further α values for low ρ's.

References

1. Atiya, A. F. (2020). Why does forecast combination work so well? *International Journal of Forecasting, 36*, 197–200. https://doi.org/10.1016/j.ijforecast.2019.03.010
2. Lichtendahl, K. C., & Winkler, R. L. (2020). Why do some combinations perform better than others? *International Journal of Forecasting, 36*, 142–149. https://doi.org/10.1016/j.ijforecast.2019.03.027
3. Schulz, F., Setzer, T., & Balla, N. (2022). Linear hybrid shrinkage of weights for forecast selection and combination. In: Proceedings of the 55th Hawaii International Conference on System Sciences, (pp. 2125–2134). https://doi.org/10.24251/HICSS.2022.267
4. Schulz, F. Non-linear hybrid shrinkage of weights for forecast selection and combination. In Wirtschaftsinformatik 2022 Proceedings, (Vol. 7). https://aisel.aisnet.org/wi2022/business_analytics/business_analytics/7
5. Stock, J. H., & Watson, M. W. (2004). Combination forecasts of output growth in a seven country data set. *Journal Forecast, 23*, 405–430. https://doi.org/10.1002/for.928

6. Timmermann, A. (2006). Forecast combinations. *Handbook of Economic Forecasting, 1*, 135–196. https://doi.org/10.1016/S1574-0706(05)01004-9
7. Yokuma, J. T., & Armstrong, J. S. (1995). Beyond accuracy: Comparison of criteria used to select forecasting methods. *International Journal of Forecasting,11*, 591–597. https://doi.org/10.1016/0169-2070(95)00615-X

Chapter 9
A Decision Support System Including Feedback to Sensitize for Certainty Interval Size

Nathalie Balla

Abstract In decision-making overconfidence and estimation biases can lead to sub-optimal outcomes and accuracy loss. A debiasing strategy presented in this work is to use feedback based on the error pattern of own previous absolute and 90% certainty (confidence) interval estimates. This is comprised in a decision support system (DSS) and applied in an experiment, where results indicate support for the key assumption that subjects are able to selectively reduce their overconfidence and their estimation bias, if present, with the help of the provided feedback.

Keywords DSS · Interval estimation · Overconfidence · Overprecision · Debiasing

Introduction

A key discussion in behavioral operations and information systems research is how to reduce overconfidence and other biases in order to achieve higher quality judgments and decisions. Frequently, experts, for example in supply management, have a sense of control that can lead to too optimistic estimations and they often cannot assess their own performance accurately [1]. Research has found that experts are often overconfident in their judgments and that this leads to low performance and detrimental judgments [1–3]. Moore and Healy [4] differentiate between three types of overconfidence: overestimation, overplacement, and overprecision. Overprecision denotes being too certain that one's estimate is more accurate than it actually is, to which this work refers to. Ren and Croson [5] investigated the newsvendor problem in inventory decisions and found overprecision leading to underestimation of the variance of demand. They measured overprecision by letting subjects answer ten general knowledge questions each with a 90% confidence interval indicating their certainty of 90% that the true answer lies within this range. In case the true answer lies within the interval in nine out of ten questions, the subject is considered well

N. Balla (✉)
Catholic University Eichstätt-Ingolstadt, Ingolstadt 85049, Germany
e-mail: NBalla@ku.de

O. Grothe et al. (eds.), *Operations Research Proceedings 2022*, Lecture Notes in Operations Research, https://doi.org/10.1007/978-3-031-24907-5_9

calibrated. A similar approach based on 90% certainty (confidence) intervals to measure overprecision is used in this work. In decision analysis, estimations regarding intervals are frequently used and Soll and Klayman [6] claim that decision makers' 90% intervals often include the true answer in under 50% of cases, which means they are too sure of themselves and not able to assess their own performance.

For this reason it is worthwhile to concentrate on interval estimation and its improvement. The method of application of certainty intervals used here corresponds to a typical method identified by Klayman [2]: asking subjects for a single numerical estimate and a 90% confidence interval, that is an upper and lower bound, for which there is a 90% probability that the correct answer lies between them.

In this short paper a novel design for a Decision Support System (DSS) is presented and applied, which aims at reducing overconfidence, particularly overprecision, and estimation biases (under- or overestimation, here meaning systematic too high or too low estimations). According to Ancarani, Di Mauro, and D'Urso [1], benchmarks should be available to experts to let them evaluate their relative performance and a decrease in overconfidence can be achieved by feedback regarding past judgments given in a timely manner. Following this notion, the subjects are provided with feedback regarding their mean percentage error (MPE) on previous absolute answers as well as feedback in which questions the correct answer lay inside the 90% certainty interval. The MPE feedback should give subjects an indication of their relative performance.

Two aspects are mainly examined: the average broadening of the intervals before and after the feedback, presumably resulting from the feedback per question if the correct answer lay within the given interval; and the shift of the intervals before and after the feedback, presumably resulting from the MPE feedback, as the MPE points in a certain direction. The assumption is that if the average interval was too narrow to include the correct answer before the feedback, subjects broaden their average interval after the feedback, indicating a reduction of overprecision. The assumption regarding the shifts is that if the average interval was too far away to include the correct answer before the feedback, subjects shift their average interval in direction of the MPE feedback, indicating a reduction of estimation bias. Additionally, the questions come from different categories and it is examined if subjects are able to selectively apply the feedback to intervals in categories in which they performed poorly. The assumption here is that subjects are able to selectively reduce overprecision as well as estimation bias, if present, with the help of the MPE and interval feedback provided. In the following, the experimental design and the assumptions are described. Subsequently, the results are presented and lastly discussed including a conclusion and an outlook.

Experimental Design

41 business students took part in the experiment (20 in treatment, 21 in control group), of which 21 are female and 20 male. Subjects were requested to answer point estimate questions and 90% certainty intervals from general knowledge categories: *number*

of residents of a country, *river length*, and *mountain height*. Example questions are: "How many residents does France have?", "How long is the Hudson River (in km)?", "How high is the Mount Everest (in meters)?". The experiment is composed of two sequences of each 15 questions, 30 questions overall. The mentioned categories are not communicated to subjects but easy to detect. All questions include visual cues for estimation support: a map of the respective country with the ten largest cities and an indication of a range of their size; a map of the river including a scale in the legend; a topographical map of the respective mountain and a reference mountain height. These cues are meant to reduce error variance. Subjects are randomly assigned to treatment or control group and the experiment starts with a welcome page and then requires subjects to answer estimation questions and to specify a 90% certainty interval by entering a lower and an upper bound. This certainty interval indicates the range of which subjects are 90% certain that the correct answer lies within it. As soon as the subject entered her answer, the next question is shown. After 15 questions either feedback is provided for 30 s (treatment group) or a blank page inviting to a 30 s break is shown (control group). For the feedback, a subject's MPE is computed and shown together with the correct answers. Additionally, the subject sees per question if the correct answer lay within her given interval. The MPE is calculated by taking the difference between the estimate and the actual (actually correct answer) per estimate, dividing this difference by the actual and multiplying it by 100. MPE is therefore the mean of these values. The MPE was selected as measure due to comprehensibility and simplicity to apply it for debiasing. For instance, a negative MPE would mean that the subject needs to place the estimation higher and to shift the certainty interval upwards. MAPE, the mean absolute percentage error, is used as performance criterion to determine improvement or deterioration between question sequences. Because the payouts depend on performance, the MAPE is also used to determine the winner subjects. In the briefing, subjects are given information on how to interpret and apply MPE for debiasing (without telling them they will receive feedback) together with information on MAPE used as performance measure. Moreover, subjects are told the meaning of the 90% certainty intervals. Following the feedback or the blank page, a subject is faced with another sequence of 15 judgments. At the end of the experiment, subjects receive performance information, are debriefed, and their MAPE is calculated. Every subject receives a payout for participation and can additionally win one of two prizes per group. If the MAPE of a subject is lower, her chance is higher to win a prize, which is ought to incentivize subjects.

Assumptions Studied

The key assumption is that subjects are able to selectively reduce their overprecision and estimation bias, if present, with the help of MPE and interval feedback. This key assumption is divided into four sub-assumptions A1–A4.

A1: The certainty intervals become broader after the feedback, if they were too narrow to include the correct answer before the feedback, more often in the treatment than in the control group. The presumption is that overprecision exists if subjects specify an interval that is too narrow to include the correct answer. A1 aims to investigate if overprecision can be reduced by, most likely, the feedback on the intervals of the questions. To test A1, the relative frequency of correct answers within the given intervals before the feedback is computed as well as the average relative size difference (between sequence one and two) of the intervals per user. This is done by dividing the average interval size in the second sequence by the one in the first sequence per category and then taking the mean over categories. It is assumed that if the correct answers lay in the given interval in less than 50% of a subject, the intervals are too narrow on average. Results for thresholds of under 33% and 25% of correct answers lying inside the interval are also presented to demonstrate stability of results. If this is the case, the average interval after the feedback should be broadened as the subjects should realize that their intervals were too narrow and they may have been overprecise. For the control group the expectation is that around 50% of subjects make their average interval broader in the second sequence, if they did not include the correct answer in the first sequence (baseline in case of randomness).

A2: The certainty intervals are shifted in the direction of the MPE feedback after the feedback more often in the treatment than in the control group. To test A2, it is analysed how the per category calculated average upper and lower bounds of the intervals are placed before and after the feedback per user. If both of the average bounds moved in the direction of the MPE feedback, that is if the MPE feedback is negative, the bounds should be larger than before the feedback and vice versa, then it can be assumed that the subject reduced her estimation bias by shifting her interval. Ratios are computed per category and treatment group from which the mean is taken over the categories per treatment group.

A3: The certainty intervals become broader more often after the feedback especially in those categories, in which the intervals were too narrow to include the correct answer before the feedback compared to no feedback given. To test A3, the category with the minimum relative frequency of correct answers within the given intervals before the feedback is determined. Then the average interval size per category and the difference of average interval size between sequence one and two per category is calculated by dividing the average interval per category of sequence two by that of sequence one. Therefore, we find the category with the maximum broadening of the average interval. The relative frequency of the matches between these two categories (minimum and maximum) indicates how often subjects selectively applied the feedback to specific categories. The objective is to show a difference in frequency of matches between the treatment groups.

A4: The certainty intervals are shifted in the direction of the MPE feedback after the feedback in those categories, in which the absolute MPE was the highest before the feedback more often in the treatment than in the control group. To test A4, the category with the maximum absolute MPE in the first sequence is determined per user and compared to the categories in which the average interval was shifted in the direction of the MPE in the second sequence. Then the relative frequency of matches between the category with the highest absolute MPE and the categories in which the average intervals were shifted in the direction of the MPE is computed. If the percentage value in the treatment group exceeds the one in the control group, selective application of the feedback to specific categories can be assumed.

For all assumptions we conduct a one-sided Fisher's exact test between the results of treatment and control group.

Results

In this section, results are presented per assumption.

A1: 19 subjects in each treatment and control group had the correct answer inside their interval in the first sequence in less than 50%. In 70% of these cases in the treatment group the average interval became broader after the feedback compared to 57.1% in the control group. For a 33% threshold for correct answers lying inside the interval, 12 subjects were in the treatment and 10 in the control group. In the treatment group 65% of subjects broadened their certainty interval after the feedback compared to 52.4% in the control group. For a 25% threshold, 9 subjects were in the treatment and 8 in the control group. Here, in the treatment group 60% of subjects broadened their certainty interval compared to 52.4% in the control group. The p-value of the Fisher's exact test is 0.45, meaning the results are not significant at a 10% level, likely due to the small sample size.

A2: In the treatment group 55% of subjects shifted the certainty interval in the direction of the MPE feedback after the feedback compared to 34.9% in the control group. The p-value of the Fisher's exact test is 0.14, which means the results are not significant at a 10% level, likely due to the small sample size.

A3: In 70% of cases the subjects broadened their average interval in that category the most after the feedback in which the least correct answers lay inside the interval in the first sequence for the treatment group. In 47.6% of cases the subjects achieved this in the control group. The p-value of the Fisher's exact test is 0.13, for which reason the results are not significant at a 10% level, again for the same reason.

A4: In 60% of the treatment group the average intervals were shifted in those categories where the absolute MPE was the highest compared to 23.8% in the control group. The p-value of the Fisher's exact test is 0.02, which means the results are significant at a 10% level.

Discussion

The shown results, despite the limited sample size, provide support for the key assumption of the combination of the feedback elements aiding subjects to reduce overprecision and estimation biases selectively. Regarding *A1*, it seems that the difference between treatment and control group becomes larger, when the threshold is higher, that is when the interval is set too narrow more often. This is consistent with expectations, because the more often subjects miss the correct answer, the stronger they are able to adapt their intervals after the feedback, whereas the control group does not have this chance. Most likely the broadening of the intervals, indicating a decrease in overprecision, originates from the interval feedback through which subjects reflect on their previously given lower and upper bounds. By realizing they were too certain in setting their bounds and therefore setting them too close to one another, they became less certain through the feedback and the reflection lead to adaptation of their future bounds, that is a decrease in overprecision. Regarding *A2*, the ratio of subjects shifting their interval in the direction of the MPE is higher for the treatment group. This shifting behavior shows that subjects do not only consider the feedback of the correct answer lying in their interval but also the MPE feedback indicating if their interval was generally too low or high. Applying this feedback to their future upper and lower bounds means they also reflect on the general position of their certainty interval independent of its size, which leads to a reduction of estimation bias, either under- or overestimation. As *A3* and *A4* also show higher ratios for the treatment group, subjects seem to be able to selectively apply the feedback to specific categories, where it is most necessary. Reducing overprecision and estimation bias selectively, leads to an overall error reduction.

Conclusion and Outlook

This work indicates that subjects are able to reduce their overprecision as well as their estimation bias, in general and for specific categories that performed poorly before the feedback. A selective consideration of feedback shows that subjects are able to recognize novel patterns and use this knowledge effectively. These two focal aspects in turn result in a general accuracy improvement in terms of MAPE reduction. This research has the limitation that, due to the Corona pandemic, it has been challenging to conduct experiments with large numbers of subjects, for which reason the sample size is rather small. However, this work is research in progress and more experiments will be conducted to support the findings. In addition, this experiment can be conducted with different categories and scenarios such as categories less obvious for subjects.

References

1. Ancarani, A., Di Mauro, C., & D'Urso, D. (2016). Measuring overconfidence in inventory management decisions. *Journal of Purchasing and Supply Management, 22*(3), 171–180. https://doi.org/10.1016/j.pursup.2016.05.001.
2. Klayman, J., Soll, J. B., Gonzalez-Vallejo, C., & Barlas, S. (1999). Overconfidence: It depends on How, What, and Whom you ask. *Organizational Behavior and Human Decision Processes, 79*(3), 216–247. https://doi.org/10.1006/obhd.1999.2847.

3. Shipman, A. S., & Mumford, M. D. (2011). When confidence is detrimental: Influence of overconfidence on leadership effectiveness. *The Leadership Quarterly, 22*(4), 649–665. https://doi.org/10.1016/j.leaqua.2011.05.006.
4. Moore, D. A., & Healy, P. J. (2008). The trouble with overconfidence. *Psychological Review, 115*(2), 502–517. https://doi.org/10.1037/0033-295X.115.2.502.
5. Ren, Y., & Croson, R. (2013). Overconfidence in newsvendor orders: An experimental study. *Management Science, 59*(11), 2502–2517. https://doi.org/10.1287/mnsc.2013.1715.
6. Soll, J. B., & Klayman, J. (2004). Overconfidence in interval estimates. *Journal of Experimental Psychology: Learning, Memory, and Cognition, 30*(2), 299–314. https://doi.org/10.1037/0278-7393.30.2.299.

Part III
Continuous and Global Optimization

Chapter 10
A Note on Matrix Reordering for Linear System Solutions by Iterative Methods in Interior Point Methods

W. Rodrigues, Marta Velazco, and A. R. L. Oliveira

Abstract The linear systems arising from interior point methods (IPM) for linear programming are solved using the preconditioned conjugate gradient method (PCG). Two preconditioners are adopted: the controlled Cholesky factorization (CCF) of the normal equations system and the splitting preconditioner. The CCF performance depends upon the previous reordering of the linear programming constraint matrix rows. A comparison among two different reordering methods is performed in order to verify the most suitable one for this approach. Variants of nested dissection (ND) and the minimum degree (MD) are among the considered heuristics. Computational experiments with large-scale linear programming problems from several collection sets are performed.

Keywords Linear programming · Interior point methods · Preconditioner · Reordering · Nested dissection

Introduction

In [1], the authors studied the influence of matrix reordering on the solution of linear systems arising from IPMs using the PCG and the hybrid preconditioner [2]. The reordering using the MD improves the performance of the hybrid preconditioner. In [1], the nested dissection (ND) heuristic was not considered. This work presents the performance of the IPM when the ND is used for reordering, and a comparison with the MD heuristic is performed. This work is organized in seven sections as follows. Section "Primal-dual Interior Point Methods" introduces the primal-dual interior point methods and the issues of the linear systems arising from such methods

W. Rodrigues · M. Velazco (✉)
UNIFACCAMP, Rua Guatemala 167, Jardim América, Campo Limpo Paulista, SP 13231-230, Brazil
e-mail: marta.velazco@gmail.com

A. R. L. Oliveira
IMECC-UNICAMP, Rua Sérgio Buarque de Holanda, 651, Campinas, SP 13083-859, Brazil

O. Grothe et al. (eds.), *Operations Research Proceedings 2022*, Lecture Notes in Operations Research, https://doi.org/10.1007/978-3-031-24907-5_10

are discussed. The hybrid preconditioner is introduced in Section "Hybrid Preconditioner". Next, in Section "Reordering Heuristics", the reordering heuristics that are used for the study are presented. In Section "Numerical Experiments", the numerical experiments using the selected linear programming problems are shown. Finally, the conclusions are presented.

Primal-dual Interior Point Methods

The Karush-Kuhn-Tucker optimality conditions for the primal and dual linear programming problems are:

$$\begin{aligned} Ax - b \quad &= 0 \\ A^t y + z - c &= 0\,, (x, y) \geq 0 \\ XZe \quad &= 0 \end{aligned} \tag{10.1}$$

where $X = diag(x)$, $Z = diag(z)$, $A \in \mathbb{R}^{m \times n}$, $c, x, z \in \mathbb{R}^n$, $b, y \in \mathbb{R}^m$ and $e \in \mathbb{R}^n$ is a vector of ones. In IPM, the system (10.1) without non-negativity constraint is solved by the Newton method. The predictor-corrector primal-dual method computes the search direction in two steps: the predictor direction and the corrector direction. For each step, a linear system (10.2) is solved with the same coefficient matrix but different right-hand sides.

$$\begin{bmatrix} A & 0 & 0 \\ 0 & A^t & I \\ Z & 0 & X \end{bmatrix} \begin{bmatrix} \Delta x \\ \Delta y \\ \Delta z \end{bmatrix} = r \tag{10.2}$$

The system (10.2) can be reduced to a system of a normal equation, by eliminating the variables Δx and Δz. The normal system equation obtained is: $\left(A\Theta^{-1}\mathrm{A}^{\mathrm{t}}\right)\Delta y = \mathrm{r}$, where $\Theta = X^{-1}Z$. It is solved by the conjugate gradient method. Its convergence depends on the condition of the system matrix then, an appropriate preconditioner is necessary. The hybrid preconditioner [2] is used in this case.

Hybrid Preconditioner

In [2], the authors consider that the optimization process occurs in two phases, and a different preconditioner is used in each one. In the first phase at the beginning of the optimization process, the CCF is used. In the second phase at the end of the process, a specific preconditioner for the IPM is used. A heuristic that was proposed by Velazco et al. [3] determines the phase change.

Controlled Cholesky Factorization (CCF): The controlled Cholesky factorization [2] (CCF) is an incomplete Cholesky factorization of the matrix $A\Theta^{-1}\mathrm{A}^{\mathrm{t}}$ with controlled fill-in. The fill-in of the factor is controlled by a parameter η. Define $E = L - \tilde{L}$, where L is the factor of the complete Cholesky factorization, $\tilde{L}$ is the factor of the incomplete factorization. Thus, the minimization of E can be solved with two heuristics: increasing η, thereby allowing more fill-ins by the column and choosing, for a fixed η, the largest entries of $\tilde{L}$ (absolute values). The initial fill-in η_0 depends on the average of the nonzero elements using column of $A\mathrm{A}^{\mathrm{t}}$.

The Splitting Preconditioner: The splitting preconditioner was proposed by Bocanegra et al. [2] for the symmetric indefinite augmented system. In this work, the splitting preconditioner version for the normal equation $A\Theta^{-1}\mathrm{A}^{\mathrm{t}}$ is used. Consider the matrix $AP = \left[\, B \ N \,\right] P$, where P is a permutation matrix such that $B \in \mathbb{R}^{m \times m}$ is nonsingular, and the remaining columns are in $N \in \mathbb{R}^{m \times n-m}$. The preconditioner for normal equations is $\Theta_{\mathrm{B}}^{\frac{1}{2}}\mathrm{B}^{-1}$.

Reordering Heuristics

Matrix reordering can improve the stability of incomplete factorization, and it can accelerate the convergence of iterative methods if some amount of fill-in, such as that which occurs in the CCF, is allowed in the incomplete factors [4]. The authors in [1] analyze the influence of matrix reordering on the solution of the linear system arising in IPMs using the PCG with the hybrid preconditioner. The reordering heuristics reduce the fill-in during the incomplete factorization, and then the CCF will be a better approximation of the complete factor because fewer significant values will be discarded. Therefore, the computational costs of the solution are reduced. Moreover, the phase change is delayed when the efficiency of the CCF is improved; thus, the performance of the splitting preconditioner will also improve. The matrix $A\Theta^{-1}A^t$ can be reordered through the permutation of the rows and columns using a permutation matrix P. P is computed using a reordering heuristic. $A\Theta^{-1}A^t$ conserves the same sparse pattern over all the optimization process because, although the diagonal matrix Θ changes at each iteration, all $\Theta_{ii} > 0$. Consequently, reordering $A\Theta^{-1}A^t$ is equivalent to reordering A by rows. That is an advantage since the reordering of the matrix A's rows is computed only once in the preprocessing step before the optimization process starts. Two reordering heuristics are presented: the MD and ND. Both heuristics represent the matrix using an undirected graph $G = (V, E)$, where V is the set of vertices, $|V| = m$, and E is the set of edges. The vertices v_i are the i rows/columns of the matrix, and the edge between the nodes v_i and v_j exists if $a_{ij} = a_{ji} \neq 0$ and $i \neq j$.

Minimum Degree (MD): In this vertex reordering heuristic [5], the initial vertex will be the vertex with the minimum degree. The minimum degree vertex in the graph will be the row/column that has the largest number of zero entries. Reordering happens by swapping the less dense rows/columns with the first position that has not yet been changed from the matrix. An already reordered row/column will no longer be considered in the reordering process. Thus, in the representation of the matrix as a graph, the permuted vertex is eliminated from the graph. The elimination of the vertex will imply that its adjacent vertices become adjacent to each other, and new edges may appear. That is, fill-in occurs.

Nested Dissection (ND): From the graph G, the algorithm finds a separator subgraph S. S is removed from G, and two disconnected graphs C^1 and C^2 remain. Next, the algorithm performs ND recursively on the subgraphs C^1 and C^2. In the final step, S is reordered using an reordering heuristic like MD [6]. Small separators result in low fill-in. The separator computing algorithm ensures the effectiveness of ND. The graph partitioning heuristic computes the vertex separator (NDV) or edge separator (NDE). The difference between the heuristic is that the NDE minimizes the number of edges between the sets, and the NDV minimizes the size of the separator S.

Numerical Experiments

This section presents the numerical experiments of the study of the performance of the predictor-corrector interior point method using the reordering heuristics. The reordering heuristics were integrated into the modified version of PCx [7] where the linear systems are solved using the PCG with the hybrid preconditioning [2] and the modification of [1]. The study addresses 63 large-scale linear programming problems from the following libraries: PDS, MNETGEN, KENNINGTON, QAPLIB, MESZAROS and NETLIB. Three reordering heuristics were used for the comparison. The implementation of the MD of the original PCx code was used for the tests, as in [1]. The code Metis [8] was used for computing the fill-reducing orderings of sparse matrices based on ND with the two separators. Three modified versions of PCx are used: PCx-Modified with the MD (PCx-MD), PCx-Modified with ND and the edge separator (PCx-NDE) and PCx-Modified with ND and the vertex separator (PCx-NDV).

Summary of the Results

The MD was faster than the other heuristics in 15 problems and the differences are small. However, the ND was faster in 33 problems with large differences, and the methods were equal for 15 problems. In 15 problems in which the density of the matrix $A\Theta^{-1}A^t$ is smaller, the differences among the reordering heuristics is very significant, and the ND heuristic is better in all cases. However, as the density increases, the MD improves the reordering time. The comparison of the approaches considers the reordering time, the total number of PCG iterations and total time. The results are summarized below.

PDS: Three large-scale and very sparse problems (sparsity bigger than 99.996%) are studied. For those problems, the CCF is used with the PCG for solving the linear systems over all the optimization processes. The ND heuristics reordered the matrix 100 times faster than the MD decreasing the total time of the IPM (Table 10.1).

MNETGEN: The problems are very sparse (bigger than 99.9936% and lower than 99.97%. In those cases, the ND heuristics reordered the matrix faster than the MD, and the fill-in of the CCF was larger. The results concerning the inner iterations of the PCG and the outer iterations of the IPM are mixed (Table 10.2).

KENNINGTON: Four large-scale and very sparse problems are studied. For those problems, the reordering times of the MD were smaller than those of ND (Table 10.3).

QAPLIB: The selected problems of this library are less sparse (sparsity up to 97%. The reordering time of ND was lower than that of the MD, but the fill-ins were similar (Table 10.4).

MESZAROS: Thirteen problems with various sparsities are studied (sparsity lower than 99.8%. The reordering times of ND were equal or lower than those of the MD. The fill-in was similar for both; however, in all problems, the number of PCG iterations increased for ND, and thus the processing time increased too (Table 10.5).

NETLIB: Twenty-five problems were studied with various sparsities (sparsity lower than 99.7%. The reordering times of ND were equal or faster than those of the MD. The fill-in was larger or equal than that of the MD. The results of the inner iterations of the PCG, the outer iterations of the IPM and processing times were similar (Table 10.6).

Table 10.1 Results for PDS problems

	PCx-MD	PCx-NDE	PCx-NDV
Reordering time	1047.692	8.752	9.284
PCG iterations	129213	12432.9	129842
IPM time	10285.07	7751.56	8246.55

Table 10.2 Results for MNETGEN problems

	PCx-MD	PCx-NDE	PCx-NDV
Reordering time	4842	1.484	1.532
PCG iterations	341055	386655	353513
IPM time	39072.88	42136.05	39973.48

Table 10.3 Results for KENNINGTON problems

	PCx-MD	PCx-NDE	PCx-NDV
Reordering time	0.78	0.952	1.056
PCG iterations	62242	59992	63604
IPM time	2642.56	2363.77	2050.88

Table 10.4 Results for QAPLIB problems

	PCx-MD	PCx-NDE	PCx-NDV
Reordering time	1.9	0.708	0.796
PCG iterations	95709	97687	94425
IPM time	1560.91	1495.85	1468.39

Table 10.5 Results for MESZAROS problems

	PCx-MD	PCx-NDE	PCx-NDV
Reordering time	0.076	0.068	0.06
PCG iterations	85029	93099	99111
IPM time	349.05	402.69	394.34

Table 10.6 Results for NETLIB problems

	PCx-MD	PCx-NDE	PCx-NDV
Reordering time	0.76	0.1	0.116
PCG iterations	70918	92908	90769
IPM time	130.15	147.97	146.93

Conclusions

This work studied the impact of matrix reordering using ND on the IPM with the systems solved using the PCG with the hybrid preconditioner. The research was carried out by comparing the results that were obtained in 63 problems from public domain libraries. Three versions of the PCx-Modified code were used: PCx-MG, PCx-NDA and PCx-NDV. In most problems, the ND reordering heuristics reduce the reordering time. The differences in the reordering times are more pronounced in larger and sparser problems, as seen in the problems of the PDS and MNETGEN libraries. However, reordering using ND has generally resulted in a larger fill-in with the CCF preconditioner. The performance of the PCG and the computational time of the predictor-corrector method have different results in the studied libraries. ND does not perform well for the MESZAROS library; however, it is the best choice for the PDS and QAPLIB libraries. The selection of the reordering heuristic depends on the sparsity of the coefficient matrix. In general, MD is a good option for generic problems.

Acknowledgements The authors are thankful for the financial support of the Brazilian National Council for Technological and Scientific Development (CNPq) and UNIFACCAMP.

References

1. Silva, D., Velazco, M., & Oliveira, A. (2017). Influence of matrix reordering on the performance of iterative methods for solving linear systems arising from interior point methods for linear programming. *Mathematical Methods of Operations Research, 85*(1), 97–112.
2. Bocanegra, S., Campos, F., & Oliveira, A. (2007). Using a hybrid preconditioner for solving large-scale linear systems arising from interior point methods. *Computational Optimization and Applications, 36*, 149–167.
3. Velazco, M., Oliveira, A., & Campos, F. (2010). A note on hybrid preconditioners for large-scale normal equations arising from interior point methods. *Optimization Methods and Software, 25*(2), 321–332.
4. Benzi, M. (2002). Preconditioning techniques for large linear systems: A survey. *Journal of Computational Physics, 182*(2), 418–477.
5. George, A., & Liu, J. (1980). A fast implementation of the minimum degree algorithm using quotient graphs. *ACM Transactions on Mathematical Software, 6*(23), 337–358.
6. George, A. (1973). Nested dissection of a regular finite element mesh. *SIAM Journal on Numerical Analysis, 10*(2), 345–363.
7. Czyzyk, J., Mehrotra, S., Wagner, M., & Wright, S. (1999). PCx an interior point code for linear programming. *Optimization Methods and Software, 11*(2), 397–430.
8. Karypis, G. (2015) METIS: A software package for partitioning unstructured graphs, partitioning meshes, and computing fill-reducing orderings of sparse matrices, version 5.0. http://glaros.dtc.umn.edu/gkhome/metis/metis/download

Chapter 11
A Tri-Level Approach for T-Criterion-Based Model Discrimination

David Mogalle, Philipp Seufert, Jan Schwientek, Michael Bortz, and Karl-Heinz Küfer

Abstract Model discrimination (MD) aims to determine the inputs, called design points, of two or more models at which these models differ most under the additional condition that the models are fitted to these points, in the case of T-optimal designs. On the one hand, nonlinear models often lead to nonconvex MD problems, on the other hand, the optimal number of design points must be determined, too. Thus, the computation of T-optimal designs is very arduous. However, if one considers finitely many design points, a well-solvable bi-level problem arises. Since the latter only represents an approximation of the original model discrimination problem, we refine the design space discretization using the equivalence theorem of MD. This yields a tri-level approach whose iterates converge to a T-optimal design. We demonstrate that the approach can outperform known solution methods on an example from chemical process engineering.

Keywords Model discrimination · Multi-level optimization · Discretization

Introduction

When modeling real processes, often several different models fit. For being able to distinguish (discriminate) which model is best suited, several criteria were developed.

A frequently used criterion is the so-called T-criterion. Let f_1 and f_2 be two competing *models* with at least continuously differentiable model functions $f_i : X \times \Theta_i \to \mathbb{R}, i = 1, 2$, a compact *design space* $X \subseteq \mathbb{R}^n$, and compact *parameter spaces* $\Theta_i \subseteq \mathbb{R}^{p_i}$. Assume that the first model f_1 is the true model having fixed parameters θ_1, $f_1(x) = f_1(x; \theta_1)$. The alternative model f_2 remains parametrized with parameters $\theta_2 \in \Theta_2$. Then, the *T-criterion* for model discrimination is defined as

D. Mogalle · P. Seufert · J. Schwientek (✉) · M. Bortz · K.-H. Küfer
Fraunhofer ITWM, Fraunhofer-Platz 1, 67663 Kaiserslautern, Germany
e-mail: Jan.Schwientek@itwm.fraunhofer.de

O. Grothe et al. (eds.), *Operations Research Proceedings 2022*, Lecture Notes in Operations Research, https://doi.org/10.1007/978-3-031-24907-5_11

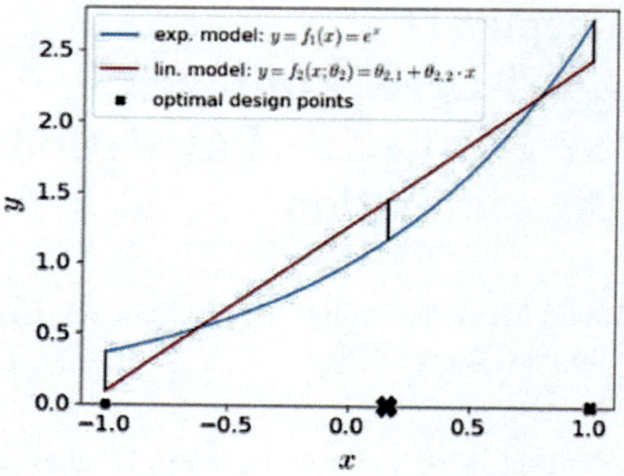

Fig. 11.1 A T-optimal design discriminating a linear alternative model from an exponential reference model. The size of the crosses corresponds to the weights, and the model differences at the design points are shown in the black lines

$$T(\xi) := \min_{\theta_2 \in \Theta_2} \int_X [f_1(x) - f_2(x; \theta_2)]^2 \xi(dx) = \min_{\theta_2 \in \Theta_2} \sum_{i=1}^{N} w_i [f_1(x_i) - f_2(x_i; \theta_2)]^2,$$

where ξ is a finite design $\xi = \begin{Bmatrix} x_1 & \dots & x_N \\ w_1 & \dots & w_N \end{Bmatrix}$ with weights $w_i \in [0, 1]$, $\sum_{i=1}^{N} w_i = 1$, and $N \in \mathbb{N}$. A design ξ^* is called *T-optimal* if it is a (global) maximizer of

$$\text{TMD} : \max_{\xi \in \Xi} T(\xi) = \max_{\xi \in \Xi} \min_{\theta_2 \in \Theta_2} \int_X [f_1(x) - f_2(x; \theta_2)]^2 \xi(\text{d}x),$$

where Ξ is the set of probability measure on X (see Fig. 11.1 for an illustration).

One benefit of the T-criterion is its relative independence from (measurement) data. For a detailed overview on model discrimination, we refer to [1].

This paper is structured as follows: In the next section we give a literature overview on solution methods for model discrimination. Section "A Tri-Level Approach" sketches our novel tri-level approach, relying on a two-fold discretization, leading to well-structured subproblems. In Section "Numerical Example: PL Versus LHHW Reaction Rate", the introduced algorithms are compared on a real-world example stemming from chemical process engineering. The paper ends with concluding remarks and future directions of research.

Existing Solution Approaches

For the computation of T-optimal designs, descent methods which were originally developed for experimental design problems (see, e.g., [1]) have been adapted. One well-known algorithm of this class is the adaptation of Fedorov's Vector Direction Method [1]. These algorithms iteratively determine new design points as maximizers of the directional derivative and update the weights by combining the current design and the new design point convexly. This yields very robust methods for computing optimal designs but they converge slowly since bad points never completely lose their weight.

On the other hand, (TMD) represents an infinite-dimensional min-max (bi-level) optimization problem which can be transformed into a semi-infinite optimization problem (SIP) via the epigraph reformulation:

$$\text{TMD}_{\text{epi}} : \max_{t \in \mathbb{R}, \xi \in \Xi} \; t \text{ s.t.}$$

Consequently, techniques from semi-infinite optimization have also been applied to solve T-criterion-based model discrimination problems. Kuczewski [2] and Duarte [3] consider a fixed number N of design points, as there always exists a finite optimal design (see [1]). The resulting SIP is solved via an adaptive discretization of the parameter space Θ_2. The lower-level problem and the discretized (upper-level) problem are solved with the global solver BARON due to their general nonconvexity in the inputs and the parameters. This corresponds to two optimization problems which must be solved multiple times alternately to global optimality. As this is repeated for several values of N, these approaches are very time consuming.

A different approach has been taken by Dette et al. [4]. They work with a discretization of the design space $X_N = \{x_1, \ldots, x_N\} \subset X$ and try to find an optimal design on this subset by linearizing the inner parameter estimation problem. The issue with Dette's algorithm is that the linearization might not be a sufficient approximation for certain model functions and, thus, the algorithm might not converge at all.

A Tri-Level Approach

Model-based design of experiments (DoE), of which model discrimination is a sub class, is a non-convex problem in general, too. By considering the DoE problem on a finite subset of the design space, one obtains a convex non-linear program which can often even be reformulated as semi-definite or second-order cone program. Thereby it can be solved to global optimality very efficiently. However, by restricting ourselves to a subset of the design space, we only obtain an approximation of the solution for the original problem. By iteratively refining the considered subset, the obtained solutions will however converge to a global optimum on the whole continuous design space. For details we refer to [5].

This solution ansatz can also be transferred to model discrimination. For that purpose, we first consider the problem (TMD) on a finite subset $\dot{X}$ of the design space X

$$\text{TMD}(\dot{X}) : \max_{\xi \in \Xi(\dot{X})} \min_{\theta_2 \in \Theta_2} \sum_{i=1}^{N} w_i \cdot [f_1(x_i) - f_2(x_i; \theta_2)]^2,$$

with N being the cardinality of the set $\dot{X}$. The semi-infinite epigraph reformulation of the problem TMD$(\dot{X})$ (see (TMD$_{\text{epi}}$)) can be solved, e.g., via a discretization approach. By considering a discrete subset $\dot{\Theta}_{2,M} = \{\theta_{2,j} | j = 1, \ldots, M\} \subset \Theta_2$, the upper-level problem is a linear problem given by

$$\text{TMD}(\dot{X}, \dot{\Theta}_{2,M}) : \max_{t \in \mathbb{R}, w \in [0,1]^N} t \text{ s.t.}$$

The only drawback of the approach is that to refine (or adapt) the discretization $\dot{\Theta}_2$, we must solve the lower-level problem

$$\text{TMD} - \text{LSQ}(\dot{X}, w, t) : \min_{\theta_2 \in \Theta_2} \sum_{i=1}^{N} w_i [f_1(x_i) - f_2(x_i, \theta_2)]^2 - t$$

to global optimality. This problem is in general non-convex and non-linear. However, it is a least-squares problem in θ_2. Such problems are well studied and can often be solved efficiently.

With this approach we obtain a (globally optimal) solution of the model discrimination problem with finitely many candidate experiments. In a second step, we must adaptively refine the discretization of the design space such that the designs obtained on the finite subsets converge to an optimal solution on the continuous design space X. For this we assume that every design ξ is *regular*, i.e., the corresponding least-squares estimator $\hat{\theta}_2(\xi)$ is unique. Based on the *Equivalence Theorem for Model Discrimination* (see [2]), a necessary and sufficient condition for a design ξ^* to be T-optimal is

$$\max_{x \in X} \left[f_1(x) - f_2\left(x, \hat{\theta}_2(\xi^*)\right) \right]^2 - T(\xi^*) \leq 0, \quad \text{(OC} - \text{MD)}$$

where $\hat{\theta}_2(\xi^*)$ is the best (weighted) least-squares estimator on the design points $x_1, \ldots, x_N$ with weights $w_1, \ldots, w_N$. This gives us a criterion for deciding whether to further refine the current discretization $\dot{X}$ or whether the design ξ^* is already optimal on the continuous design space X.

Starting with an initial discretization $\dot{X}_0 \subseteq X$, we iteratively solve the linear semi-infinite program TMD$(\dot{X}_k)$, $k \in N_0$, and check whether the returned design ξ_k^* is optimal. We then either stop the routine or refine the discretization $\dot{X}_k$ by adding the globally optimal solution x_k^* of

$$\text{TMD} - \text{OC}\left(\xi_k^*, \hat{\theta}_2\right) : \max_{x \in X} \left[f_1(x) - f_2\left(x, \hat{\theta}_2(\xi_k^*)\right) \right]^2 - T\left(\xi_k^*\right)$$

to obtain $\dot{X}_{k+1} = \dot{X}_k \cup \{x_k^*\}$. The problem TMD $-$ OC$\left(\xi_k^*, \hat{\theta}_2\right)$ is a global optimization problem, too. Nevertheless, for some problem classes, e.g., polynomials, it can be solved efficiently. Additionally, if a sufficiently good initial discretization $\dot{X}_0$ is selected it must not be solved very often.

Algorithm 1: Tri-level approach 2-ADAPT-MD

Input: finitely many design point candidates $\dot{X}_0 = \{x_1, \ldots, x_N\}$, initial design ξ^0, initial parameter space discretization $\dot{\Theta}_{2,M_0} \subset \Theta_2$, tolerances $\delta, \varepsilon > 0$, $k = 0$

while True: $l = 0$

    while True:

        solve $\text{TMD}(\dot{X}_k, \dot{\Theta}_{2,M_l})$ with optimal solution w_l^*, t_l^*

        solve $\text{TMD} - \text{LSQ}(\dot{X}_k, w_l^*, t_l^*)$ with opt. solution $\theta_{2,l}^*$ and value $LSQ(\theta_{2,l}^*)$

        if $LSQ(\theta_{2,l}^*) \leq \delta$: break, **else:** set $\dot{\Theta}_{2,M_{l+1}} = \dot{\Theta}_{2,M_l} \cup \{\theta_{2,l}^*\}$ and $l \leftarrow l + 1$

    combine $\dot{X}_k$ and weights w_l^* to obtain ξ_k^* and set $\hat{\theta}_2 = \theta_{2,l}^*$

    solve $\text{TMD} - OC(\xi_k^*, \hat{\theta}_2)$ with optimal solution x_k^* and value $OC(x_k^*)$

    if $OC(x_k^*) \leq \varepsilon$: break, **else:** set $\dot{X}_{k+1} = \dot{X}_k \cup \{x_k^*\}$ and $k \leftarrow k + 1$

return $\dot{X}_k, \xi_k^*$ and $\hat{\theta}_2$

This approach was developed in [6], where it was also proven that the iterates converge towards arbitrary good approximations of T-optimal designs. Notice that the inner loop basically solves a (linear) semi-infinite program by adaptive discretization. It is straightforward to show that the initial feasible set of this SIP is compact, so we have guaranteed convergence in the inner loop (see, e.g., [7]). However, in finite time we may only find an approximate solution. Nevertheless, one can control this error from the inner loop and extend the proofs by Fedorov [1] and Dette [4] to guarantee overall convergence towards an approximate T-optimal design. This approximation can be arbitrarily good depending on the convergence tolerance in the inner loop.

Numerical Example: PL Versus LHHW Reaction Rate

We now apply the tri-level approach 2-ADAPT-MD to a discrimination task in chemical process engineering and compare it to two methods for model discrimination mentioned above, Fedorov's and Dette's algorithm. Further numerical examples can be found in [6].

First, we give some information on how we implemented the algorithms. We used Python such that we can use different solvers to solve the various subproblems:

- We try to compute the global solutions of the least-squares problems using KNITRO in a multi-start approach.

- For the determination of a new design point, we consider a fine reference discretization of the input space and evaluate the squared distance at each point.
- We compute the optimal weights by solving the corresponding linear program using MOSEK via the Pyomo interface.
- We check optimality of the current design $(OC - MD)$ up to some given tolerances δ in the inner loop and ε in the outer loop.

For a reaction of two gases A and B to a gas C, one can use either the *power-law* (*PL*) or the *Langmuir-Hinshelwood-Hougen-Watson* (*LHHW*) expression to model the reaction rate. They are given by

$$r_{PL} = kp_A^n p_B^m \quad \text{and} \quad r_{LHHW} = \frac{kK_A p_A K_B p_B}{1 + K_A p_A + K_B p_B + K_C p_C},$$

where p_A, p_B, and p_C are the partial pressures of each component, k is the reaction rate constant, n and m are the orders of reaction, and K_A, K_B, and K_C are adsorption coefficients. To discriminate r_{PL} and r_{LHHW}, we take the LHHW type expression as our reference model and the power-law expression as our alternative model.

For the reference model, we set the parameters to $k = 2$, $K_A = 0.5$, $K_B = 2$, and $K_C = 1.2$. The design space consists of the partial pressures which we assume to be the discretization of the set $[1, 10] \times [1, 10] \times [0, 10]$ with step size 1, i.e., $(p_A, p_B, p_C) \in X := \{1, 2, \ldots, 10\} \times \{1, 2, \ldots, 10\} \times \{0, 1, \ldots, 10\}$. The parameter space for the power-law expression is assumed to be $(k, n, m) \in \Theta_2 := [0.1, 1]^3$, and we choose the initial design

$$\xi^{(0)} = \begin{Bmatrix} (1,1,0) & (10,1,0) & (1,10,0) & (5,5,0) & (10,10,10) & (5,5,5) \\ 1/6 & 1/6 & 1/6 & 1/6 & 1/6 & 1/6 \end{Bmatrix}.$$

As outer stopping tolerance we take $\varepsilon = 10^{-4}$. All computations were performed on a laptop with an Intel Core i7-5600U CPU and 8 GB RAM. As Table 11.1 shows, our algorithm outperforms Fedorov's and Dette's algorithm in solution quality and run time. The "…" in the optimal designs hint at more design points with neglectable weight.

Conclusions

We have introduced a novel tri-level algorithm to compute T-optimal experimental designs. The algorithm utilizes a two-fold adaptive discretization of the parameters space Θ_2 as well as of the design space X. The main advantage of the 2-ADAPT-MD algorithm is that it converges under mild assumptions on the considered models. By using an increasing discretization in design points x and parameters θ, we obtain linear programs and least-square programs as most frequent sub-problems to solve. These problem classes are well studied, and specialized solvers can be used to

Table 11.1 Comparison of algorithms

Algorithm	Reached accuracy	Computed design ξ^*	Run time [s]
Fedorov	0.0048	$\begin{Bmatrix} (10,5,0) & (10,2,10) & (10,10,10) & \dots \\ 0.4426 & 0.3077 & 0.2060 & \dots \end{Bmatrix}$	125.5
Dette	52.16	$\begin{Bmatrix} (10,1,10) & (10,10,10) \\ 0.9845 & 0.0155 \end{Bmatrix}$	25.37
2-ADAPT-MD	2.65×10^{-5}	$\begin{Bmatrix} (10,5,0) & (10,2,10) & (10,10,10) \\ 0.4835 & 0.3154 & 0.2011 \end{Bmatrix}$	17.82

compute corresponding solutions efficiently. Via an example from chemical engineering, we have shown that 2-ADAPT-MD can compete and even outperform state of the art model discrimination methods.

References

1. Fedorov, V. V., & Leonov, S. L. (2013). *Optimal design for nonlinear response models.* CRC Press.
2. Kuczewski, B. (2006). *Computational aspects of discrimination between models of dynamic systems.* Ph.D. Thesis, University of Zielona Góra.
3. Duarte, B. P., Wong, W. K., Atkinson, A. C. (2015). A semi-infinite programming based algorithm for determining T-optimum designs for model discrimination. *Journal of Multivariate Analysis, 135.*
4. Dette, H., Melas, V. B., Guchenko, R. (2015). Bayesian T-optimal discriminating designs. *Annals of Statistics, 43*(5).
5. Seufert, P., Schwientek, J., & Bortz, M. (2021). Model-based design of experiments for high-dimensional inputs supported by machine-learning methods. *Processes, 9*, 508.
6. Mogalle, D. (2021). *Computing T-optimal designs for model discrimination by approximating the inner optimization problem.* Master's Thesis, TU Kaiserslautern.
7. López, M., Still, G. (2007). Semi-infinite programming. *European Journal of Operational Research, 180*(2).

Part IV
Decision Analysis and Support

Chapter 12
A Decision Support Method to Assess Energy Policy Impacts on Different Household Types for a Socially Just Energy Transition in Germany

Audrey Dobbins and Ulrich Fahl

Abstract Households are responsible for a third of the final energy consumption in Germany in 2018 with the average household meeting 60% of household energy service needs with fossil fuels. Energy transition targets to increase renewables and energy efficiency will require high upfront capital investments into building renovations, heater and appliance upgrades. Overall less than 17% of households have sufficient capital and are in the decision-making power to undertake investments. The household sector is disaggregated into 56 profiles by key socio-economic parameters (income, building type, tenure status, urbanization) and, together with a budget constraint limiting the total available capital on all investments and consumption per profile, incorporated into an energy system optimization model to account for the differentiated needs and financial capabilities of diverse household types. This method provides a platform from which to evaluate the impact of various policies and measures. Selected scenarios include a comparison between a reference case and two carbon revenue redistribution schemes. The results yield insights into the energy-related investment and consumption patterns for different household types emphasizing fuel types, emissions and quantification of suppressed demand (expressed through household budget deficit for unmet household service demands), and show that to evaluate the energy welfare of lower income households several aspects must be considered.

Keywords Energy system model · Suppressed demand · Energy welfare

Challenges for a Socially Just Energy Transition in Germany

The household sector has a key role to play in achieving the German energy transition targets to decarbonize by 2045 since it is responsible for 27.5% of the total final energy consumption and 10.1% of direct greenhouse gas emissions in 2018 [1–3].

A. Dobbins (✉) · U. Fahl
Institute of Energy Economics and Rational Energy Use, University of Stuttgart, Heßbrühlstr. 49a, 70565 Stuttgart, Germany
e-mail: audrey.dobbins@ier.uni-stuttgart.de

O. Grothe et al. (eds.), *Operations Research Proceedings 2022*, Lecture Notes in Operations Research, https://doi.org/10.1007/978-3-031-24907-5_12

Households are expected to contribute to targets by increasing renewable energy in heating and energy efficiency, but recent assessments have found that the building sector did not meet the 2020 targets nor is it expected they will meet the targets for 2030 [4]. Around 60% of household energy service demands are met with fossil fuels [3] indicating that a shift towards increasing energy efficiency and renewable energy will require significant investments to changing the underlying household energy infrastructure. The challenge will be to mobilize investment in the lower income and rental sector, where there is often a lack of financial capital for the high upfront costs of investments and the decision-making power as tenants. Therefore, there is a need to consider the variation in investment capabilities of households undertaken in energy planning within the energy system modelling.

Carbon taxes have long been implemented as a means to reflect the environmental damage incurred through the combustion of fossil fuels, but these can disproportionately impact lower income households and tenants who lack the financial capacity or decision-making power to alter the structure designating the types of fuels and amount of energy necessary to meet household energy service demands. In pursuit of balancing social inequities experienced by some households as a result of the carbon tax, revenues derived from the carbon tax should fund the reduction of the EEG levy on electricity as a means to dampen these distributional impacts [4], or alternatively to redistribute to the population through a climate bonus [5]. General acceptance for the climate bonus is at risk if there is no steering mechanism in place to ensure that the funds go towards decreasing emissions [6]. Lower income households are prone to rebound when households invest in energy efficiency, or similarly consumption may increase by increasing income because households are now able to afford previously unmet household service demands (suppressed demand) [7]. Therefore, it is critical to understand the impacts policies aimed at discouraging fossil fuel consumption and the progressivity redistribution policies have on different households to better estimate the energy welfare of households in addition to the overall energy and emissions.

Methods

Characterization of the Household Sector

To explore the realistic impacts of policies on households it is necessary to account for the differences in the investment and consumption patterns of various households according to socio-economic and built environment characteristics [8]. The share of income spent on energy is often used to measure the affordability of energy and is a key indicator for energy poverty [9]. In 2018, the average household in Germany spent 3.3% of their income expenditure on household energy, whereas a lower income household spent around 11.3% [10]. These disparities are further reflected in Fig. 12.1

which shows the monthly energy-related expenditures on consumption and investments expressed as magnitude per household and shares of income by income groups in Germany in 2018 [11]. The magnitude of expenditure increases with income due to increasing drivers of demand (e.g. size of the home, number of occupants), but the share of income spent decreases [11, 12]. For expenditures on investment, the magnitude increases with income with shares representing less than 1% of the total income across all income groups. However, such that consumption costs can be reduced. These expenditure patterns reveal that these increasing expenditures on investment enable higher income households to invest in more efficient and renewable thereby reducing energy bills.

The potential to afford the high upfront investment costs is examined by compiling the potential monthly financial savings accumulated per income group. Less than a quarter (22.2%) of all households save more than the average household with approximately 539€ per month in 2018 which could be considered theoretically available for investments in renewable energy and energy efficient technologies or building upgrades [11]. While, the share of homeowners increases with income, only a total of 16.7% of all households have higher-than-average savings and are at the same time homeowners with decision-making power. This underscores the limitations in the potential of the household sector to be able to undertake the necessary investments to shift the household energy infrastructure to achieve the energy transition targets.

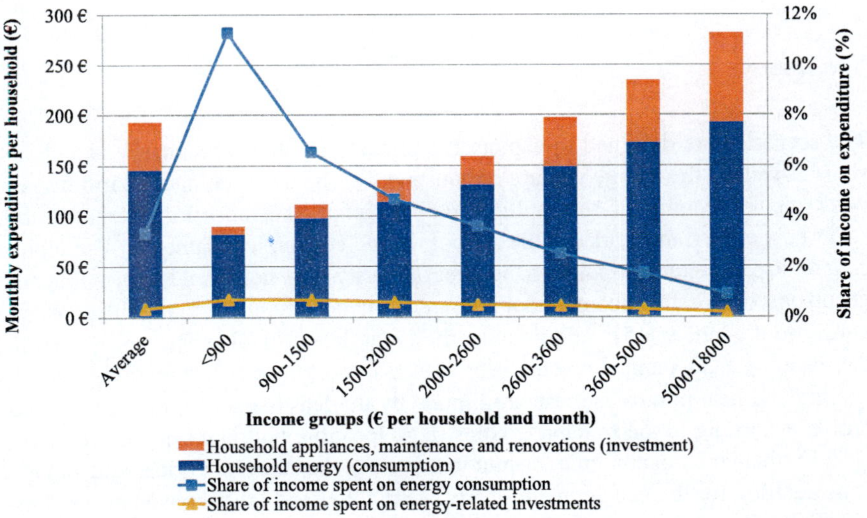

Fig. 12.1 Income and expenditure by income group per household in Germany, 2018 [own graph based on 11 as given in 12]

TIMES-Actors-Model-Households Energy System Optimization Model

Energy system optimization models are applied to assess the least-cost path to achieving specific objectives, e.g. climate, energy. However, the household sector is typically represented in an aggregated manner with differentiation only by building type or location [13]. To better characterize the differentiated needs and capabilities of various households, the TIMES-Actors-Model-Households (TAM-HHs) disaggregates the population into 56 different profiles according to socio-economic and built environment parameters including building type (multi-family, single-family), tenure (owner, tenant), location (urban, rural) and income group (7 statistical groups according to [11]) (see [14, 15] for more details). Each profile has its own distinct energy demand profile based on the distribution of end-uses and fuel types according to the demand drivers unique to each [15]. A budget constraint is included to express the financial capacity of each of these defined profiles to afford all energy-related investment and consumption costs, which is [14, 15]. This addition converts the model into a dual objective model to maximize the benefits to the consumer in addition to minimizing the overall costs and is a tabulation of the total statistically available savings and actual energy-related investment and consumption expenditures for all households within a defined profile [11, 16].

Scenarios

The scenarios are designed to explore the impact of policies on energy and emissions as well as the energy welfare of households. The overarching scenario framework includes common assumptions for the input data and projections of the socio-economic framework and the various techno-economic parameters. The socio-economic parameters (population, number of households, household size, number of occupants, disposable income, floor area and climatic variables) are provided exogenously to the model [15]. Additionally, the input data also takes into consideration the resource and technical potential for biomass, rooftop solar installations, district heating and heat pumps as these are limited by the feasible access of each defined profile according to their characteristics (e.g., location, building type, useful area) [15]. Further techno-economic assumptions include the consumer prices and include consideration for the development of EEG levy relief and other consumer taxes [3, 17]. The carbon price is defined in line with current policy developments according to the tax scheme [18] with a CO_2 tax in 2035 with an upper bound of 200€ per ton of CO_2-equivalent [4], and reaching 250€ per ton of CO_2-equivalent in 2050. This accounts for the expected damage cost for climate change, thereafter with a linear extrapolation to 260€ per ton of CO_2-equivalent in 2060. Two carbon revenue redistribution schemes are explored with either an annual climate bonus distributed

Table 12.1 Scenario descriptions

Scenario	Description
Reference (REF)	Disaggregated, all expected policies implemented, budget constraints
CB	Climate bonus; 100€ per capita annually; carbon tax; EEG surcharge without levy relief
CBLI	Climate bonus only for half the population (lower incomes); €200 per capita annually; carbon tax; EEG surcharge without levy relief

as 100€ given per capita (CB) or with a climate bonus of 200€ provided to 50% of the lower income population (CBLI), see Table 12.1.

Results

Energy and Emissions

Overall, the environmental impact varies little between the two distribution variations where in 2030 the final energy consumption across all households in the CB scenario results in 1781 PJ (52 Mt CO_2-eq) and 1788 PJ (54 Mt CO_2-eq) in the CBLI scenario, compared to the Reference scenario with 1736 PJ (54 Mt CO_2-eq). This top-down (average) result is insufficient to understand the complexities of energy consumption in the household sector. Due to the disaggregation of the household sector in the TAM-HHs model, the results can also be analyzed from the household perspective. When comparing the average energy consumption per household in the lower four income groups, shifts in the total consumption can be observed with an average of 31 GJ, 34 GJ and 35 GJ per household in the REF, CB and CBLI scenarios, respectively, as shown in Fig. 12.2. The REF scenario incorporates shares of renewables and fossil fuels of 7.2% and 63.2%, respectively, while in the CB scenario the shares are 5.6% and 62.1%, respectively, and 7.9% and 58.8% in the CBLI scenario, respectively. This indicates a greater shift for the lower income households in the CBLI scenario towards renewables and away from fossil fuels compared to both other scenarios. In 2030, the majority of space heaters reach the end of their technological lifetime and require replacing. However, since the lower four income groups have been unable to accumulate sufficient budget to afford investment into infrastructural changes, these households opt instead to extend the use of the existing infrastructure and technologies by instead shifting the types of fuels to supply demand in order to comply with climate objectives. For example, the key bridging solution is blending fuels such as hydrogen or biofuels into the existing infrastructure until such a time when sufficient budget is accumulated to afford new investments and the infrastructural changes. The magnitude of fossil fuel consumption is greater in both the CB and CBLI scenarios compared to the reference scenario, which runs counter to the mandate to fund emission reductions. To understand these underlying causes of

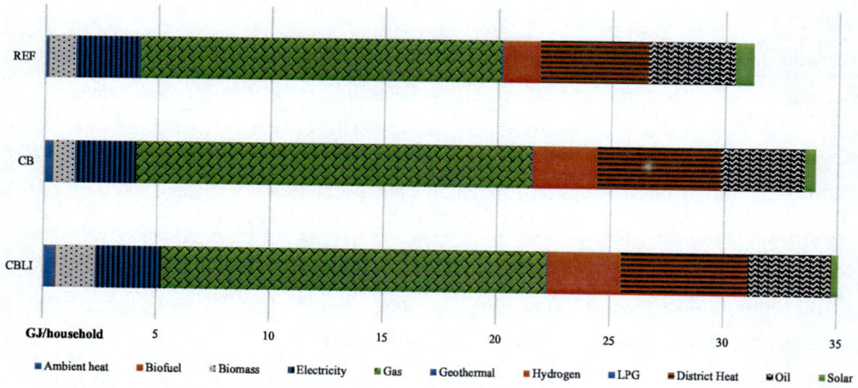

Fig. 12.2 Average energy consumption per household for the lower four income groups, 2030 [own graph as given in 15]

household energy consumption patterns, it is also necessary to assess the influence of suppressed demand on these different levels of consumption.

Suppressed Demand

The inclusion of the budget constraints in the model allows the analysis of the suppressed, which can provide a more holistic picture of the energy welfare of households. In both the REF and the two Climate Bonus scenarios, energy consumption was suppressed in the lower four income groups of the population restricting consumption to varying degrees in 2030. In 2030 in the REF scenario, 11.4 million people require an additional 84€ per capita. The addition of the climate bonus of 100€ per person in the CB scenario yielded an increase in the short-term investment expenditure in lieu of consumption expenditure, reduced both the number of people affected and the budget deficit to 5.7 million people requiring an additional 52€ per capita. For these households, the budget remains restricted despite the additional cash flow, and the cost-optimal solution is to continue to suppress demand and save budget for future investments. The investment and consumption patterns in higher income groups do not change with the addition of the climate bonus. By increasing the distribution to 200€ per capita to the lower 50% of the population (CBLI scenario), the suppressed demand is reduced to the equivalent of about 118€ per capita for approximately 131,000 people affected, such that the severity and the extent of suppressed demand is reduced significantly and focused to specific households.

Discussion

The debate about the types of policies to support a just energy transition is ongoing. In order to consider the energy welfare of households, it is also necessary to understand the implications and distributional impacts of policies. The carbon tax acts as a price signal to encourage a shift in investment towards renewable technologies and energy efficiency, however this can also result in negative consequences for lower income households and tenants, who lack either the capital or decision-making power to make structural changes to avert the effect of the carbon tax. An assessment of energy use and emissions, or expenditure on consumption alone do not provide a holistic view on the energy welfare of households. Lower income households already have negative savings and may suppress energy demands. The presented methodology allows the comparison of the expected suppressed demand experienced by households across the various scenarios in addition to an evaluation of the investment and consumption patterns across all households, or specific household types. This highlights that due to the existing suppressed demands, lower income households must decide between investments that will improve the long-term energy consumption patterns, or expenditure to meet current unmet needs. The more a lower income household can put towards their available budgets, the more they are able to do both of these as was shown with the CBLI policy approach.

Conclusion

This paper showed that an analysis of only the energy consumption does not give adequate insights into the energy welfare of households. By incorporating the budget constraint in the method, the suppressed demands can be monetized and deeper insights are given on the trade-offs between investment and consumption decisions, which can support the discourse on support schemes towards achieving the energy transition targets and the energy welfare of households. The methodology facilitates further assessments to find the right redistribution scheme, and can be expanded to include other socio-economic parameters, such as household composition (e.g. single-parent households, elderly). Due to the long-term nature of investment decisions and how these affect consumption patterns, an assessment of the energy welfare of households should consider not only the affordability of energy, but also view this within the context of suppressed demands. Considering these multiple aspects will support the energy welfare of households for a socially just energy transition.

References

1. BMU. (2018). Climate action in figures. Facts, trends and incentives for German Climate Policy 2018 edition.
2. BMU. (2019). Climate action plan 2050—Germany's long-term low greenhouse gas emission development strategy.
3. BMWi. (2021). *Energy data: Complete edition.* https://www.bmwk.de/Redaktion/EN/Artikel/Energy/energy-data.html
4. Harthan, R. O., et al. (2020). *Abschätzung der Treibhausgasminderungswirkung des Klimaschutzprogramms 2030 der Bundesregierung.* BMU.
5. BMF. (2019). *Financing climate action.* https://www.bundesregierung.de/breg-en/issues/climate-action/klimaschutzziele-finanzieren-1674072
6. Barckhausen, A., Becker, J., Dütschke, E., Piria, R., Preuß, S. (2022). Akzeptanz und Kommunikation eines CO2-Bepreisungssystems, Berlin.
7. Sorrell, S. (2007). *The rebound effect: An assessment of the evidence for economy-wide energy savings from improved energy efficiency.* UKERC.
8. Cayla, J.-M., Maizi, N., & Marchand, C. (2011). The role of income in energy consumption behaviour: Evidence from French households data. *Energy Policy, 39*, 7874–7883.
9. EPOV. (2021). *European energy poverty observatory: Indicators and data.* https://www.energypoverty.eu/indicators-data
10. BMWi. (2021). The energy of the future. In *8th Monitoring report on the energy transition—Reporting years 2018 and 2019.* Federal Ministry for Economic Affairs and Climate Action, Berlin.
11. Destatis. (2018). Einkommens- und Verbrauchsstichprobe. Fachserie 15, Heft 4, Wiesbaden.
12. BMWSB. (2020). Buildings Energy Act. Federal Ministry for Housing, Urban Development and Construction.
13. Haasz, T. (2017). *Entwicklung von Methoden zur Abbildung von Demand Side Management in einem optimierenden Energiesystemmodell. Fallbeispiele für Deutschland in den Sektoren Industrie, Gewerbe, Handel, Dienstleistungen und Haushalte.* Doctoral thesis, University of Stuttgart.
14. Dobbins, A., Fahl, U. Energy poverty or vulnerable consumers? An energy-economic method to compare the policy approaches to addressing vulnerabilities in the energy system in Germany (pp. 317–326).
15. Dobbins, A. (2022) *System analysis of the significance of energy poverty on household energy use and emissions in Germany.* Doctoral thesis, University of Stuttgart. http://dx.doi.org/10.18419/opus-12471
16. IMF. (2019). *World economic outlook database.* https://www.imf.org/external/pubs/ft/weo/2019/02/weodata/index.aspx. Accessed 20 Jan 2020.
17. BMU. (2019). Projektionsbericht-der-Bundesregierung 2019 für Deutschland gemäß Verordnung (EU) Nr. 525/2013.
18. Vermittlungsausschuss. (2019). *Gesetz zur Umsetzung des Klimaschutzprogramms 2030 im Steuerrecht.* Abgeschlossene Vermittlungsverfahren. Ergebnis.

Chapter 13
A Multi-Perspective Approach for Exploring the Scenario Space of Future Power Systems

Ulrich Frey, Karl-Kiên Cao, Thomas Breuer, Manuel Wetzel, Shima Sasanpour, Jan Buschmann, Kai von Krbek, and Aileen Böhme

Abstract There are many possible future energy systems. We explore the range of parameter uncertainty and quantify parameter interrelations to generate hundreds of scenarios for Germany. Only sensible parameter combinations remain as inputs to an energy system optimization and coupled models. In the past, computational limitations have been a major obstacle to calculate such an enormous space of scenarios. In contrast, we use high-performance computing. To utilize high-performance computing (HPC) efficiently, the parallel solver for linear programs PIPS-IPM++ is applied. We integrate it into a tool chain of different components including scenario generation, energy system optimization and results evaluation and couple a large diversity of software packages in a fully automated HPC workflow. This enables the calculation of all scenarios in a matter of days. Furthermore, we use a set of 37 indicators to provide a comprehensive assessment of the simulated energy systems. In this way, we cover multiple perspectives, such as system adequacy, security of supply or behavior of market actors. Whereas scenarios with already expanded capacities do not lead to clear results, a "green field" approach or higher spatial resolution do. Yet, we identified three clusters of scenarios. This allows to study disruptive events like price shocks in a vast parameter space and to identify countermeasures for the long-term.

Keywords Energy systems analysis · PIPS-IPM · High-performance computing

U. Frey · K.-K. Cao (✉) · M. Wetzel · S. Sasanpour · J. Buschmann · K. von Krbek
German Aerospace Center (DLR), Institute of Networked Energy Systems, Stuttgart and Oldenburg, Germany
e-mail: karl-kien.cao@dlr.de

T. Breuer
Jülich Supercomputing Centre, Forschungszentrum Jülich (FZJ), Jülich, Germany

A. Böhme
GAMS Software GmbH, Frechen, Germany

O. Grothe et al. (eds.), *Operations Research Proceedings 2022*, Lecture Notes in Operations Research, https://doi.org/10.1007/978-3-031-24907-5_13

The Three Challenges of Energy Scenario Analysis

Despite great progress in energy systems analysis, three key challenges are still apparent in future energy system modelling. The first challenge are computational limitations. To analyze a multitude of scenarios, parallelization and high-performance computers (HPC) are necessary. Many projects have, in contrast, used shortcuts like analyzing only typical days instead of complete timeseries [1]. The second challenge is that future pathways are highly contingent on assumptions. Different assumptions have led to very different scenarios [2]. As a consequence, comparisons and evaluations are highly problematic. The traditional solution, while scientifically not satisfactory, has been to take a very selective set of assumptions. The last challenge is that, given a certain methodology, only certain aspects of future scenarios are typically analyzed. For example, optimization models tend to concentrate on system costs [3], whereas agent-based simulations focus on individual strategies of actors. Hence, there is a certain blindness to some aspects just by the methodology the researcher has chosen.

Taken together, these problems substantially reduce the trust in energy systems modeling. Hence, this paper tries to address all three problems by answering: "If it would be possible to explore the full possibility space of future energy scenarios, could we select those that are near optimal from a multitude of perspectives?" This allows us to get nearer to our goal, a comprehensive assessment of future energy systems.

Our contribution to addressing the first challenge is the further application of the parallel solver—PIPS-IPM++ [4]. It allows us to solve Energy System Optimization Models (ESOMs) on HPCs by exploiting the block-structure of the corresponding linear programs (LPs). Our solution to the second problem is to sample from a huge parameter space. The third problem is addressed by coupling different tools, e.g., the ESOM, REMix [5] and the agent-based simulation, AMIRIS [6]. The resulting indicators provide a comprehensive assessment of energy scenarios including security of supply and market impact. Therefore, we are able to find *points of interests* (POIs), defined as a special space where many indicators evaluated are above average.

Methods

Our modeling goals require the integration of various complex steps. Therefore, we developed a workflow of multiple software packages within a HPC environment (see Fig. 13.1).

We start with a basic parameterization of REMix, which is already useable for energy scenario analysis without any uncertainty consideration. We automatically generate a large variety of scenarios with a newly developed parameter sampling tool (Scenario Generator) to describe the parametric uncertainty of various instances of the basic REMix model. These model instances are LPs generated by GAMS,

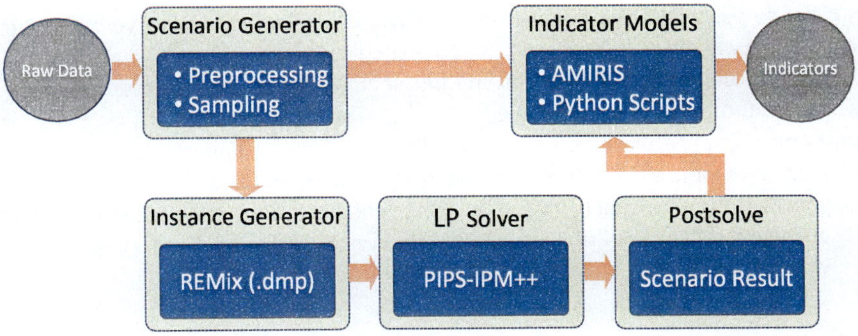

Fig. 13.1 Software components and data flow pipeline that is executed for each scenario on HPCs

which are passed to the parallel solver PIPS-IPM++ for scenario solving. The optimal solution of each LP can be interpreted as one possible scenario in terms of power infrastructure required in the future. Each can be evaluated by calculating numerous indicators. The HPC workflow consists of executing this process for each scenario, as well as providing the data structure to exchange data between individual components. Thus, we are able to evaluate all scenario indicators to observe the points of interests by statistical analyses.

Basic Energy System Optimization Model and Scenario Generation

The REMix parameterization represents a high-resolution network of the German power system on transmission grid level. The 479 nodes represent unique locations of transformer substations. Additionally, nine neighboring countries are included with fixed imports and exports to Germany. The model focuses on the power sector with several power plant, storage and grid technologies included. In the "green field" approach no pre-existing power plants are available. Lignite and coal power plants can be expanded with their current capacities as upper limit. Gas-fired power plants can be further expanded as transition technologies. CO_2 prices drive investments into CO_2-poor technologies.

The input parameters include historical weather profiles for the dispatch of the renewable energies for the years 1995–2018. Additionally, techno-economic parameters such as capital expenditures, fuel cost, CO_2 allowance cost, efficiencies, fixed and variable operations and maintenance costs are included, which have been subject to our parameter sampling approach. Drivers are varied randomly to create different instances of the model. For consistency reasons (e.g., coupled oil and gas prices), we need (i) a collection of possible parameter values, (ii) information about their probability distributions and (iii) information about possible interrelations. For this, we define pseudo-correlations (from negative to positive) of the drivers based on expert assessments.

Thus, a literature research considering about 50 sources [7] including energy scenario studies on both Europe, e.g., [8] and Germany, e.g., [9] derives statistical descriptors of the drivers' values. A statistical derivation of a probability distribution of parameter values from different studies is impossible. Instead, we use truncated normal distributions, which are defined by the collected statistical descriptors, which results in consistent REMix instances and thus, various scenarios to be passed to the solver.

Parallel Computing of Multiple Optimization Models Using PIPS-IPM++

Each model instance to be solved by PIPS-IPM++ has to be annotated, which means that variables and constraints are assigned to independent blocks to be treated in parallel by the solver. Despite a large variety of conceivable criteria to define these blocks, we annotate each model instance into time blocks, which represent predefined time slices in the modeled operation horizon. We need to stress that the available computing hardware determines how model instances are annotated. In our case, the corresponding limitations are a maximal total wall-clock time of 24 h and a maximum of 192 GB RAM per compute node. We annotate the problem instances into 730 blocks to be solved in 216 MPI tasks distributed across 18 compute nodes with 4 cores per task. PIPS-IPM++ is executed using the hierarchical approach, which is for our application the best setup to avoid memory issues. As a result, we are able to solve model instances with about 94.6 M variables (including 3713 linking variables and thereof 3356 globally) and 91.2 M constraints (ca. 367 k locally and 693 linking globally) in about 14 h. After a successful solve, a post-solving process creates a solution as GDX file to be used by subsequent workflow steps (indicator models).

Indicator Assessment

To assess the various aspects of future energy system scenarios, we coupled several models (see Section "Workflow Automatization") for a more comprehensive analysis of the solved ESOM instances. For that, indicators [10] are defined, which are computed by indicator models (e.g., an agent-based model that simulates the behavior of stakeholders at the electricity market for each scenario). For some indicators, the interpretation is clear, e.g., system costs or CO_2 emissions—lower is better. However, for some indicators this is less clear. Therefore, indicators are scored in respect to the overall mean of all scenarios. If it is above or below one standard deviation in the desired direction (if possible), it is considered for further investigation. Scenarios that score on a lot of these indicators are selected and dubbed points of interest.

Workflow Automatization

The overarching goal of our workflow automatization is to provide a basis for analyzing a large number of scenarios, but also to allow massively parallel implementation on HPC with automatic data exchange. The challenge is to maintain a bug-free workflow consisting of dozens of scripts or program calls, which are linked in a serial manner and are subject to continuous development. Hence, any change in just one component might break the whole workflow. Parser scripts used for data transformation require exception handlings, which were not implemented initially. Nevertheless, replacing broken workflow components is not always possible, e.g., replacing PIPS-IPM++ by a commercial solver.

To keep an overview of this complex workflow we extended the software JUBE [11], consisting of ~ 6.200 directories and ~ 140.000 files and ~ 1.8 TB of data in total, cumulative across several scenario runs. The JUBE extension introduces another layer of parallelism to the workflow besides the solver resulting in a reduction of the total workflow runtime. Implementing this exchange of independent and highly specialized software in a stable manner took a team of 10 about 1.5 years, calculations took about 1.100.000 core hours.

Results

First, we cluster indicators and inputs. For robustness reasons, both k-means and k-medoids are employed, using both the BIC and GAP method to determine the optimal number of clusters. For 1000 scenarios with pre-existing power plant capacities and low spatial resolution this does not lead to discernible clusters. Hence, we evaluate a lower number of highly resolved and green field ESOMs. In both cases, we observe three clearly delineated clusters (nine indicators are collapsed into two dimensions via PCA, i.e. main correlations). Figure 13.2 shows the clusters for the green field approach. The first cluster (red) is in-between extremes for most indicators. The second cluster (green) subsumes power systems with higher shares of renewables (RE-share), low CO_2-emissions and medium capability for flexible load-balancing. The third one (blue) is opposite to the first one with high dependency on natural gas, more CO_2-emissions, a high electricity price and some shortages for providing enough energy reserves.

Correlations between indicators are as expected, e.g., a high RE-share corresponds to low CO_2-emissions, etc. This lends credibility to the scenario analyses.

Points of interest are all scenarios where a majority of indicators show values one standard deviation above (e.g. RE-share) or below (e.g., CO_2-emissions) the mean

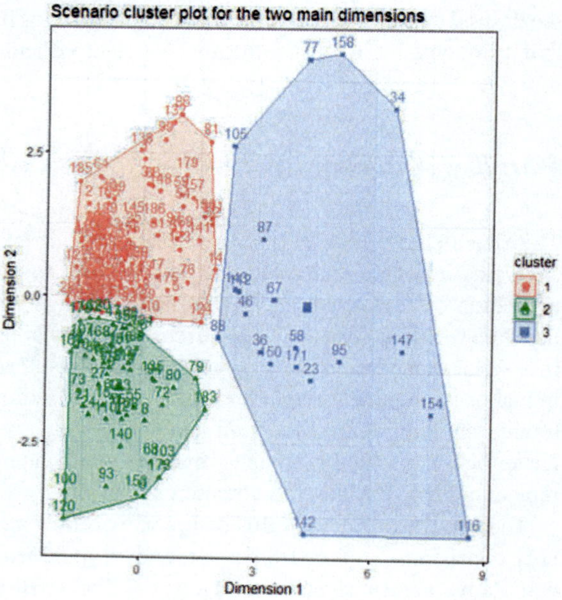

Fig. 13.2 Scenario clusters for the two main dimensions

of all scenarios. Overall, there are few points of interest, i.e. systems where many indicators would point to a system that is satisfactory concerning system adequacy, security of supply, and economic performance. Differences between scenarios are small, i.e., t-tests between potential “good” and “bad” systems are not significant.

Discussion

This paper addresses three problems of current energy systems analysis, i.e. computational limitations, model results that are highly dependent on varying assumptions and the limited perspectives of single models on only some aspects. By implementing a complex and scalable HPC work flow through coupling a number of specialized models, application of PIPS-IPM++ , and a comprehensive set of indicators, this paper proposes a solution to these problems. However, some limitations remain. Initially, a high number of scenarios did not show distinguishable clusters due to low spatial resolution and limited options due to a large size of already available power plants capacities. With the green field approach or highly resolved ESOMs relevant bottlenecks could be identified. However, only a selected number of scenarios could be calculated, yet. Points of interests are defined statistically, not from a system perspective.

Conclusion

Our results pave the way to more robust energy system modeling since they cover a large range of assumptions and future pathways. We found a few scenarios that seem to satisfy a number of desiderata for a near-optimal energy system. Our indicator set is easily reusable and allows a comprehensive assessment of energy systems, most notably system adequacy, security of supply, sustainability, and economic performance.

Thus, from an operations research perspective, lessons learned concern the scale-up of the HPC workflow maintained by different developers. It is crucial to find a trade-off between (a) an early prototype and (b) the early preparation of all workflows.

The established broad scale analysis can be reused for future analyses which evaluate new indicators (e.g., for sustainability assessments) and also put emphasis on systems beyond the power sector and solving of mixed-integer linear programs. Due to the HPC capability and automation this workflow provides full scalability, which can be further improved by making the parallel solver PIPS-IPM++ more robust and computationally more performant.

Acknowledgements We thank our UNSEEN colleagues, a project funded by the German Federal Ministry for Economic Affairs and Climate Action, grant FKZ 03EI1004. We gratefully acknowledge the Gauss Centre for Supercomputing e.V. for providing computing time on the GCS Supercomputer JUWELS at Jülich Supercomputing Centre (JSC).

References

1. Cao, K. -K., von Krbek, K., Wetzel, M., Cebulla, F., & Schreck, S. (2019). Classification and evaluation of concepts for improving the performance of applied energy system optimization models. *Energies, 12*(24).
2. Gils, H. C., Gardian, H., Kittel, M., Schill, W. P., Murmann, A., Launer, J., Gaumitz F., van Ouwerkerk J., Mikurda J., & Torralba-Díaz, L. (2022). Model-related outcome differences in power system models with sector coupling—Quantification and drivers. *Renewable and Sustainable Energy Reviews, 159.*
3. Ringkjøb, H. K., Haugan, P. M., & Solbrekke, I. M. (2018). A review of modelling tools for energy and electricity systems with large shares of variable renewables. *Renewable and Sustainable Energy Reviews, 96.*
4. Rehfeldt, D., Hobbie, H., Schönheit, D., Koch, T., Möst, D., & Gleixner, A. (2022). A massively parallel interior-point solver for LPs with generalized arrowhead structure, and applications to energy system models. *European Journal of Operational Research, 296*(1).
5. Gils, H. C., Scholz, Y., Pregger, T., de Tena, D. L., & Heide, D. (2017) Integrated modelling of variable renewable energy-based power supply in Europe. *Energy, 123.*
6. Deissenroth, M., Klein, M., Nienhaus, K., & Reeg, M. (2017) Assessing the plurality of actors and policy interactions: agent-based modelling of renewable energy market integration. *Complexity, 2017.*
7. A multi-perspective approach for exploring the scenario space of future power systems: input data. https://doi.org/10.23728/B2SHARE.4E5E2D11B8224FB8809CDC2D07EEFF04

8. Ruiz, P., Nijs, W., Tarvydas, D., Sgobbi, A., Zucker, A., Pilli, R., Jonsson, R., Camia, A., Thiel, C., Hoyer-Klick, C., Dalla Longa, F., Kober, T., Badger, J., Volker, P., Elbersen, B. S., Brosowski, A., & Thrän, D. (2019). ENSPRESO-an open, EU-28 wide, transparent and coherent database of wind, solar and biomass energy potentials. *Energy Strategy Reviews, 26*.
9. Bernath, C., Boßmann, T., Deac, G., Elsland, R., Fleiter, T., Franke, B., Kühn, A., Pfluger, B., Ragwitz, M., Rehfeldt, M., & Tersteegen, B. (2017). *Langfristszenarien für die Transformation des Energiesystems in Deutschland—Modul 3: Referenzszenario und Basisszenario*. Karlsruhe, Germany.
10. Energy system indicators. https://doi.org/10.23728/b2share.fe70b138419243c0817425a0d2d5ae32
11. JUBE benchmarking environment 2008–2022. Retrieved from http://www.fz-juelich.de/jsc/jube

Chapter 14
A Quantum Computing Approach for the Unit Commitment Problem

Pascal Halffmann, Patrick Holzer, Kai Plociennik, and Michael Trebing

Abstract Planning energy production is a challenging task due to its cost-sensitivity, fast-moving energy markets, uncertainties in demand, and technical constraints of power plants. Thus, more complex models of this so-called *unit commitment problem (UCP)* have to be solved more rapidly, a task that probably can be solved more efficiently via quantum computing. In this article, we model a UCP with minimum running and idle times as a quadratic unconstrained optimization problem to solve it on quantum computing hardware. First experiments confirm the advantages of our formulation in terms of qubit usage and connectivity and most importantly solution quality.

Keywords Quantum computing · Unit commitment problem · Quadratic programming · Energy planning

Introduction

An energy supply that is both stable and environmentally sustainable is vital for viable economic growth and social welfare. However, with the recent substantial increases in energy market prices and price fluctuations, energy generation is a highly cost-sensitive field comprising a complex system of power units and grids. The complexity of the so-called *unit commitment problem (UCP)* presents challenges when it comes to planning energy generation due to technical constraints regarding power units and the power grid—a task that, with increasing reliance on renewable energy, is exacerbated further by weather-induced uncertainties. In conclusion, there is a real incentive to calculate the optimal solution to a realistic UCP model in a short period

P. Halffmann (✉) · P. Holzer · M. Trebing
Department of Financial Mathematics, Fraunhofer Institute for Industrial Mathematics, Am Fraunhofer Platz 1, 67663 Kaiserslautern, Germany
e-mail: pascal.halffmann@itwm.fraunhofer.de

K. Plociennik
Competence Center High Performance Computing, Fraunhofer Institute for Industrial Mathematics, Am Fraunhofer Platz 1, 67663 Kaiserslautern, Germany

O. Grothe et al. (eds.), *Operations Research Proceedings 2022*, Lecture Notes in Operations Research, https://doi.org/10.1007/978-3-031-24907-5_14

of time. However, solvers on classical computers often cannot accomplish this task, especially when uncertainties from renewable energy supply are considered.

In contrast, *quantum computing (QC)* provides a new computational paradigm based on counterintuitive phenomena in quantum mechanics, such as considering all possible solutions at once in a state of superposition. This has the potential of rapidly solving various tasks, from optimization problems to simulations and even communication. Admittedly, in the current, so called *noise intermediate scale quantum (NISQ)* era, only few quantum computers are available with a limited amount of qubits, a rather restricted number of connections between them, and low fault-tolerance. Suitable use cases have just been discovered and algorithms are in development. Nevertheless, QC is a promising technology of the near-term future and has already achieved some remarkable results for example in chemistry [3] and finance [10]. Therefore, the question arises whether QC will eventually be able to provide solutions to the UCP faster and, due to its inherent capability of coping with uncertain and stochastic parameters, has advantages when considering e.g. uncertain supply from renewable energies.

Previous Work

The unit commitment problem has been studied extensively in the last decades. There exist numerous versions with varying complexity and purpose. We refer to [9] for an overview. It is beyond the scope of this article to provide a full literature overview on the extensively published topic of quantum computing. So far, only two contributions try to solve the UCP via quantum computing: In [1], a UCP problem with quadratic cost function, demand satisfaction and minimum and maximum power generation is discussed. They use a generic method to transform this model to a quadratic unconstrained problem. A computational study with three to twelve power units shows that while D-Wave returns near-optimal solutions, the computation via Gurobi is 30,000 times faster for the largest instance. Recently, the authors of [8] considered a quantum computing approach for a distributed UCP, where power units are concentrated in a connected hub which allows a decomposition into subproblems. They apply a quantum version of the decomposition and coordination alternate direction method of multipliers (ADMM) to this problem. While they report that demand satisfaction, minimum running and idle times, ramping, and power grid constraints among others are considered in the model using the same generic method for the problem transformation as before, a full formulation of their problem is not given.

Our Contribution

In order to solve an optimization problem via a quantum computer it can be transformed into a so-called *Ising Hamiltonian*, which is an operator measuring the total energy of a physical system. This operator has a one-to-one correspondence to a QUBO. Generically, a linear optimization problem with constraints is transformed into a QUBO by transforming inequality constraints to equality constraints via slack variables and adding equality constraints as quadratic penalty terms to the objective function. At last, continuous and integer variables are encoded and replaced by binary variables. Clearly, this method is not problem-specific and has some drawbacks: Each slack variable (or its binary encoding, if it is non-binary) requires additional qubits. The quadratic penalty term results in quadratic terms using every combination of variables used for this penalty term, thus, in an all-to-all connection between the corresponding qubits. Current NISQ devices have a limited number of qubits with restricted interconnectivity and in particular no all-to-all connectivity, which requires additional qubits to facilitate the connection. It is therefore crucial to avoid introducing large numbers of qubits with many interactions. Even in the future, the degree of interconnectivity between qubits can be expected to present a limiting factor.

We propose a novel formulation for a unit commitment problem modeling minimal variable and starting costs, demand satisfaction as well as minimum running and idle times as a QUBO problem. This formulation is explicitly designed for reducing the number of and the connectivity between qubits, while ensuring that all constraints are satisfied even though they are transformed to penalty terms. This has been especially achieved by avoiding slack variables and, largely, squared sums of variables in our method. As pointed out, quantum computing is still in its infancy. We do not expect that solving a UCP with current QC hardware and algorithms can cope with real-world sized problem instances or is competitive against classical approaches. We merely present a proof-of-concept for an alternative direction: given the current rapid development of QC and its present successes, our model may provide faster or better solving of the UCP in the future, especially for large instances with e.g. uncertain demand.

The remainder of this article is organized as follows: In Section "Classicform", we formally introduce the unit commitment problem and present a formulation as mixed-integer linear problem. In the next section, our formulation as quadratic unconstrained binary problem is given. An illustrative example of our formulation, together with a comparison to resource requirements for the generic method, and solving the model both via a quantum computer simulator and a quantum annealer, follows in Section "Example". We conclude this article with an outlook on future directions of research.

Linear Formulation of the Unit Commitment Problem

In this section we present the classical mixed-integer linear formulation of the unit commitment problem. In general, the UCP deals with finding a cost-minimal operation schedule for a set of thermal units $i \in \{1, \ldots, N\}$ to meet a given demand for electricity over a distinct set of time steps $t \in \{1, \ldots T\}$. Further, technical properties of both the thermal units and the underlying power grid have to be respected. Take note that there does not exist one single UCP, but several variants are present in the literature. For an overview on different formulations, we refer to [6]. In the remainder of this article, we propose a model with the following assumptions. Each thermal unit i has the following properties: linear, production dependent costs $varcost_i$ and fixed costs $startcost_i$ for starting the unit, minimum and maximum power generation output, $mingen_i$, $maxgen_i$, and minimum running time and minimum idle time, $minup_i$, $mindown_i$. At each time step t the residual demand rd_t, demand minus supply by renewable energies plus spinning reserve, has to be met. Limitations due to the power grid are omitted. This type of unit commitment problem is commonly modeled as follows:

$$
\begin{aligned}
\min \ & \sum_{t=1}^{T}\sum_{i=1}^{I}\left(varcost_i \cdot gen_{t,i} + startcost_i \cdot start_{t,i}\right) \\
\text{s.t.} \ & \sum_{i} gen_{t,i} = rd_t, && \forall t = 1,\dots,T, \\
& on_{t,i} \cdot mingen_i \le gen_{t,i}, && \forall t = 1,\dots,T,\ i = 1,\dots,I, \\
& on_{t,i} \cdot maxgen_i \ge gen_{t,i}, && \forall t = 1,\dots,T,\ i = 1,\dots,I \\
& on_{t,i} - on_{t-1,i} \le start_{t,i}, && \forall t = 1,\dots,T,\ i = 1,\dots,I, \\
& \sum_{\tau=t}^{t-1+minup_i} on_{\tau,i} \ge start_{t,i} \cdot minup_i, && \forall t = 1,\dots,T,\ i = 1,\dots,I, \\
& \sum_{\tau=t+1-mindown_i}^{t} start_{\tau,i} \le 1 - on_{t-mindown_i,i}, && \forall t = 1,\dots,T,\ i = 1,\dots,I, \\
& on_{t,i}, start_{t,i} \in \mathbb{B} && \forall t = 1,\dots,T,\ i = 1,\dots,I, \\
& gen_{t,i} \in \mathbb{R}_{\ge 0} && \forall t = 1,\dots,T,\ i = 1,\dots,I.
\end{aligned}
$$

Here we have three sets of decision variables $on_{t,i}$, $gen_{t,i}$ and $start_{t,i}$, $t \in \{1, \dots T\}$, $i \in \{1, \dots, N\}$. The binary variable $on_{t,i}$ observes whether unit i is running at time t. With $gen_{t,i} \in \mathbb{R}_{\ge 0}$ we denote the power generated by unit i at time step t. The variables $start_{t,i} \in [0, 1]$ track the starting of power units.

Quadratic Unconstrained Binary Formulation of the Unit Commitment Problem

In quantum physics, the *Hamiltonian*, an operator corresponding to the total energy of the system it refers to, is used to calculate the energy-minimal state of that system. One of the most famous Hamiltonians is the *Ising Hamiltonian* describing the energy of a solid in a ferromagnetic field using n spins $s_i = \pm 1$:

$$H(s_1, \dots, s_n) = -\sum_{i<j} J_{ij} \cdot s_i \cdot s_j - \sum_{i=1}^{N} h_i \cdot s_i .$$

Due to its close resemblance to a quadratic unconstrained binary problem $\min_{x\in\mathbb{B}^n} x^\top Qx$, optimization problems are transformed to QUBOs and then to Ising Hamiltonians in order to solve these problems on a quantum computer, see [5] for methods for the transformation. Spins can be transformed to binary variables via $x_i = {(s_i+1)}/{2}$. We concisely state the UCP as QUBO as follows:

$$
\begin{aligned}
\min \ & \sum_{t=1}^{T}\sum_{i=1}^{I}(varcost_i \cdot maxgen_i \cdot on_{t,i} + startcost_i \cdot start_{t,i}) \\
& +A\cdot\sum_{t=1}^{T}\left(\sum_{i=1}^{I} maxgen_i \cdot on_{t,i} - rd_t\right)^2 \\
& +B\cdot\sum_{t=1}^{T}\sum_{i=1}^{I}(on_{t,i}\cdot(1-on_{t-1,i}) + 2\cdot start_{t,i}\cdot(on_{t-1,i}+1-on_{t,i}) - start_{t,i}) \\
& +C\cdot\sum_{t=1}^{T}\sum_{i=1}^{I}\left(start_{t,i}\cdot minup_i - \sum_{\tau=t}^{t-1+minup_i} start_{t,i}\cdot on_{\tau,i}\right) \\
& +D\cdot\sum_{t=1}^{T}\sum_{i=1}^{I}\left(\sum_{\tau=t}^{t-1+mindown_i}(start_{t,i}+on_{t-1,i}-on_{t,i})\cdot on_{\tau,i}\right) \\
\text{s.t.} \ & on_{t,i}, start_{t,i} \in \mathbb{B} \qquad \forall\, t=1,\ldots,T,\ i=1,\ldots,I.
\end{aligned}
$$

The penalty terms are in the same order as the constraints for the classical formulation. By a closer look one can identify that if the $start$-variable is not properly set (i.e. it is zero although it has to be one), the last penalty term provides a bonus for an infeasible solution. Hence the penalty for wrongly setting $start$-variables has to be higher: $B > D \cdot \max_{i=1,\ldots,I} mindown_i$. It is also possible to model a UCP with energy generation between $mingen_i$ and $maxgen_i$ using discrete power generation steps. Given a step size $step_i$, we need $d_i := \lfloor \log_2 ({(maxgen_i - mingen_i)}/{step_i}) \rfloor + 1$ variables using a logarithmic encoding of the steps ([7]). In total, our model needs $T \cdot I \cdot \left(2 + \sum_{i=1}^{I} d_i\right)$ variables. Then we can replace the power output $maxgen_i \cdot on_{t,i}$ by

$$
mingen_i \cdot on_{t,i} + \left(\sum_{k=1}^{d_i-1} 2^k \cdot gen_{t,i,k} + \left(d_i + 1 - 2^{d_i}\right)\cdot gen_{t,i,d_i}\right)\cdot step_i .
$$

Further, we have to add a coupling constraint between the on-variables and the discrete power generation variables $gen_{t,i,k}$:

$$
A\cdot\sum_{t=1}^{T}\left((1-on_{t,i})\cdot\sum_{k=1}^{d_i} gen_{t,i,k}\right).
$$

Table 14.1 Input data of our example

Name	$mingen_i$	$maxgen_i$	$minup_i$	$mindown_i$	$startcost_i$	$varcost_i$
Unit 0	1	1	1	1	50	30
Unit 1	1	1	2	1	25	45

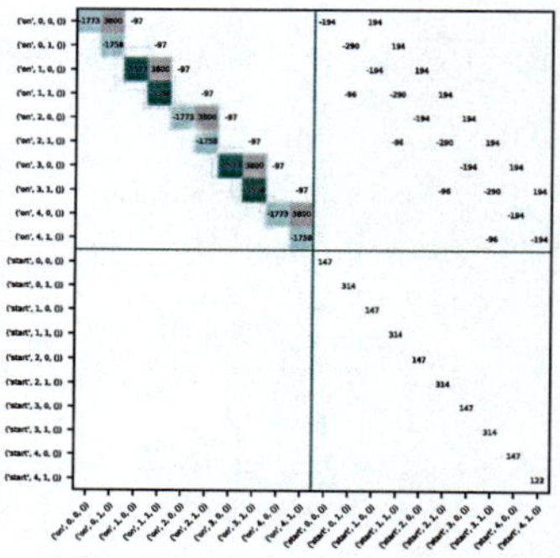

Fig. 14.1 The QUBO matrix corresponding to our example. The first index counts time steps, the second the unit

Illustrative Example

In order to illustrate our model, we have built the QUBO problem for an instance with two thermal units and five time steps. Further, we assume that each unit can only operate on maximal power generation if turned on. The parameter values are given in Table 14.1. Further, the residual demand is given by $\{1, 2, 1, 2, 1\}$. The upper triangular matrix of the QUBO problem with penalty factors $A = 1900$, $B = 97$, $C = 96$, $D = 96$ is visualized in Fig. 14.1.

Clearly, we have obtained a sparse matrix, as only 55 of 400 entries (13.75%) are nonzero. This number rapidly decreases to 1 to 2% for larger instances. Further, the maximum number of nonzero entries per row/column is 5. Hence, corresponding qubits are less connected and the embedding onto real hardware needs only few extra qubits. In contrast, for the generic formulation, we get the following penalty term controlling the relation between $start$- and on-variables: $(start_{t,i} - (on_{t,i} - on_{t-1,i} + slack_{t,i}))^2$. This results in an additional qubit for each $start$-variable with a necessary connection to the qubits responding to $start_{t,i}$,$on_{t,i}$, and $on_{t-1,i}$. The effect of one slack variable for every power plant and time step with only three interactions each may appear relatively harmless. Yet, for the minimum running time constraint this requires the introduction of at least $\lfloor \log_2(minup_i) \rfloor + 1$ slack variables for every unit and time step. Due to squaring, these interact with one $start$-variable and a number of on-variables each and also with each other. The same holds true for the minimum idle time.[1] In total, the generic model has 50 variables with 106 interactions between different variables, while our tailor-made

[1] Remark that in our example this parameter is set to 1 and does not need any binarization, increased parameter values worsen the generic model.

approach calls for only 38 interactions of 20 variables thus makes better use of system resources.

We have solved this model using the *Gurobi* solver (Version 9.5.1) on classical hardware, the *Quantum Approximate Optimization Algorithm (QAOA)* with warm start [4] on the *IBM Qiskit* 0.37.1 QASM simulator, and the *D-Wave Quantum Annealer* (Version Advantage_system 5.2 with over 5000 qubits) using 23 physical qubits instead of the 20 virtual qubits. The Gurobi solver used on the classical formulation finds the optimal solution $\{0, 1, 1, 1, 1, 0, 1, 1, 0, 1, 0, 1, 1, 0, 0, 0, 0, 1, 0, 0\}$ with objective value 370 (Ising objective value -20530). Here, unit 0 is on in time steps 1 to 3. Unit 1 is running in time steps 0, 1, 3, and 4. For 1000 shots, this is the solution occurring most often on D-Wave. The QASM simulator with warm started QAOA, however, does not find the optimal solution but finds $\{0, 1, 1, 1, 0, 1, 1, 1, 0, 1, 0, 1, 1, 0, 0, 0, 1, 0, 0, 0\}$ with objective value 410 (Ising objective value -20490). This solution is feasible and just switches the occupancy of the units in time step 2, which is also reflected in changed states for both $start$ variables in time step 3.

Conclusion

In this article, we have presented a new formulation for the unit commitment problem as a QUBO such that it can be solved using quantum computers. Our model correctly penalizes infeasible solution while providing a compact matrix with fewer necessary qubits and connections between qubits than generic translation methods. First tests show the advantages of our model in practice. While our model still has a binary encoding of the continuous variables occupying valuable qubits, a recent publication show that it may possible to encode these variables without binary encoding on some QC devices [2].

Besides the introduction of uncertainty to the residual demand, immediate future research will obviously be to carry out an extensive computational study of our model on various systems (e.g. Gurobi vs. quantum computer) comparing solution quality and running times. Due to the limited availability of qubits and connectivity in gate-based quantum computers, we will focus on the quantum annealers from D-Wave. This study promises to return answers to two important questions: can current quantum computing devices outperform classical solvers on classical hardware for practical problems and, if not, what is necessary in the future to ascertain "quantum supremacy" in practice?

Acknowledgements This publication has been funded by the German Federal Ministry for Economic Affairs and Climate Action (grant no. 03EI1025A).

References

1. Ajagekar, A., & You, F. (2019). Quantum computing for energy systems optimization: Challenges and opportunities. *Energy, 179*. https://doi.org/10.1109/TPWRS.2022.3141794
2. Bermejo, P., & Orus, R. Variational quantum continuous optimization: A cornerstone of quantum mathematical analysis. https://doi.org/10.48550/arXiv.2210.03136
3. Cao, Y., Romero, J., Olson, J., Degroote, M., Johnson, P., Kieferová, M., Kivlichan, I., Menke, T., Peropadre, B., Sawaya, N., Sim, S., Veis, L., & Aspuru-Guzik, A. (2019). Quantum chemistry in the age of quantum computing. *Chemical Reviews, 119*(19), 10856–10915. https://doi.org/10.1021/acs.chemrev.8b00803
4. Egger, D., Mareček, J., & Woerner, S. (2021). Warm-starting quantum optimization. *Quantum 5*, 479. https://doi.org/10.1287/ijoc.2019.0944
5. Glover, F., Kochenberger, G., Hennig, R. et al. (2022). Quantum bridge analytics I: A tutorial on formulating and using QUBO models. *Annals of Operations Research, 314*, 141–183. https://doi.org/10.1007/s10479-022-04634-2
6. Knueven, B., Ostrowski, J., & Watson, J.-P. (2020). On mixed-integer programming formulations for the unit commitment problem. *INFORMS Journal on Computing, 32*(4), 857–876. https://doi.org/10.1287/ijoc.2019.0944
7. Lucas, A. (2014, February 12). Ising formulations of many NP problems. *Frontiers in Physics* Sec. Interdisciplinary Physics. https://doi.org/10.3389/fphy.2014.00005
8. Nikmehr, N., Zhang, P., & Bragin, M. (2022). Quantum distributed unit commitment. *IEEE Transactions on Power Systems*. https://doi.org/10.1016/j.energy.2019.04.186
9. Padhy, N. (2004). Unit commitment—A bibliographical survey. *IEEE Transactions on Power Systems, 19*(2), 1196-1205. https://doi.org/10.1109/TPWRS.2003.821611
10. Stamatopoulos, N., Mazzola, G., Woerner, S., & Zeng, W. (2022). Towards quantum advantage in financial market risk using quantum gradient algorithms. *Quantum, 6*, 770. https://doi.org/10.22331/q-2022-07-20-770

Chapter 15
The Sales Force Deployment Problem for Teams of Sales Representatives Within Sales Territories

Tobias Vlćek

Abstract We address the sales force deployment problem and its four subproblems for teams of sales representatives within sales territories. We contribute by showing that the problem can be reduced to the uncapacitated facility location problem under convex profit contribution functions with a unique maximum. In our numerical study, we show that instance sizes considered difficult for the sales force deployment problem of individual representatives can be solved optimally in minutes for teams of representatives. In our largest instance, with 2020 potential locations and sales coverage units, it took on average 140.73 s.

Keywords Salesforce · Territory alignment · Uncapacitated facility location problem

Introduction

By addressing the sales force deployment problem (SFDP), decision makers are assigning customer accounts and their selling potential to sales representatives in order to maximize profit [7]. The SFDP is composed of four interrelated subproblems: sales territory alignment, sales force sizing, sales resource allocation and sales force location. Solving these, decision makers are grouping small geographic sales coverage units (SCUs) in a preset area into sales territories according to managerial criteria while determining the size of the sales force, the allocation of the selling time to each individual SCU, and the location of a sales office for each territory [2]. Past literature considered sales representatives working independent of each other in separate territories. We contribute by solving the SFDP for teams of sales representatives within sales territories. This can have benefits, as teams can support and

T. Vlćek (✉)
University of Hamburg, Institute of Logistics, Transport and Production, Moorweidenstraße 18, 20148 Hamburg, Germany
e-mail: tobias.vlcek@uni-hamburg.de

O. Grothe et al. (eds.), *Operations Research Proceedings 2022*, Lecture Notes in Operations Research, https://doi.org/10.1007/978-3-031-24907-5_15

represent each other while sharing their knowledge to improve sales and the onboarding of new members. In addition, sales teams can lead to higher profits as they allow solutions with a lower number of office locations and their associated fixed costs.

Related Research

Due to the interrelatedness of the subproblems, past research has often focused on solving the subproblems of the SFDP separately. The profit was usually the central criterion, additional criteria included contiguity, equal profits and equal workloads per sales representative [6]. We refer interested readers to the overviews by [4] and [7] and subsequently focus on literature directly related to our work. Skiera and Albers [5] proposed an approach based on aggregated concave sales response functions to solve two of the four subproblems of the SFDP for individual sales representatives. Drexl and Haase [2] considered all four subproblems simultaneously. They developed a non-linear mixed-integer programming model and used a heuristic to approximate solutions. Later, [3] addressed the problem by solving the linear relaxation of the model with column generation combined with Branch-and-Price. We contribute to the literature by explicitly addressing all four subproblems of the SFDP for teams simultaneously, while showing that the problem can be reduced to the uncapacitated facility location problem (UFLP) under certain conditions.

Profit Contribution

Our aim is a maximization of the overall profit over all sales territories. We employ a known and widely accepted concave aggregated sales response function where the selling time is a constant proportion of the travel time within each SCU while considering the incurred selling costs [2, 3, 5]. We formalize the profit contribution function by defining the following sets and symbols:

$\mathcal{J}$: set of SCUs, indexed by j,
$\mathcal{I}$: set of potential office locations ($\mathcal{I} \subseteq \mathcal{J}$) indexed by i,
α : per-unit profit contribution of sales,
t : allocated selling time to SCU j,
b : calling time elasticity ($0 < b < 1$),
n_j : calling time profitability, e.g., potential accounts in j ($n_j > 0$),
β_{ij} : ratio of travel time to the selling time allocated to SCU j by the sales team located in SCU i ($0 < \beta_{ij} < 1$),
μ : productivity scaling parameter of the salesforce ($\mu > 0$),
h : cost per hour of travel time (e.g., gas, car insurance, ...) ($h > 0$),

g : hourly cost per sales representative (e.g., salary, variable location costs, ...) ($g > 0$) and
f_i : fixed costs of allocating a sales office in SCU i.

Note, the hourly costs per sales representative g were integrated into the fixed costs associated with each opened office in former approaches [2, 3, 5]. Now, we can formulate the sales response function

$$s_{ij}(t) = \mu \times n_j \times ((1 - \beta_{ij}) \times t)^b \tag{15.1}$$

and the selling cost function

$$k_{ij}(t) = (h \times \beta_{ij} + g) \times t, \tag{15.2}$$

both of which are dependent on the associated sales team location i and the allocated selling time t to SCU j. The farther the travel distance between the sales team location i of a sales territory and the SCU j, the higher the ratio of the travel time to the selling time β_{ij}. Now, we can calculate the profit contribution of each allocated SCU j to a sales team location i with (15.3).

$$p_{ij}(t) = \alpha \times s_{ij}(t) - k_{ij}(t) \tag{15.3}$$

In contrast to [2, 3, 5] we don't consider the selling costs as negligible. They comprise not only the hourly travel costs h, but also the hourly costs per sales person g (e.g., salary, variable costs of the office location, ...) to account for multiple representatives per location. This is important, as the profit in the application of the former approaches was only bound by the fixed costs f_i associated with each realized location. Now, it is also bound by the selling costs in each SCU j. We use this characteristic and differentiate the profit contribution function (15.3) by t while setting the derivate to $\hat{p}$.

$$\frac{\partial p_{ij}(t)}{\partial t} = \hat{p} \tag{15.4}$$

Now, (15.4) is strictly monotonically decreasing; thus there exists a unique solution for each $\hat{p}$. This allows us to solve (15.4) for t to determine the selling time $t_{ij}(\hat{p})$ for all combinations of i and j for different marginal profits $\hat{p}$ with (15.5).

$$t_{ij}(\hat{p}) = \left(\frac{\beta_{ij} \times h + g + \hat{p}}{\alpha \times \mu \times n_j \times b \times (1 - \beta_{ij})^b} \right)^{\frac{1}{b-1}} \tag{15.5}$$

As our selling time is unconstrained, the optimal selling time $t_{ij}(\hat{p})$ for each possible combination of i and j has marginal profits $\hat{p} = 0$. We know these selling times are finite, as $\lim_{t \to \infty} \frac{\partial p_{ij}(t)}{\partial t} = -(\beta_{ij} \times h + g)$. Hence, we can calculate the optimal selling time $t_{ij}(0)$ with (15.5) to determine the maximal profit $p_{ij}(t_{ij}(0))$ of each possible allocation with equation (15.3).

Model

Based on the pre-calculated maximal profit $p_{ij}(t_{ij}(0))$, we can reduce the problem to the UFLP to solve the four interconnected subproblems. Note, that the UFLP is nonetheless $\mathcal{NP}$-hard as it can be reduced from the set covering problem [1]. First, we define the decision variable: $X_{ij} = 1$, if SCU j is assigned to the office in SCU i, else 0. Now, we can maximize the overall profit P of our sales territories with the UFLP:

$$\text{maximize} \quad \sum_{i\in\mathcal{I}}\sum_{j\in\mathcal{J}} p_{ij}(t_{ij}(0)) \times X_{ij} - \sum_{i\in\mathcal{I}} f_i \times X_{ii} \tag{15.6}$$

subject to:

$$\sum_{i\in\mathcal{I}} X_{ij} \leq 1 \qquad \forall j \in \mathcal{J} \tag{15.7}$$

$$X_{ij} \leq X_{ii} \qquad \forall i \in \mathcal{I}, \forall j \in \mathcal{J} \tag{15.8}$$

$$X_{ij} \in \{0, 1\} \qquad \forall i \in \mathcal{I}, \forall j \in \mathcal{J} \tag{15.9}$$

In our objective function (15.6), we maximize the total profit contribution. (15.7) ensures that each SCU j is allocated to at most one realized sales force location. (15.8) guarantees that SCUs can only be assigned to realized locations, while (15.9) defines the domain of our decision variable. Recall that β_{ij} describes the ratio of the travel time to the selling time by the team allocated in SCU i. Let a_{ij} $\forall i \in \mathcal{I}, \forall j \in \mathcal{J}$ describe the air-distance between the centroids of all possible SCU combinations i and j in kilometres. As long as $a_{ij} < a_{iv} \implies \beta_{ij} < \beta_{iv}$ $\forall i \in \mathcal{I}, \forall j \in \mathcal{J}, \forall v \in \mathcal{J}$ and $\mathcal{I} \subseteq \mathcal{J}$, the consideration of contiguity constraints is not necessary, as each SCU will be allocated to the nearest realized sales-team location. In case these conditions do not hold, we can enforce contiguity by introducing additional constraints, but this is beyond the scope of this paper.

After solving the model, we can answer all four subproblems: the sales territory alignment is represented by the optimal SCU assignment X_{ij}. The sales force teams are located in each SCU i where $X_{ii} = 1$. The sales resource allocation s_i (expressed in selling time) of each location can be calculated with (15.10).

$$s_i = \sum_{j\in J} t_{ij}(0) \times X_{ij} \;\; \forall i \in \mathcal{I} \tag{15.10}$$

Note, that $s_i = 0$ $\forall i \in \mathcal{I} : X_{ii} = 0$ as no sales team is allocated in these SCUs. The overall size of the sales force S (in selling time) is then just the sum of the sales resource allocations s_i of all sales team locations. If we consider the total available selling time T per sales person, the number of sales representatives per district r_i can be calculated by $r_i = s_i/T$ $\forall i \in \mathcal{I}$. Naturally, $r_i \in \mathbb{R}_{\geq 0}$ $\forall i \in \mathcal{I}$ as we do not restrict the selling time to fixed intervals.

Numerical Study

To evaluate the solvability of our model, we implement it on an 8-Core AMD Ryzen 5700G CPU with 32 GB RAM using Julia 1.72, JuMP 2.0 and the solver Gurobi 9.5. We solve all instances till optimality. For our numerical study, we replicate the setting of [3] and assume a fictional company in the German market for medical products. Following their structure, the sales representatives have a ratio of travel time to selling time by

$$\beta_{ij} = 0.1 + \min\{0.9, 2 \times a_{ij}/800\}.$$

This assumes a travel time proportional to the air-distance a_{ij}, sales representatives with an average travel speed of 100 km/h by car and an average 8-h work day. Note, that the travel speed is 100 km/h, as representatives are driving on highways most of the time. This leads to the observation, that 90% of the selling time will be calling time in case $i = j$ and that SCUs with an air-distance $a_{ij} \geq 360$ km cannot be profitable due to the lack of calling time. To ensure that $\beta_{ij} < 1$ we enforce a maximal air-distance $R = 360$ km and replace $\mathcal{J}$ by $\mathcal{J}_i$ where $j \in \mathcal{J}_i | a_{ij} < R$. We benchmark per-unit profit contributions $\alpha \in \{5, 10, 15, 20, 25\}$ while setting $\mu = 1$ and $b = 0.3$. To keep the data set public, we estimate the number of potential customer accounts n_j by dividing the number of residents from the *ZENSUS 2011* data set[1] by 1000, to approximate the number of medical practices within each SCU j. For the selling cost function, we vary the hourly cost per sales person $g \in \{40, 50, 60, 70, 80\}$ and the costs per hour of travel time $h \in \{10, 20, 30, 40, 50\}$. Moreover, we assume the yearly fixed costs f_i to be equal for all SCUs i while varying them between 10,000 and 50,000 in steps of 10,000. The total available selling time T per sales person is 1600 h per year if we assume an 8-h day and 200 working days per year. In addition, we vary the size of the SCU sets $|\mathcal{J}|$ between 401, 758, 1168 and 2020 and assume that all SCUs j represent a potential location, thus $\mathcal{I} = \mathcal{J}$. An example of a resulting territory alignment is shown in Fig. 15.1. In total, we benchmarked 3125 different variations. All data sets and our code is available on GitHub[2].

Our benchmark shows that the number of territories and the number of sales representatives per territory depend a lot on the parameters of the profit contribution function. For example, high fixed costs per office location combined with low driving costs lead to a few office locations with multiple representatives per location that make use of the entire maximal air-distance R. Moreover, we observe that the managerial criteria of equal profits and workload per sales representative compared across territories are fulfilled in all optimal solutions of the optimization due to the identical marginal profits. Table 15.1 shows the average computation time depending on the number of SCUs $|\mathcal{J}|$ and the fixed costs f_i. It is apparent, that the computation time depends on the problem size and the fixed costs.

---

1 https://atlas.zensus2011.de/

2 https://github.com/beyondatlas/salesforce

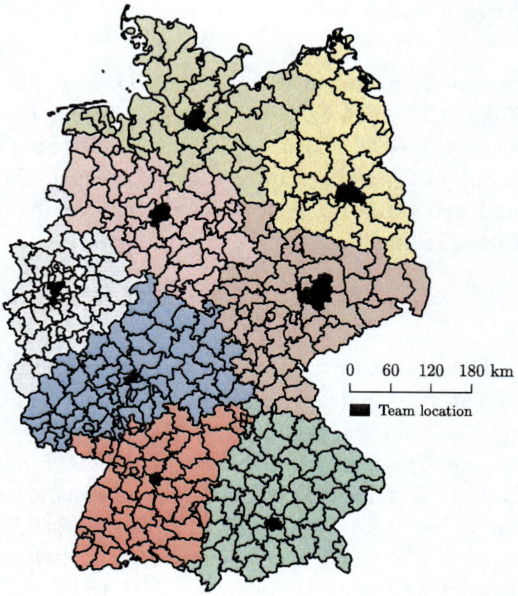

Fig. 15.1 Example of an optimal territory alignment in our numerical study for $\alpha = 20$, $g = 60$, $h = 20$, $f_i = 40,000$ with $|\mathcal{J}| = 401$ solved in 0.59 s

Table 15.1 Computation time (seconds)

	f_i				
$\|\mathcal{J}\|$	10,000	20,000	30,000	40,000	50,000
401	2.1	2.3	2.4	2.4	2.5
758	7.1	8.6	9.5	10.1	10.5
1168	26.2	34.0	37.6	39.5	41.2
2020	116.7	139.9	143.4	150.0	153.5

Note: The time excludes the model-building time and shows the time Gurobi 9.5 needed to solve the instances with a gap of 0%

Nonetheless, the solver could always determine the optimal solution within 2 min. Other parameters are not displayed, as they had no notable influence. This is an improvement compared to past SFDP implementations of individual sales representatives. For example, in the computational study of [3] no solution on instances with $|\mathcal{I}| = 50$ and $|\mathcal{J}| = 2140$ could be proved to be optimal within two hours. Our lower computation time is the result of the reduced model formulation due to the pre-calculated profits and the allowance of fractional representatives r_i per territory.

Conclusion

Overall, our results are promising. By differentiating the profit contribution function and setting the marginal profits to 0, we could determine the optimal selling time in each SCU, since the utilised profit contribution function has a unique maximum by its structure if the selling costs are not neglected. This allowed us to reduce the sales force deployment problem for teams to the uncapacitated facility location problem. Moreover, we showed in our numerical study that optimal contiguous solutions for the SFDP for teams can be determined within minutes for problem sizes considered difficult in the past for the SFDP of individual representatives. Still, we also have to acknowledge the limits of our approach. First, the profit contribution function has to be convex with a unique maximum. Second, contiguous optimal solutions without explicit contiguity constraints can only be guaranteed if the ratio of the travel time to selling time β_{ij} fulfils certain, albeit realistic, conditions. Future research could look at the problem if the number of representatives per sales territory has to be integral and how the computation time changes if contiguity constraints have to be included.

Acknowledgements This research did not receive any specific grant from funding agencies in the public, commercial, or not-for-profit sectors. The author would like to thank Jan Pape, who worked as a student assistant at the project. All other support is gratefully acknowledged.

References

1. Cornuéejols, G., Nemhauser, G., & Wolsey, L. (1983). The uncapacitated facility location problem (tech. rep.). *Cornell University Operations Research and Industrial Engineering*.
2. Drexl, A., & Haase, K. (1999). Fast approximation methods for sales force deployment. *Management Science, 45*(10), 1307–1323. https://doi.org/10.1287/mnsc.45.10.1307
3. Haase, K., & Müuller, S. (2014). Upper and lower bounds for the sales force de-ployment problem with explicit contiguity constraints. *European Journal of Operational Research, 237*(2), 677–689. https://doi.org/10.1016/j.ejor.2014.01.061
4. Howick, R., & Pidd, M. (1990). Sales force deployment models. *European Journal of Operational Research, 48*(3), 295–310. https://doi.org/10.1016/0377-2217(90)90413-6
5. Skiera, B., & Albers, S. (1998). Costa: Contribution optimizing sales territory alignment. Marketing Science, 17 (3), 196-213. https://doi.org/10.1287/mksc.17.3.196
6. Zoltners, A. A., & Sinha, P. (1983). Sales territory alignment: A review and model. *Management Science, 29*(11), 1237-1256. https://doi.org/10.1287/mnsc.29.11.1237
7. Zoltners, A. A., & Sinha, P. (2005). The 2004 isms practice prize winner-sales territory design: Thirty years of modeling and implementation. *Marketing Science, 24*(3), 313-331. https://doi.org/10.1287/mksc.1050.0133

Part V
Discrete and Combinatorial Optimization

Chapter 16
A Heuristic Column Generation Approach for the Stochastic Bin Packing Problem

John Martinovic, Nico Strasdat, Jean-François Côté, and Vinícius Loti de Lima

Abstract The stochastic bin packing problem (SBPP) is an extension of the well-studied classical bin packing problem with respect to imperfect information on the item sizes. From a practical point of view, the latter are typically represented by (stochastically independent) normally distributed random variables with given means and variances. In this scenario, the SBPP requires to determine the minimum number of bins needed to pack all the items, with the risk of overloading a bin not exceeding a certain tolerable limit. Such computations are of high relevance in server consolidation applications, where decisions have to be made before witnessing the true item characteristics. The resulting integer optimization problems are generally nonlinear and therefore difficult to solve. For this reason, previous approaches from the literature can only handle small instance sizes exactly. In this work, we present a column generation algorithm using heuristic information and near-optimal solutions of the associated (challenging) pricing problems. Based on numerical tests, we show that in most cases this heuristic approach already leads to an optimal solution, so that much larger instance sizes can now be dealt with in reasonable time.

Keywords Cutting and packing · Stochastic bin packing problem · Normal distribution · Column generation · Heuristics

---

This work does not relate to the author's position at Amazon.

J. Martinovic (✉) · N. Strasdat
Institut für Numerische Mathematik, Technische Universität Dresden, Dresden, Germany
e-mail: john.martinovic@tu-dresden.de

N. Strasdat
e-mail: nico.strasdat@tu-dresden.de

J.-F. Côté
Département d'opérations et systèmes de décision, Université Laval, Laval, Canada
e-mail: jean-francois.cote@fsa.ulaval.ca

V. L. de Lima
Amazon.com, Seattle, USA
e-mail: vloti@amazon.com

O. Grothe et al. (eds.), *Operations Research Proceedings 2022*, Lecture Notes in Operations Research, https://doi.org/10.1007/978-3-031-24907-5_16

Introduction and Preliminaries

In this article, we consider a single-stage stochastic optimization problem, hereinafter referred to as the *stochastic bin packing problem (SBPP)*. The SBPP generalizes the classical BPP, see [3], in a sense that decisions have to be made based on imperfect information on the item sizes, which is the usual source of uncertainty [7], and cannot be changed later when the true characteristics have been revealed. Such problems (and generalizations of it) arise in data center workload management to keep operational costs low, see [1, 2, 9] for some application-oriented research. More formally, given a set $I := \{1, \ldots, n\}$ of *items (jobs)*, each being characterized by a random size c_i, the aim is to find the minimum number of unit capacity *bins* needed for a valid packing. In this context, an assignment is called *feasible* if the items of any bin respect its capacity with high probability (specified by some *error bound* $\varepsilon > 0$). Here we just consider normally distributed item sizes $c_i \sim \mathcal{N}(\mu_i, \sigma_i^2)$ with given mean $\mu_i \in (0, 1)$ and variance $\sigma_i^2 > 0, i \in I$. This does not represent an actual restriction, since many real-world workloads can be approximated well by that type of distribution, see [2, 7, 9]. For completeness, we highlight that the problem we consider also appeared under different names in the literature, among which the *bin packing problem with chance constraints* is the most relevant, see [2, 5] for some publications. In addition, according to Perboli et al. [8], bin packing problems with other random input data can also be referred to as a stochastic BPP.

Although having some obvious parallels to the well-studied BPP, an exact solution of the SBPP has not yet been addressed very successfully in the literature. This is due to the fact that associated compact (assignment) models generally possess nonlinear constraints, many symmetric solutions, and a very poor LP bound, leading to challenging integer programs even if additional information (like heuristic solutions) are available, see [7]. Moreover, transferring promising concepts like flow-based approaches did not lead to an efficient solution framework either, because of the exponential state space, see [6]. For these reasons, the optimal solution of many (even moderately-sized) benchmark instances is still unknown. However, given the substantial progress in terms of lower and upper bounds for the SBPP in recent years, see [2, 7], approximate solutions of reasonable quality are available in many cases. In this paper, we would therefore like to present a new heuristic approach based on *column generation* (CG) using the latter information. In that algorithm, the arising pricing problems (that are, *stochastic knapsack problems*) are very challenging and can only be dealt with approximately in reasonable time. Computational tests show, however, that numerous larger instances can be solved to proven optimality for the first time.

Solution Approaches from the Literature

Let $E = (n, \boldsymbol{c}, \varepsilon)$ with $\boldsymbol{c} := (c_1, \dots, c_n)$ denote an instance of the SBPP consisting of n items with normally distributed workloads $c_i \sim \mathcal{N}(\mu_i, \sigma_i^2)$, $i \in I$, and an error bound $\varepsilon > 0$ limiting the probability of overloading a bin. More generally, any vector $\boldsymbol{a} \in \mathbb{B}^n$ (with $\mathbb{B} := \{0, 1\}$) satisfying

$$\mathbb{P}\left[\boldsymbol{c}^\top \boldsymbol{a} > 1\right] \leq \varepsilon \quad \text{or equivalently} \quad w(\boldsymbol{a}) := \sum_{i \in I} \mu_i a_i + q_{1-\varepsilon} \cdot \sqrt{\sum_{i \in I} \sigma_i^2 a_i} \leq 1, \tag{16.1}$$

with $\mathbb{P}[\cdot]$ referring to the probability of an event and $q_{1-\varepsilon}$ denoting the $(1-\varepsilon)$-quantile of a standard normal distribution $\mathcal{N}(0, 1)$, is called a *(feasible) pattern.* The set of all patterns will be referred to as $\mathcal{P}$. To ensure solvability, we demand $\mathbb{P}[c_i > 1] \leq \varepsilon$, that is, $\mu_i + q_{1-\varepsilon} \cdot \sigma_i \leq 1$, for any $i \in I$.

Remark 1 Of course, real workloads cannot be negative, so that (ideally) truncated normal distributions should be used. However, for the benchmark sets appearing in Sect. "Computational Results", the probability of negative workloads is considerably small (below 10^{-4}), so that the normal distribution is an almost exact approximation.

Let u denote an upper bound for the number of bins required in an optimal solution and let $K := \{1, \dots, u\}$ denote the set of all bins, then we obtain the

Nonlinear Assignment Model for the SBPP (from [7])

$$z = \sum_{k \in K} y_k \to \min$$

$$\text{s.t.} \quad \sum_{k \in K} x_{ik} = 1, \qquad i \in I, \tag{16.2}$$

$$\sum_{i \in I} \mu_i x_{ik} + q_{1-\varepsilon} \cdot \sqrt{\sum_{i \in I} \sigma_i^2 x_{ik}} \leq 1, \qquad k \in K, \tag{16.3}$$

$$x_{ik} \leq y_k, \qquad i \in I, k \in K, \tag{16.4}$$

$$y_k \in \mathbb{B}, \qquad k \in K, \tag{16.5}$$

$$x_{ik} \in \mathbb{B}, \qquad (i, k) \in I \times K. \tag{16.6}$$

In this formulation, y_k models the decision whether bin $k \in K$ is used or not, whereas the x_{ik} are classic assignment variables stating which item will be packed into which bin. Besides the already known capacity constraints (16.3), any item has to be packed exactly once (see (16.2)) and the variable types have to be linked by (16.4) to obtain a consistent decision.

Remark 2 As shown in [6], for $0 < \varepsilon \leq 0.5$ and any $k \in K$ the nonlinear capacity constraint (16.3) can be expressed in a root-free manner by demanding

$$\sum_{i\in I}\left(q_{1-\varepsilon}^2\cdot\sigma_i^2+2\mu_i-\mu_i^2\right)x_{ik}-2\sum_{i\in I}\sum_{j>i}\mu_i\mu_j x_{ik}x_{jk}\le 1,\quad \sum_{i\in I}\mu_i x_{ik}\le 1.$$

A similar idea is given by converting (16.3) into *second-order cone* constraints via

$$r_k=1-\sum_{i\in I}\mu_i x_{ik}\ge 0,\quad q_{1-\varepsilon}^2\sum_{i\in I}\sigma_i^2 x_{ik}^2\le r_k^2$$

for any $k\in K$. Either way, these reformulations still possess lots of symmetries and a bad LP bound, so that even some instances with only $n\approx 30$ items cannot be solved to optimality within a time limit of ten minutes, see [7, Sect. 4].

Interestingly, the nonlinear model presented above can be solved in $\mathcal{O}(n\cdot\log(n))$ time if the integrality of the x-variables is relaxed, leading to a lower bound $lb^\star\in\mathbb{Z}_+$ satisfying $OPT\le 2\cdot lb^\star-1$, see [7, Subsect. 3.1]. The solution of that relaxation can be obtained by a *fractional next-fit decreasing (FNFD)* algorithm, where the items are sorted with respect to non-increasing values of σ_i^2/μ_i. Then, in a specific iteration (belonging to some fixed item $i\in I$), the largest possible portion $x\in[0,1]$ of item i is assigned to the current bin, so that (16.1) still holds. If the entire item does not fit into this bin, a new bin is opened to accommodate the remaining $1-x$ units of item i. As a second main result, a *first-fit decreasing (FFD)* heuristic based on the same sorting criterion is shown to establish an upper bound $ub^\star$ with $ub^\star\le 2\cdot OPT-1$ in $\mathcal{O}(n\cdot\log(n))$ time, see [7, Theorem 11]. Both these approximations are near-optimal and time-efficient, and hence they will be part of the new solution approach presented in Sect. "A Column Generation Approach".

A Column Generation Approach

For convenience, we identify each pattern with its unique subset of items, so that we have $\mathcal{P}:=\{J\subseteq I\mid w(J)\le 1\}=\{J_q\mid q\in Q\}$, where $w(J)$ represents the "size" of the pattern, similar to (16.1). Let $\xi_q\in\mathbb{B}$ denote whether pattern J_q is used or not, then a pattern-based model for the SBPP is given by

$$\sum_{q\in Q}\xi_q\to\min\quad\text{s.t.}\quad\sum_{q\in Q:i\in J_q}\xi_q\ge 1,\ i\in I,\ \xi_q\in\mathbb{B},q\in Q.\tag{16.7}$$

The LP relaxation of (16.7) can be tackled by standard CG. More precisely, for some (already constructed) pattern pool $\widetilde{\mathcal{P}}\subset\mathcal{P}$, we let $\pi\in\mathbb{R}_+^n$ denote the prices associated with the covering constraints (of (16.7)) in the *restricted master problem (RMP)* and consider the *pricing problem* $\pi^\star:=\pi(J)\to\max_{J\in\mathcal{P}}$ with $\pi(J):=\sum_{i\in I}\pi_i$ for all $J\in\mathcal{P}$. A pattern $J\in\mathcal{P}\setminus\widetilde{\mathcal{P}}$ should be added to the RMP if it has negative reduced costs, i.e., if $c(J):=1-\pi(J)<0$ or, equivalently, $\pi(J)>1$ holds. If no such pattern exists the solution is optimal. Note that the pricing problem is a challenging stochastic knapsack problem, and classic *dynamic programming (DP)* will lead to an exponential state space. To overcome this issue, we will use the (inexact) dominance rule given in Alg. 1 to considerably reduce the number of states.

Algorithm 1: DP-based heuristic

Set $S := \{\emptyset\}$ and sort the items in non-increasing order of π_i/μ_i.
foreach $i \in I$ **do** `// try to pack items in a particular order`
  foreach $J \in S$ **do** `// check every possible state`
    Define $J^+ := J \cup \{i\}$
    if $w(J^+) \leq 1$ **then** `// check feasibility`
      if *J^+ is approx. dominated by some $\tilde{J} \in S$* **then**
        Ignore J^+.
      else `// remove approx. dominated states and add` J^+
        Define $S_{\text{nd}} := \{\tilde{J} \in S \mid \tilde{J}$ is not approx. dominated by $J^+\}$.
        Update $S := S_{\text{nd}} \cup \{J^+\}$.
Add a suitable subset of patterns (from S) with $\pi(J) > 1$ to the RMP.

More precisely, a pattern $J \in \mathcal{P}$ is *approximately dominated* by another pattern $\widetilde{J} \in \mathcal{P}$ if $\pi(J) \leq \pi(\widetilde{J})$ and $w(J) \geq w(\widetilde{J})$ hold. By that, we no longer guarantee that all required patterns will be represented. Note that the set S in Alg. 1 can be sorted because of the monotonicity of the values $(w(J), \pi(J))$. Thus, an efficient implementation is possible by means of sorted lists.

Remark 3 Another heuristic way to deal with the pricing problems is given by using *polynomial-time approximation schemes (PTAS)*, see [4] for an examplary algorithm and related theoretical results. However, for the instances considered later, the practical applicability of this specific approach (from [4]) is rather restricted, because an exponential number (in fact, $n^{1/\delta}$) of subproblems has to be solved to obtain a certain bound (depending on some sufficiently small $\delta > 0$) for the optimal profit. Among others, this is also due to the fact that positive prices close to zero strongly increase the required number of iterations (see parameter $N_1 := \lfloor \log_{1+\delta} \min_{i \in I} \pi_i \rfloor$ in Algorithm $\mathcal{A}$ proposed in [4]).

Our overall algorithm now consists of two levels (L0 and L1), see Alg. 2. In the first one, we compute the best lower and upper bound from [7] and perform a simple optimality test. Effectively, for larger values of n, this phase gathers all instances whose optimal solution was already known from previous results in the literature. As a preparation for L1, we use the columns obtained from the FFD heuristic and run CG with our inexact dominance rule. With the patterns generated during that procedure, in L1 we then solve the associated RMP with binary variables leading to a new heuristic solution. Afterwards, we again check for optimality by comparing its objective value to $lb^\star$. However, the latter steps are only promising when $\lceil z_{CG} \rceil \leq lb^\star$ holds after having performed CG, since in the opposite case the RMP with binary variables cannot generate a solution attaining $lb^\star$. In addition, we cannot use the value $\lceil z_{CG} \rceil$ as a lower bound either because the pricing problems in CG are just tackled heuristically.

Algorithm 2: solution method

```
Compute lb* and apply the FFD heuristic to obtain ub*.
if lb* = ub* then                                        // Level 0 (L0)
|  Stop: The packing is optimal.

Perform CG (starting with patterns from the FFD heuristic) where the pricing
problems are "solved" by Alg. 1. Let z_CG denote the objective value after CG
has finished.
if ⌈z_CG⌉ ≤ lb* then                                     // Level 1 (L1)
|  Solve the binary RMP to get a feasible solution with value ub_RMP.
|  if lb* = ub_RMP then
|  |  Stop: The packing is optimal.
```

Computational Results

We coded Alg. 2 in Python (with adding up to 100 promising patterns to the RMP per iteration of Alg. 1, if possible) and solved the appearing binary problems by Gurobi 9.5 on an AMD A10-5800K processor with 16 GB RAM with time limit $\bar{t} = 600\,s$. The benchmark set we consider consists of 320 randomly generated instances from [7] with parameter values $n \in \{30, 50, 100, 200\}$ and $\varepsilon \in \{0.01, 0.05, 0.1, 0.2\}$, i.e., for each fixed pair (n, ε) a total number of 20 instances is dealt with. The results are displayed in Table 16.1 and can be summarized as follows: Our heuristic approach is able to solve 292 (out of 320) instances to proven optimality, and even the hardest subset just requires less than seven minutes of computation time on average. In particular, 229 of these optimal solutions have been obtained by the new main component (that is, L1) of Alg. 2, which leads to convincing results even for large values of n. Moreover, only nine instances (all with $n = 200$) reached the time limit, two of these already during CG due to the time required to solve the pricing subproblems. For the remaining 19 instances, Alg. 2 was successfully applied, but optimality of the heuristic solution could not be manifested in L1. By and large, we also see that Alg. 2 needs more time for larger choices of ε, because there we have a less restricted assignment policy and much more feasible patterns to check. Among other things, this is supported by the number of CG steps which averages roughly $1.5n$, but increases slightly if ε grows.

Table 16.1 Number of instances solved in Level 0 and Level 1 (abbreviated by L0 and L1), as well as total number of optimally solved instances (*opt*) and average runtime (*t*) of Alg. 2

n	ε	L0	no L1	L1	*opt*	t
30	0.01	10	3	7	17	0.1
	0.05	9	0	11	20	0.2
	0.10	9	1	10	19	0.2
	0.20	12	1	7	19	0.2
50	0.01	3	0	17	20	1.0
	0.05	4	2	14	18	1.1
	0.10	8	0	12	20	1.0
	0.20	6	1	12	18	1.7
100	0.01	0	2	18	18	12.6
	0.05	0	1	19	19	19.3
	0.10	1	1	18	19	25.6
	0.20	1	0	19	20	38.9
200	0.01	0	1	17	17	180.1
	0.05	0	2	18	18	200.3
	0.10	0	1	16	16	309.5
	0.20	0	0	14	14	406.7

The column "no L1" counts the number of instances for which level 1 was not executed due to $\lceil z_{CG} \rceil > lb^\star$ (see line 7 in Alg. 2)

Conclusions

In this article, we presented a new CG-based heuristic for the SBPP which benefits from (i) recent theoretical and algorithmic progress in terms of lower and upper bounds, see [7], and (ii) a new idea to approximately solve the very challenging pricing problems. Although not representing an exact solution framework, numerical tests show that almost all instances can be solved to proven optimality (in reasonably short time) – about 75% of them for the first time ever. In the future, we will mainly focus on alternative strategies to (approximately) tackle the pricing problems, e.g., by appropriate state-space relaxations or some more tailored PTAS.

References

1. Chen, M., Zhang, H., Su, Y.-Y., Wang, X., Jiang, G., Yoshihira, K. (2011). Effective VM sizing in virtualized data centers. In *Proceedings of the 12th IFIP/IEEE International Symposium on Integrated Network Management and Workshops*, 594–601.
2. Cohen, M. C., Keller, P. W., Mirrokni, V., & Zadimoghaddam, M. (2019). Overcommitment in Cloud Services: Bin Packing with Chance Constraints. *Management Science, 65*(7), 3255–3271.

3. Delorme, M., Iori, M., & Martello, S. (2016). Bin packing and Cutting Stock Problems: Mathematical Models and Exact Algorithms. *European Journal of Operational Research, 255*, 1–20.
4. Goyal, V., & Ravi, R. (2010). A PTAS for the chance-constrained knapsack problem with random item sizes. *Operations Research Letters, 38*(3), 161–164.
5. Klopfenstein, O., & Nace, D. (2008). A robust approach to the chance-constrained knapsack problem. *Operations Research Letters, 36*, 628–632.
6. Martinovic, J., Hähnel, M., Scheithauer, G., Dargie, W., Fischer, A. (2019). Cutting stock problems with nondeterministic item lengths: A new approach to server consolidation. *4OR 17*(2), 173–200.
7. Martinovic, J., Selch, M. (2022). Mathematical models and approximate solution approaches for the stochastic bin packing problem. *Computers and Operations Research 135*, Article 105439.
8. Perboli, G., Tadei, R., & Baldi, M. (2012). The Stochastic Generalized Bin Packing Problem. *Discrete Applied Mathematics, 160*(7–8), 1291–1297.
9. Wang, M., Meng, X., Zhang, L. (2011). Consolidating virtual machines with dynamic bandwidth demand in data centers. In *Proceedings IEEE INFOCOM*, pp. 71–75.

Chapter 17
A Penalty Branch-and-Bound Method for Mixed-Integer Quadratic Bilevel Problems. Part I: Key Ideas and a Fixed Parameter Setting

Andreas Horländer and Martin Schmidt

Abstract We propose an algorithm for solving bilevel problems with mixed-integer convex-quadratic upper level as well as convex-quadratic and continuous lower level. The method is based on a classic branch-and-bound procedure, where branching is performed on the integer constraints and on the complementarity constraints resulting from the Karush–Kuhn–Tucker reformulation of the lower-level problem. However, instead of branching on constraints as usual, suitably chosen penalty terms are added to the objective function to create new subproblems in the tree. In this first part, we consider a fixed penalty parameter, derive the main ideas, and prove the correctness of the method for this setting.

Keywords Bilevel optimization · Branch-and-bound · Penalty methods · Mixed-integer optimization

Introduction

Bilevel optimization gained increasing attention over the last years and decades mainly because bilevel models form a powerful tool for hierarchical decision making as it appears in the energy sector [8] or in security applications [7]. At the same time, bilevel problems are very hard to solve both in theory and practice [4]. Most of the solution techniques rely on branch-and-bound (B&B) or branch-and-cut techniques; see [3] for a state-of-the-art method and [6] for a recent survey.

In [1], a novel algorithm has been developed for solving mixed-integer linear complementarity problems (MILCPs). This B&B method uses penalizations of violated

The first author thanks the DFG for their support within RTG 2126 "Algorithmic Optimization".

A. Horländer (✉) · M. Schmidt
Department of Mathematics, Trier University, Universitätsring 15, 54296 Trier, Germany
e-mail: horlaender@uni-trier.de

M. Schmidt
e-mail: martin.schmidt@uni-trier.de

O. Grothe et al. (eds.), *Operations Research Proceedings 2022*, Lecture Notes in Operations Research, https://doi.org/10.1007/978-3-031-24907-5_17

constraints instead of constraint branching. It has been shown that this algorithmic idea leads to good numerical results for MILCPs. Since the latter class of problems is highly related to certain bilevel problems, the goal of this paper is to apply and extend the ideas in [1] to the case of bilevel optimization. To be more specific, we discuss bilevel problems with convex mixed-integer quadratic problems in the upper and convex-quadratic problems in the lower level, i.e., we consider mixed-integer bilevel problems of the form

$$\min_{x,y} \quad Q(x,y) := \frac{1}{2}x^\top H_x x + \frac{1}{2}y^\top H_y y + c_x^\top x + c_y^\top y \tag{17.1a}$$

$$\text{s.t.} \quad Ax + By \geq a, \; x_i \in \{0,1\}, \; i \in I \subseteq [n_x] := \{1,\dots,n_x\}, \; y \in S(x), \tag{17.1b}$$

where $S(x)$ is the set of optimal solutions of the x-parameterized lower level

$$\min_{y} \quad \frac{1}{2}y^\top G_y y + x^\top G_{xy} y + d_y^\top y \quad \text{s.t.} \quad Cx + Dy \geq b \tag{17.2}$$

with $A \in \mathbb{R}^{m\times n_x}$, $B \in \mathbb{R}^{m\times n_y}$, $C \in \mathbb{R}^{\ell\times n_x}$, $D \in \mathbb{R}^{\ell\times n_y}$, $G_{xy} \in \mathbb{R}^{n_x\times n_y}$, $c_x \in \mathbb{R}^{n_x}$, c_y, $d_y \in \mathbb{R}^{n_y}$, $a \in \mathbb{R}^m$, and $b \in \mathbb{R}^\ell$. The matrices $H_x \in \mathbb{R}^{n_x\times n_x}$ and $H_y, G_y \in \mathbb{R}^{n_y\times n_y}$ are symmetric and positive semidefinite. The upper-level problem (17.1) thus has a convex-quadratic objective function, linear constraints, and mixed-integer variables. The objective function of the lower level is also convex and quadratic because the term $x^\top G_{xy} y$ is linear in y. The constraints of the lower level are linear and the variables are continuous. Since the upper level is a mixed-integer quadratic problem and the lower level is a quadratic problem, we call Problem (17.1) an MIQP-QP bilevel problem.

We reformulate these problems using the classic Karush–Kuhn–Tucker (KKT) reformulation and tackle the integrality constraints as well as the KKT complementarity conditions using a novel penalty B&B method. By doing so, we introduce a new class of algorithms for solving bilevel optimization problems. In this first part of the paper, we consider the case of a fixed penalty parameter, develop the key ideas and prove the correctness of our method for this setting. The case of iteratively adapted penalty parameters, the respective convergence theory, and some preliminary numerical results are presented and discussed in the second part of the paper [5].

A Penalty Branch-and-Bound Method

Main Ideas and Derivation of the Algorithm

We first apply the KKT reformulation to the bilevel problem (17.1), i.e., we replace the lower level with its KKT conditions, which are necessary and sufficient for global optimality in our setup. They are given by

$$G_y y + G_{xy}^\top x + d_y - D^\top \lambda = 0, \quad 0 \le \lambda \perp Cx + Dy - b \ge 0 \tag{17.3}$$

and the KKT reformulation of the bilevel problem (17.1) thus reads

$$\min_{x,y,\lambda} \quad Q(x, y) \quad \text{s.t.} \quad Ax + By \ge a, \quad G_y y + G_{xy}^\top x + d_y - D^\top \lambda = 0, \tag{17.4a}$$

$$0 \le \lambda \perp Cx + Dy - b \ge 0, \quad x_i \in \{0, 1\}, \quad i \in I. \tag{17.4b}$$

From Theorem 2.3 in [2], it follows that (x^*, y^*) is a global optimal solution of the problem (17.1) if (x^*, y^*, λ^*) is a global optimal solution of (17.4). Hence, we will work with the latter problem in what follows. Note that Problem (17.4) contains two complicating aspects: the integrality and the KKT complementarity constraints. To address these two aspects, we again reformulate the problem. We remove the problematic constraints and penalize their violation by extending the objective function with additional piecewise-linear penalty terms and obtain the continuous problem

$$\min_{x,y,\lambda} \quad Q(x, y) + \alpha \sum_{i \in I} \min\{x_i, 1 - x_i\} + \beta \sum_{j=1}^{\ell} \min\{\lambda_j, (Cx + Dy - b)_j\}, \tag{17.5a}$$

$$\text{s.t.} \quad Ax + By \ge a, \quad Cx + Dy \ge b, \tag{17.5b}$$

$$G_y y + G_{xy}^\top x + d_y - D^\top \lambda = 0, \quad \lambda \ge 0, \quad 0 \le x_i \le 1, \quad i \in I, \tag{17.5c}$$

where $\alpha, \beta > 0$ are suitably chosen penalty parameters. For what follows, let $\pi(x, y, \lambda)$ denote the objective function of the penalty reformulation (17.5) and let $\mathcal{P}$ denote its feasible set. Problem (17.5) can thus be written as $\min\{\pi(x, y, \lambda) : (x, y, \lambda) \in \mathcal{P}\}$. In order to measure the violation of the integrality and complementarity constraints, we make use of the following definition.

Definition 1 For a given point $(x, y, \lambda) \in \mathcal{P}$, we call

$$F_w(x, y, \lambda; \alpha, \beta) := \alpha \sum_{i \in I} \min\{x_i, 1 - x_i\} + \beta \sum_{j=1}^{\ell} \min\left\{\lambda_j, (Cx + Dy - b)_j\right\}$$

the *weighted infeasibility measure*. Moreover, $F_u(x, y, \lambda) := F_w(x, y, \lambda; 1, 1)$ is called the *unweighted infeasibility measure*.

The following theorem establishes the connection between the penalty reformulation (17.5) and the bilevel problem (17.1).

Theorem 1 *Let (x^*, y^*, λ^*) be a global optimal solution of the penalty reformulation (17.5) for arbitrarily chosen $\alpha, \beta > 0$. If the solution is feasible for the KKT reformulation (17.4), i.e., $F_w(x^*, y^*, \lambda^*; \alpha, \beta) = 0$, then it is also globally optimal for Problem (17.4). Furthermore, (x^*, y^*) is globally optimal for the bilevel problem (17.1).*

Proof Suppose that (x^*, y^*, λ^*) is not a global optimum of the KKT reformulation (17.4), i.e., there exists a point $(\bar{x}, \bar{y}, \bar{\lambda})$ with $Q(\bar{x}, \bar{y}) < Q(x^*, y^*)$. Then,

$$\pi(\bar{x}, \bar{y}, \bar{\lambda}) = Q(\bar{x}, \bar{y}) < Q(x^*, y^*) = \pi(x^*, y^*, \lambda^*)$$

holds, which contradicts the global optimality of (x^*, y^*, λ^*) w.r.t. the penalty reformulation (17.5). According to Theorem 2.3 in [2], (x^*, y^*) is also globally optimal for Problem (17.1). □

According to Theorem 1, we can obtain globally optimal solutions of Problem (17.1) by solving the penalty reformulation (17.5). However, this problem is still hard to solve. Compared to the KKT reformulation (17.4), Problem (17.5) has only linear constraints but has a nonsmooth and nonconvex objective function due to the penalty terms. This seems to be obstructive w.r.t. Theorem 1, which requires a global solution of Problem (17.5). Fortunately, we will show that the method described in the following can tackle these aspects.

A Branch-and-Bound Method For Fixed Penalty Parameters

We now derive a B&B method for solving (17.5) that branches by adding suitably chosen penalty terms instead of adding constraints or refining variable bounds. As in classic B&B, we first solve the root node relaxation of Problem (17.5), i.e., $\min\{Q(x, y) \colon (x, y, \lambda) \in \mathcal{P}\}$, for which we assume that it is feasible and bounded from below. If integrality or complementarity constraints are violated in the obtained solution, we choose one among them for branching and construct the subproblems

$$\begin{aligned} S_1(x, y, \lambda) &:= \min_{(x,y,\lambda)\in\mathcal{P}} \; Q(x, y) + \alpha x_i, \\ S_2(x, y, \lambda) &:= \min_{(x,y,\lambda)\in\mathcal{P}} \; Q(x, y) + \alpha(1 - x_i), \end{aligned}$$

if we choose an integrality constraint or

$$\begin{aligned} \tilde{S}_1(x, y, \lambda) &:= \min_{(x,y,\lambda)\in\mathcal{P}} \; Q(x, y) + \beta\lambda_j, \\ \tilde{S}_2(x, y, \lambda) &:= \min_{(x,y,\lambda)\in\mathcal{P}} \; Q(x, y) + \beta(Cx + Dy - b)_j \end{aligned}$$

otherwise. By doing so, the penalty terms in the objective of (17.5) are split in linear parts, which leads to convex-quadratic subproblems S_1, S_2, $\tilde{S}_1$, and $\tilde{S}_2$. An arbitrary node problem in the B&B tree is then defined as

$$\min_{(x,y,\lambda)\in\mathcal{P}} \; \pi_N(x, y, \lambda) \tag{17.6}$$

with

$$\pi_N(x, y, \lambda) := Q(x, y) + \alpha \left(\sum_{i \in Z} x_i + \sum_{i \in O} (1 - x_i) \right) + \beta \left(\sum_{j \in D} \lambda_j + \sum_{j \in P} (Cx + Dy - b)_j \right)$$

and the tuple $N := (Z, O, D, P)$ of index sets containing the indices for which binary variables are driven to 0 (Z) or to 1 (O) as well as for which either the dual variable λ_j is driven to 0 (D) or for which the primal constraint is driven to 0 (P). The overall method is given in Algorithm 1. In Line 1 of Algorithm 1 the set $\mathcal{N}$ of

---

Algorithm 1: A penalty B&B method to solve Problem (17.5)

Data: A bilevel problem of the form (17.1) and $\alpha, \beta > 0$.
Result: A globally optimal solution (x^*, y^*, λ^*) of Problem (17.5).
1 Set $\mathcal{N} \leftarrow \{(\emptyset, \emptyset, \emptyset, \emptyset)\}$ and $u \leftarrow \infty$.
2 **while** $\mathcal{N} \neq \emptyset$ **do**
3   Choose an $N \in \mathcal{N}$ and set $\mathcal{N} \leftarrow \mathcal{N} \setminus \{N\}$.
4   Solve Problem (17.6) for N and obtain the solution (x_N, y_N, λ_N).
5   **if** $\pi(x_N, y_N, \lambda_N) < u$ **then**
6   Set $(x^*, y^*, \lambda^*) \leftarrow (x_N, y_N, \lambda_N)$ and $u \leftarrow \pi(x_N, y_N, \lambda_N)$.
7   **if** $\pi_N(x_N, y_N, \lambda_N) < u$ ***and*** ($\exists i \in I \setminus (Z \cup O)$ ***or*** $\exists j \in [\ell] \setminus (D \cup P)$) **then**
8   Choose either an $i \in I \setminus (Z \cup O)$ and set
        $\mathcal{N} \leftarrow \mathcal{N} \cup \{(Z \cup \{i\}, O, D, P), (Z, O \cup \{i\}, D, P)\}$ or a $j \in [\ell] \setminus (D \cup P)$ and set
        $\mathcal{N} \leftarrow \mathcal{N} \cup \{(Z, O, D \cup \{j\}, P), (Z, O, D, P \cup \{j\})\}$.
9 **end**
10 **return** (x^*, y^*, λ^*).

---

open nodes is initialized with the root node and the incumbent u is set to infinity. Every time a node in the B&B tree is solved, the objective function of the penalty reformulation (17.5) is evaluated at the solution to check if a new incumbent is found (Line 5–6). The branching step is performed in Lines 7–8. Note that we, of course, only choose integer variables or complementarity constraints as branching candidates that are fractional or violated.

The following two lemmas and the theorem are direct extensions of the corresponding results in [1] to the setting considered here. The proofs can be done analogously. To verify the correctness of the method, we first show that a complete evaluation of the B&B tree yields an optimal solution of Problem (17.5).

Lemma 1 *Let (x^*, y^*, λ^*) be a globally optimal solution of Problem (17.5). Then, $\pi(x^*, y^*, \lambda^*) = \min\{\pi_N(x_N, y_N, \lambda_N) \colon N \in \mathcal{N}'\}$ holds, where the minimum is taken over all tuples in $\mathcal{N}' := \{N = (Z, O, D, P) \colon Z \cup O = I, Z \cap O = \emptyset, D \cup P = [\ell], D \cap P = \emptyset\}$.*

Furthermore, we state that the implicit pruning in Line 7 of Algorithm 1 is correct. To this end, we show that the optimal objective value of a node is not larger than the optimal objective value of any node below in the tree.

Lemma 2 *Let $N' = (Z', O', D', P')$ be a successor node of $N = (Z, O, D, P)$ in the B&B tree, i.e., $Z \subseteq Z'$, $O \subseteq O'$, $D \subseteq D'$, and $P \subseteq P'$ holds. For an optimal solution (x_N, y_N, λ_N) of the problem in node N, it holds $\pi_N(x_N, y_N, \lambda_N) \leq \pi_{N'}(x_{N'}, y_{N'}, \lambda_{N'})$.*

We can now establish a correctness theorem for Algorithm 1.

Theorem 2 *Suppose that the root node relaxation of Problem (17.5) is feasible and bounded from below. Then, Algorithm 1 terminates after finitely many steps with a globally optimal solution of Problem (17.5).*

Conclusion

We have just seen that Algorithm 1 computes globally optimal points for the penalty reformulation (17.5). However, the point is only optimal for the bilevel problem (17.1) if $F_u(x^*, y^*, \lambda^*) = 0$ holds, i.e., if the solution satisfies all integrality and complementarity constraints. To address this issue, we will show in the second part of this paper [5] that we can set up an iterative penalty method so that we converge to points that satisfy all integrality and complementarity constraints and which are, thus, optimal solutions of the original bilevel problem.

References

1. de Santis, M., de Vries, S., Schmidt, M., & Winkel, L. (2022). A penalty branch-and-bound method for mixed-binary linear complementarity problems. *INFORMS Journal on Computing*.
2. Dempe, S., & Dutta, J. (2012). Is bilevel programming a special case of a mathematical program with complementarity constraints? *Mathematical Programming, 131*(1), 37–48.
3. Fischetti, M., Ljubić, I., Monaci, M., & Sinnl, M. (2017). A new general-purpose algorithm for mixed-integer bilevel linear programs. *Operations Research, 65*(6), 1615–1637.
4. Hansen, P., Jaumard, B., & Savard, G. (1992). New branch-and-bound rules for linear bilevel programming. *SIAM Journal on Scientific and Statistical Computing, 13*(5), 1194–1217.
5. Horländer, A., & Schmidt, M. (2022). *A penalty branch-and-bound method for mixed-integer quadratic bilevel problems*. Part II: Penalty parameter updates and numerical results.
6. Kleinert, T., Labbé, M., Ljubić, I., & M. Schmidt. A survey on mixed-integer programming techniques in bilevel optimization. *EURO Journal on Computational Optimization*, 9, 100007.

7. Smith, J. C., & Song, Y. (2020). A survey of network interdiction models and algorithms. *European Journal of Operational Research, 283*(3), 797–811.
8. Wogrin, S., Pineda, S., & Tejada-Arango, D. A. (2020). Applications of bilevel optimization in energy and electricity markets. In S. Dempe & A. Zemkoho (Eds), *Bilevel Optimization: Advances and Next Challenges*, pp. 139–168. Springer International Publishing.

Chapter 18
A Penalty Branch-and-Bound Method for Mixed-Integer Quadratic Bilevel Problems. Part II: Penalty Updates and Numerical Results

Andreas Horländer and Martin Schmidt

Abstract In the first part of this paper, we propose a penalty branch-and-bound method for solving bilevel problems with mixed-integer convex-quadratic upper level as well as convex-quadratic and continuous lower level and analyze the method for a fixed penalty parameter. In this second part, we extend the algorithm and its analysis towards iteratively adapted penalty parameters, prove the correctness of this extended method, and show its applicability by some first numerical results.

Keywords Bilevel optimization · Branch-and-bound · Penalty methods · Mixed-integer optimization

Introduction

In Part I of this paper, we studied bilevel problems of the form

$$\min_{x,y} \quad Q(x, y) := \frac{1}{2}x^\top H_x x + \frac{1}{2}y^\top H_y y + c_x^\top x + c_y^\top y \tag{18.1a}$$

$$\text{s.t.} \quad Ax + By \geq a, \; x_i \in \{0, 1\}, \; i \in I \subseteq [n_x] := \{1, \dots, n_x\}, \tag{18.1b}$$

$$y \in \arg\min_{y'} \left\{ \frac{1}{2}(y')^\top G_y y' + x^\top G_{xy} y' + d_y^\top y' : Cx + Dy' \geq b \right\}. \tag{18.1c}$$

We then used the KKT reformulation (see Problem (4) in [1]) to derive the penalty reformulation

---

The first author thanks the DFG for their support within RTG 2126 "Algorithmic Optimization". Both authors kindly acknowledge the support of RHRK.

A. Horländer (✉) · M. Schmidt
Department of Mathematics, Trier University, Universitätsring 15, 54296 Trier, Germany
e-mail: horlaender@uni-trier.de

M. Schmidt
e-mail: martin.schmidt@uni-trier.de

O. Grothe et al. (eds.), *Operations Research Proceedings 2022*, Lecture Notes in Operations Research, https://doi.org/10.1007/978-3-031-24907-5_18

$$\min_{x,y,\lambda} \quad Q(x, y) + \alpha \sum_{i \in I} \min\{x_i, 1 - x_i\} + \beta \sum_{j=1}^{\ell} \min\{\lambda_j, (Cx + Dy - b)_j\} \quad (18.2a)$$

$$\text{s.t.} \quad Ax + By \geq a, \quad Cx + Dy \geq b, \quad (18.2b)$$

$$G_y y + G_{xy}^\top x + d_y - D^\top \lambda = 0, \quad \lambda \geq 0, \quad 0 \leq x_i \leq 1, \quad i \in I. \quad (18.2c)$$

Moreover, we derived a penalty B&B method for solving this reformulation and proved its correctness for the case of a fixed penalty parameter. However, the penalty parameter might not be large enough so that the resulting solution of the reformulation does not satisfy all integrality and complementarity constraints. In this part, we answer the question on how to adapt the penalty parameter so that we provably obtain a solution of the original bilevel problem.

The key theorems and the resulting algorithm are presented and discussed in the Section "Iterative Penalty Parameter Updates". Afterward, we discuss some preliminary numerical results in the Section "Numerical Results".

Iterative Penalty Parameter Updates

We first derive the result that the outcome of Algorithm 1 in [1] converges to points having an unweighted infeasibility measure F_u of zero if α and β tend to infinity.

Theorem 1 *Let the root node relaxation of (18.2) be feasible and bounded from below and let the KKT reformulation of the original bilevel problem be solvable. Furthermore, let $(\bar{x}(\alpha, \beta), \bar{y}(\alpha, \beta), \bar{\lambda}(\alpha, \beta))$ be a global solution of Problem (18.2) for given penalty parameters α and β. Then, $F_u(\bar{x}(\alpha, \beta), \bar{y}(\alpha, \beta), \bar{\lambda}(\alpha, \beta)) \to 0$ for $\alpha \to \infty$ and $\beta \to \infty$. Furthermore, for a given $t > 0$, there exist finite penalty parameters α and β with $F_u(\bar{x}(\alpha, \beta), \bar{y}(\alpha, \beta), \bar{\lambda}(\alpha, \beta)) \leq t$.*

Proof Let (x^*, y^*) be a globally optimal solution of the KKT reformulation of (18.1). Due to optimality, it holds

$$\begin{aligned} &\pi(\alpha, \beta, \bar{x}(\alpha, \beta), \bar{y}(\alpha, \beta), \bar{\lambda}(\alpha, \beta)) \\ &\leq \pi(x^*, y^*, \lambda^*) = Q(x^*, y^*) + F_w(\alpha, \beta, x^*, y^*, \lambda^*) = Q(x^*, y^*). \end{aligned} \quad (18.3)$$

Let (x_r, y_r, λ_r) be the optimal solution and $Q_r := Q(x_r, y_r) > -\infty$ be the optimal objective value of the root node relaxation. Note that the latter is independent of the penalty parameters α and β. Using Inequality (18.3), we get

$$\begin{aligned} Q(x^*, y^*) &\geq \pi(\alpha, \beta, \bar{x}(\alpha, \beta), \bar{y}(\alpha, \beta), \bar{\lambda}(\alpha, \beta)) \\ &= Q(\bar{x}(\alpha, \beta), \bar{y}(\alpha, \beta)) + F_w(\alpha, \beta, \bar{x}(\alpha, \beta), \bar{y}(\alpha, \beta), \bar{\lambda}(\alpha, \beta)) \\ &\geq Q_r + F_w(\alpha, \beta, \bar{x}(\alpha, \beta), \bar{y}(\alpha, \beta), \bar{\lambda}(\alpha, \beta)). \end{aligned}$$

The last inequality follows from the optimality of (x_r, y_r, λ_r) for the root node. By using the definition of F_w, we get

$$\alpha \sum_{i\in I} \min\{\bar{x}_i, 1-\bar{x}_i\} + \beta \sum_{j\in[\ell]} \min\{\bar{\lambda}_j, (C\bar{x}+D\bar{y}-b)_j\} \le Q(x^*, y^*) - Q_r$$

with constant right-hand side $Q(x^*, y^*) - Q_r$. Now, we have

$$\begin{aligned}\sum_{i\in I} \min\{\bar{x}_i, 1-\bar{x}_i\} &\le \frac{Q(x^*, y^*) - Q_r}{\alpha},\\ \sum_{j\in[\ell]} \min\{\bar{\lambda}_j, (C\bar{x}+D\bar{y}-b)_j\} &\le \frac{Q(x^*, y^*) - Q_r}{\beta}\end{aligned} \tag{18.4}$$

and taking the limits yields

$$\lim_{\alpha\to\infty,\beta\to\infty} \Big(\sum_{i\in I} \min\{\bar{x}_i, 1-\bar{x}_i\} + \sum_{j\in[\ell]} \min\{\bar{\lambda}_j, (C\bar{x}+D\bar{y}-b)_j\}\Big) = 0.$$

This proves the first claim. For the second claim, we use (18.4) to see that for $\alpha, \beta \ge 2(Q(x^*, y^*) - Q_r)/t$, we get

$$\sum_{i\in I} \min\{\bar{x}_i, 1-\bar{x}_i\} \le \frac{(Q(x^*, y^*) - Q_r)t}{2(Q(x^*, y^*) - Q_r)} = \frac{t}{2}$$

as well as the analogous inequality for the complementarities and, thus, $F_u(\bar{x}(\alpha,\beta), \bar{y}(\alpha,\beta), \bar{\lambda}(\alpha,\beta)) \le t$. □

To make use of Theorem 1, we modify Algorithm 1 in [1] so that the penalty parameters are increased after every evaluation of the B&B tree. Furthermore, we relax the condition $F_u(x^*, y^*, \lambda^*) = 0$ to $F_u(x^*, y^*, \lambda^*) \le t$ for a given tolerance $t > 0$.

Since the value of $Q(x^*, y^*)$ is unknown in practice, one can also use the value $Q(\tilde{x}, \tilde{y})$ of an already obtained point $(\tilde{x}, \tilde{y})$, which is feasible for the KKT reformulation of the bilevel problem. This yields a valid and finite, but also worse, upper bound for the unweighted infeasibility F_u since $Q(x^*, y^*) \le Q(\tilde{x}, \tilde{y})$ holds.

We can now present a multi-tree penalty B&B method that computes an approximate solution to the bilevel problem (18.1) by repeatedly applying Algorithm 1 in [1] for increasing penalty parameters α and β; see Algorithm 1. Here, a point $(x^*, y^*, \lambda^*) \in \mathcal{P}$ is called *approximate solution* of Problem (18.1) if it minimizes $Q(x, y)$ and if it satisfies the condition $F_u(x^*, y^*, \lambda^*) \le t$.

In Line 3, Algorithm 1 in [1] is applied to obtain a globally optimal solution of Problem (18.2) for α^k and β^k. If the solution is not an approximate solution of Problem (18.1), the penalty parameters are increased and the procedure is repeated (Lines 10 and 11). Further steps are added in Lines 6–8, where every node solution that yields a better objective value than the incumbent is checked for feasibility

Algorithm 1: A penalty B&B method for MIQP-QP bilevel problems

Data: A bilevel problem of the form (18.1), $\alpha^1 > 0$, $\beta^1 > 0$, $t > 0$, and $\tau > 0$.
Result: An approximate solution (x^*, y^*, λ^*) of the bilevel problem (18.1).

1 Set $u_{\max} \leftarrow \infty$, $\alpha_{\max} \leftarrow \infty$, $\beta_{\max} \leftarrow \infty$, and $k \leftarrow 1$.
2 **while** $\alpha^k \le \alpha_{max}$, $\beta^k \le \beta_{max}$ **do**
3 Apply Algorithm 1 in [1] with $u \leftarrow u_{\max} + \tau$, $\alpha \leftarrow \alpha^k$, and $\beta \leftarrow \beta^k$.
4 **if** $\pi^k(x_N^k, y_N^k, \lambda_N^k) < u$ *for any node N of the last execution of Algorithm 1* **then**
5 Set $(x_{\text{opt}}^k, y_{\text{opt}}^k, \lambda_{\text{opt}}^k) \leftarrow (x_N^k, y_N^k, \lambda_N^k)$ and $u \leftarrow \pi^k(x_N^k, y_N^k, \lambda_N^k)$.
6 **if** $F_u(x_N^k, y_N^k, \lambda_N^k) \le t$ **then**
7 Set $\alpha_{\max}, \beta_{\max} \leftarrow 2(Q(x_N^k, y_N^k) - Q_r)/t$.
8 **if** $F_u(x_N^k, y_N^k, \lambda_N^k) = 0$ **then** set $u_{\max} \leftarrow u$.
9 **end**
10 **if** $F_u(x_{opt}^k, y_{opt}^k, \lambda_{opt}^k) > t$ **then**
11 Choose new penalty parameters $\alpha^{k+1} > \alpha^k$ and $\beta^{k+1} > \beta^k$.
12 **else**
13 Set $\alpha^{k+1} > \alpha_{\max}$ and $\beta^{k+1} > \beta_{\max}$ to terminate the algorithm.
14 **end**
15 Set $k \leftarrow k + 1$.
16 **end**
17 **return** $(x^*, y^*, \lambda^*) := (x_{\text{opt}}^k, y_{\text{opt}}^k, \lambda_{\text{opt}}^k)$

w.r.t. the KKT reformulation of the bilevel problem. If the unweighted infeasibility measure F_u for this point is within the tolerance t, we update the upper bounds of the penalty parameters according to the proof of Theorem 1. Moreover, if $F_u = 0$ holds, we update the upper bound of the incumbent, which can be used in further evaluations of the B&B tree to initialize the incumbent instead of setting it to infinity; see Line 3. Therefore, it has to be modified by the value τ, which is also shown in the following theorem.

Theorem 2 *Let the KKT reformulation of the bilevel problem be solvable, the root node relaxation of (18.2) be feasible and bounded from below, and let $t, \tau > 0$ be given. Furthermore, suppose that $\alpha^{k+1} - \alpha^k \ge \varepsilon$ and $\beta^{k+1} - \beta^k \ge \varepsilon$ holds for a fixed $\varepsilon > 0$ and every k. Then, Algorithm 1 terminates after finitely many steps in an iteration k^* with a globally optimal solution $(x_{\text{opt}}^{k^*}, y_{\text{opt}}^{k^*}, \lambda_{\text{opt}}^{k^*})$ of the penalty reformulation with α^{k^*} and β^{k^*} that satisfy the condition $F_u(x_{\text{opt}}^{k^*}, y_{\text{opt}}^{k^*}, \lambda_{\text{opt}}^{k^*}) \le t$.*

Proof According to Theorem 1, the algorithm terminates with finite penalty parameters $\alpha^k \le \alpha_{\max}$ and $\beta^k \le \beta_{\max}$ with a solution that satisfies the condition of the theorem. Using the assumption that the increase of the penalty parameters is not arbitrarily small, we obtain $k_{\max} \le \max\{(\alpha_{\max} - \alpha^1)/\epsilon, (\beta_{\max} - \beta^1)/\epsilon\} < \infty$, where $k_{\max}$ denotes the number of maximally required penalty parameter updates. According to Theorem 2 in [1], a complete evaluation of the B&B tree takes only finitely many steps so that Algorithm 1 also terminates in finite time. It remains to show that the update $u \leftarrow u_{\max} + \tau$ at the end of iteration k cannot lead to overlooking a globally optimal solution of Problem (18.2) with new penalty parameters α^{k+1} and β^{k+1} in Line 4 of the next iteration $k+1$. We obtain $u_{\max}$ in iteration k by the objective value π^k of a node's solution that satisfies all integrality and complementarity conditions. Let $(\tilde{x}^k, \tilde{y}^k, \tilde{\lambda}^k)$ be this solution, i.e., $(\tilde{x}^k, \tilde{y}^k, \tilde{\lambda}^k)$ is the optimal solution of a node N for iteration k for which $u_{\max} = \pi^k(\tilde{x}^k, \tilde{y}^k, \tilde{\lambda}^k)$ holds. Additionally, it is feasible for the KKT reformulation of (18.1), i.e., it holds

$$F_u(\tilde{x}^k, \tilde{y}^k, \tilde{\lambda}^k) = \sum_{i \in I} \min\{\tilde{x}_i^k, 1 - \tilde{x}_i^k\} + \sum_{j \in [\ell]} \min\left\{\tilde{\lambda}_i^k, (C\tilde{x} + D\tilde{y} - b)_i\right\} = 0.$$

Then, the point $(\tilde{x}^k, \tilde{y}^k, \tilde{\lambda}^k)$ is also feasible for the node $\tilde{N} = (\tilde{Z}, \tilde{O}, \tilde{D}, \tilde{P})$ with $\tilde{Z} := \{i \in I : \tilde{x}_i^k \leq 1 - \tilde{x}_i^k\}$, $\tilde{O} := \{i \in I : \tilde{x}_i^k > 1 - \tilde{x}_i^k\}$, $\tilde{D} := \{j \in \{1, \ldots, \ell\} : \tilde{\lambda}_j^k \leq (C\tilde{x}^k + D\tilde{y}^k - b)_j\}$, and $\tilde{P} := \{j \in \{1, \ldots, \ell\} : \tilde{\lambda}_j^k > (C\tilde{x}^k + D\tilde{y}^k - b)_j\}$. Furthermore, it holds

$$\alpha^k \left(\sum_{i \in \tilde{N}} \tilde{x}_i^k + \sum_{i \in \tilde{O}} (1 - \tilde{x}_i^k) \right) + \beta^k \left(\sum_{j \in \tilde{D}} \tilde{\lambda}_j^k + \sum_{j \in \tilde{P}} (C\tilde{x}^k + D\tilde{y}^k - b)_j \right) = 0,$$

which is why the point $(\tilde{x}^k, \tilde{y}^k, \tilde{\lambda}^k)$ is also an optimal solution of the node $\tilde{N}$ for every other $\alpha^{k+1} > \alpha^k$ and $\beta^{k+1} > \beta^k$. If we use the incumbent $u = u_{\max} + \tau > \pi^k(\tilde{x}^k, \tilde{y}^k, \tilde{\lambda}^k) = \pi_{\tilde{N}}^k(\tilde{x}^k, \tilde{y}^k, \tilde{\lambda}^k)$, in Algorithm 1 in iteration $k+1$, then for an optimal solution $(x_{\text{opt}}^{k+1}, y_{\text{opt}}^{k+1}, \lambda_{\text{opt}}^{k+1})$ of Problem (18.2) with penalty parameters α^{k+1} and β^{k+1}, it holds $\pi^{k+1}(x_{\text{opt}}^{k+1}, y_{\text{opt}}^{k+1}, \lambda_{\text{opt}}^{k+1}) \leq \pi_{\tilde{N}}^{k+1}(\tilde{x}^{k+1}, \tilde{y}^{k+1}, \tilde{\lambda}^{k+1}) = \pi_{\tilde{N}}^k(\tilde{x}^k, \tilde{y}^k, \tilde{\lambda}^k) < u$, where the first inequality follows from Lemma 1 in [1]. Due to the strict inequality in the last estimation, the optimal solution $(x_{\text{opt}}^{k+1}, y_{\text{opt}}^{k+1}, \lambda_{\text{opt}}^{k+1})$ cannot be overlooked in Line 4 of iteration $k+1$ of Algorithm 1. To guarantee this, τ has to be chosen strictly positive. □

Let us close this section with two remarks. First, we also derived a single-tree version of Algorithm 1 in which we check for every node if the obtained solution fulfills all constraints that have been added by branching so far, i.e., if

$$\sum_{i \in Z} x_{N,i} + \sum_{i \in O} (1 - x_{N,i}) + \sum_{j \in D} \lambda_{N,j} + \sum_{j \in P} (Cx_N + Dy_N - b)_j = 0 \qquad (18.5)$$

holds. If this is not the case, we set $Q(x, y) = 0$ and solve the node again to check if the node is infeasible w.r.t. the integrality and complementarity constraints. If it is, we prune the node and, otherwise, we increase the parameters until Condition (18.5) is satisfied in the associated solution. By doing so, we can prune nodes more efficiently and only a single B&B tree is required. However, we have to solve nodes multiple times so that it is not clear a priori if the single-tree is performing better. We will discuss this in more detail in the next section.

Second, we derived a modified version of Algorithm 1 in which we do not branch on the integrality conditions but only on the complementarity constraints and let the MILP solver handle the other ones. We compare these two different variants in the next section as well.

Numerical Results

In this section, we present a brief and preliminary numerical comparison of the penalty B&B method with solving the classic KKT reformulation of the given bilevel problem, which is still the most frequently used approach to solve bilevel problems in practice. We choose Gurobi 9.5.1 (with cuts, presolve, and heuristics deactivated to get a fair comparison) as the solver applied to the KKT formulation, where we handle complementarity constraints via SOS1 conditions; see [2].

Our test set consists of a subset of the instance collections Denegre Denegre, Int0Sum, MIPLIB2010, MIPLIB2017, MIPLIB3, Small, and Xuwang as used in [3]. Moreover, we excluded all instances that all methods can solve in less than 1 s and all instances that no method can solve within the time limit of 1 h; leading to 157 remaining instances in total. Our algorithm has been implemented in Python 3.9.7 and all occurring subproblems have been solved using Gurobi 9.5.1. All computations were executed on the high performance cluster "Elwetritsch" at the TU Kaiserslautern, which is part of the "Alliance of High Performance Computing Rheinland-Pfalz" (AHRP). We used a single Intel XEON SP 6126 core with 2.6 GHz and 32 GB RAM.

As discussed at the end of the last section, we tested both a multi-tree (MT) and a single-tree (ST) variant of our method as well as a version in which we branch on complementarity and integrality constraints (BOTH; in this case, we first branch on integrality constraints and then on complementarity constraints) and on complementarity constraints only (COMP). While doing so, we always choose the most violated integrality or complementarity constraint for branching in a depth-first search. Moreover, we set $\alpha^1 = \beta^1 = 100$ with update factor 100 and $\tau = 10^{-1}$ and $t = 10^{-4}$. The results are given in Fig. 18.1.

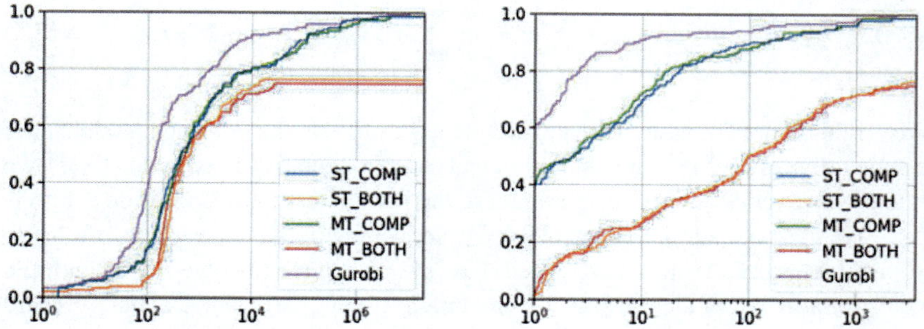

Fig. 18.1 Node counts (left) and running times (in s; right) versus percentage of instances

Conclusion

The numerical results show that the penalty B&B method is more effective if we only apply the penalty branching to complementarity conditions and let Gurobi handle the integrality constraints. The proposed method is rather competitive w.r.t. the classic KKT reformulation when it comes to node counts, although it is still slightly outperformed by it. The running times are worse, which can be devoted to our prototypical implementation that is not comparable to a commercial software. Based on the results for the node counts, we see a clear potential of the new method. However, it needs to be improved—both w.r.t. further algorithmic techniques (such as cutting planes or tailored heuristics) as well as w.r.t. its implementation—in order to obtain a method that is more powerful than the commercial state-of-the-art.

References

1. Horländer, A., & Schmidt, M. (2022). *A penalty branch-and-bound method for mixed-integer quadratic bilevel problems*. Part I: Key ideas and a fixed parameter setting.
2. Kleinert, T., & Schmidt, M. (2023). Why there is no need to use a big-*M* in linear bilevel optimization: A computational study of two ready-to-use approaches. *Computational Management Science*.
3. Kleinert, T., & Schmidt, M. (2021). Computing feasible points of bilevel problems with a penalty alternating direction method. *INFORMS Journal on Computing, 33*(1), 198–215.

Chapter 19
Aircraft Fleet Planning: An Optimization Model with Integrated CO_2 Trading Systems

Lisa-Marie Manke and Imke Joormann

Abstract The pressure on airlines to reduce their emitted CO_2 emissions is continuously increasing. Due to new fuel-efficient air frame and engine technologies, aircraft modernization offers an opportunity to reduce annual emissions. But with limited production capacities of the manufacturers and related costs, it is not possible to renew a whole fleet at once. At the same time, CO_2 trading systems lead to higher costs for older fleets. Therefore, there needs to be long term planning regarding the fleet composition of an airline. For this, the fleet planning problem determines economically optimal replacement times for aircraft. The fleet planning problem can be formulated as a MILP and combines fleet composition and fleet assignment. The fleet composition problem considers different aircraft types as well as retrofits; here, retrofits are modifications of aircraft to lower the amount of emitted emissions without replacing the whole aircraft. These aircraft are then used within the fleet assignment problem to fulfill a given demand of flights. The emissions are then considered within CO_2 trading systems as that is currently the only affecting restriction for airlines concerning CO_2 emissions.

Keywords Mixed-integer programming · Transportation · Airline applications

Introduction

In 2019, the aviation sector was responsible for at least 3% of the worldwide emitted CO_2 emissions [2], and the pressure on airlines to reduce the emitted emissions is continuously increasing, as seen for example in the "Fit for 55"-initiative of the EU [1]. Due to new fuel-efficient air frame and engine technologies, see, e.g., [3], aircraft modernization offers an opportunity to reduce annual emissions. Moreover,

L.-M. Manke (✉) · I. Joormann
Cluster of Excellence SE^{2}A—Sustainable and Energy-Efficient Aviation, Technische Universität Braunschweig, 38106 Braunschweig, Germany
e-mail: lisa-marie.manke@tu-braunschweig.de

Institute of Automotive Management and Industrial Production, Technische Universität Braunschweig, Mühlenpfordtstr. 23, 38106 Braunschweig, Germany

O. Grothe et al. (eds.), *Operations Research Proceedings 2022*, Lecture Notes in Operations Research, https://doi.org/10.1007/978-3-031-24907-5_19

there are possibilities to retrofit an old aircraft so that the emitted CO_2 emissions decrease. For example, it is possible to re-engining an aircraft which reduces the emitted emissions per flight by 12.5% or to reduce the cabin weight which would result in 1.2–2.1% reduced emitted emissions per life cycle, see [5]. With limited production capacities of the manufacturers and related costs, it is not possible to renew a whole fleet at once. At the same time, CO_2 trading systems lead to higher costs for older fleets. Therefore, there needs to be long term planning regarding the fleet composition of an airline. For this, the fleet planning problem (FPP) determines economically optimal replacement times for aircraft.

The FPP models the composition of the fleet under the restriction that the operation of the considered airline is not affected. For this purpose a given demand of flight-hours will be fulfilled under the restriction that the aircraft-specific working hours will not be exceeded. This model was first introduced in [4], but gets rather big even for small instances. We will introduce an alternative formulation for the FPP with beneficial properties for the computation of the solution: it is possible to decompose the FPP into two different optimization problems, the composition of the fleet and the assignment of aircraft to flights. Afterwards, we can model the fleet composition via a network flow problem, leaving the assignment of aircraft to flights as the only part adopted from [4]. Additionally, we model trading systems for CO_2 emissions, as these are the current political instruments affecting airlines most regarding CO_2 emissions; see [6] for a complete introduction of the considered systems.

Model

We will first introduce the two planning problems separately, describe the relevant trading systems for the CO_2 emissions and then combine them to the mixed-integer programming formulation of the FPP.

Fleet Composition

The first problem contained in the FPP is the fleet composition over time (FCP). As movements in a fleet's stock can be represented by flows in a graph, it is possible to describe the FCP as a network flow problem. For the FCP, we are given the number of considered periods $T \in \mathbb{Z}_{>0}$ and the set of different aircraft types C. The set C consists of base aircraft types B as they are available on the market, and retrofitted aircraft types R, so that $C = B \cup R$. For retrofitted aircraft, some modification is done on a base type, leading to lower emitted CO_2 emissions than for the original type. We consider a set of retrofit options ρ; if the retrofit $r \in \rho$ is available for $b \in B$, there is a retrofitted aircraft type $r_b \in R_b$ so that $R = \bigcup_{b \in B} R_b$. In our computational study, we use different retrofit options where the emission reduction potential ranges from 1.2 to 12.5%.

For each $b \in B$, the maximum service time is given by $\overline{T}_b \in \mathbb{Z}_{>0}$; retrofitted aircraft types have the same maximum service time as their original aircraft type. The first period in which $b \in B$ is available for sale is $\underline{t}_b \in \mathbb{Z}$, $\underline{t}_r \in \mathbb{Z}$ is the period from when on a retrofit r is available. A negative $\underline{t}_b$ or $\underline{t}_r$ here means that the market launch was before the beginning of the optimization. An aircraft bought in period t will be delivered in $t + 1$.

For $n \in \mathbb{Z}$, let $[n] := \{0, 1, 2, \dots, n\}$ if $n \geq 0$ and $[n] := \emptyset$ otherwise. Let $N = (V, A)$ be the network graph. To construct the graph, we add one vertex $v_{b,\theta,t}$ to V for all base aircraft $b \in B$, $t \in \{\max\{0, \underline{t}_b\}, \dots, T\}$ and $\theta \in [\min\{\overline{T}_b, t - \underline{t}_b\}]$. Let $\mathcal{T} := \{\max\{1, \underline{t}_r, \underline{t}_b\}, \dots, T\}$ with $\underline{t}_r := 0$ for terms without any retrofitted type. Let $\Theta_b := [\min\{\overline{T}_b, t - \underline{t}_b\}] \setminus \{0\}$ and $\Theta_r := [\min\{\overline{T}_b, t - \underline{t}_b\}] \setminus \{0, 1\}$—note that a retrofit cannot take place in the same period as the purchase of an aircraft. Then, for the retrofit options, we add vertices $v_{r_b,\theta,t}$ for all $r_b \in R_b$, $t \in \mathcal{T}$ and $\theta \in \Theta_r$. Lastly we add a vertex for the purchase market v_{pm} and one for the retail market v_{rm} to V.

Each arc of the graph represents a different kind of movement within the fleet: purchasing, selling, retrofitting and owning. The integer flow of every arc is equal to the number of aircraft in stock performing this particular movement. Aircraft are bought at the purchase market, which offers only non-retrofitted aircraft types, leading, for each potential purchase action, to the arc $a^{\text{pur}}_{b,t} = (v_{\text{pm}}, v_{b,0,t}) \in A$ for all $b \in B$ and $t \in \mathcal{T}$. Regarding aircraft sales, it is prohibited to sell any aircraft in period T, otherwise part of an optimal solution would always be to sell off the whole fleet in the last period. Furthermore, it is not possible to sell aircraft in the period directly after they were purchased; since normally, an airline has no reason to buy an aircraft only to resell it directly afterwards, this serves the purpose of complexity reduction. Hence, we add selling arcs $a^{\text{sell}}_{c,\theta,t} = (v_{c,\theta,t-1}, v_{\text{rm}})$ to A for all $c \in C$, $t \in \mathcal{T}$ and $\theta \in \Theta_c$. Here $\Theta_c := \Theta_b$ when $c = b \in B$ and $\Theta_c := \Theta_r$ when $c = r \in R$. Since it is not allowed to retrofit a base aircraft type within its purchase period, this leads to the arcs $a^{\text{retro}}_{b,r_b,\theta,t} = (v_{b,\theta-1,t-1}, v_{r_b,\theta,t}) \in A$ for all $b \in B$, $r_b \in R_b$, $t \in \mathcal{T}$ and $\theta \in \Theta_b$. Older aircraft consume more fuel and the blocked time for a single flight increases due to wear and tear. Hence, we need to keep track of the aging process of the aircraft and add the aging arcs $a^{\text{own}}_{c,\theta,t} = (v_{c,\theta-1,t-1}, v_{c,\theta,t}) \in A$ for all $c \in C$, $t \in \mathcal{T}$ and $\theta \in \Theta_c$. An example for the resulting graph is shown in Fig. 19.1.

Let $\delta^+ : V \to \mathcal{P}(A)$ and $\delta^- : V \to \mathcal{P}(A)$ be the functions assigning each vertex its outgoing and incoming arcs, respectively, where $\mathcal{P}(A)$ is defined as the power set of A. Let $f : \mathcal{P}(A) \to \mathbb{Z}_{\geq 0}$ be the function which assigns sets of arcs their summed up flow values. This enables us to formulate the IP for the FCP as

$$\min NPV(f(a^{\text{sell}}), f(a^{\text{pur}}), f(a^{\text{retro}})) \tag{19.1a}$$

$$\text{s.t. } f(\delta^+(v_{c,\theta,0})) = EF_{c,\theta} \quad \forall c \in B \quad \forall \theta \in \Theta_c \tag{19.1b}$$

$$f(\delta^-(v_{c,\theta,t})) - f(\delta^+(v_{c,\theta,t})) = 0 \quad \forall c \in C \quad \forall t \in \mathcal{T} \setminus \{T\} \quad \forall \theta \in \Theta_c \tag{19.1c}$$

$$f(a^{\text{pur}}_{b,t}) \leq UB^{\text{pur}}_{b,t} \quad \forall b \in B \quad \forall t \in \mathcal{T} \tag{19.1d}$$

$$\sum_{\substack{b \in B: \\ \exists r_b \in R_r}} \sum_{\theta \in \Theta_b} f(a^{\text{retro}}_{b,r_b,\theta,t}) \leq UB^{\text{retro}}_{r,t} \quad \forall r \in \rho \quad \forall t \in \mathcal{T} \tag{19.1e}$$

$$f(\alpha) \in \mathbb{Z}_{\geq 0} \quad \forall \alpha \in \mathcal{P}(A) \tag{19.1f}$$

The objective function (19.1a) is the net present value (NPV) of future costs, composed of the expenses for the purchase of new aircraft and for the implementation of retrofits (positive values) and the revenue from the sale of aircraft (negative values). Equation (19.1b) sets the aircraft in stock as initial flow, where $EF_{c,\theta}$ is the number of aircraft of type $c \in C$ and age $\theta \in \Theta_c$ in stock. Equation (19.1c) are the flow conversation constraints. Lastly (19.1d) and (19.1e) give upper bounds on the purchase of new aircraft and retrofits as there are limitations from the manufacturers side. Here $UB^{\text{pur}}_{b,t}$ is the maximum number of aircraft of type b that can be purchased by an airline per period and $UB^{\text{retro}}_{r,t}$ is the maximum number a retrofit $r \in \rho$ can be done; R_r is the set of retrofitted aircraft types originating from $r \in \rho$.

Flight Assignment Under CO_2-Certificate Trading Systems

The flight assignment problem (FAP) is the problem of assigning a set of aircraft to a given set of flights. As this might be very complicated when the numbers of flights and aircraft are large, we consider some flights and aircraft to be equivalent. For

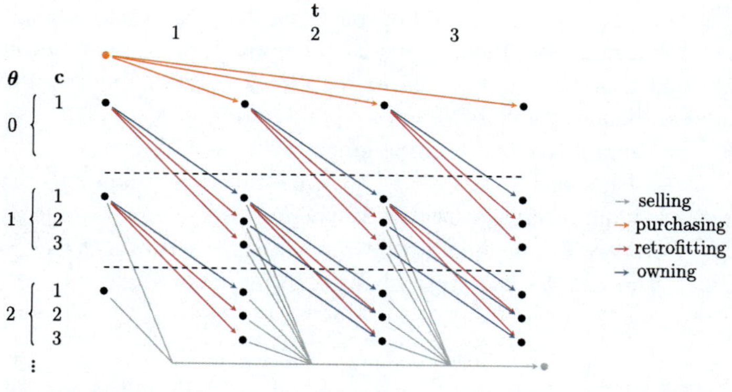

Fig. 19.1 Fleet composition graph with $T = 3$ periods, one base aircraft type $B = \{1\}$ and two retrofitted aircraft types $R = \{2, 3\}$ with θ as the age of the aircraft; name of nodes $v_{c,\theta,t}$ given on the border of the network with $c \in C = B \cup R$ and $t \in [T]$

aircraft, the equivalence classes consist of all aircraft of the same type and age. For flights, the equivalence classes are called net classes L and a flight belongs to a certain net class $\ell \in L$ if its length and seat demand is within the respective ranges of ℓ.

The main political instrument to limit the amount of emitted CO_2 emissions within the aviation sector active currently are trading schemes, where airlines need to buy certificates for their emitted CO_2. As of 2022, there are two main trading schemes relevant for European airlines, one from the European Union, applicable for all flights within the EU, and one from the International Civil Aviation Organisation (ICAO), applicable on flights between members states of the ICAO. To model this, we split L into the set L_I, consisting of net classes considering international flights, and L_E, the set of net classes considering flights within the European air traffic, so that $L = L_I \cup L_E$ holds.

To provide a linear program (LP) for the FAP together with the trading systems, we need flight and emission variables. The flight variables are $z_{c,\theta,\ell} \in \mathbb{R}_{\geq 0}$, capturing the number of flights in net class $\ell \in L$ to be operated by the aircraft $c \in C$ at age $\theta \in \Theta_c$, where C and Θ_c are from the FCP formulation with $t := 0$ for the Θ_c formula. For complexity reduction we assume that the flight variables are continuous, as they are typically large so that the possible rounding error should be relatively small. The emission variables $e_I, e_E \in \mathbb{R}_{\geq 0}$ capture the amount of emitted CO_2 emissions that must be compensated by certificates within the considered period on international flights (I) and within Europe (E), respectively.

The *supply constraints* ensure that the number of assigned flight hours does not exceed the number of flight hours given by the specific set of aircraft. For each aircraft of type c and age θ there is a fixed number of flight hours $MU_{c,\theta} \in \mathbb{Z}$ allowed per period. The number of aircraft in stock with these characteristics is given by $S_{c,\theta}$. The time needed for a flight of net class ℓ performed by an aircraft of type c and age θ is $BT_{c,\theta,\ell}$. The set $L_{c,\theta,q} \subseteq L$ is the subset of net classes which can be assigned to an aircraft of type c and age θ located in $q \in \{I, E\}$:

$$\sum_{q \in \{I,E\}} \sum_{\ell \in L_{c,y,q}} BT_{c,\theta,\ell} z_{c,\theta,\ell} \leq MU_{c,\theta} S_{c,\theta} \quad \forall c \in C \quad \forall \theta \in \Theta_c.$$

The *demand constraints* guarantee that the number of demanded flights d_ℓ in a net class ℓ is not larger than the total number of flights assigned to this net class by the set of aircraft:

$$\sum_{c \in C} \sum_{\theta \in \Theta_c} z_{c,\theta,\ell} \geq d_\ell \quad \forall \ell \in L.$$

The *emission constraints* collect the CO_2 emissions emitted and to be compensated in the considered period. Here we need to differentiate between the system of the EU and the one from ICAO. Within the EU, it is mandatory to compensate the whole amount of emitted CO_2 emissions per period from 2027 on, while for the ICAO, airlines just need to compensate the amount of emitted CO_2 emissions above the levels from 2019, so a smaller percentage of the whole amount. This difference leads to the constraints

$$\sum_{c\in C}\sum_{\theta\in\Theta_c}\sum_{\ell\in L_E} 9.75 FU_{c,y,\ell} z_{c,\theta,\ell} = e_E,$$

$$p_I \sum_{c\in C}\sum_{\theta\in\Theta_c}\sum_{\ell\in L_I} 9.75 FU_{c,\theta,\ell} z_{c,\theta,\ell} = e_I,$$

where $FU_{c,\theta,\ell} \in \mathbb{R}$ is the amount of fuel needed to perform a flight of net class ℓ with an aircraft of type c and age θ. The growth rate p_I of the aviation emissions within the considered period relative to 2019 base values is a scalar, computed and published by the ICAO for the international aviation system. The amount of fuel use is multiplied by the CO_2 emission factor of jet fuel, 9.75 kgCO_2/(gallon jet fuel) and the number of assignments $z_{c,\theta,\ell}$ to obtain the total emissions.

Fleet Planning Problem

Until now, we have described the flight assignment problem for one period in time. To obtain an integrated formulation for the FPP, the FAP needs to be solved for every period. To then be able to identify the corresponding period for the solutions, we add an additional index to the variables of the FAP. Hence, e_q becomes $e_{q,t}$ and $z_{c,\theta,\ell}$ becomes $z_{c,\theta,\ell,t}$ for all $t \in \mathcal{T}$. Furthermore, the data of the FAP depends on the solution of the FCP: the supply of available aircraft is given by the number of aircraft determined in the FCP. Therefore, $S_{c,\theta}$ from the supply constraints will be replaced by the incoming flow of the vertex $v_{c,\theta,t}$ from the flow network describing the FCP. The new objective function is the sum of the costs from the FCP and FAP.

Together, this yields the mixed integer formulation of the FPP:

$$\min NPV(f(a^{\text{sell}}), f(a^{\text{pur}}), f(a^{\text{retro}}), z, e)$$

$$\text{s.t. } f(\delta^+(v_{c,\theta,0})) = EF_{c,\theta} \quad \forall c \in C \quad \forall \theta \in \Theta_c$$

$$f(\delta^-(v_{c,\theta,t})) - f(\delta^+(v_{c,\theta,t})) = 0 \quad \forall c \in C \quad \forall t \in \mathcal{T} \setminus \{T\} \quad \forall \theta \in \Theta_c$$

$$f(a^{\text{pur}}_{b,t}) \leq UB^{\text{pur}}_{b,t} \quad \forall c \in B \quad \forall t \in \mathcal{T}$$

$$\sum_{\substack{b\in B:\\ \exists r_b \in R_r}} \sum_{\theta\in\Theta_b} f(a^{\text{retro}}_{b,r_b,\theta,t}) \leq UB^{\text{retro}}_{r,t} \quad \forall r \in \rho \quad \forall t \in \mathcal{T}$$

$$\sum_{q\in\{I,E\}} \sum_{\ell\in L_{c,\theta,q}} BT_{c,\theta,\ell} z_{c,\theta,\ell,t} \leq MU_{c,\theta} f(\delta^+(v_{c,\theta,t})) \quad \forall c \in C \quad \forall t \in \mathcal{T} \quad \forall \theta \in \Theta_c$$

$$\sum_{c\in C}\sum_{\theta\in\Theta_c} z_{c,\theta,\ell,t} \geq d_{\ell,t} \quad \forall \ell \in L \quad \forall t \in \mathcal{T}$$

$$\sum_{c\in C}\sum_{\theta\in\Theta_c}\sum_{\ell\in L_E} 9.75 FU_{c,\theta,\ell} z_{c,\theta,\ell,t} = e_{E,t} \quad \forall t \in [T]$$

$$\sum_{c\in C}\sum_{\theta\in\Theta_c}\sum_{l\in L_I} p_I (9.75 FU_{c,\theta,\ell} z_{c,\theta,\ell,t}) = e_{I,t} \quad \forall t \in [T]$$

$$e_{q,t},\ z_{c,\theta,\ell,t} \in \mathbb{R}_{\geq 0},\ f(\alpha) \in \mathbb{Z}_{\geq 0} \quad \forall \alpha \in \mathcal{P}(A).$$

Conclusion and Outlook

After setting up the MILP formulation of the FPP we implemented the model via SCIP. For performance tests we also set up a random instance generator. Comparing the computational results with the original formulation from [4], we were able to solve small instances within about 1 s versus 30 s in the referenced paper. For some large instances, we were also able to cut the solving time to roughly 35 min from 450 min. These preliminary computational suggest that the reformulation as a flow problem is beneficial. Next steps in our work will be finding a specialized algorithm solving the FPP by analyzing the structure of the problem and doing a full computational study to verify the given preliminary results.

Acknowledgements The authors would like to acknowledge the funding by the Deutsche Forschungsgemeinschaft (DFG, German Research Foundation) under Germany's Excellence Strategy—EXC 2163/1-Sustainable and Energy Efficient Aviation—Project-ID 390881007.

References

1. European Commission. (2021). *Fit for 55—Delivering the EU's 2030 climate target on the way to climate neutrality*. European Commission, Brussels, Belgium. https://www.eesc.europa.eu/en/our-work/opinions-information-reports/opinions/fit-55-delivering-eus-2030-climate-target-way-climate-neutrality
2. IEA. (2021). *Aviation*. IEA, Paris. https://www.iea.org/reports/aviation
3. Karpuk, S., Radespiel, R., & Elham, A. (2022). Assessment of future airframe and propulsion technologies on sustainability of next-generation mid-range aircraft. *Aerospace, 9*(5), 279. https://www.mdpi.com/2226-4310/9/5/279
4. Müller, C., Kieckhäfer, K., & Spengler, T. S. (2018). The influence of emission thresholds and retrofit options on airline fleet planning: an optimization approach. *Energy Policy, 112*, 242–257. https://doi.org/10.1016/j.enpol.2017.10.022
5. Schäfer, A. W., Evans, A. D., Reynolds, T. G., & Dray, L. (2016). Costs of mitigating CO_2 emissions from passenger aircraft. *Nature Climate Change, 6*(4), 412–417.
6. Scheelhaase, J., Maertens, S., Grimme, W., & Jung, M. (2018). EU ETS versus CORSIA—A critical assessment of two approaches to limit air transport's CO_2 emissions by market-based measures. *Journal of Air Transport Management, 67,* 55–62.

Chapter 20
PaMILO: A Solver for Multi-objective Mixed Integer Linear Optimization and Beyond

Fritz Bökler, Levin Nemesch, and Mirko H. Wagner

Abstract In multi-objective optimization, several potentially conflicting objective functions need to be optimized. Instead of one optimal solution, we look for the set of so called non-dominated solutions. An important subset is the set of non-dominated extreme points. Finding it is a computationally hard problem in general. While solvers for similar problems exist, there are none known for multi-objective mixed integer linear programs (MOMILPs) or multi-objective mixed integer quadratically constrained quadratic programs (MOMIQCQPs). We present PaMILO, the first solver for finding non-dominated extreme points of MOMILPs and MOMIQCQPs. It can be found on github under https://github.com/FritzBo/PaMILO. PaMILO provides an easy-to-use interface and is implemented in C++17. It solves occurring subproblems employing either CPLEX or Gurobi. PaMILO adapts the Dual-Benson algorithm for multi-objective linear programming (MOLP). As it was previously only defined for MOLPs, we describe how it can be adapted for MOMILPs, MOMIQCQPs and even more problem classes in the future.

Keywords Software · Multi-objective · Non-dominated extreme points · Mixed integer linear programming · Mixed integer quadratically constrained quadratic programming · Dual-Benson

Introduction

Many optimization problems can be seen as *multi-objective problems*, where several potentially conflicting objective functions need to be optimized. The solution of such a problem consists not of one single optimal solution value, but instead a

F. Bökler · L. Nemesch (✉) · M. H. Wagner
Theoretical Computer Science: https://tcs.uos.de/, Universität Osnabrück, Osnabrück, Germany
e-mail: lnemesch@uni-osnabrueck.de

F. Bökler
e-mail: fboekler@uni-osnabrueck.de

M. H. Wagner
e-mail: mirwagner@uni-osnabrueck.de

O. Grothe et al. (eds.), *Operations Research Proceedings 2022*, Lecture Notes in Operations Research, https://doi.org/10.1007/978-3-031-24907-5_20

set of so called *Pareto-optimal* solution values. A solution is Pareto-optimal, if an improvement in one objective has to worsen at least one other objective. Subclasses of multi-objective problems include *multi-objective linear programming* (MOLP), *multi-objective integer linear programming* (MOILP), *multi-objective mixed integer linear programming* (MOMILP), and *multi-objective mixed integer quadratically constrained quadratic programming* (MOMIQCQP). In this paper, we are interested in the so-called extreme points of MOMILPs and MOMIQCQPs. As MOLPs and MOILPs are special cases of MOMILPs, this also covers finding extreme points for those.

Especially MOMILPs are relevant problems in practice. For example in [1], a model for power generation expansion planning is formulated as a MOMILP. The objectives consider the economic cost of expanding the infrastructure, the environmental impact and the cost of resulting economic damage. In [9], a MOMILP is used in a completely different field to optimize financial portfolios. A set of portfolios is calculated from which an investor can choose which one fits best to their preferences.

Extreme points are an especially interesting subset of the Pareto-optimal solution: If a decision process mirrors reducing multiple criteria into one dimension through a concave function, an optimal solution can always be represented by exactly one specific extreme point [2]. Searching for extreme points can be seen as equal to parametric optimization, extreme points are called *break points* there. There are no extreme point solvers for MOMILPs or MOMIQCQPs and only few for MOLPs or MOILPs: *bensolve* [13] and *inner* [7] are solvers for MOLPs, *PolySCIP* [6] is able to find extreme points of both MOLPs and MOILPs.

Contribution. We introduce *PaMILO* (Parametric Mixed Integer Linear Optimization), a new solver for MOMILPs and MOMIQCQPs. It can read most common input formats and provides an easy-to-use console interface for practitioners. The only requirement is the existence of a tight lower bound for each objective, a so-called *ideal point*. The existence of it is an assumption also made by the aforementioned state-of-the-art solvers for MOLPs and MOILPs. Which is reasonable, as practical instances usually have an ideal point. The algorithmic approach of PaMILO is based on the Dual-Benson algorithm. We explain, how the Dual-Benson algorithm is adapted for MOMILPs and MOMIQCQPs. Furthermore, through this technique the Dual-Benson algorithm can be adapted to all kinds of multi-objective optimization problems.

Definitions

In MOMIQCQP, problems are of the form:

$$\begin{aligned} \min \quad & f_i(x) := x^\mathsf{T} P_i x + c_i^\mathsf{T} x && i = 1, \ldots, d \\ \text{s.t} \quad & x^\mathsf{T} Q_j x + a_j^\mathsf{T} x \leq b && j = 1, \ldots, m \\ & x \in \mathbb{R}^\ell \times \mathbb{Z}^{\ell'}, \end{aligned}$$

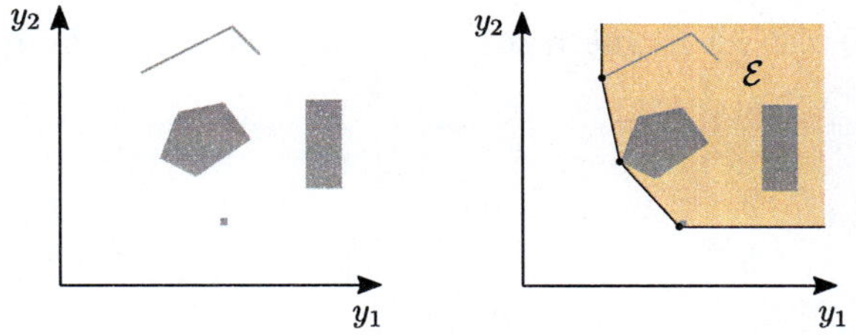

Fig. 20.1 Example of a MOMILP and its Edgeworth-Pareto hull

where $d, \ell, \ell', m \in \mathbb{N}$ and $n = \ell + \ell'$. For an index $i = 1, \ldots, d$, we denote by $P_i \in \mathbb{Q}^{n \times n}$ the *quadratic objective function matrix* i, and by $c_i \in \mathbb{Q}^n$ the *objective function vector* i. For an index $j = 1, \ldots, m$, we denote by $Q_j \in \mathbb{Q}^{n \times n}$ the *quadratic constraint matrix* j, and by $a_j \in \mathbb{Q}^n$ the *constraint vector* j. A single solution x is *feasible* if it fulfills all constraints, and we denote the set of all feasible solutions by $\mathcal{X}$. The set of all feasible solutions in objective space $\{f(x) : x \in \mathcal{X}\}$ is called $\mathcal{Y}$. For MOLP, MOILP and MOMILP all quadratic matrices $P_{i=1,\ldots,d}$ and $Q_{i=1,\ldots,m}$ are $\mathbf{0}$. The objective function then becomes Cx with $c_{i=1,\ldots,d}^{\mathsf{T}}$ as the rows of C, and the constraints can be described through $Ax \leq b$ with $a_{j=1,\ldots,m}^{\mathsf{T}}$ as the rows of A. Furthermore, in MOLP all variables are continuous and in MOILP all variables are integers. Without loss of generality, we only consider minimization problems here. But PaMILO is also able to solve problems with maximization objectives.

A point $y^* \in \mathcal{Y}$ is called *non-dominated* if there is no $y \in \mathcal{Y} \setminus \{y^*\}$ with $y_i \leq y_i^* \; \forall i = 1, \ldots, d$. In general, there is not a single non-dominated point, but a (not necessarily finite) set of non-dominated points. We call this the *non-dominated set*, or $\mathcal{Y}_N$.

PaMILO searches for a subset of $\mathcal{Y}_N$, the non-dominated extreme points. To properly define an extreme point, we first define the *Edgeworth-Pareto hull* of a MOMIQCQP: $\mathcal{E} := \text{conv}(\mathcal{Y}) + \mathbb{R}^d_{\geq 0}$. Then, the *non-dominated extreme points* are the vertices of $\mathcal{E}$. A vertex of the (convex) set $\mathcal{E}$ is a point $y \in \mathcal{E}$ such that there is a valid hyperplane H with $H \cap \mathcal{E} = \{y\}$. PaMILO finds a finite representation of $\mathcal{E}$. If the number of extreme points is finite itself, this representation is the exact set of extreme points. In MOMILP, this set is always finite. But there are MOMIQCQPs in which it may be infinite. Figure 20.1 shows a MOMILP and its Edgeworth-Pareto hull.

A special case of the Edgeworth-Pareto hull is the *upper image* for MOLPs. It is defined as $\mathcal{P} := \mathcal{Y} + \mathbb{R}^d_{\geq 0}$ [11]. In MOLP, $\mathcal{Y}$ is convex by itself and, thus, every point of $\mathcal{Y}_N$ lies on the boundary of $\mathcal{P}$. The same does not hold for MOMIQCQP, points of $\mathcal{Y}_N$ can lie inside $\mathcal{E}$.

The extreme points can also be characterized through the *weighted-sum problem* (WSP) of a MOMILP. Given a weight vector $w \in \mathcal{W} = \{w \in \mathbb{R}_{\geq 0} : \sum_{i=1}^d w_i = 1\}$, it is $\text{WSP}(w) := \min_{x \in \mathcal{X}} w^{\mathsf{T}} f(x)$. An extreme point is a point $y \in \mathcal{Y}$ where a $w \in \mathcal{W}$ exists, so that the weighted sum in objective space is only minimal for this point. Simply solving $\text{WSP}(w)$ might result in a dominated point in some special cases.

While PaMILO is able to handle these special cases, here we simply assume they do not appear.

Algorithm

PaMILO adapts the Dual-Benson algorithm, which we briefly describe in this section. Originally, the Dual-Benson algorithm is only defined for MOLPs. In Sect. 20.3.1 we explain how to also apply it to MOMILPs and MOMIQCQPs. A detailed description of the algorithm and its theoretical preliminaries can be found in [8, 10, 11]. The algorithm operates on the *lower image* of the problem. The lower image $\mathcal{D}$ is defined in [11] as

$$\mathcal{D} := \left\{ \left(w_1, \ldots, w_{d-1}, b^{\mathsf{T}} u \right) : w \in \mathcal{W}, A^{\mathsf{T}} u = C^{\mathsf{T}} w, u \in \mathbb{R}^m_{\geq 0} \right\}.$$

Heyde and Löhne showed in [11] that the lower image is geometrically dual to the upper image. Particularly, every facet in the lower image corresponds to an extreme point in the upper image and vice versa. Hence, by enumerating the facets of the lower image, we obtain the extreme points of the MOLP.

The algorithm works by iteratively refining an outer approximation of $\mathcal{D}$. The initial outer approximation consists of one trivial facet of $\mathcal{D}$ and the boundaries given by $w \in \mathcal{W}$. The algorithm then improves this approximation by subsequently enumerating unvisited vertices of the approximation. A vertex is *unvisited* if it was not previously used in an iteration. In each iteration, the algorithm picks an unvisited vertex v of the approximation and shoots a ray down onto the boundary of $\mathcal{D}$. Shooting down a ray from v is the same as solving WSP$(v_1, \ldots, v_{d-1}, 1 - \sum_i^{d-1} v_i)$. Either, the ray hits $\mathcal{D}$ directly at v and confirms that v is an extreme point of $\mathcal{D}$, or the ray hits a new point on the boundary of $\mathcal{D}$. If such a point is hit, a new facet supporting inequality of $\mathcal{D}$ is constructed that cuts off v. This improved approximation of $\mathcal{D}$ has new unvisited vertices for further iterations, so a vertex enumeration is done. When no unvisited vertex remains, the final approximation is the lower image itself. Because of geometric duality, this also gives us the vertices and facets of the upper image. Thus, we have found all non-dominated extreme points (Fig. 20.2).

Generalizing the Dual-Benson Algorithm

As mentioned before, we can adapt the Dual-Benson algorithm to also find extreme points of MOMILPs and MOMIQCQPs. A previous generalization to multi-objective combinatorial problems was done by Bökler and Mutzel in [5]. Our adaption is based upon the ideas developed in [4].

The Edgeworth-Pareto hull $\mathcal{E}_{\mathcal{I}}$ of a MOMILP $\mathcal{I}$ (under assumption of an ideal point) is a convex polyhedron with a recession cone equal to the positive orthant.

The number of its vertices is finite. Thus, there is a MOLP $\mathcal{J}$ so that its image $\mathcal{Y}_{\mathcal{J}}$ is equal to $\mathcal{E}_{\mathcal{I}}$. $\mathcal{J}$ can be constructed by simply using the facets of $\mathcal{E}_{\mathcal{I}}$ as constraints and making $C_{\mathcal{J}}$ the identity matrix. This MOLP has an upper image $\mathcal{P}_{\mathcal{J}}$ and a corresponding lower image $\mathcal{D}_{\mathcal{J}}$. As $\mathcal{P}_{\mathcal{J}}$ has the same vertices as $\mathcal{E}_{\mathcal{I}}$, a Dual-Benson algorithm operating on $\mathcal{J}$ finds the extreme points of $\mathcal{I}$.

There is a catch though, since we do not actually know $\mathcal{J}$. But we do not need to: The only time the Dual-Benson algorithm interacts with $\mathcal{J}$ is when shooting down rays onto the boundary of $\mathcal{D}_{\mathcal{J}}$. But ray shooting is the same as solving weighted-sum problems. Thus, a weighted-sum oracle for $\mathcal{I}$ is sufficient to realize ray shooting onto $\mathcal{D}_{\mathcal{J}}$. With such a weighted-sum oracle, the Dual-Benson algorithm is able to find the extreme points of the MOMILP $\mathcal{I}$.

For MOMIQCQPs, the same arguments can directly be applied if the number of extreme points is finite. But there also are MOMIQCQPs with an infinite number of extreme points. In this case, the Dual-Benson algorithm would not terminate. This can be avoided by only adding dual facets when the length of the ray hitting them is above a threshold value $\varepsilon > 0$. The resulting outer approximation of the lower image corresponds to an inner approximation of the actual Edgeworth-Pareto hull [14]. The extreme points of this inner approximation are a finite subset of the extreme points of $\mathcal{E}$.

As we never use distinct characteristics of MOMILPs or MOMIQCQPs, this argumentation can be generalized for all possible multi-objective problems. As long as a weighted-sum is computable, a Dual-Benson algorithm using a weighted-sum oracle is able to find the extreme points or at least the extreme points of an approximation of $\mathcal{E}$.

Implementation

PaMILO is implemented in C++17. Installation can be easily done by using cmake. After installation, PAMILO provides an easy-to-use command-line inter-

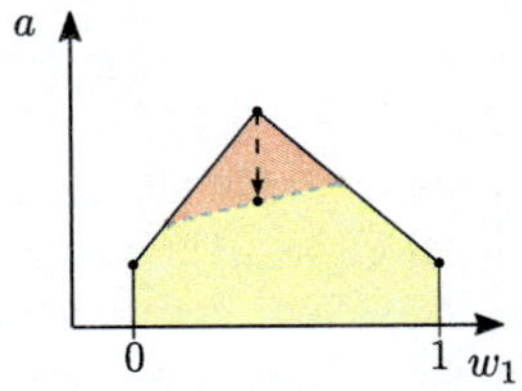

(a) A ray is shot down from a dual vertex and finds a new dual facet.

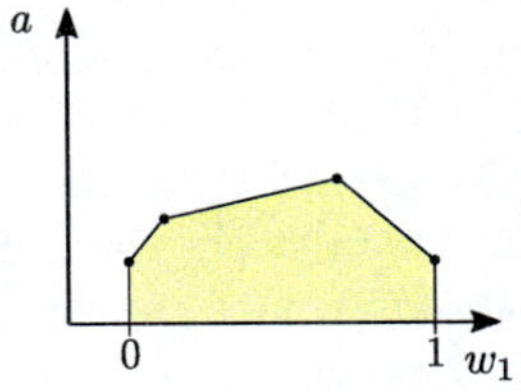

(b) The new facet is added to the dual approximation.

Fig. 20.2 Example of one iteration of the Dual-Benson algorithm. The approximation (red) of $\mathcal{D}$ (yellow) becomes more accurate with the new facet face. PaMILO can be found on github under https://github.com/FritzBo/PaMILO. For academic purposes, it can be used subject to the MIT software license.

The weighted-sum problems are solved through CPLEX or Gurobi. Due to the need to solve multi-objective problems, recent versions of the solvers are required. For technical reasons, only the Gurobi version can handle MOMIQCQPs.

To minimize numerical error, instances undergo preprocessing. If the value ranges of objectives in the non-dominated points differ strongly, the number of numerical errors increases. PaMILOs preprocessing tries to normalize these ranges for each objective function. Through this, we observe a significant decrease of numerical issues in practice.

Many formats are supported for input files, the most well known are the `.lp` and the `.mps` formats. The output consists of three files, all beginning with a user defined output name as prefix. The `*_sol` file contains all extreme points and one solution corresponding to each in `.json` format, the `*_log` file contains logging information of PaMILO, and the `*_(cplex|gurobi)` file contains the logging output from the respective solver.

Computational Results

To get an impression of PaMILOs performance on established benchmark instances, we compared it to two other start-of-the-art solvers: bensolve [13] and PolySCIP [6]. PolySCIP is used with CPLEX internally and PaMILO with Gurobi, the best variant for the respective solver. We used the Dual-Benson algorithm as bensolves algorithm. Computations were done on an Intel Xeon GOLD 6134 CPU with 256 GB RAM, and all instances had a time limit of 30 minutes.

Experiments for MOLPs use instances from [5]. They show a very clear superiority of bensolve over PolySCIP and PaMILO on all instances. For MOILPs, we generated instances according to the scheme for general MOILPs described in [12]. On those, PaMILO is often faster than PolySCIP while finding similar sets of extreme points. For MOMILPs, we generated instances according to the scheme described in [15]. The instances we generated were much bigger than the ones used in [15], but PaMILO is still able to find extreme points. A brief look on some of the computational results for MOILPs and MOMILPs is given in Table 20.1.

In general, we observe that the most cost expensive part of the calculations are the vertex enumerations. In theory, the computational effort to enumerate the vertices can grow exponentially in the number of objectives. We observe such growth with PaMILO for many instances. But for a fixed number of objectives, we also observe a delay between subsequent outputs that is only incremental polynomial in the oracle calls consistent with [3].

To the best of our knowledge, there are no benchmark instance sets for MOMIQCQPs described in literature. Our experiments aim at giving a first impression. As there are now several extreme point solvers available, a comprehensive computational study on their capabilities is a goal for future research.

Table 20.1 Computational results for MOILPs and MOMILPs

Instances			PaMILO			PolySCIP		
Type	d	n	Time [s]	$\lvert\mathcal{Y}_{\mathrm{Ex}}\rvert$	#	Time [s]	$\lvert\mathcal{Y}_{\mathrm{Ex}}\rvert$	#
MOILP	3	100	**184.19**	(28)	(27/30)	712.66	(24)	(22/30)
	4	60	**191.8**	(90.5)	(18/30)	828.4	(67.5)	(18/30)
	5	50	**570.6**	(139)	(26/30)	–	(61)	(13/30)
MOMILP	3	300	**412.29**	10,074.5	30	–	–	–
	4	200	**781.79**	(24,930.0)	(29/30)	–	–	–
	5	40	**208.67**	21,546.0	30	–	–	–

Time and number of extreme points ($\mathcal{Y}_{\mathrm{Ex}}$) are median. Entries in () are reduced data because of timeouts
Bold entries indicate fastest time for the instance class

Conclusion

PaMILO is the first solver able to find the extreme points of MOMILPs and MOMIQCQPs and proves to be comparable to start-of-the-art solver PolySCIP for MOILPs. Thus, it can be an important new addition to the repertoire of practitioners.

References

1. Antunes, C. H., Martins, A. G., & Brito, I. S. (2004). A multiple objective mixed integer linear programming model for power generation expansion planning. *Energy, 29*(4), 613–627.
2. Benson, H. P. (1995). Concave minimization: Theory, applications and algorithms. In *Handbook of global optimization*, pp. 43–148. Springer.
3. Bökler, F. (2018). *Output-sensitive complexity of multiobjective combinatorial optimization with an application to the multiobjective shortest path problem*. Ph.D. thesis
4. Bökler, F., Parragh, S.N., Sinnl, M., & Tricoire, F. (2021). An outer approximation algorithm for multi-objective mixed-integer linear and non-linear programming. Preprint. https://doi.org/10.48550/ARXIV.2103.16647
5. Bökler, F., & Mutzel, P. (2015). Output-sensitive algorithms for enumerating the extreme nondominated points of multiobjective combinatorial optimization problems. In *Algorithms—ESA 2015*, LNCS, Vol. 9294, pp. 288–299
6. Borndörfer, R., Schenker, S., Skutella, M., & Strunk, T. (2016). Polyscip. In *Mathematical Software—ICMS 2016*. LNCS, Vol. 9725, pp. 259–264
7. Csirmaz, L. (2021). Inner approximation algorithm for solving linear multiobjective optimization problems. *Optimization, 70*(7), 1487–1511.
8. Ehrgott, M., Löhne, A., & Shao, L. (2012). A dual variant of Benson's "outer approximation algorithm for multiple objective linear programming. *Journal of Global Optimization, 52*(4), 757–778.
9. Ehrgott, M., Waters, C., Kasimbeyli, R., & Ustun, O. (2009). Multiobjective programming and multiattribute utility functions in portfolio optimization. *INFOR: Information Systems and Operational Research, 47*(1), 31–42.
10. Hamel, A. H., Löhne, A., & Rudloff, B. (2014). Benson type algorithms for linear vector optimization and applications. *Journal of Global Optimization, 59*(4), 811–836.
11. Heyde, F., & Löhne, A. (2008). Geometric duality in multiple objective linear programming. *SIAM Journal on Optimization, 19*(2), 836–845.

12. Kirlik, G., & Sayın, S. (2015). Computing the nadir point for multiobjective discrete optimization problems. *Journal of Global Optimization, 62*(1), 79–99.
13. Löhne, A., & Weißing, B. (2017). The vector linear program solver Bensolve—notes on theoretical background. *European Journal of Operational Research, 260*(3), 807–813.
14. Löhne, A., Rudloff, B., & Ulus, F. (2014). Primal and dual approximation algorithms for convex vector optimization problems. *Journal of Global Optimization, 60*(4), 713–736.
15. Mavrotas, G., & Diakoulaki, D. (1998). A branch and bound algorithm for mixed zero-one multiple objective linear programming. *European Journal of Operational Research, 107*(3), 530–541.

Chapter 21
Vehicle Routing with Heterogeneous Time Windows

Petra Mutzel, Tim Niemann, Lukas Schürmann, Sebastian Stiller, and Andreas M. Tillmann

Abstract We consider a novel variant of the heterogeneous vehicle routing problem (VRP) in which each customer has different availability time windows for every vehicle. In particular, this covers our motivating application of planning daily delivery tours for a single vehicle, where customers can be available at different times each day. The existing literature on heterogeneous VRPs typically distinguishes properties of the vehicle fleet such as costs or capacities, but apparently, windows of customers have only been regarded in a homogeneous fashion thus far. To solve the problem, we employ a branch-and-price framework based on a set partitioning formulation together with a parallelizable labeling algorithm. The heterogeneous time window structure yields notable computational gains by allowing to decompose the pricing problem as well as to utilize a customer-vehicle assignment branching rule. We show that this branching rule leads to more balanced search trees than the usual arc flow branching, and demonstrate its efficiency in numerical experiments.

Keywords Vehicle routing problem · Heterogeneous time windows · Branch and price

Introduction

The vehicle routing problem (VRP) is a well-studied problem class with a long history and numerous variants inspired by practical applications, cf., e.g., [2, 7]. In the basic (capacitated) VRP, we are given a directed graph $\mathcal{G} = (\mathcal{V}, \mathcal{E})$ with $\mathcal{V} = \mathcal{C} \cup \{v_0\}$, where $\mathcal{C}$ represents customers and v_0 is the depot of a set of vehicles $\mathcal{K}$, each with capacity Q. Each arc $e \in \mathcal{E}$ has an associated cost $c_e \geq 0$ (typically, driving time),

P. Mutzel · L. Schürmann
Institute for Computer Science, Universität Bonn, Friedrich-Hirzebruch-Allee 8, 53115 Bonn, Germany

T. Niemann (✉) · S. Stiller · A. M. Tillmann
Institute for Mathematical Optimization, Technische Universität Braunschweig, Universitätsplatz 2, 38106 Braunschweig, Germany
e-mail: tim.niemann@tu-braunschweig.de

O. Grothe et al. (eds.), *Operations Research Proceedings 2022*, Lecture Notes in Operations Research, https://doi.org/10.1007/978-3-031-24907-5_21

and each customer $c \in \mathcal{C}$ has a demand q_c. We are looking for a cost-minimal set of tours in $\mathcal{G}$ that each start and end at the depot such that every customer $c \in \mathcal{C}$ is visited exactly once and the sum of all demands on each tour does not exceed the vehicle capacity.

We extend the palette of VRP variants by adding *heterogeneous time windows*: Customers can only be visited during individual availability time windows, which can be different for each vehicle. Our study of this model is motivated by a real-world practical application in which we need to schedule delivery tours for one vehicle over a given time horizon (e.g., a few weeks) and a fixed set of customers whose availability times differ from day to day. Thus, each day, the vehicle travels a different route so that at the end of the time horizon, all customers have been visited. In our application, vehicles are tantamount to days, but it is conceivable that different availability times may also directly relate to delivery or vehicle types—for instance, while recipients can be assumed to handle small packages by themselves, they may need to coordinate with helpers when receiving large pieces of furniture.

Time window constraints in VRPs have been well-known since the work of Solomon [6], but to the best of our knowledge, the case of time windows which differ for different vehicles (or days, or deliveries) has not been examined thus far. Previous works on heterogeneous VRP variants (cf. [3]) considered different aspects such as the cost and capacities on a vehicle-specific basis, but different time windows appear to be a novel concept. In particular, our approach also allows for customers to be completely unavailable for arbitrary time spans. Moreover, heterogeneous time windows not only enable addressing special customer needs, but also opens a way to reduce search tree sizes and runtime of dedicated VRP branch-and-price solvers. Notably, the heterogeneity gives rise to an effective branching rule that we shall call *vehicle assignment branching* and which compares favorably with the classical arc flow branching.

We will briefly explain our model and the problem-specific branch-and-price solution procedure in Section "Model and Solution Method", including a theoretical analysis of the two branching rules. In Section "Experimental Results", we discuss computational experiments that illustrate the practical impact of heterogeneity-based model properties and branching on different test instances. Some final remarks in Section "Concluding Remarks" conclude the paper.

Model and Solution Method

For ease of presentation, we assume henceforth that each customer has (at most) one time window per vehicle. Nevertheless, we point out that it is possible to handle multiple (disjoint) time windows as well as, in case of daily route planning as in our motivating application, multiple vehicles per day; space limitations prevent us from providing the technical details of the corresponding modifications.

Model

Let $\mathcal{P}$ be the set of all feasible tours (w.r.t. time-window, capacity and possible further constraints), and $\mathcal{P}_k \subset \mathcal{P}$ those performed by vehicle $k \in \mathcal{K}$. For each tour p, we define its cost $C_p := \sum_{e \in p} c_e$. With binary variables x_p to decide whether tour p partakes in the overall solution, the *set-partitioning formulation* of our VRP is then given by

$$\min \sum_{p \in \mathcal{P}} C_p x_p$$
$$\text{s.t.} \sum_{p \in \mathcal{P}: i \in p} x_p = 1 \ \forall i \in \mathcal{C}, \sum_{p \in \mathcal{P}_k} x_p = 1 \ \forall k \in \mathcal{K}, x_p \in \{0, 1\} \ \forall p \in \mathcal{P}. \tag{SPF}$$

The first constraint guarantees that each customer is visited exactly once, and the second one that each vehicle is used on at most one tour (for technical reasons, we include empty tours with cost 0 in each $\mathcal{P}_k$ to write this in equality form); the latter stems from our daily route planning application and may take a different form in other cases.

Due to the exponential number of tours and the resulting impracticality of directly solving (SPF), we adopt the established VRP solution approach of *branch-and-price* with a dedicated labeling algorithm; see, e.g., [2].

Pricing Problem

The general idea of branch-and-price is to start with few tours $\mathcal{P}' \subset \mathcal{P}$, and then iteratively add new tours to $\mathcal{P}'$ and re-optimize the reduced problem (with variables x_p, $p \in \mathcal{P}'$) until a provably optimal solution to the original problem is found. To find promising new tours $p \in \mathcal{P} \setminus \mathcal{P}'$, we consider the following *reduced-cost pricing problem*: Given an optimal dual solution (π^*, τ^*) for the LP relaxation of the reduced problem, solve

$$r(\pi^*, \tau^*) = \min_{k \in \mathcal{K}} \Big\{ -\tau_k^* + \min_{p \in \mathcal{P}_k} \sum_{(i,j) \in p} c_{ij} - \pi_i^* \Big\}. \tag{RCPP}$$

Since the depot node $v_0 \notin \mathcal{C}$ has no dual variable, we define $\pi_{v_0}^* := 0$ to simplify notation. If we find a tour with negative reduced cost ($r(\pi^*, \tau^*) < 0$), we add it to $\mathcal{P}'$; otherwise, the reduced LP was solved to optimality (w.r.t. all tours, including omitted ones).

This pricing problem amounts to an *elementary shortest path problem with resource constraints*, which is strongly NP-hard and usually solved using a labeling algorithm, cf. [2]. Due to space limitations, we omit the relatively straightforward details on adapting the labeling scheme to our present setting, but note that vehicle capacities and customer time windows are translated to resource constraints. Since (RCPP) naturally decomposes into independent pricing subproblems for every vehicle $k \in K$, we only need to consider the customers who have a time window for the corresponding vehicle $k \in K$ for each subproblem. Depending on the instance, this significantly reduces the size of the relevant subgraph, which in turn speeds up the labeling algorithm. Furthermore, the subproblems can be solved in parallel.

Branching Rules

In a branch-and-price framework, branching on the standard problem variables often interferes with the pricing mechanism. Several alternative branching rules for VRPs have been proposed to deal with this issue, see [2] for an overview of the most common ones. Branching on *arc flow variables* $y_e := \sum_{p \in \mathcal{P}: e \in p} x_p$, $e \in \mathcal{E}$, appears to be the most popular strategy, even though the branching decisions, especially setting $y_e = 0$ (removing an arc), have a rather low overall impact.

Our heterogeneous structure allows us adapt the branching idea originally introduced by Ryan and Foster in [4], and formulated in [5] for the general assignment problem. The idea is to branch on assignment decisions; in the present setting, these translate to decisions whether to let specific customers get served by certain vehicles. To that end, we define auxiliary *vehicle assignment variables*

$$d_i^k := \sum_{p \in \mathcal{P}_k : i \in p} x_p \in [0, 1], \quad \forall\, i \in \mathcal{C}, k \in \mathcal{K}.$$

Then, we can branch on a (fractional) d_i^k, setting it to 0 or 1, respectively. These *vehicle assignment branching* decisions can easily be accounted for in subsequent pricing steps: $d_i^k = 0$ leads to the exclusion of customer i from pricing subproblem for vehicle k, and $d_i^k = 1$ removes i from all pricing subproblems except for vehicle k. Moreover, any exact pricing scheme can be adapted to ensure the condition $d_i^k = 1$ remains valid, cf. [5], and in leaves of the resulting decision tree, there exists an integral optimal solution for the LP relaxation of (SPF). The latter arises from the fact that for each solution x such that $d_i^k \in \{0, 1\}$ for all $i \in \mathcal{C}$, $k \in \mathcal{K}$, all tours in $\mathcal{P}_k^{>0} := \{p \in \mathcal{P}_k : x_p > 0\}$ visit the same set of customers for each $k \in \mathcal{K}$, as $\sum_{p \in \mathcal{P}_k} x_p = 1$ must be satisfied. Consequently, the C_p values are identical for each $p \in \mathcal{P}_k^{>0}$, and any solution with $x_{p'} = 1$ for an arbitrary $p' \in \mathcal{P}_k^{>0}$ and $x_p = 0$ for each $p \in \mathcal{P}_k^{>0} \setminus \{p'\}$ is feasible and optimal. Applying this transformation for every $k \in \mathcal{K}$ yields an integral optimal solution. We remark that the variables d_i^k are not actually included in the problem, but are handled implicitly during pricing.

Table 21.1, which gives the potential branching tree depths for all previously mentioned branching rules, exhibits a theoretical advantage of our rule. Clearly, vehicle assignment branching leads to more balanced and possibly much smaller

Table 21.1 Maximal number of branchings for different branching rules

branching variables	number of 0-branches	number of 1-branches
tour variables (x_p)	$\mathcal{O}(2^{\lvert\mathcal{V}\rvert})$	$\lvert\mathcal{K}\rvert$
arc flow variables (y_e)	$\lvert\mathcal{E}\rvert - \lvert\mathcal{V}\rvert$	$\lvert\mathcal{V}\rvert$
vehicle assignment variables (d_i^k)	$(\lvert\mathcal{K}\rvert - 1)\lvert\mathcal{V}\rvert$	$\lvert\mathcal{V}\rvert$

tree, especially when the instance has many arcs but relatively few vehicles, as in this case $|\mathcal{K}| \times |\mathcal{V}| \ll |\mathcal{E}|$. The practical impact of our branching rule will be assessed by means of the computational experiments discussed in the following.

Experimental Results

To obtain test instances for our VRP variant, we modified Solomon's data set [6], which includes only homogeneous time windows, for our purposes. Based on the three instances R101, C101 and RC101, we generated new instances by taking the first $|\mathcal{C}| \in \{40, 50, \ldots, 100\}$ customers and inheriting travel and service times, customer demands and vehicle capacities. We then add a time window for each customer-vehicle pair $i \in \mathcal{C}, k \in \mathcal{K}$ with probability $\alpha_{\text{TW}} \in \{0.25, 0.5, 0.75, 1\}$, retaining the lengths of time windows from the respective base instance but randomly choosing their starting times (such that all time windows lie within depot working hours), where the size of $\mathcal{K}$ was set heuristically to ensure feasibility.

We implemented our approach in C++ with SCIP 8.0 [1] and tested it on a 24-core machine; on average, parallelization of the pricing subproblems saved about half the computation time with this setup, compared to single-thread execution. We solved each instance twice: once with our vehicle assignment branching, and once using arc flow branching. In both cases, we chose the variable with value closest to 0.5 for branching. The results in Table 21.2 focus on the instances based on R101; those based on the other two base instances yielded similar results. Besides instance parameters, we report the runtime in seconds ("DNF" marks cases for which the solution process did not finish after one hour), number of branching nodes, and optimality gap at termination for both variants. We also state the best objective value found by the vehicle assignment branching variant; the objective value for the arc flow branching variant was mostly very similar, so differences in gap values are predominantly due to dual bounds. All decimals were rounded to two significant digits.

From Table 21.2, we can make some key observations: Almost all instances can be solved faster and with fewer search nodes using vehicle assignment branching than with arc flow branching, especially those with many time windows per customer (larger α). For instances with many customers, the vehicle assignment branching variant can solve more instances and often significantly reduce the final gap when

Table 21.2 Results for test instances based on R101, comparing two branching rules

			vehicle assignment branching				arc flow branching		
$\|\mathcal{C}\|$	$\|\mathcal{K}\|$	α_{TW}	best obj.	time	nodes	% gap	time	nodes	% gap
40	11	0.25	1231.45	0.01	1	0	0.01	1	0
40	8	0.50	1043.80	0.05	3	0	0.05	3	0
40	7	0.75	811.64	0.96	37	0	1.19	59	0
40	12	1.00	741.23	6.79	573	0	12.44	1239	0
50	13	0.25	1604.27	0.01	1	0	0.01	1	0
50	10	0.50	1271.13	0.06	1	0	0.06	1	0
50	14	0.75	920.46	2.23	129	0	2.10	103	0
50	12	1.00	853.98	14.28	265	0	14.80	299	0
60	14	0.25	1724.36	0.08	13	0	0.13	25	0
60	17	0.50	1119.48	0.54	43	0	1.16	70	0
60	14	0.75	983.33	49.05	1490	0	125.34	4718	0
60	15	1.00	943.59	1590.46	20503	0	DNF	69128	0.71
70	16	0.25	2080.74	0.03	1	0	0.03	1	0
70	17	0.50	1335.39	23.85	1286	0	67.52	4251	0
70	17	0.75	1174.52	3527.82	51265	0	DNF	70380	1.15
70	16	1.00	1096.63	DNF	7718	1.17	DNF	7294	8.35
80	18	0.25	2213.67	3.65	265	0	12.33	1145	0
80	21	0.50	1418.48	485.65	23473	0	2435.35	114433	0
80	20	0.75	1221.76	DNF	26970	0.3	DNF	36499	1.27
80	18	1.00	1148.86	DNF	2609	0.77	DNF	3010	0.91
90	25	0.25	2024.82	68.85	8177	0	396.55	47164	0
90	21	0.50	1534.87	DNF	68465	1.4	DNF	77536	2.21
90	20	0.75	1602.31	DNF	8807	21.37	DNF	8744	21.86
90	19	1.00	1504.87	DNF	360	24.71	DNF	397	24.80
100	25	0.25	2124.97	263.48	18639	0	DNF	218423	0.76
100	23	0.50	1540.94	508.18	6227	0	DNF	54166	0.57
100	19	0.75	1570.67	DNF	2325	13.67	DNF	3637	13.69
100	19	1.00	1581.19	DNF	16	25.89	DNF	15	25.97

hitting the time limit. In general, solving times increase notably with the number of time windows; for example, with 70 customers, the instance with $\alpha_{TW} = 0.25$ can be solved in well under a second, whereas more than an hour is needed if every customer has a (different) time window for each vehicle. However, increasing customer flexibility naturally allows for more efficient routings, as reflected by the primal bounds decreasing significantly the larger α_{TW} gets.

Finally, it is worth mentioning another empirical observation that is not apparent from Table 21.2: If a lot of customer-vehicle pairs have similar time windows, i.e., the instance "approaches" time-window homogeneity, then arc flow branching is preferable to vehicle assignment branching, which can likely be explained by the former then affecting several vehicles simultaneously.

Concluding Remarks

We introduced the concept of heterogeneous time windows for the VRP to account for modern-day customer availability requirements, and demonstrated that it allows for often significant computational gains by decomposition of pricing problems and the specialized vehicle assignment branching rule. In ongoing work, we also successfully adapted our approach to multiple time windows per day in a daily delivery tour planning application. As future research, we plan to assess the algorithmic efficiency in yet more general variants with, e.g., multiple vehicles per day and incorporating robustness w.r.t. travel delays. Furthermore, it is of interest to construct problem-specific primal and pricing heuristics to further speed up computations and enable solving larger or more complicated instances.

Acknowledgements The authors gratefully acknowledge funding by the German Federal Ministry for Education and Research, grant nos. 05M20MBA-LeoPlan and 05M20PDA-LeoPlan.

References

1. Bestuzheva, K., Besançon, M., Chen, W.-K., Chmiela, A., Donkiewicz, T., van Doornmalen, J., Eifler, L., Gaul, O., Gamrath, G., Gleixner, A., Gottwald, L., Graczyk, C., Halbig, K., Hoen, A., Hojny, C., van der Hulst, R., Koch, T., Lübbecke, M., Maher, S. J., . . . , Witzig, J. (2021). The SCIP optimization suite 8.0. ZIB-report 21-41, Zuse Institute Berlin. http://nbn-resolving.de/urn:nbn:de:0297-zib-85309
2. Costa, L., Contardo, C., & Desaulniers, G. (2019). Exact branch-price-and-cut algorithms for vehicle routing. *Transportation Science, 53*(4), 946–985. https://doi.org/10.1287/trsc.2018.0878
3. Koç, Ç., Bektaş, T., Jabali, O., & Laporte, G. (2016). Thirty years of heterogeneous vehicle routing. *European Journal of Operational Research, 249*(1), 1–21. https://doi.org/10.1016/j.ejor.2015.07.020
4. Ryan, D. M., & Foster, B. A. (1981). An integer programming approach to scheduling. In A. Wren (Ed.), *Computer scheduling of public transport* (pp. 269–280).
5. Sol, M. (1994). *Column generation techniques for pickup and delivery problems*. Ph.D. thesis, Technische Universiteit Eindhoven, The Netherlands.
6. Solomon, M. M. (1984). *Vehicle routing and scheduling with time window constraints: Models and algorithms (Heuristics)*. Ph.D. thesis, University of Pennsylvania, PA, USA.
7. Vidal, T., Laporte, G., & Matl, P. (2020). A concise guide to existing and emerging vehicle routing problem variants. *European Journal of Operational Research, 286*(2), 401–416. https://doi.org/10.1016/j.ejor.2019.10.010

Part VI
Energy and Environment

Chapter 22
A Bicriteria Almost Equal Minimum Cost Flow Model for Day-Ahead Trading

E. Finhold, T. Heller, S. O. Krumke, and N. Leithäuser

Abstract As the share of renewable energy and therefore the fluctuation in power generation increases, using a battery to market the power saved/used by a flexible process becomes more and more important and attractive. The charging and discharging process of a battery can be modeled as a flow model in a time-expanded graph. We model the costs for this process as edge costs between nodes that correspond to consecutive points in time. An optimal battery strategy be obtained by computing a minimum cost flow in the given network. If the charging and discharging of a battery takes place as a coupled process of a production process, a steady and even charging and discharging is often important. In this paper, we describe a bicriteria flow model based on the Almost-Equal-Minimum Cost Flow Problem (AEMCFP) in which both trading profits and a steady flow of energy are considered as the two objectives. In the AEMCFP one is given additional sets of edges on which the flow values differ at most by a given constant. We obtain a strongly polynomial algorithm based on a parametric search approach. Furthermore, we present a case study for bidding on the German day ahead market.

Keywords Network flows · Minimum cost flow · Parametric search · Energy trading

Introduction

In the course of the transformation of the energy system towards a system with a high share of renewable energies, the marketing of flexible storage is becoming increasingly important. The flexibility should be used to be able to react to times of high or low energy—these are often reflected by price signals on different markets.

E. Finhold · T. Heller (✉) · N. Leithäuser
Department of Optimization, Fraunhofer Institute for Industrial Mathematics, ITWM, 67663 Kaiserslautern, Germany
e-mail: till.heller@itwm.fraunhofer.de

S. O. Krumke
Department of Mathematics, University of Kaiserslautern, 67663 Kaiserslautern, Germany

O. Grothe et al. (eds.), *Operations Research Proceedings 2022*, Lecture Notes in Operations Research, https://doi.org/10.1007/978-3-031-24907-5_22

One way to market this flexibility is the day ahead (DA) market which is a daily auction where bids can be submitted for hourly products of the following day. The German DA market, the EPEX Spot, allows bidding on all hours of the following day, i.e. each participant can submit one bid per hour.[1] Trading strategies for the DA market are widely discussed and often arise from demand side management (e.g. [9]). In this paper we focus on bidding strategies that indicate in which hour energy is sold (or bought, respectively). Thus, a bidding strategy can be modeled as an inflow—outflow problem of the battery. Modeling energy flows as flows in a network graph has a rather long history (cf. [1]). If the charging and discharging of a battery takes place as a coupled process of a production process, a steady and even charging and discharging is often important. The *Almost Equal Maximum Flow Problem (AEMFP)* tries to find a maximum flow in a given network graph with the additional property that the flow value on edges of so called *homologous edge sets* do not differ by more than a given value, i.e. the minimum and maximum flow value are not too far apart. The allowed deviation can be either given by a constant, or depends on a function with the minimum flow value as input. This problem was formulated in [3] and generalized to the *Almost Equal Minimum Cost Flow Problem (AEMCFP)* the setting of minimum cost flows in [4]. In this work, we present a strongly polynomial algorithm for solving the AEMCFP as well as a numerical study on computed day ahead trading strategies.

The Model

The German DA market at the EPEX Spot allows the buying and selling of energy in hourly resolution of the following day. Besides the rule set allowing complex bids, we consider only one price forecast and are only interested in the exact amount to be traded in an hour. We call the maximal amount of energy that can be bought or sold in an hour the *flexibility*. Now, we are interested in finding an optimal trading strategy for the day ahead market under consideration of limitations from the ramp-up (or cool-down) process. Note that this problem can be formulated as a linear program and hence solved in polynomial time. However, we are interested in a formulation as a network flow model in order to obtain a combinatorial polynomial time algorithm.

The Network Graph

In order to construct the underlying network graph, we are given an initial battery level b^{in}, a minimal battery level $b^{\min}$, a maximal battery level $b^{\max}$ and a desired end battery level b^{out}. Furthermore, the flexibility is denoted by δ. For each hour t, a price forecast p_t is given. By adding an edge e from t to s with $l_e = 0$, $u_e = \infty$

---

[1] A more in-depth explanation of the rules on the German DA market can be found on www.epexspot.com.

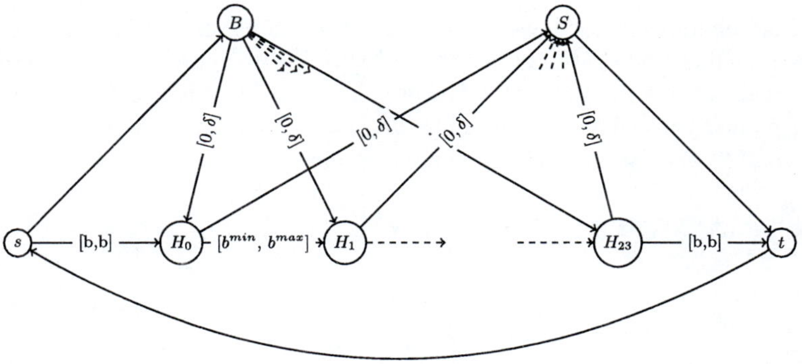

Fig. 22.1 An exemplary DA trading network graph for consecutive hours $H_0, H_1, \ldots, H_{23}$, a battery level b, minimum and maximum battery level $b^{\min}$, $b^{\max}$ and a flexibility δ

and $c(e) = 0$, we obtain a circulation network. Thus, by setting all the balances of nodes to zero, we can find the flow with minimum cost among all possible flows. We add homologous sets $R_i := \{(H_{i-1}, H_i), (H_i, H_{i+1})\}$ for $i = 1, \ldots, 22$ and $R_0 := \{(s, H_0), (H_0, H_1\}, R_{23} := \{(H_{22}, H_{23}), (H_{23}, t)\}$ with deviation Δ. Note that by subdividing the edges which are contained in two homologous sets one can obtain disjoint homologous sets. We define the DA trading strategy da_t for hour t as $\text{da}_t := f(B, H_t) - f(H_t, S)$. See Fig. 22.1 for an example of the constructed network.

Thus, a DA trading strategy can be computed by solving (22.8) on this network graph.

The Almost Equal Minimum Cost Flow Problem

We start with a formal definition of the AEMCFP. Let $G = (V, E)$ denote a graph with a source $s \in V$ and a sink $t \in V$. Furthermore, we are giving a balance for each node by $b : V \to \mathbb{Z}$, a cost c_e for each edge $e \in E$ and lower and upper bounds for each edge by $l, u : E \to \mathbb{N}$. In addition to that, so called *homologous sets* $R_i \subseteq E$ as subset of the edge sets with a corresponding value $\Delta_i \in \mathbb{N}$ are given. The AEMCFP can be formulated as an optimization problem in the flow variables f_e $(e \in E)$ and $f_i, i, \ldots, k$:

$$\textbf{(AEMCFP)} \quad \min \sum_{e\in E} c_e f_e \tag{22.1}$$

$$\text{s.t.} \quad \sum_{e\in\delta^+(v)} f_e - \sum_{e\in\delta^-(v)} f_e = b_v \quad \forall v \in V \tag{22.2}$$

$$l_e \le f_e \le u_e \quad \forall e \in E \tag{22.3}$$

$$f_i \le f_{r_i} \le f_i + \Delta_i \quad \forall r_i \in R_i, \forall R_i \tag{22.4}$$

Problem definitions of different variants of the AEMCFP as well as a complexity analysis can be found in [4]. For the rest of the section we consider only one homologous edge set R with a constant deviation Δ. Note that given a lower bound λ for the homologous edge set, one can find a flow that fulfills the homologous edge constraint (22.4) by solving the following problem:

$$\textbf{(AEMCFP}(\lambda)\textbf{)} \qquad \min \sum_{e \in E} c_e f_e \tag{22.5}$$
$$\text{s.t.} \quad \sum_{e \in \delta^+(v)} f_e - \sum_{e \in \delta^-(v)} f_e = b_v \quad \forall v \in V$$
$$l_e \leq f_e \leq u_e \quad \forall e \in E$$
$$\lambda \leq f_e \leq \lambda + \Delta \quad \forall e \in R$$

We obtain the dual problem of the problem (22.5) as

$$\textbf{(D-AEMCFP}(\lambda)\textbf{)} \qquad \max \quad b^T \pi - u^T \mu + \lambda^T \alpha - (\lambda + \Delta)^T \beta \tag{22.6}$$
$$\text{s.t.} \quad A^T \pi - \mu + \alpha - \beta \leq c. \tag{22.7}$$

Thus, this problem is equivalent to finding the maximum value of (22.6) over all nodes of the polyhedron defined by (22.7). Since polyhedra are convex and taking the maximum over convex functions is a convex function, the function (22.6) is convex in λ. Thus, the AEMCFP can be solved by computing the minimum of the function (22.6).

Theorem 1 *The function*

$$F(\lambda) := \max_{(\pi,\mu,\alpha,\beta) \in P} b^T \pi - u^T \mu + \lambda^T \alpha - (\lambda + \Delta)^T \beta$$

is a piecewise convex function and the AEMCFP can be solved by solving

$$\min \left\{ F(\lambda) : 0 \leq \lambda \leq \min_{r \in R_\Delta} u_r \right\}. \tag{22.8}$$

In the next section we will describe how to solve (22.8).

A Parametric Algorithm Approach

In order to obtain a strongly polynomial algorithm, we use the parametric search technique by Megiddo (e.g. [7]). The idea is to run an algorithm on a parametric instance, i.e. an instance where some input values are handled as parameters. Each time the algorithm has to solve a comparison on a term containing a parametric value, a subroutine has to resolve it. Given the optimal value λ, (22.8) can be solved by a minimum cost flow computation. We take λ as a parametric input value and denote the corresponding parametric network by G_λ. As base algorithm we take the *minimum mean cycle canceling algorithm*[2] which runs in $\mathcal{O}(n^2m^3\log(n))$ time on a graph with n nodes, m edges and arbitrary real valued edge costs (cf. [2]). A minimum mean cycle can be found in $\mathcal{O}(n^2)$ (cf. [5]). In a straight forward implementation, the minimum mean cycle algorithm runs in $\mathcal{O}(n^2)$ iterations, and in each of these a comparison between two edge capacities has to be computed. Since we run the algorithm on a parametric instance, each of the comparisons has to be resolved, i.e. we have to decide for a given test value $\lambda \lesseqgtr \lambda^*$. We consider the following problems.

$$
\begin{aligned}
\mathbf{P}(\lambda)\quad & \min\ c^Tx \\
& \text{s.t. } Nx(\lambda) = b \\
& 0 \le x(\lambda) \le u \\
& \lambda \le x(\lambda) \le \lambda + \Delta
\end{aligned}
\qquad
\begin{aligned}
\mathbf{P}(\lambda+\epsilon)\quad & \min\ c^Tx(\lambda+\epsilon) \\
& \text{s.t. } Nx(\lambda+\epsilon) = b \\
& 0 \le x(\lambda+\epsilon) \le u \\
& \lambda+\epsilon \le x(\lambda+\epsilon) \le \lambda+\epsilon+\Delta
\end{aligned}
$$

where N denotes the incidence matrix and we deviate λ by ϵ. Let y denote the change of the flow value, i.e. $y := x(\lambda+\epsilon) - x(\lambda)$. Thus, for a given $x(\lambda)$, y can be computed by solving

$$
\begin{aligned}
\mathbf{C}(x(\lambda))\qquad \min\ & c^Ty \\
\text{s.t. } & Ny = 0 \\
& 0 \le x(\lambda) + y \le u \\
& \lambda+\epsilon \le x(\lambda) + y \le \lambda+\epsilon+\Delta.
\end{aligned}
$$

This problem is a circulation problem in the residual network $G_{x(\lambda)}$ and therefore can be solved by a minimum cost computation. With this, the problem of deciding whether $\lambda \lesseqgtr \lambda^*$ can be solved in $\mathcal{O}(T_{\text{MC}}(n,m))$, where $T_{\text{MC}}(n,m)$ denotes the time needed for computing a minimum cost flow in a graph with n nodes and m edges. By keeping already computed comparison results, the total number of comparison calls is $\mathcal{O}(\log(m))$, and, thus, the parametric mean cycle canceling algorithm runs in $\mathcal{O}(n^2 + \log(m)\cdot T_{\text{MC}}(n,m))$ time. We summarize this in the following theorem.

Theorem 2 *Given a graph G with n nodes and m edges,* (22.8) *can be computed in* $\mathcal{O}(nm^2\log(n)\cdot\{n^2+\log(m)\cdot T_{MC}(n,m)\})$. □

Note that the algorithm described above can be used iteratively to solve the AEM-CFP for multiple disjoint homologous sets. While we aimed for a strongly polynomial algorithm one can use a binary search on the λ value in order to obtain a weakly polynomial algorithm with running time $\mathcal{O}(\log(U)\cdot T_{\text{MC}}(n,m))$.

[2] Note that there might be other, better suited algorithms for this. We will elaborate this in the future.

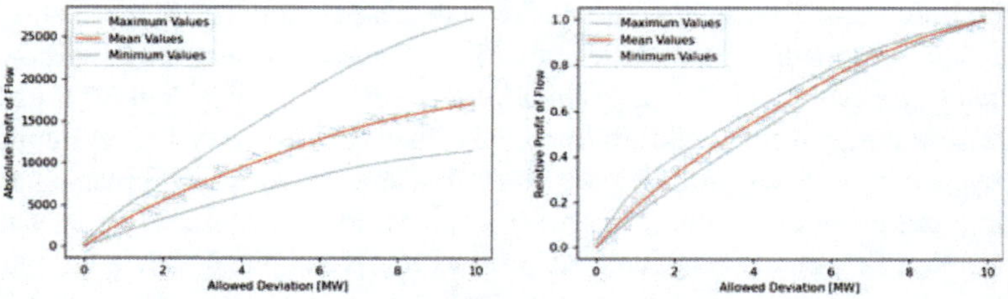

Fig. 22.2 Evaluation on absolute (left) and relative (right) flow values for different allowed deviation values. On the x-axis the allowed deviation is represented, whereas on the y-axis the absolute (left) respectively relative (right) costs are represented

Numerical Results

In this section we present a case study based on the German DA market where we aim to market flexibility from a virtual battery.[3] We consider the trading strategy as a bicriteria problem. The first objective is given by the allowed deviation on consecutive hours, whereas the second objective is given by the forecasted[4] profit. We run our analysis on days between 01/01/2020 and 30/06/2020. For a given deviation, we defined a network model and iterate over different deviation values. The battery level at any point has to lie in the interval [0, 100 MWh]. The initial battery level is given as 50 MWh and the total flexibility at any point is 10 MW. See Fig. 22.2 for an evaluation.

For the relative values as shown in the right chart, we observe that while the profit function for mean deviation values can be assumed to be linear, a difference from this behaviour is seen for deviations in the interval [0, 2] MW and [7, 10] MW where the maximal relative profits increase from 0 to 0.4 and only from 0.8 to 1 respectively. In this intervals, trade-offs between an allowed deviation and obtained profit allow a relatively large gain in profit with a relatively small additional deviation. With this, one can find trading strategies that incorporate the allowed deviation on consecutive points in time as second objective. For absolute values, we obtain similar results as shown in the left chart of Fig. 22.2.

Conclusion

In this paper we proposed a flow based model for DA trading strategies. By using the parametric search method of Megiddo, we obtained a strongly polynomial algorithm to solve the AEMCFP with one homologous edge set. The worst case running time

[3] See [6] for a description of a similar setting.

[4] For the forecasted prices of the DA prices we use a neural network model which was trained on 5 years of electricity prices as well as solar and wind forecasts. For more details we refer to [8] for an in-depth explanation of the neural network model.

of the algorithm could be further improved by using other minimum cost flow algorithms and by using refinements of the parametric search approach e.g. as introduced in [10]. This is left open for future research. We gave a construction of a suitable network graph in which the trading strategy can be obtained by solving a related flow problem. By applying the *Almost Equal Minimum Cost Flow Problem* to this graph, we were able to incorporate additional properties regarding the charging (discharging) difference between consecutive hours into the flow model. We presented a case study on the German DA market in a bicriteria fashion where we considered both profits and allowed difference as separate objectives.

References

1. Ahuja, R. K., Magnanti, T. L., & Orlin, J. B. (1988). *Network flows*. Alfred P. Sloan School of Management.
2. Goldberg, A. V., & Tarjan, R. E. (1989). Finding minimum-cost circulations by canceling negative cycles. *Journal of the ACM (JACM), 36*(4), 873–886.
3. Haese, R., Heller, T., & Krumke, S. O. (2020). Algorithms and complexity for the almost equal maximum flow problem. In *Operations Research Proceedings 2019* (pp. 323–329). Springer.
4. Heller, T. (2021). *Virtual prosumer consortia: A game theoretic optimization approach*. Ph.D. thesis, TU Kaiserslautern.
5. Karp, R. M. (1978). A characterization of the minimum cycle mean in a digraph. *Discrete Mathematics, 23*(3), 309–311.
6. Leithäuser, N., Heller, T., Finhold, E., & Schirra, F. (2022). Optimal trading of flexible power consumption on the day-ahead market. In *International Conference on Operations Research* (pp. 175–181). Springer.
7. Megiddo, N. (1978). Combinatorial optimization with rational objective functions. In *Proceedings of the Tenth Annual ACM Symposium on Theory of Computing* (pp. 1–12). ACM.
8. Ramentol, E., Schirra, F., & Wagner, A. (2020). Short- and long-term forecasting of electricity prices using embedding of calendar information in neural networks.
9. Schäfer, P., Westerholt, H. G., Schweidtmann, A. M., Ilieva, S., & Mitsos, A. (2019). Model-based bidding strategies on the primary balancing market for energy-intense processes. *Computers and Chemical Engineering, 120*, 4–14.
10. Toledo, S. (1993). Maximizing non-linear concave functions in fixed dimension. In *Complexity in numerical optimization* (pp. 429–447). World Scientific.

Chapter 23
A Real Options Analysis of the Siting and Cost-Efficient Layout of Charging Infrastructure for Fuel Cell and Battery Electric Vehicles

Lars Wohlan and Reinhard Madlener

Abstract The German federal government has set up the goal of 10 million electric cars on German roads by 2030 and is aiming for the complete electrification of road traffic by 2050. Currently, the focus is particularly on the development of charging infrastructure for battery electric vehicles (BEVs), with 1 million publicly accessible charging stations to be built by 2030. However, the expansion of hydrogen infrastructure is also being supported with large subsidies. Although there are numerous studies on either the cost of charging infrastructure for BEVs or fuel cell electric vehicles (FCEVs), there are few comparative research results available so far. In this study, therefore, a model for the spatial distribution of charging infrastructure for BEVs and FCEVs is first developed for the district of Steinburg in the German federal state of Schleswig–Holstein. In a second step, the results and corresponding economic costs are compared in a real options analysis for the period 2021 to 2050, considering different charging infrastructure expansion curves. In addition to the hardware infrastructure costs, operations and maintenance and electricity or hydrogen costs for the charging infrastructure are explicitly taken into account.

Keywords Charging infrastructure · Real options analysis · Electric vehicles

L. Wohlan
RWTH Aachen University, Aachen, Germany
e-mail: Lars.Wohlan@rwth-aachen.de

R. Madlener (✉)
Institute for Future Energy Consumer Needs and Behavior (FCN), School of Business and Economics / E.ON Energy Research Center, RWTH Aachen University, Aachen, Germany
e-mail: RMadlener@eonerc.rwth-aachen.de

Department of Industrial Economics and Technology Management, Norwegian University of Science and Technology (NTNU), Trondheim, Norway

O. Grothe et al. (eds.), *Operations Research Proceedings 2022*, Lecture Notes in Operations Research, https://doi.org/10.1007/978-3-031-24907-5_23

Introduction

The German federal government has set a target of having 10 million electric cars on German roads by 2030 and has decided to implement an extensive support program. Under this master plan, 50,000 public charging points for BEVs will be available by the end of 2022 and a total of one million by 2030 [1]. In September 2022, Germany had overachieved the 2020 target with 68,000 available public charging points [2]. In addition, the German federal government has launched the National Hydrogen and Fuel Cell Technology Innovation Program to nurture the market for FCEVs. The refueling process for hydrogen is similar to that for gasoline and is characterized by high throughputs at the pump, but it also features high costs for the construction of a single refueling station. In contrast, the charging process for battery vehicles takes longer. Additionally, the cost of a single charging point for BEVs is lower. To date, there is insufficient research regarding which technology a policy-maker should promote from an economic point of view. Technical drawbacks of FCEVs, such as high efficiency losses in the energy conversion, are typically not part of such analysis. Large efficiency losses imply a higher demand for renewable electricity, implying larger installed capacities than for BEV. Based on these caveats, some experts have rated FCEVs as non-competitive (e.g., [3]).

To solve the above-mentioned policy-maker's problem, the two competing technologies, BEV and FCEV charging infrastructure, are evaluated by means of a real options analysis with a focus on a model region in the district of Steinburg in the state of Schleswig–Holstein. In addition to the hardware infrastructure costs, operation and maintenance and electricity or hydrogen fuel costs for the charging infrastructure are explicitly taken into account. Cost projections were considered as well, but note that these input parameters need to be updated regularly in order to reflect the dynamic development. A decision tree is defined, considering learning/experience curves, economic data, capacity expansion pathways, and a stochastic cost function that is based on a β-PERT non-normal distribution, where the upper and lower quartiles are taken for the up- and down-factors in the binomial tree used for the real options analysis. The results of this decision tree can support policy-makers in deciding which alternative vehicle technology they should support, and to systematically re-evaluate their decisions over time.

In the context of this evaluation, two research questions are addressed:

1. What is an optimal spatial distribution of the public charging infrastructure for BEVs or FCEVs in the model region for different penetration rates of e-mobility?
2. What economic costs and options for action result from the spatial distribution of the charging infrastructures for BEVs and FCEVs from the policy-maker's perspective, considering the predefined policy goals for e-mobility?

Spatial Distribution Modeling of Charging Infrastructure

The developed spatial distribution model calculates the charging infrastructure demand for a node i such that the charging demand is covered within a maximum distance to the station and the costs for the charging infrastructure operator are minimized. The maximum distance is based on the reachability of a conventional fuel station and the range of the vehicle type (BEV or FCEV). Node i represents the center of one of the 36 km^2 of clustered squares in the district of Steinburg. To classify the results for the charging infrastructure for BEVs and FCEVs, the clustering of the district is shown in Fig. 23.1.

The squares were divided into 4 categories, applying the single linkage method, a clustering algorithm based on the similarity coefficient method [4]: 'country', 'small municipality', 'large municipality', and 'city'. As an example, the results from the developed model are shown below for four penetration shares: 5, 10, 50, and 100%. These penetration shares imply that the corresponding share of all vehicles in the Steinburg district is either a BEV (Fig. 23.2) or an FCEV (Fig. 23.3). When looking at the results for BEVs, it is noticeable that the expansion is approximately linear. Hence, a doubling of the penetration share is accompanied by a doubling of charging stations. In addition, the highest charging infrastructure demand is in the squares of the "city" category. On the one hand, this is due to the high population density in these squares and, on the other hand, to the low proportion of home charging. The home charging share in this model is assumed to be 90% in the country and 45% in the city category. It should be noted, however, that a higher population density does not always go hand in hand with a higher need for charging infrastructure in the same square. This is mainly due to the fact that a conveniently located square can ideally cover the needs of several neighborhoods.

The spatial distribution results for public charging infrastructure for FCEVs are very different from those obtained for the charging infrastructure needs of BEVs. First, a significantly lower number of dispensers is required than charging points. On the one hand, this is due to the shorter charging time and the associated larger maximum dispensing quantity per hour. On the other hand, it is due to the higher usage rates for FCEV charging infrastructure, since a dispenser is generally only

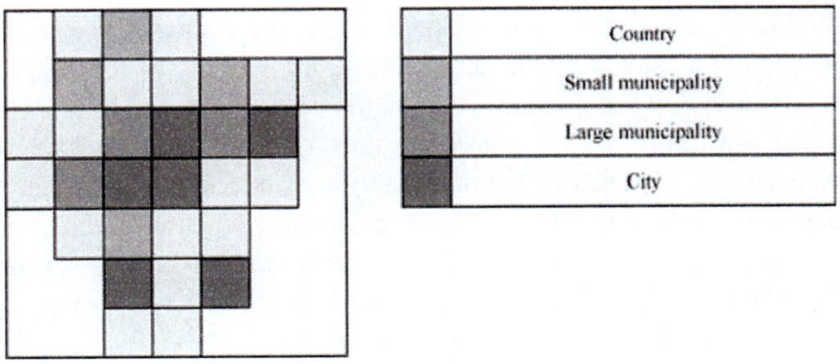

Fig. 23.1 Clustered topology of the district of Steinburg in Schleswig Holstein

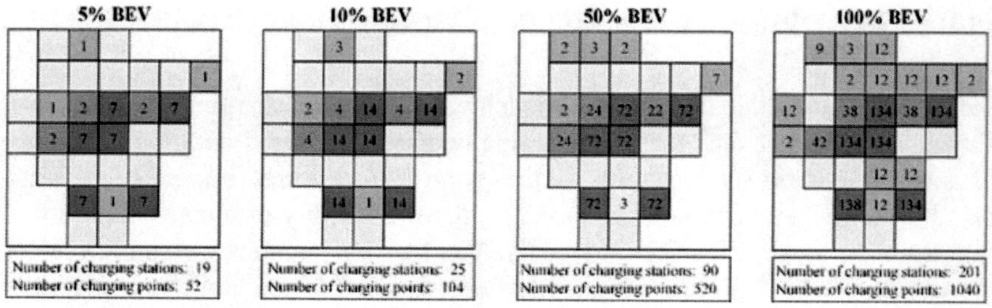

Fig. 23.2 Spatial distribution results for public charging at various penetration shares

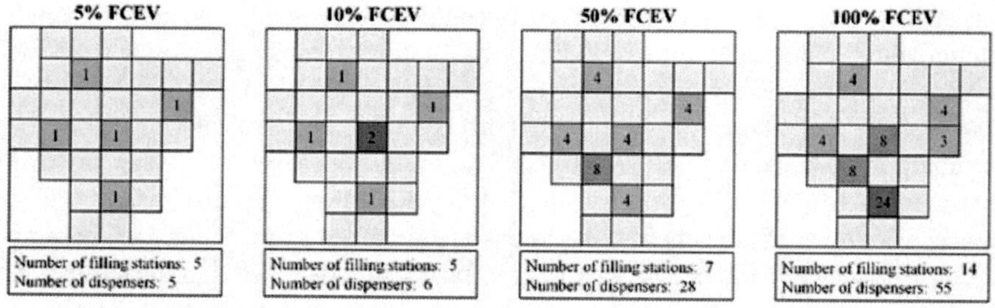

Fig. 23.3 Spatial distribution results for dispenser for various penetration shares

occupied when people are fueling their vehicles. A BEV charging station may be occupied by a parked car whose charging process was already completed several hours ago.

In addition, the development of the required charging infrastructure is not linear. If the penetration share doubles from 5 to 10%, only one additional fuel pump is needed at one of the existing charging stations. This shows that at the 5% penetration level, the 5 charging stations are not fully utilized. However, they are necessary to meet the charging demand of users within the maximum distance. Thus, there is a high demand for charging infrastructure for FCEVs in this model, even at low penetration shares, to meet the user needs based on the accessibility to conventional fueling stations as a yardstick.

Although the location of only one hydrogen refueling station is in a square of the 'city' category, overall, the locations are found in the vicinity of the 'city' squares. This can be explained by the greater accepted maximum distance by the user to the nearest charging option for FCEVs. A FCEV has a higher range and, therefore the user accepts a greater distance to the next charging option. For example, the node with 24 fuel pumps in the 100% FCEV diagram can serve the two squares to its right and left in the city category. In total, there are 14 hydrogen fueling stations in the 100% FCEV scenario. This corresponds to about half of the 30 conventional fueling stations currently available in the Steinburg district [5].

A Real Options Analysis of BEV/FCEV Charging Infrastructures

Based on a net present value (NPV) calculation, a decision tree for the real options analysis was calculated (cf. Fig. 23.4) for three defined scenarios based on a 50% target achievement by 2030, a 100% target achievement by 2030, and a 150% target achievement by 2030 (16 branches à 3 scenarios = 48 branches). On the basis of this decision tree the value of the real option to switch is calculated. It is assumed that depending on the stochastic price and cost developments the infrastructure technology built from 2021–2030 will either be retained from 2031–2050 or abandoned in favor of the alternative infrastructure option (i.e. not renewed at the end of the expected lifetime).

A favorable "cost low" and an unfavorable "cost high" development were calculated using Monte Carlo simulation (for details cf. [6]). Figure 23.5 shows that the favorable cost development for charging infrastructure for FCEVs is always higher than the unfavorable one for charging infrastructure for BEVs. While for the period 2021 to 2025 the favorable FCEV yearly cost is 136% larger than the unfavorable BEV yearly cost, this difference is only 30% for the period 2046 to 2050.

The total cost of BEV charging infrastructure considering annual cost and the costing interest rate of 0.7% is between €508 million and €598 million for the period 2021 to 2050 in the 100% scenario. The total costs for FCEV charging infrastructures, on the other hand, amount to between €777 million and €874 million, depending on the assumed cost development. Note that due to the strict cost dominance of charging infrastructure for BEVs, exercising the option to switch the technology focus in 2030

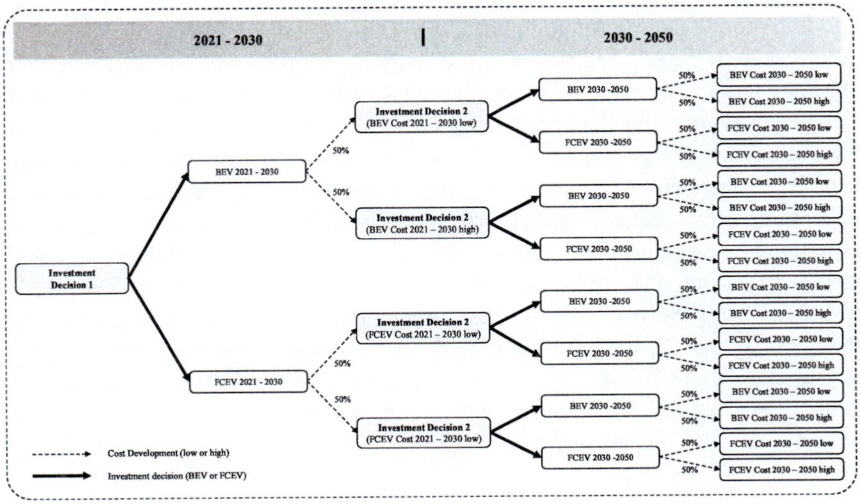

Fig. 23.4 Binomial decision tree for the real options analysis, 2021–2030 and 2030–2050

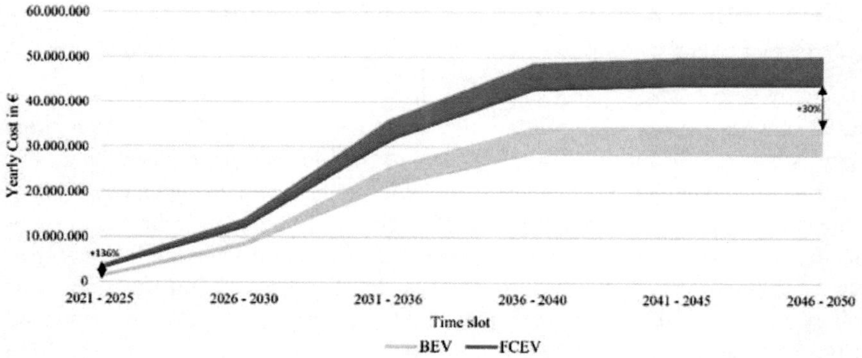

Fig. 23.5 Corridors for possible future cost development from ROA for the 100% scenario

from BEV to FCEV infrastructure is not a realistic one, as even the worst-case BEV infrastructure cost scenario is preferable to the best-case FCEV scenario.

For using the developed real options model and calculating the value of the option to switch, the input parameter values are varied. As the costs are dominated by electricity and hydrogen costs, these two cost factors are crucial. Thus, the option to switch would be exercised if the electricity price from 2031 to 2050 were between 64 and 70% higher than the specified input parameters. This would correspond to an electricity price of between €248 and €257 per MWh. Similarly, a significant decrease in the hydrogen price would make it worthwhile to exercise the option to switch. Thus, depending on the scenario, the hydrogen price would have to be between 36 and 40% lower from 2030 onwards, so that exercising the option would create value. This would correspond to a hydrogen price of between €2.80 and €3.00 including production, transportation and distribution from 2030 onwards.

As shown in Fig. 23.6 the costs of the charging infrastructure are dominated by electricity costs and hydrogen costs, respectively, and the shares of electricity and hydrogen costs both increase over time. Due to the high-cost shares of electricity and hydrogen, respectively, the model results react very sensitively to adjustments of the input parameters for electricity and hydrogen prices. In contrast, a change in the charging infrastructure requirements has only a minor impact on the results.

In summary, an expansion of charging infrastructure for BEVs has lower total costs than the expansion of charging infrastructure for FCEVs. However, the reason for this are not the investment costs for the charging infrastructure per se, but primarily the electricity and hydrogen prices. While at high penetration rates the charging infrastructure for FCEVs is even cheaper than the charging infrastructure for BEVs, the total costs are largely driven up by the hydrogen price.

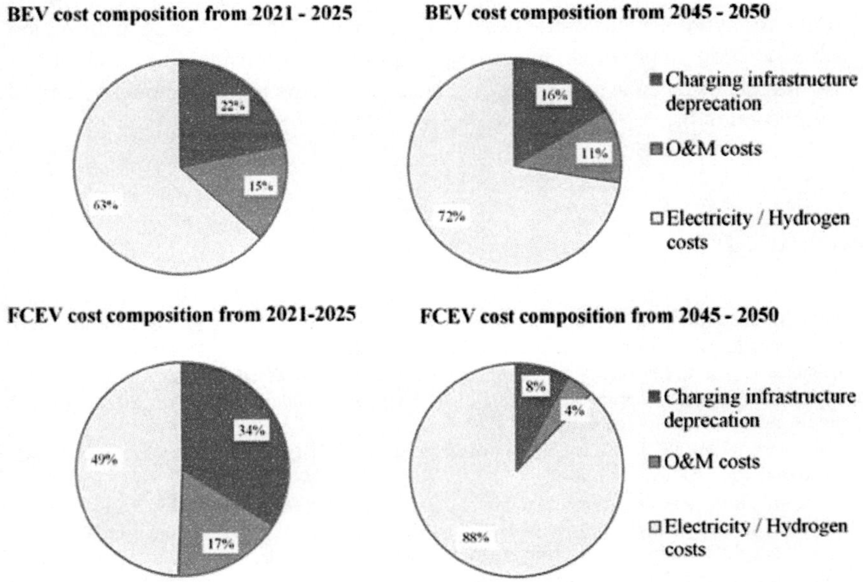

Fig. 23.6 Development of cost shares of BEV (upper plots) and FCEV (lower plots) charging infrastructure

Conclusion

The discussion of results and insights gained from the developed spatial distribution model and real options analysis are reported in much more detail in [6], and summarized here only briefly by the following three key statements:

1. *Influence of penetration rates.* The model for the spatial distribution of the charging infrastructure shows that high initial investments in FCEV charging infrastructure are necessary to sufficiently cover user needs even at low FCEV penetration rates. When considering public charging infrastructure, BEV charging infrastructure turns out to be always less expensive. Including the costs of private home charging points, however, the pure infrastructure costs for FCEVs are lower at high penetration rates.
2. *Electricity and hydrogen as cost drivers.* The analysis of the cost composition reveals that with increasing penetration rates, electricity and hydrogen costs dominate the total costs for the charging infrastructure. The reasons for this can be found in the learning curve and the associated decreasing costs for the hardware. The cost composition shows that the research focus when assessing the total charging infrastructure costs should be especially on electricity and hydrogen cost developments.
3. *Necessary cost development of hydrogen.* Finally, the analysis of the additional scenario provides evidence which factors can lead to the use of the option to switch being a real decision option. For this to happen, the average production

costs for hydrogen must fall into the range from €1.2 to €1.7 per kg in the period 2031 to 2050. In principle, selected optimistic studies consider this development to be realistic but, due to the long time horizon, such projections should be treated with caution [4].

References

1. Bundesministerium für Umwelt, Naturschutz, nukleare Sicherheit und Verbraucherschutz Homepage https://www.bmuv.de. Accessed July 21, 2021.
2. Bundesnetzagentur https://www.bundesnetzagentur.de/DE. Accessed October 27, 2022.
3. Michael Liebreich/Liebreich Associates: Clean Hydrogen Ladder, Version 4.1, 2021. Concept credit: Adrian Hiel, Energy Cities.
4. Anderberg, M. R. (1973). The broad view of cluster analysis. In *Cluster Analysis for Applications*, Academic Press/Elsevier, pp. 1–9.
5. Steinburg Homepage https://www.steinburg.de/startseite. Accessed August 13, 2021.
6. Wohlan L., Madlener R., Specht M. (2021). A real options analysis of the siting and cost-efficient layout of charging infrastructure for fuel cell and battery electric vehicles, FCN Working Paper No. 15/2021, Institute for Future Energy Consumer Needs and Behavior, RWTH Aachen University, December 2021.

Chapter 24
A Tabu Search Approach to the Short-Term Operational Planning of Power Systems

Ionela Knospe, Roman Stainko, Anna Gattinger, Michael Bögl, Katharina Rafetseder, and Dominik Falkner

Abstract The accelerating transition towards clean energy is raising the need to consider electric power systems with an increased integration of renewable energy sources, energy storage systems and flexible loads, in addition to the classical programmable generators and electricity demand. Within this setting, we consider the optimal day-ahead operational planning of power systems with a quarter-hourly time resolution and with the objective of minimizing its total operating cost. The electricity demand and photovoltaic production data is provided by forecast models that are based on historical and real-world data sets. For solving this optimization problem we use a hierarchical approach based on tabu search.

Keywords Metaheuristics · Prescriptive analytics · Optimal power flow

Introduction

Power system operational planning aims to provide a reliable and efficient supply of electricity at any time and deals with formulations of the Unit Commitment Problem (UCP) and Optimal Power Flow (OPF). An overview of UCP formulations, both deterministic and stochastic, can be found in [1] and for a study of OPF methods we refer to [2], both references containing also a comprehensive literature review. We study here the day-ahead operational planning of power systems with a 15 min time resolution, by considering the forecasts for the power consumption and renewable generation and the day-ahead market price.

Overview of the components of the power system. We consider here power systems with the following components and technical details:

- Programmable generators (G): minimum and maximum capacity, minimum up time and down time, ramp up and ramp down rates, maximum reserve limit, generation costs, startup costs, shutdown costs and reserve costs;

I. Knospe (✉) · R. Stainko · A. Gattinger · M. Bögl · K. Rafetseder · D. Falkner
RISC Software GmbH, Softwarepark 32a, 4232 Hagenberg, Austria
e-mail: ionela.knospe@risc-software.at

O. Grothe et al. (eds.), *Operations Research Proceedings 2022*, Lecture Notes in Operations Research, https://doi.org/10.1007/978-3-031-24907-5_24

- Energy Storage Systems (S): maximum capacity, minimum and maximum charging and discharging power, minimum charging and discharging time, charging and discharging efficiency factors, initial and final state of charge (SOC);
- Photovoltaics (PV): installed capacity per site;
- Loads (D): power consumption;
- Transmission lines (L): reactance, maximum flow, and the tap ratio if the branch is a transformer; they are modelled here by means of DC power flows.

The load and photovoltaics generation is provided by forecast models, whose data basis consists of historical data and real-world datasets. The forecast models are based on non-linear regression like Gradient Boosting [3] utilizing state of the art clear sky models and additional features from weather forecast providers.

Energy storage arbitrage is a technique where power is bought and stored during off-peak hours, when the grid prices are cheapest, and is then used during peak hours, when grid electricity prices are highest. The planning strategy used here for all energy storage units in the system is to achieve arbitrage. Furthermore, we assume that they have at most one charging/discharging cycle per day and that their initial and final state of charge is 0.

In addition to the above components, two parameters are included in our model, which address and control the self-sufficiency of the system, see also [4]:

- Seamless index (SI): indicates how much from the total load can be covered in a self-sufficient manner over the entire planning horizon. SI takes values in the interval $[0, 1]$, with 1 meaning complete self-sufficiency from the external grid and respectively 0, complete dependency to it, see (24.17) below for its definition.
- Reserve factor (RF): enables capacity reserve allocation for the programmable generators in each time step of the planning horizon, in order for the system to be able to manage the forecast errors for load and photovoltaic generation. RF takes values in the interval $[0, 1)$, it influences the power output and creates an additional component in the total operational costs, see (24.18)–(24.20) below.

Formulation of the model. For solving the operational planning of the power system we determine the commitment status of all $i \in G$ and $s \in S$ for all time steps $t \in T$, i.e., the values of the following binary and real variables:

$z_{i,t} \in \{0, 1\}$ 1, if i is committed at time step t and 0, otherwise
$u_{i,t} \in \{0, 1\}$ 1, if i is turned on at time step t and 0, otherwise
$v_{i,t} \in \{0, 1\}$ 1, if i is turned off at time step t and 0, otherwise
$z^{ch}_{s,t} \in \{0, 1\}$ 1, if s is charging at time step t and 0, otherwise
$z^{dch}_{s,t} \in \{0, 1\}$ 1, if s is discharging at time step t and 0, otherwise
$P_{i,t} \geq 0$ the real power output of i at time step t
$R_{i,t} \geq 0$ the real power reserve allocation of i at time step t
$P_{ext,t}$ the real power transfer from the external grid at time step t
$P^{ch}_{s,t} \geq 0$ the real power dispatch when charging of s at time step t
$P^{dch}_{s,t} \leq 0$ the real power dispatch when discharging of s at time step t.

We mention that $P_{ext,t}$ is ≤ 0 at export and ≥ 0 at import. Moreover, results are also the optimal power flows on all branches in the power system. With the above notations, we want to minimize the total operational costs of the system:

$$f = \sum_{t \in T} \left[\sum_{i \in G} (a_i P_{i,t} \Delta t + SU_i u_{i,t} + SD_i v_{i,t}) + \rho_t P_{ext,t} \Delta t + \sum_{i \in G} c_i^{res} R_{i,t} \Delta t \right]$$

where a_i are the generating costs, c_i^{res} the reserve costs, SU_i and SD_i are the start up and the shut down costs of the generator $i \in G$, ρ_t is the hourly market price, Δt is the time resolution, such that the technical constraints of the components and of the system are satisfied, see (24.2)–(24.6) and (24.8)–(24.20) below.

The Solution Approach

Meta-heuristics have been widely used in solving complex combinatorial optimizations problems, because they can provide acceptable solutions in reasonable computational time and are thus good substitutes for exact algorithms, see [5]. In recent years there has been an increased research interest in integrating machine learning into meta-heuristics, in different parts like algorithm selection, initialization, evolution or parameter setting, see e.g. [6] and the references therein for a comprehensive review.

Our solution approach to the problem described in Section "Introduction", called the hierarchical approach, is based on tabu search with adaptive neighborhood selector, which learns the success rate of the neighborhood operators and applies it in subsequent iterations. As its name suggests, the solution approach is a hierarchical one, the generation of one solution involving three steps detailed in Sections "Assignment of Commitment State", "Energy Storage System Dispatcher" and "DC-OPF" below, see also [4]. The subproblems in the second and third step are linear programs, which we solve using the open-source LP solver CLP through Google's optimization suite OR-Tools. The tabu search uses neighborhood operators tailored to the current problem formulation. Similarly to the ideas in [7], the operators are used in the first iterations without any special selection, in order to collect their initial experience. After the "ramp-up phase", the operators are applied according to their success rate, measured by the decrease of the objective function value. The credit assignment has a fading memory, the credits assigned decreasing over time. According to the nomenclature in [6], this approach can be seen as a hybrid approach combining score-based, average and extreme value-based credit assignments.

This hierarchical approach provides acceptable solutions in reasonable computational time, and although not necessarily optimal, they constitute a valuable alternative to the equivalent all-in-one formulation as mixed integer linear program, named here the integrated approach. The hierarchical approach was proposed with the intention that the underlying software becomes open-source, enabling thus other learning components to be integrated. The integrated approach with the use an open-source MIP solver turned out to be unsuccessful in most test instances for larger power systems, see also Section "Assessment of the Solution Approach". Behind the proposed hierarchical approach was also the motivation of having a model which can be easily extended with other technical and operational constraints of the power systems, which are not necessarily linear. Thus, in the final step of generating a solution, the AC-OPF problem could be also solved with an external software, in case the more accurate solution of the AC version is needed.

Assignment of Commitment State

In this step, all programmable generators and energy storage systems are assigned a commitment state over the entire planning horizon, more precisely, values are assigned to the binary variables $z_{i,t}$, $u_{i,t}$, $v_{i,t}$, for all $i \in G$ and $z^{ch}_{s,t}$, $z^{dch}_{s,t}$ for all $s \in S$ and $t \in T$, see the formulation of the model in Section "Introduction".

Neighborhood operators change the commitment state of the programmable generators and of the energy storage systems during the tabu search, e.g., they introduce a commitment block for $i \in G$ or shift it, extend or shorten the charging or discharging cycle of $s \in S$. All generated assignments take into consideration the minimum up time and down time of the programmable generators, the minimum and maximum charging/discharging times of the energy storage systems, as well as the fact that their charging interval precedes the discharging interval.

Energy Storage System Dispatcher

For each $s \in S$, we denote by $SOC^{\max}_s$ its maximum capacity, by $P^{ch,\min}_s$ and $P^{ch,\max}_s$ the minimum and maximum charging power, by $P^{dch,\min}_s$, $P^{dch,\max}_s$ the minimum and maximum discharging power and finally by η^{ch}_s, $\eta^{dch}_s \in (0, 1]$ the energy storage system's charging and discharging efficiency factors. In addition, we define the following real decision variables: $P_{s,t}$, $\forall t \in T$ for the power dispatch of s, which is negative (charging), positive (discharging), or zero otherwise, and $SOC_{s,t}$, $\forall t \in T$ for the state of charge of s at each time step t for all $s \in S$. The energy storage system dispatcher solves for each $s \in S$ the following linear optimization problem:

$$\text{Maximize } f(s) = \sum_{t \in T} \left[\rho_t P_{s,t} z^{dch}_{s,t} \Delta t - \rho_t P_{s,t} z^{ch}_{s,t} \Delta t\right] \tag{24.1}$$

subject to

$$P_{s,t} \leq P_s^{dch,\max} z_{s,t}^{dch} - P_s^{ch,\min} z_{s,t}^{ch}, \quad \forall t \in T \tag{24.2}$$

$$P_{s,t} \geq P_s^{dch,\min} z_{s,t}^{dch} - P_s^{ch,\max} z_{s,t}^{ch}, \quad \forall t \in T \tag{24.3}$$

$$SOC_{s,t} = SOC_{s,t-1} - \eta_s^{ch} P_{s,t} z_{s,t}^{ch} \Delta t - \frac{1}{\eta_s^{dch}} P_{s,t} z_{s,t}^{dch} \Delta t, \quad \forall t \in T \tag{24.4}$$

$$0 \leq SOC_{s,t} \leq SOC_s^{\max}, \quad \forall t \in T \tag{24.5}$$

$$SOC_{s,0} = SOC_{s,T-1} = 0 \tag{24.6}$$

where $z_{s,t}^{ch}$, $z_{s,t}^{dch}$, for all $t \in T$ are fixed inputs from Section "Assignment of Commitment State".

DC-OPF

In the final step of the solution generation, we solve for each time step $t \in T$ a DC optimal power flow problem in order to obtain the real power output $P_{i,t}$ for all $i \in G$, $P_{ext,t}$ and $F_{i,j,t}$, the power flows through the branches. The model of the DC-OPF problem can be summarized as minimization of

$$f_t = \sum_{i \in G} (a_i P_{i,t} \Delta t + SU_i u_{i,t} + SD_i v_{i,t}) + \rho_t P_{ext,t} \Delta t + \sum_{i \in G} c_i^{res} R_{i,t} \Delta t \tag{24.7}$$

subject to

$$P_i^{\min} z_{i,t} \leq P_{i,t} \leq P_i^{\max} z_{i,t}, \quad \forall i \in G \tag{24.8}$$

$$P_{i,t-1} - RD_i \leq P_{i,t} \leq P_{i,t-1} + RU_i, \quad \forall i \in G \tag{24.9}$$

$$\theta_{n_s,t} = 0 \tag{24.10}$$

$$F_{i,j,t} = BaseMVA \times 1/x_{i,j}(\theta_{i,t} - \theta_{j,t}), \quad \forall (i,j) \in L \tag{24.11}$$

$$F_{j,i,t} = -F_{i,j,t}, \quad \forall (i,j) \in L \tag{24.12}$$

$$F_{i,j,t} = 0, \quad \forall i, j \in N, (i,j) \notin L \tag{24.13}$$

$$F_{i,j,t} \leq F_{i,j}^{\max}, F_{j,i,t} \leq F_{j,i}^{\max}, \quad \forall (i,j) \in L \tag{24.14}$$

$$\sum_{i \in G_n} P_{i,t} + \sum_{s \in S_n} P_{s,t} + \sum_{j \in PV_n} P_{j,t} + \delta_{n,n_s} P_{ext,t} \tag{24.15}$$

$$= \sum_{d \in D_n} D_{d,t} + \sum_{(i,n) \in L} F_{n,i,t} + \sum_{(n,i) \in L} F_{n,i,t}, \quad \forall n \in N \tag{24.16}$$

$$-(1 - SI) \sum_{d \in D} D_{d,t} \leq P_{ext,t} \leq (1 - SI) \sum_{d \in D} D_{d,t} \tag{24.17}$$

$$\sum_{i \in G} R_{i,t} \geq RF \sum_{i \in G} P_i^{\max} \tag{24.18}$$

$$R_{i,t} \leq \min(R_i^{\max}, RU_i) z_{i,t}, \quad \forall i \in G \tag{24.19}$$

$$P_{i,t} + R_{i,t} \leq P_i^{\max} z_{i,t}, \quad \forall i \in G \tag{24.20}$$

In the above linear program $u_{i,t}$, $v_{i,t}$, $i \in G, t \in T$ are obtained in Section "Assignment of Commitment State" and $P_{s,t}$, $s \in S, t \in T$, are results of Section "Energy Storage System Dispatcher". The other notations used are defined as follows: N denotes the set of nodes, n_s is the slack bus or reference bus from which relative voltage angles at all other buses are calculated, $\theta_{i,t}$ is the voltage phase angle at bus i relative to the slack bus, $F_{i,j,t}$ is the power flow on the line from bus i to bus j, δ_{n,n_s} is equal 1 if the n is the slack bus and 0 otherwise, $BaseMVA$ is the base power for the network, $x_{i,j}$ and $F_{i,j}^{\max}$ are the reactance and maximum allowable flow for the line connecting the nodes i and j, $P_i^{\min}$, $P_i^{\max}$ the minimum and maximum real power output of $i \in G$ and RU_i and RD_i the ramp up and ramp down rates. Finally, G_n, S_n, PV_n, D_n are the subsets of G, S, PV and D respectively connected to bus $n \in N$.

Assessment of the Solution Approach

The basis for testing our proposed approach for solution quality and performance is the IEEE 5-, 14-, 30- and 300-Bus Systems. In addition, we use as benchmark for our approach the objective function value of the integrated approach. We emphasise here, that a one-to-one comparison between the approaches is not in all cases possible, as by breaking down the generation of a solution in several steps in the first approach, the optimal solution from the equivalent MIP formulation may not always be obtained.

The hierarchical approach starts from a configuration with all programmable generators committed over the entire planning horizon. Furthermore, the charging and discharging intervals of the energy storage systems are assigned with a sequence of consecutive time steps where in average the market price is lowest for charging and highest for discharging. The number of neighborhood operators used in the tabu search is equal to 8 and the tabu list size is set to 100.

For the first three smaller test power systems, our proposed approach with tabu search is fast and generates good solutions in relatively few iterations. Figure 24.1 shows the result of our solution approach on a test instance based on IEEE 5-Bus System with additionally one energy storage system and one photovoltaic unit. Without giving any details on the underlying test instance data, we just point here to the fact that the solution captures the change around noon in the relation between the generation costs of the programmable generators and the market price, and thus the system imports power from the external grid in the first half of the day and exports power in the second half of the day, as much as the value of the seamless index allows it.

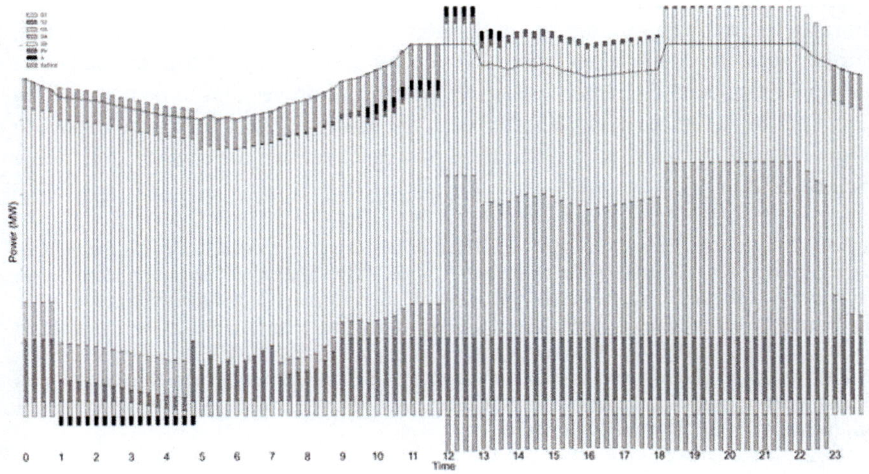

Fig. 24.1 Result of our approach for a test instance based on the IEEE 5-bus system

Table 24.1 Numerical results for the IEEE 300-bus system

	Hierarchical approach	Iterations	Runtime (min)	Integrated approach	Relative gap (%)	Absolute gap
Test 1	137,608,164.69	720	82	137,607,524.78	0.0004	639.91
Test 2	3,944,286.25	400	37	3,931,869.91	0.3147	12,416.34

For the larger IEEE 300-Bus System, the proposed method also generates good solutions for the test instances considered. The benchmark values were obtained in this case with a commercial solver, as the open-source MIP solver SCIP did not converge satisfactorily after approximatively one hour. Table 24.1 contains some numerical results for this power system with additionally 5 energy storage systems and 5 photovoltaic units. The two test instances differ in the generating costs of the programmable generators, load and photovoltaic generation profiles and hourly market price. The financial benefits of the energy storage systems depend on their capacity and market price values, but they are absolutely necessary to shift energy horizontally.

Conclusions

We considered here the day-ahead operational planning of power systems with a basis of power system components and technical details which can be further extended, for example with other renewable energy sources like wind farms or with flexible loads. The proposed approach can be easily adapted to cover these cases and can thus find different applications, like in the planning of microgrids or energy communities. Furthermore, the adaptive neighborhood selector used within the tabu search generates good results on the larger test power system and proves thus to be a promising direction for future research, together with other ways of learning in the evolution of tabu search, or in the initialization and parameter setting, the classification being the one used in [6].

Acknowledgements Parts of the work described in this paper have been funded by the State of Upper Austria, Austria, as part of the program "#upperVISION2030" within the project "Secure Prescriptive Analytics". Project number: Wi-2021-305601/18-Au. Parts of the work described in this paper have been funded by the European Union from the European Regional Development Fund (EFRE) and the State of Upper Austria, Austria, within the project "RESINET—Resilienzsteigerung in Energienetzen". Project number: Wi-2020-702019/19-Cz. Funding content: Structural Funds Operational Program "Investments for Growth and Jobs 2020", Program Priority 7, Investment Priority REACT-EU, Measure 02-REACT.

References

1. Abdou, I., & Tkiouat, M. (2018). Unit commitment problem in electrical power system: A literature review. *International Journal of Electrical and Computer Engineering (IJECE), 8*(3), 1357–1372.
2. Khan, B., & Singh, P. (2017). Optimal power flow techniques under characterization of conventional and renewable energy sources: A comprehensive analysis. *Journal of Engineering, 2017.*
3. Ke, G., Meng, Q., Finley, T., Wang, T., Chen, W., Ma, W., Ye, Q., & Liu, T. Y. (2017). LightGBM: A highly efficient gradient boosting decision tree. *Advances in Neural Information Processing Systems, 30,* 3149–3157.
4. Hosseinnezhad, V., Rafiee, M., Ahmadian, M., & Siano, P. (2016). Optimal day-ahead operational planning of microgrids. *Energy Conversion and Management, 126,* 142–157.
5. Blum, C., & Roli, A. (2003). Metaheuristics in combinatorial optimization: Overview and conceptual comparison. *ACM Computing Surveys (CSUR), 35,* 268–308.
6. Karimi-Mamaghan, M., Mohammadi, M., Meyer, P., Karimi-Mamaghan, A. M., & Talbi, E.-G. (2022). Machine learning at the service of meta-heuristics for solving combinatorial optimization problems: A state-of-the-art. *European Journal of Operational Research, 296,* 393–422.
7. Achterberg, T., Koch, T., & Martin, A. (2005). Branching rules revisited. *Operations Research Letters, 33*(1), 42–54.

Chapter 25
Eco-Energy-Efficient Simultaneous Lot-Sizing and Scheduling: A Tri-criteria Problem

Markus Hilbert, Andreas Dellnitz, and Andreas Kleine

Abstract In the context of lot-sizing and scheduling, minimizing energy consumption is a typical criterion for improving a company's environmental footprint. In the literature, however, it is insufficiently questioned whether minimizing energy consumption actually also minimizes energy-related emissions. In this paper, we show that such a positive one-to-one relationship does not always hold. In fact, when energy prices fluctuate over time, energy costs may additionally conflict with these two goals. To demonstrate this, we develop a three-criteria lot-sizing and scheduling problem and derive the three-dimensional Pareto front using the elastic constraint method.

Keywords Energy-efficiency · Multiobjective optimization · Elastic constraint

Motivation

In energy-efficient production planning, i.e., lot-sizing and/or scheduling, dynamic electricity price tariffs are well-established concepts to strive for a reduction in energy costs. Typically, Time-Of-Use tariffs (TOU) or Real-Time-Pricing (RTP) are considered, and tradeoffs between the cost of electricity and another criterion, such as makespan, are examined; cf. [1]. In this context, minimization of energy-related carbon emissions—so-called Scope 2 emissions (see https://ghgprotocol.org)—is rarely considered. In [2] and [3], for example, the authors minimize energy-related carbon emissions, multiplying electricity consumption by a constant emission factor; then, minimizing electricity consumption and energy-related carbon emissions essentially coincide. A recently conducted study in 2021 commissioned by the German Fed-

M. Hilbert (✉) · A. Kleine
Department of Operations Research, University of Hagen,Universitätsstraße 41, 58097 Hagen, Germany
e-mail: markus-andre.hilbert@fernuni-hagen.de

A. Dellnitz
Chair of Quantitative Methods, Leibniz School of Business, Expo Plaza 11, 30539 Hannover, Germany

O. Grothe et al. (eds.), *Operations Research Proceedings 2022*, Lecture Notes in Operations Research, https://doi.org/10.1007/978-3-031-24907-5_25

eral Network Agency found that time-related carbon emission factors in the German electricity mix fluctuate significantly over the course of the year, in particular during the night; see [4]. A first step towards time-based emission optimization can be found in [5] and [6] under TOU and in [7] under RTP. These articles examined tradeoffs between carbon emissions and other criteria, but not electricity consumption.

To the best of our knowledge, no one has yet studied the tradeoffs between the three criteria—electricity consumption, electricity costs, and carbon emissions—under RTP. To this end, we develop a tri-criteria integrated lot-sizing and scheduling program and compute the corresponding Pareto fronts for selected instances using the elastic constraint method. This method is an extension of the well-known ε-constraint method, but significantly less frequently used to tackle multiobjective problems.

After introducing the theoretical concepts in Section "Theoretical Concepts", Section "Tri-criteria MIP" is devoted to model development and problem description. Section "Results" presents the results of the computational study. Section "Conclusion and Road Ahead" summarizes and provides an outlook.

Theoretical Concepts

Real-Time-Pricing and Fluctuating Emission Factors

Demand response (DR) can be understood as electricity demand that responds to economic signals. If these signals are electricity prices and an end-user, e.g. a company, reduces its electricity consumption in response to high market prices, we call it price-based DR. In the presence of an RTP-based DR program, the electricity price fluctuates on an hourly basis or even shorter, e.g. a quarter hourly basis; cf. [8]. As mentioned in the prior section, emission factors of the electricity mix vary greatly throughout the day, especially during nighttime hours. Moreover, emission factors and spot market electricity prices are not perfectly positive correlated. While corresponding planning approaches that exploit fluctuating emissions [5–7] also consider nighttime hours, we show that emission factors also fluctuate significantly during a weekday, e.g., between 8 AM and 8 PM. To justify this, we consider the electricity mix in the 6th calendar week of 2022 in Germany, corresponding trajectories of quarter hourly electricity prices and emission factors of the electricity mix are illustrated in Fig. 25.1.

Here, quarter hourly continuous intraday average spot market prices are used (Mon–Fri, 8 AM to 8 PM for each day) and the corresponding emission factors are calculated based on the data at https://www.ipcc.ch and www.uba.de. To illustrate the tradeoffs between electricity prices and emission factors, we normalize the data and present the transformed data in Fig. 25.2. A simple grid classification is used to cluster the data. It can be seen that many points are in the white areas, i.e., normalized electricity prices and normalized emission factors are similar: the smaller the price, the smaller the emission factors, and vice versa. However, this is not the case in the

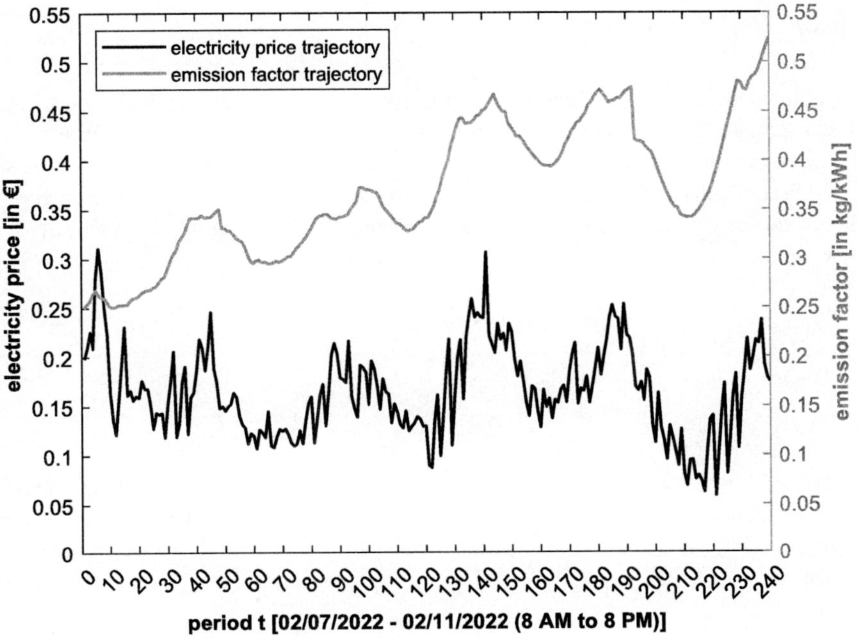

Fig. 25.1 Price and factor trajectory

Fig. 25.2 Normalized prices and factors

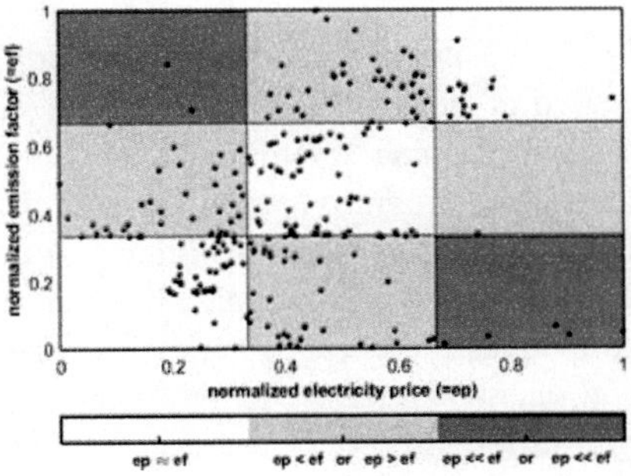

light gray shaded and dark gray shaded areas where many normalized data points are displayed. In these areas, there are significant tradeoffs between electricity prices and emission factors. In the dark gray shaded southeastern area, for example, we see high normalized electricity prices but the corresponding normalized emission factors are low.

Elastic Constraint Method

The main idea of the elastic constraint method is to relax the ε-constraints by allowing these ε-bounds to be violated, but penalizing the violations in the objective function. Since we are dealing with a tri-criteria optimization problem and want to obtain a three-dimensional Pareto front by applying exact rather than heuristic approaches, the elastic constraint method might be less expensive from a computational point of view, which is also emphasized in [9]. The scalarization is as follows:

$$\begin{aligned} \min_{x\in S} \quad & f_j(x) + \sum_{k\neq j} \mu_k \cdot s_k \\ \text{s.t.} \quad & f_k(x) + l_k - s_k = \varepsilon_k \; k \neq j \\ & s_k, l_k \geq 0 \qquad\qquad k \neq j \end{aligned} \tag{25.1}$$

where $\mu_k \geq 0, \forall k \neq j$, are penalty factors, $f_j(x)$, $f_k(x)$ being the jth and kth objective functions, respectively, and S being the feasible region. In general, weakly efficient solutions can be found using this scalarization approach. For further details, see [9].

Tri-criteria MIP

In [1], it is pointed out that multicriteria planning approaches have been little investigated in the context of energy-efficient lot-sizing and/or scheduling considering a parallel machine environment. Therefore, we propose an integrated lot-sizing and scheduling program comprising energy costs, energy-related emissions and energy consumption—without involving setup costs and warehousing. Due to the ecological nature of the criteria—at least for energy consumption and emissions—we refer to the problem as eco-energy-efficient tri-criteria problem. As levers for energy consumption, energy costs and energy-related emissions, we consider different machine states with different power levels and a speed-scaling approach for the machines, which are common concepts in the literature; see [1]. Table 25.1 contains the symbols used in our mixed-integer program (MIP) given by (25.2)–(25.11). For an overview of the setting, see Fig. 25.3.

$$\min \quad \left\{ \mathrm{C} = \sum_{t=1}^{T} c_t^{elec} \cdot s_t^{buy}; \mathrm{E} = \sum_{t=1}^{T} e_t^{elec} \cdot s_t^{buy}; \mathrm{P} = \sum_{t=1}^{T} s_t^{buy} \right\} \tag{25.2}$$

s.t.

$$\sum_{v=1}^{N} \sum_{t=1}^{T} \sum_{m=1}^{M} a_{vj}^{prod} \cdot x_{jmtv} = d_j \quad \forall\, j \in \mathcal{J} \tag{25.3}$$

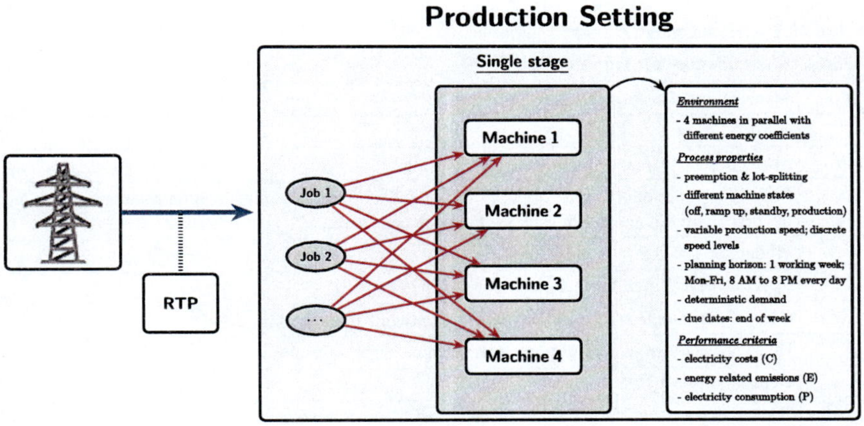

Fig. 25.3 Setting overview

$$\sum_{i=0}^{I} \delta_{imt}^{state} = 1 \quad \forall\, m \in \mathcal{M},\ t \in \mathcal{T} \tag{25.4}$$

$$\widehat{\delta}_{mtv}^{state\_I} - \sum_{j=1}^{J} x_{jmtv} = 0 \quad \forall\, t \in \mathcal{T},\ m \in \mathcal{M},\ v \in \mathcal{N} \tag{25.5}$$

$$\delta_{Imt}^{state} - \sum_{v=1}^{N} \widehat{\delta}_{mtv}^{state\_I} = 0 \quad \forall\, t \in \mathcal{T},\ m \in \mathcal{M} \tag{25.6}$$

$$\delta_{imt}^{state} + \delta_{hm,t+1}^{state} \le 1 + \gamma_{ih}^{tran} \quad \forall\, i, h \in \mathcal{I},\ m \in \mathcal{M},\ t \in \mathcal{T}\backslash\{T\} \tag{25.7}$$

$$\sum_{m=1}^{M} \left(\sum_{v=1}^{N} \widehat{a}_{vm}^{elec\_I} \cdot \widehat{\delta}_{mtv}^{state\_I} + \sum_{i=0}^{I-1} a_{im}^{elec} \cdot \delta_{imt}^{state} \right) = s_t^{buy} \quad \forall\, t \in \mathcal{T} \tag{25.8}$$

$$\delta_{0mt}^{state} + \delta_{1mt}^{state} = 1 \quad \forall\, m \in \mathcal{M},\ t \in \mathcal{T}^{init} \tag{25.9}$$

$$x_{jmtv},\ \delta_{imt}^{state},\ \widehat{\delta}_{mtv}^{state\_I} \in \{0, 1\} \quad \forall\, t \in \mathcal{T},\ j \in \mathcal{J},\ i \in \mathcal{I},\ m \in \mathcal{M},\ v \in \mathcal{N} \tag{25.10}$$

$$s_t^{buy} \ge 0 \quad \forall\, t \in \mathcal{T} \tag{25.11}$$

We minimize electricity costs C, emissions E, and consumption P simultaneously by multiplying the electricity cost rate c_t^{elec}, emission factor e_t^{elec} or the neutral factor '1' for each period t by the energy consumption s_t^{buy} in period t and aggregating accordingly, see (25.2). Using (25.3), we model equality conditions for demand fulfillment d_j for each job j. Equation (25.4) ensure that a machine has only one state and never becomes stateless. (25.5) and (25.6) control the production state in tandem with the speed level, i.e., $\delta_{mtv}^{state\_I}$ equals 1 if any x_{jmtv} equals 1 in (25.5). The latter is only the case when a job j is assigned to machine m at speed level

Table 25.1 Indices, parameters and variables

Indices, parameters and decision variables	
m	Machine $m \in \mathcal{M} = \{1, \ldots, M\}$
j	Job $j \in \mathcal{J} = \{1, \ldots, J\}$
i, h	Machine states $i, h \in \mathcal{I} = \{0, \ldots, I\}$
t	Period (a quarter hour) $t \in \mathcal{T} = \{1, \ldots, T\}$. $t \in \mathcal{T}^{init}$ serves as initialization and corresponds to the first quarter of an hour of each working day considered
ν	Production speed level $\nu \in \mathcal{N} = \{1, \ldots, N\}$
c_t^{elec}	Cost rate [€/kWh] of the electricity purchased in period t
e_t^{elec}	Carbon emission factor [kg/kWh] of the electricity mix purchased in period t
$a_{\nu j}^{prod}$	Hourly production rate of job j on each machine at speed level ν
d_j	Demand of job j
γ_{ih}^{tran}	Transition parameter from state i to h (1 if possible, 0 otherwise)
a_{im}^{elec}	Electricity consumption of machine m in state i without $i = I$
$\widehat{a}_{\nu m}^{elec\_I}$	Electricity consumption of machine m in production state I at speed level ν
$x_{jmt\nu}$	Binary variable equals 1 if job j will be processed on machine m in t at speed level ν, otherwise 0
δ_{imt}^{state}	Binary variable equals 1 if machine m has state i in period t, otherwise 0
$\widehat{\delta}_{mt\nu}^{state\_I}$	Binary variable equals 1 if machine m is in production state I at speed level ν in t, otherwise 0
s_t^{buy}	Nonnegative variable equals the amount of electricity [in kWh] to be purchased in period t
C	Free variable equals the energy costs to be minimized
E	Free variable equals the carbon emissions to be minimized
P	Free variable equals the total power consumption to be minimized

ν in period t. Equalities (25.6) couple the production mode $i = I$ of a machine to one speed level exclusively. Constraints (25.7) control the state transitions of a machine. Here, a machine can either maintain a state $\delta_{imt}^{state} + \delta_{hm,t+1}^{state} = 2$, with $h = i$ and $1 + \gamma_{ih}^{tran} = 2$, or it can change it by choosing $h \neq i$ if the transition from state i to h is feasible. (25.8) balance the electricity consumption and (25.9) serve as initialization. In the first period of each working day, a machine m is either in off mode ($i = 0$) or in ramp up mode ($i = 1$). (25.10)–(25.11) are binary and non-negativity conditions.

Results

We conduct a simulation study over five different demand scenarios ($S1$–$S5$) for four machines and five jobs. The quantities d_j are randomly generated and the total demand of all jobs ranges from $\approx 50-90\%$ of the machine utilization (= cap. ut.) at highest production speed. The calculations are performed via GAMS using CPLEX, applying both the elastic constraint method (= elCM) with $\mu_1 = \mu_2 = 0.1$ and the ε-constraint method (= εCM). Figure 25.4 shows a Pareto front plot for a selected

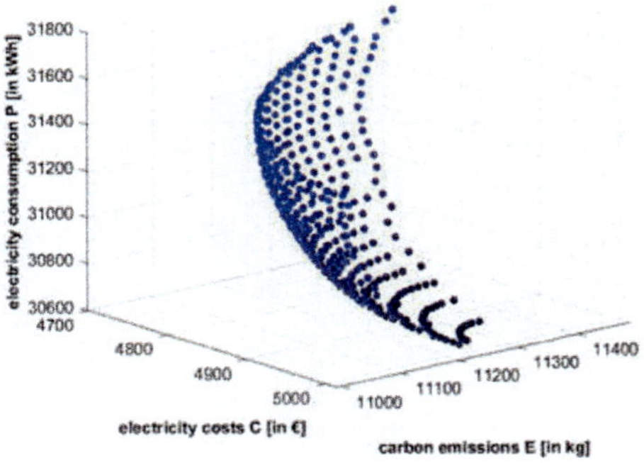

Fig. 25.4 Pareto front representation

scenario (‘S3’ for M = 4 and J = 5). The implications of the results are also representative for the other scenarios.

It can be seen that the three criteria are in conflict with each other, resulting in a Pareto front representation in three-dimensional space. This is also true if only the bi-criteria subproblems of the MIP are considered (the corresponding figures are skipped). In the context of the RTP, electricity costs are not only in conflict with energy-related emissions, but also with electricity consumption. Consequently, the question arises whether electricity costs are an environmental criterion, at least when considering dynamic price tariffs. Furthermore, minimizing both electricity consumption and carbon emissions are not concurrent when market-based emission factors are considered, i.e., these are in conflict with each other. It is therefore questionable whether approaches—that minimize emissions via the carbon cost of energy consumption using a constant emission factor—really minimize energy-related emissions. Of course, one could add a more classical criterion such as makespan. However, the interactions shown between electricity costs, energy-related emissions, and energy consumption would still remain.

As secondary results, we have compared the ε-constraint method with the elastic constraint method in terms of computational effort, see Table 25.2. For each scenario, the elastic constraint method leads to significantly lower computation times for calculating a Pareto front representation (see the corresponding Δ-values in Table 25.2). Thus, this method seems to be a real alternative in multiobjective settings when exact Pareto fronts are to be determined.

Conclusion and Road Ahead

In this paper, we have developed a tri-criteria lot-sizing and scheduling problem, minimizing electricity costs, energy-related emissions, and electricity consumption. We apply the elastic constraint method to obtain a three-dimensional Pareto front,

Table 25.2 Computation times in seconds

$M = 4, J = 5$			
Scenario; cap. ut.	εCM	elCM	Δ
$S1; \approx 93\%$	3154.69	1097.36	2057.33
$S2; \approx 75\%$	2401.09	1055.67	1345.42
$S3: \approx 88\%$	3147.04	1016.88	2130.16
$S4; \approx 59\%$	2029.36	1100.95	928.41
$S5; \approx 70\%$	2459.20	999.10	1460.10

showing that all three criteria are in conflict. Our results show that focusing solely on energy consumption is not enough; at least energy-related emissions should be taken into account. However, considering additional criteria can lead to a significant increase in computational effort. Therefore, the development of heuristics could be fruitful research.

References

1. Bänsch, K., Busse, J., Meisel, F., Rieck, J., Scholz, S., Volling, T., & Wichmann, M. (2021). Energy-aware decision support models in production environments: A systematic literature review. *Computers and Industrial Engineering, 159*, 1–27. https://doi.org/10.1016/j.cie.2021.107456
2. Ding, J., Song, S., & Wu, C. (2016). Carbon-efficient scheduling of flow shops by multi-objective optimization. *European Journal of Operational Research, 248*, 758–771. https://doi.org/10.1016/j.ejor.2015.05.019
3. Gu, W., Li, Z., Dai, M., & Yuan, M. (2021). An energy-efficient multi-objective permutation flow shop scheduling problem using an improved hybrid cuckoo search algorithm. *Advances in Mechanical Engineering, 13*, 1–15. https://doi.org/10.1177/16878140211023603
4. EUPD Research. https://www.eupd-research.com/co2-emissionen
5. Schulz, S., & Linß, F. (2020). Time-dependent emission minimization in sustainable flow shop scheduling. In *Operations Research Proceedings 2019* (pp. 583–589). https://doi.org/10.1007/978-3-030-48439-2_71
6. Zhang, H., Zhao, F., Fang, K., & Sutherland, J. (2014). Energy-conscious flow shop scheduling under time-of-use electricity tariffs. *CIRP Annals, 63*, 37–40. https://doi.org/10.1016/j.cirp.2014.03.011
7. Dellnitz A., Braschczok D., Ostmeyer J., Hilbert M., & Kleine A. (2020). Energy costs vs. carbon dioxide emissions in short-term production planning: A business case study. *Journal of Business Economics, 90*, 1383–1407. https://doi.org/10.1007/s11573-020-01000-1
8. Albadi, M., & El-Saadany, E. (2008). A summary of demand response in electricity markets. *Electric Power Systems Research, 78*, 1989–1996. https://doi.org/10.1016/j.epsr.2008.04.002
9. Ehrgott, M. (2005). *Multicriteria optimization*. Springer.

Chapter 26
Energy-Efficient Driving Model by Clustering of GPS Information

Michael Breuß, Ali Sharifi Boroujerdi, and Ashkan Mansouri Yarahmadi

Abstract In this paper we propose a novel approach to distinguish the style of drivers with respect to their energy efficiency. A unique property of the proposed method is that it relies exclusively on Global Positioning System (GPS) data. This setting is highly robust and available in practice as these GPS logs can easily be obtained. To rely on positional data alone means that all possible derived features from it will be highly correlated, so we have to consider a single feature. Here, we propose to explore the use of acceleration differences of a movement. Our strategy relies on agglomerative hierarchical clustering. The approach can be easily implemented to perform fast, even on huge amount of real-world data logs.

Keywords Energy efficiency · Driving style analysis · Clustering · GPS data

Introduction

The driving style significantly impacts the fuel consumption of a moving car. Intelligent hybrid cars could be adapted to the driving style of a human to maximise their mileage. Such possibility of optimisation could be manifold, ranging from effectively utilizing the electric engine to suggesting energy-optimal paths [8, 10]. In current study we aim to provide a major step towards including the driving style as an additional constraint into this objective. Thus an automated way of classifying driving style concerning the energy consumption is presented in this work. This may lead to a data-oriented model to describe the efficiency of the driving styles.

M. Breuß · A. Mansouri Yarahmadi (✉)
Institute for Mathematics, BTU Cottbus-Senftenberg, Platz der Deutschen Einheit 1, 03046 Cottbus, Germany
e-mail: yarahmadi@b-tu.de

M. Breuß
e-mail: breuss@b-tu.de

A. Sharifi Boroujerdi
Volkswagen Infotainment GmbH, Bochum, Germany
e-mail: ali.sharifi.boroujerdi@volkswagen-infotainment.com

O. Grothe et al. (eds.), *Operations Research Proceedings 2022*, Lecture Notes in Operations Research, https://doi.org/10.1007/978-3-031-24907-5_26

To develop a robust method in terms of applicability, we solely rely on data that can be easily acquired in any vehicle, namely Global Positioning System (GPS) logs. These setups are also very popular due to their plenty of corresponding devices on consumer market with relatively affordable prices. However, one has to admit that benefits come with certain drawbacks, namely obtained noisy samples with insufficient accuracy or even total loss of the logs because of hardware failures. Our modeling is based on an unsupervised approach aiming to infer hidden structures from a large volumes of GPS logs. By the unsupervised approach, we avoid the labeling task to assign pre-defined driving style to GPS data as it is a cumbersome task concerning the huge volume of data.

Now we turn our attention to show the importance of utilizing GPS data in classification and recognition of the drivers in real world scenarios, namely public transportation sector. The coal transportation pattern as shown in [14] is very much influenced by behaviour of the truck drivers. The average velocity maintained by the truck drivers acquired based on their GPS logs was chosen as an index to model the truck movement patterns. The risky taxi drivers are identified in [7] after their velocities within the regions that they pass were analysed. Another study concerning reckless taxi driver identification is reported in [15], to label a cab with a highly deviated route compared to those traveled by his colleagues. The task to recognize individual drivers from a given set of drivers is also tackled in [2], making use of a ranking of statistically extracted features related to car movement yet obtained from a variety of sensor information. Let us also emphasis on the robustness of our chosen GPS logs in contrast to other data modalities, namely videos and pictures. More specifically, the law constraint that needs to be enforced on the collected data while the driver crosses the national borders of countries when it comes to the captured visual data by auto sensors is much more restricted compared to GPS data. Within the context of the already referred literature, let us highlight that we extract a single feature that statistically processes neighbor GPS logs of a moving car in contrast to [7, 14, 15] with their focus to be solely on velocity or the traveled routes. Specifically, the practicability in number of used features comes to the picture, as we use only one against 137 features adopted in [2].

Nowadays modern vehicles are supplied with the necessary embedded hardware systems for the purpose of processing a collection of high volume acquired sensor data, especially cameras and radar [9]. One significant advantage of our GPS based method is a low volume of data as only few GPS logs are required to classify a driver on board. This consequently can be realized within the context of hybrid cars, allowing to switch the driving mode according to the analysis of the driving style of the operator [4].

On the Proposed Method to Analyse the GPS Data

The spatio-temporal positions of a moving car are collected in form of coordinate pairs $(x, y) \in \mathbb{R}^2$, with each pair to be accompanied with a label containing the time stamp t on which the position has been recorded. Such discrete set of collected samples is sufficient to describe the movement of a car. In context of current study, we speak of a movement pattern if the car does not stop during a considered time interval.

In order to apply our method we remove all GPS logs with the same longitude and latitude values irrespective if they are sampled on a same or a varying time interval. Such cases may represent a hardware failure or a still standing car, respectively. To allow a meaningful and unbiased comparison we additionally drop movement patterns with less than 10 samples and split them if they exceed a length of 24. This is done to avoid an unbiased comparison among the drivers. After the preprocessing we obtain about 3225 movement patterns from the real-world data set [13].

Clustering of Movement Patterns and Novel Feature

Feature construction The core of our proposed method is a refined formulation of differences in acceleration performed by a driver. We construct our feature as

$$\omega(t) := \exp\left(-1/\sqrt{\sigma\left(j(t)\right)}\right) \tag{26.1}$$

with $j(t)$ as the jerk value for each position in the movement pattern at the time of t, and the function σ to denote the standard deviation. We recall that the jerk j of a movement can be deduced as $\dddot{x}(t) = \ddot{v}(t) = \dot{a}(t) = j(t)$, while the derivatives are taken from the position, velocity and acceleration with respect to the time t.

In practice and to obtain a measure of $j(t)$, we start by populating a one dimensional discrete grid Φ, with its point values representing the velocities between adjacent spatio-temporal coordinates $(x, y) \in \mathbb{R}^2$ of subsequent GPS logs. Let us note that it is not crucial by which value Φ_t we start in this grid, since we will finally consider just differences in positions. Next we apply a second-order centered difference approximation to the computed velocities on Φ

$$|j(t)| \approx \sqrt{\left(\frac{(\Phi_{t+1} - 2\Phi_t + \Phi_{t-1})}{\Delta t^2}\right)^2} \tag{26.2}$$

leading us to approximated jerk values. Note that, we used dummy grid points on both ends of ϕ grid concerning each movement pattern to account for boundary points and while adopting the central finite difference operator. The dummy points have the same values equal to their corresponding boundaries. With this, we conclude the construction of our feature ω.

Feature properties The motivation for proposing this feature is as follows. The exponential part of (26.1) discriminates those movement patterns having a very high amount of fluctuation rate around their mean jerk values. They occur due to the existence of at least one noisy GPS log inside the movement pattern or a dubious driving style with a large number of strong accelerations and decelerations. Further, in order to discern drivers with lower jerk fluctuation rates we consider the square root of the standard deviation inside the exponential function.

Small feature values correspond to those movement patterns representing drivers with less accelerations and decelerations in their driving patterns. An energy-saving driving style can therefore be identified by small feature values, whereas more racy drivers will usually exhibit larger feature values.

Algorithmic details An agglomerative clustering method developed by The MathWorks [11] finds the optimal number of clusters by taking singleton clusters at the lowest level and keeps on merging them pairwise to establish a bottom-up hierarchy of clusters. The Ward criterion proposed by Wishart [12] was used as the core of the agglomerative clustering. The criterion is used to decide if merging two clusters C_a and C_b is optimal, in a sense of computing a standard sum of squared error (SSE) among each individual component c_{ax} and all components c_{bx} from C_a and C_b, respectively:

$$\frac{|C_a|\,|C_b|}{|C_a| + |C_b|} \sum_{\nu=1}^{\ell} \sum_{\eta=1}^{k} \left(c_{a\nu} - c_{b\eta}\right)^2 . \tag{26.3}$$

Since the cluster elements are comprised of (26.1), measuring the rate of acceleration change, a harmonic mean type weighting term published by Aldous [1] between cardinalities $|C_x|$ of clusters C_a and C_b is used to weight the SSE difference between C_a and C_b in (26.3). The harmonic mean is near the arithmetic mean if the two numbers $|C_a|$ and $|C_b|$ are close, but smaller if they differ a lot.

To have an insight about the optimal cluster numbers, an L-curve heuristic method was adopted according to [5, 6]. The L-curve method plots the WCSS as function of cluster numbers and suggests an optimum cluster number as WCSS reaches below a threshold value. In our case four number of clusters were found to be optimal. The WCSS is given by:

$$\sum_{a=1}^{M} \sum_{x \in C_a} \|x - c_a\|_2^2 \tag{26.4}$$

where M is the total number of clusters and c_a is the centroid of cluster C_a.

Experimental Results and Validation

Our dataset at hand contains 10,000 number of trajectories comprised of GPS logs obtained within a distinct areas of Beijing city [13]. Total number of logs in dataset is more than 17 million representing around 9 million kilometres of taxi travels across the city. Our expectation is to see a certain variety of driving styles by taxi drivers.

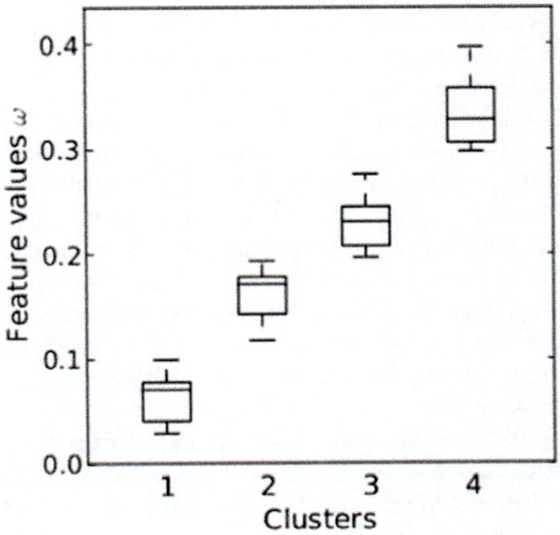

Fig. 26.1 A whisker plot representation of ω values and their distributions among different established clusters concerning the centre city areas. More precisely, Clusters 2 and 3 represent a spectrum of the drivers with an average fuel consuming driving style. The 1st cluster contains solely the energy saving drivers and the last cluster has the drivers with a high fuel consumption, containing also noisy patterns

The region under current study is chosen to cover a vast area of Beijing city centre. The logs are preprocessed in accordance to the description in Section "On the Proposed Method to Analyse the GPS Data", which resulted in 3225 number of movement patterns. All the feature values are obtained based on (26.1).

Our clustering result based on agglomerative approach is visualized in Fig. 26.1. Here, each box marks the boundaries of the first and third quartiles of a cluster and the bars indicate the full extent of the feature values in that cluster. The horizontal line indicates the cluster median. As we expected, drivers are distributed across four distinct different clusters, concerning the chosen region. The energy saving drivers are located in first cluster, where as the fourth cluster contains mostly noise and also highly fuel consuming drivers. A majority of average fuel consuming drivers lies within the scope of second and third drivers.

As WCSS did not have a major reduction after adding the 5th cluster, according to L-curve heuristic the optimal number of clusters is found to be four. We recall that the central city area might be enforced by maximum velocity limitations leading the reckless drivers also to drive more carefully. This means, the fourth cluster contains with a high probability only noisy samples. The other three clusters represent a spectrum of drivers from optimum to more fuel consuming ones.

Conclusions

In this paper, we proposed an approach to classify the drivers based on the energy consumption inherent to their driving styles. Our algorithm benefits from a novel feature developed based on jerk capable of discerning the style of car driving based on only GPS logs. As we perform an unsupervised classification, the cumbersome task of labeling is not required here, which means our method can build data-driven models on even huge amount of GPS logs. The more number of samples are provided, it will be more likely to clearly distinguish noise and classified individual drivers.

At the technical level, let us comment that our jerk-based feature considers velocity differences but not the absolute size of the underlying velocities. To account for the absolute velocities, the GPS data needs to be augmented with their corresponding class of streets e.g. distinguishing highways from other streets where typically lower velocities are enforced. We refer to [3] as a more advanced and technically oriented work, that robustly clusters the driver style concerning different classes of streets. This is achieved by retaining the same feature setting explained in current work along with a boosted dataset collected on different streets enforced by different level of velocity limit.

We also conjecture that within the scope of industry, our work is a realistic solution to be used in fuel consumption related scenarios and also as a measure to avoid other negative consequences, namely micro-particles emitted from tires as a concern for human health, because of driving in a not energy saving style. In a deeper perspective, our proposed work still stay relevant by the emerge of next generation of electric vehicles as they become more heavy and strong in terms of weight and power and their increased torque capabilities resulting to more produced airborne pollutants emanated from tires.

In brief, our research aims to deduce deep insights from pure unlabeled data to model the driver behaviour. The model can consequently be brought in connection with environment friendly measures, namely dealing with airborne pollutants.

Acknowledgements This work was supported by the German Federal Ministry of Education and Research (BMBF), funding code 05M13ICC.

References

1. Aldous, D. (1989). The harmonic mean formula for probabilities of unions: Applications to sparse random graphs. *Discrete Mathematics, 76*, 167–176.
2. Banerjee, T., Chowdhury, A., Chakravarty, T., & Ghose, A. (2020). Driver authentication by quantifying driving style using GPS only. In *2020 IEEE International Conference on Pervasive Computing and Communications Workshops (PerCom Workshops)* (pp. 1–6). IEEE.
3. Breuß, M., Sharifi Boroujerdi, A., & Mansouri Yarahmadi, A. (2022). Modelling the energy consumption of driving styles based on clustering of GPS information. *Modelling, 3*(3), 385–399.
4. Ding, Y. (2018). Eco-route: Recommending economical driving routes for plug-in hybrid electric vehicles.
5. Hansen, P. C. (1992). Analysis of discrete ill-posed problems by means of the L-curve. *SIAM Review, 34*(4), 561–580.
6. Hansen, P. C., & O'Leary, D. P. (1993). The use of the l-curve in the regularization of discrete ill-posed problems. *SIAM Journal on Scientific Computing, 14*, 1487–1503.

7. Liao, Z., Yu, Y., & Chen, B. (2010). Anomaly detection in GPS data based on visual analytics. In *Proceedings of the IEEE Symposium on Visual Analytics and Science Technology* (pp. 51–58).
8. Merting, S., Schwan, C., & Strehler, M. (2015). Routing of electric vehicles: Constrained shortest path problems with resource recovering nodes. In 15th Workshop on Algorithmic Approaches for Transportation Modelling, Optimization, and Systems (Vol. 48, pp. 29–41).
9. Notz, D., Becker, F., Kühbeck, T., & Watzenig, D. (2020). Extraction and assessment of naturalistic human driving trajectories from infrastructure camera and radar sensors.
10. Pickenhain, S., & Burtchen, A. (2015). Optimal energy control of hybrid vehicles. In *Proceedings of 6th International Conference on High Performance Scientific Computing* (accepted for publication).
11. The MathWorks. (2020). *Statistics and machine learning toolbox*. Natick.
12. Wishart, D. (1969). An algorithm for hierarchical classifications. *Biometrics, 25*(1), 165–170.
13. Yuan, J., Zheng, Y., Zhang, C., Xie, W., Xie, X., & Huang, Y. (2010). T-drive: Driving directions based on taxi trajectories. In *ACM SIGSPATIAL GIS 2010* (pp. 99–108).
14. Yuniar, D., Djakfar, L., Wicaksono, A., & Efendi, A. (2020). Truck driver behavior and travel time effectiveness using smart GPS. *Civil Engineering Journal, 6*(4), 724–732.
15. Zhang, D., Li, N., Zhou, Z.H., Chen, C., Sun, L., & Li, S. (2011). iBAT: Detecting anomalous taxi trajectories from GPS traces. In *Proceedings of the 13th International Conference on Ubiquitous Computing* (pp. 99–108). ACM.

Chapter 27
Identifying Critical Demand Periods in Capacity Planning for Networks Including Storage

Andreas Bley and Philipp Hahn

Abstract We consider a capacity planning problem for networks including storage. Given a graph and a time series of demands and supplies, we seek for integer link and storage capacities that permit a single commodity flow with valid storage in- and outtakes over all time steps. This problem arises, for example, in power systems planning, where storage can be used to buffer peaks of varying supplies and demands. For typical time series spanning a full year at hourly resolution, this leads to huge optimization models. To reduce the model size, time series aggregation is commonly used. The time horizon is sliced into fixed size periods, e.g. days or weeks, a small set of representative periods is chosen via clustering methods, and a much smaller model involving only the chosen periods is solved. Representative periods, however, typically do not contain the situations with the most extreme demands and supplies and the strongest effects on storage. In this paper, we show how to identify such critical periods using principal component analysis (PCA) and convex hull computations and we compare the quality and solution time of the reduced models to the original ones for benchmark instances derived from power systems planning.

Keywords Time series analysis · Capacity planning · Model reduction

Introduction

We consider a single-commodity capacitated network design problem with storage, which arises, for example, in the planning of power and transport networks where storage can be used to balance loads among consecutive load scenarios. Given a graph and a time series of load scenarios, the task is to find minimum cost edge and storage capacities that permit single-commodity flows in each time step without exceeding

A. Bley · P. Hahn (✉)
Institut für Mathematik, Universität Kassel, Heinrich-Plett-Straße 40, 34132 Kassel, Germany
e-mail: philipp.hahn@uni-kassel.de

A. Bley
e-mail: andreas.bley@uni-kassel.de

O. Grothe et al. (eds.), *Operations Research Proceedings 2022*, Lecture Notes in Operations Research, https://doi.org/10.1007/978-3-031-24907-5_27

the edge and storage capacities. For instances with many time steps, the resulting models are huge and techniques to reduce their size are needed, c.f. [6]. As storage adds dependencies among consecutive time steps, reductions based on isolated time steps do not work well. Instead, short periods of consecutive time steps need to be considered. In practice, typically few representative periods are chosen via clustering techniques and then a model involving only these is solved [4, 6, 10]. Representative periods, however, often do not cover the extreme scenarios governing the capacity installation. In this paper, we extend the technique presented in [1], which is based on principal component analysis and convex hull computations, to find such extreme periods.

Formally, in the (single-commodity) capacitated network design problem with storage (CNDS) we are given a graph $G = (V, E)$ and, for each edge $ij \in E$, a capacity unit u_{ij}, that can be installed in integer multiples at cost c_{ij} per unit. Storage capacity can be installed in integer multiples of u_i at cost c_i at each node $i \in V_S \subseteq V$. Let $A = \{(i, j), (j, i) \mid ij \in E\}$. All arcs in A except for a subset $A^0 \subset A$ can carry flow. Furthermore, we are given a collection of time periods, each containing several consecutive time steps. Time-dependencies exist only between time steps within the same period. Using this scheme, a full time series is given as a single period containing all time steps, while a reduced model contains only few short periods taken from the time series. We let $[k] := \{1, \ldots, k\}$ for each $k \in \mathbb{N}$. We denote by P the number of given periods and, for each $p \in [P]$, by S_p the number of time steps in period p. The set of all time steps is $T = \{(p, s) \mid p \in [P], s \in [S_p]\}$. Let $T^0 = T \cup \{(p, 0) \mid p \in [P]\}$. For each time step $(p, s) \in T$, we are given a vector $d^{p,s} = (d_j^{p,s})_{j \in V}$, where $d_j^{p,s}$ is the supply or demand at node j in time step (p, s). Our goal is to find minimum cost edge and storage capacities, such that, for each $(p, s) \in T$, there are storage in- and outtakes and a flow that respect the storage and edge capacities and satisfy the resulting node balances.

Using variables $x_{ij}, y_i \in \mathbb{Z}_+$ for the number of capacity and storage units installed on edge $ij \in E$ and node $i \in V$, variables $l_i^{p,s,+}, l_i^{p,s,-}, \bar{l}_i^{p,s}, \bar{l}_i^{p,0} \in \mathbb{R}_+$ for the storage in- and outtake and level at node $i \in V$ in time step $(p, s) \in T$, and variable $f_{(i,j)}^{p,s} \in \mathbb{R}_+$ for the flow sent via arc $(i, j) \in A$ in time step $(p, s) \in T$, we obtain the following MILP-model for CNDS:

$$\min \sum_{ij \in E} c_{ij} x_{ij} + \sum_{i \in V} c_i y_i \qquad \text{(CNDS-IP)}$$

$$\text{s.t.} \quad f_{(i,j)}^{p,s} + f_{(j,i)}^{p,s} \leq x_{ij} u_{ij} \quad ij \in E,\ (p, s) \in T \qquad (27.1)$$

$$\bar{l}_i^{p,s} \leq y_i u_i \quad i \in V,\ (p, s) \in T^0 \qquad (27.2)$$

$$l_i^{p,s,+} - l_i^{p,s,-} + \bar{l}_i^{p,s-1} = \bar{l}_i^{p,s} \quad i \in V_S,\ (p, s) \in T \qquad (27.3)$$

$$\sum_{(j,i) \in \delta^-(i)} f_{(j,i)}^{p,s} - \sum_{(i,j) \in \delta^+(i)} f_{(i,j)}^{p,s} - l_i^{p,s,+} + l_i^{p,s,-} = d_i^{p,s} \quad i \in V,\ (p, s) \in T \qquad (27.4)$$

$$f_{(i,j)}^{p,s} = 0 \quad (i,j) \in A_0,\ (p,s) \in T \tag{27.5}$$

$$l_i^{p,s,+}, l_i^{p,s,-}, \bar{l}_i^{p,s} = 0 \quad i \in V \setminus V_S,\ (p,s) \in T \tag{27.6}$$

$$l_i^{p,s,+}, l_i^{p,s,-}, \bar{l}_i^{p,s} \geq 0 \quad i \in V,\ (p,s) \in T$$

$$f_{(i,j)}^{p,s} \geq 0 \quad (i,j) \in A,\ (p,s) \in T$$

$$x_{ij} \geq 0, \quad x_{ij} \in \mathbb{Z} \quad ij \in E$$

$$y_i \geq 0, \quad y_i \in \mathbb{Z} \quad i \in V_S$$

Inequalities (27.1) and (27.2) ensure that in each time step the flow does not exceed the edge capacities and the storage levels do not exceed storage capacities. Constraints (27.3) link the storage levels of consecutive time steps within a period and the corresponding in- and outtakes. Equalities (27.4) ensure that the flow in each time step satisfies all node balances including storage in- and outtakes. For simplicity, flow and storage variables are defined for all arcs and nodes but fixed to zero for all non-flow arcs and non-storage nodes in (27.5) and (27.6), respectively. The objective is to minimize the sum of all capacity installation costs.

Note that the capacity variables x_{ij} and y_i do not depend on the time steps, while flow and storage variables do. Also note that the initial storage level in each period is unrestricted. Thus, initially full storage can be used to help satisfy demands. The storage capacity needed for this, however, causes costs.

Reducing the Model

To model the CNDS problem for a full time series and all dependencies between consecutive time steps, a single period containing all time steps is given as input to the model (CNDS-IP). Our goal is to construct a smaller, approximate model that involves only few shorter periods and less time steps in total. For this, we first generate a large collection of short periods that cover the original time series and then pick a small sub-collection of these to remain in the reduced model.

Generating Periods

Assume the full time series is given as a single period containing all m time steps and we are given a desired sub-period length S, e.g. $S = 168$ for a period of 1 week and hourly time steps, and a desired shift Q between two consecutive periods, e.g. $Q = 24$ for a shift of 1 day. For simplicity, assume that $P = (m - S)/Q \in \mathbb{N}$. We then construct P sub-periods of size S, each shifted by a multiple of Q. More precisely, the S consecutive time steps $(p-1)Q + 1$ to $(p-1)Q + S$ of the original full time series are contained in period $p \in [P]$ and we have $d^{p,s} = d_{\text{orig}}^{1,(p-1)Q+s}$, where d_{orig} denotes the demand vectors indexed according to the original full time series and d the demand (re-)indexed according to the time steps s in the created period $p \in [P]$.

Passing the periods in $[P]$ (or a sub-collection thereof) with the corresponding demand vectors d as input to (CNDS-IP), one obtains a relaxation of the model for the full time series. Edge and storage capacities that permit valid flows and storage in- and outtakes for the full times series' model also permit valid flows and storage in- and outtakes for the model involving the periods in $[P]$ instead.

Identifying Critical Periods

In order to identify a small sub-collection of the generated periods, we extend the method presented in [1] to select critical isolated scenarios in robust network design (without dependencies among time steps) to periods spanning multiple time steps. That method first uses principal component analysis (PCA) to identify the directions of the statistically largest variation among the demand vectors. In the second step, all demand vectors are projected onto subspaces spanned by one or more of the most important principal components. For each of these subspaces, the vertex set of the convex hull covering the projected demand vectors is computed. Eventually, the scenarios corresponding to the vertices are selected as critical and chosen to remain in the reduced problem. In a symmetrized variant of this method, for each scenario s both the original demand vector d^s and its negative $-d^s$ are projected onto the subspaces, which may lead to fewer scenarios corresponding to vertices. For subspace dimension b, we denote the non-symmetrized and the symmetrized variants by CH_b and SCH_b, respectively.

To extend that method to time periods, we first apply a slicing procedure to the original demand vectors in order to create a new matrix whose rows represent the demands of periods of length S. More precisely, with $n = |V|$, the given node demands form a matrix $D_{\text{orig}} \in \mathbb{R}^{m \times n}$, whose sth row is the demand vector $d_{\text{orig}}^{1,s}$ for time step s. Concatenating the demand vectors $d^{p,s}$ of all time steps $s \in [S_p]$ in a sub-period $p \in [P]$ to a single row, we obtain the matrix $D \in \mathbb{R}^{P \times S \cdot n}$. The pth row D_p of D can be regarded as the demand or 'feature' vector of the entire period p. Eventually, we apply the methods described in [1] to the feature vectors D_p for periods $p \in [P]$ to choose a small subset of extreme periods $P' \subseteq [P]$, that will be finally kept in the reduced model.

Time series, however, typically feature some natural periodicities, such as daily and weekly patterns. If the shift Q is no multiple of these periodicities, seemingly strong variations among different periods' demands are artificially introduced by the different offsets. To eliminate these artificial variations, we can realign the periods' demand data to a uniform pattern before applying the PCA. For weekly periods, for example, we cyclically reorder the time steps $s \in [S_p]$ in each row p in such a way that the time step corresponding to Sunday 0:00 is in the same column in each row.

Note that, in contrast to the approach presented in [2], our approach does not require to first solve (CNDS-IP) for each period $p \in [P]$ in order to decide which ones to use in the reduced model. It solely relies on analyzing the extremality of the (adjusted) demand vectors, which is computationally much less demanding.

Instances and Computational Experiments

To assess the effectiveness of our approach, we compare the solution times and the objective values of (CNDS-IP) for the full time series and for the chosen subset of periods for some benchmark instances. The method is implemented in Python 3.7. We use Gurobi 9.5.0 [3] as ILP solver, except for the Benders decomposition approach, where CPLEX 12.9 [5] with its built-in Benders algorithm is used. In both cases, we admit a mip-gap of 0.1% to keep solution times relatively small. The reported results are obtained on a machine with two 14-core Intel Xeon (R) E5-2690 v4, 2.6 GHz processors and 256 GB of RAM. The presented plots are created using grblogtools [8] and plotly [9].

In the following, we report on two relatively small benchmark instances derived from a power grid planning problem including storage. In both instances, the network consists of a 6 node densely meshed bidirectional core, resembling countries and their exchange capacities, and 50 nodes representing 26 renewable and 24 thermal generators, which are unidirectionally connected to only one country node and an artificial sink each. The demands and supplies of the renewable generators and countries are given as a time series, while the thermal generators are flexible within their capacity bounds. The time series of the smaller instance I1 contains 8736 hourly time steps covering one year, that of instance I2 a total of 61,320 covering seven years. We generate periods of length 168 (1 week) and step size 168, so periods do not overlap. In the PCA-based period selection from Section "Identifying Critical Periods", we always standardize the data prior applying PCA. This turned out to yield better results than using the unscaled demand values, contrarily to the results obtained in [1] for critical scenarios in robust network design. The reasons are not yet clear to us.

Figure 27.1 shows the objective values obtained with the subset of periods chosen with our approach depending on the number of principle components (PCs) consid-

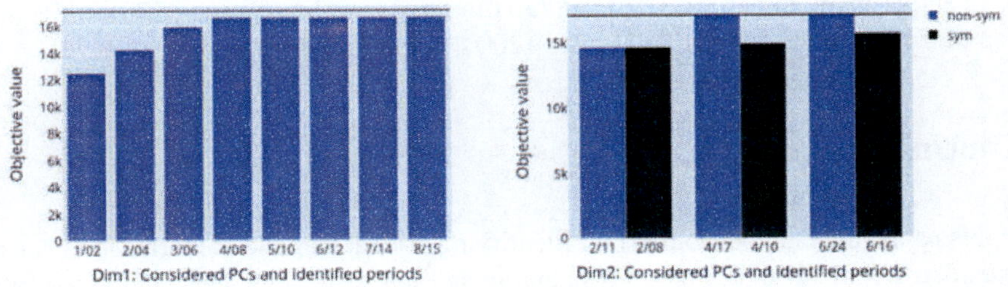

Fig. 27.1 Objective of reduced models for varying number of PCs for instance I1

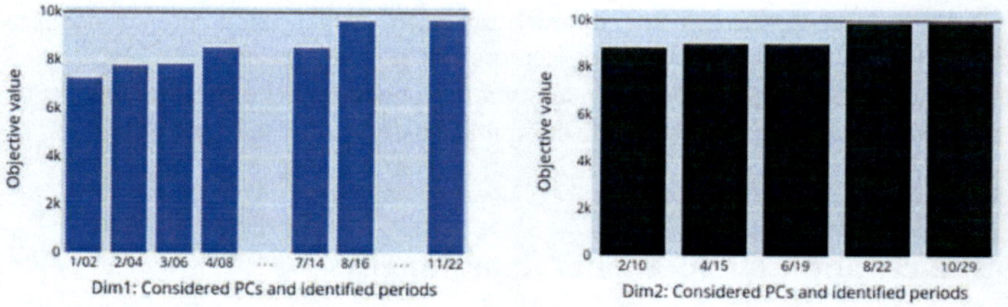

Fig. 27.2 Objective of reduced models for varying number of PCs for instance I2

ered for the smaller instance I1. The left plot shows the results for method variant CH_1, which just chooses the two extreme periods per PC. The right plot shows the results for variants CH_2 and SCH_2. The horizontal line shows the objective for the full time series. Method CH_1 yields a very good approximation already for 4 PCs, choosing 8 periods. Unfortunately, considering more PCs with this variant does not close the remaining gap. Considering the first 4 PCs, variant CH_2 identifies 17 (out of 52) periods, leading to a reduced model that actually achieves the full model's optimal value. With the symmetric variant we miss one of the critical periods. The model for the full time series needs 11,011 s to be solved to optimality, a gap of 0.1% is reached after 2180 s. The solution times for the reduced models range from 10 s for 2 periods to roughly 250 s for 17 and 500 s for 24 periods. The generation and selection of the periods, including PCA and convex hull computations, requires just a few seconds.

The results for instance I2 are shown in Fig. 27.2. For this instance, the full time series model could not be solved in reasonable time. After 14,881 s the computation was aborted with a gap of 0.12%. Again, the non-symmetric variant for dimension 1 is shown left and the symmetric variant for dimension 2 right. As before, we obtain a very good approximation with variant CH_1, choosing only 16 out of 365 periods when considering the first 8 PCs, but we cannot close the remaining gap. For CH_2, this gap closes almost with the first 6 and completely with the first 8 PCs considered, leading to 22 or 29 out of 365 periods chosen, respectively. The runtimes of the reduced models range from 17 s for 2 to roughly 1000 s for 29 considered periods.

Using Benders algorithm with the 22 and 17 identified critical periods in the initial master problem and generating Benders cuts for the others, we solve the full models of I2 and I1 involving all 61,320 and 8736 time steps in 11,740 s and 3770 s while a gap of 0.1% is reached after 9400 s and 1200 s, respectively.

Conclusion

In this paper, the method from [1] for identifying critical demand scenarios has been extended to critical demand periods spanning multiple time steps. The method solely works on the raw demand data. Neither the topology, the capacity, nor any other structural properties are considered. Also, the method is very fast. In relation to solving (the reduced) CNDS-IP the times for PCA and convex hull computations are negligible. In our experiments, the resulting reduced models deliver good approximations of the original ones. Further, using the critical periods in the master problem of a Benders decomposition leads to a significant speed up for finding an exact solution. In the future, we plan to apply this method to a large scale sector coupled energy systems planning model where the model is reduced using both representative and critical periods.

Acknowledgements This work was supported by the German Federal Ministry for Economic Affairs and Energy (BMWi), Project *NSON II*, [7].

References

1. Bley, A., & Hahn, P. (2021). *Identifying critical demand scenarios for the robust capacitated network design problem using principal component analysis*. Technical report. Institute for Mathematics, University of Kassel.
2. Bukenberger, J. P., & Webster, M. D. (2020). Approximate latent factor algorithm for scenario selection and weighting in transmission expansion planning. *IEEE Transactions on Power Systems, 35*(2), 1099–1108.
3. Gurobi Optimization, LLC. (2022). Gurobi optimizer reference manual. https://www.gurobi.com
4. Hoffmann, M., Kotzur, L., Stolten, D., & Robinius, M. (2020). A review on time series aggregation methods for energy system models. *Energies, 13*, 641.
5. IBM Ilog CPLEX. (2020). User's manual for CPLEX (Vol. 12, p. 9).
6. Mahdavi, M., Antunez, C. S., Ajalli, M., & Romero, R. (2019). Transmission expansion planning: Literature review and classification. *IEEE Systems Journal, 13*(3), 3129–3140.
7. Mende, D., Harms, Y., Härtel, P., Frischmuth, F., Stock, D. S., Braun, M., Herrmann, M., Hofmann, L., Valois, M., Bley, A., Hahn, P., Jurczyk, J., & Rathke, C. (2020). NSON II: Next steps in economical connection and international integration of offshore wind energy in the north seas. In *Proceedings of the 19th Wind Integration Workshop*, Ljubljana, Slovenia.
8. Miltenberger, M., Oberdieck, R., Siefen, K., Aramon, M., & Bowly, S. grblogtools. https://github.com/Gurobi/grblogtools
9. Plotly Technologies Inc. (2015). *Collaborative data science*. https://plot.ly
10. Saxena, A., Prasad, M., Gupta, A., Bharill, N., Patel, O. P., Tiwari, A., Er, M. J., Ding, W., & Lin, C. T. (2017). A review of clustering techniques and developments. *Neurocomputing, 267*, 664–681.

Chapter 28
Industrial Use or Storage of CO_2? A Compound Real Options Valuation for the Retrofitting of Coal-Fired Power Plants

Qinghan Yu and Reinhard Madlener

Abstract We investigate sequential investment in carbon capture and storage (CCS), i.e., the case of retrofitting a coal-fired power plant, and then invest in carbon capture and utilization (CCU) for methanol production. A (nested) compound real options model based on a backward recursive dynamic programming algorithm is used for the analysis, which seems helpful for decision-makers who have to make capital-intensive irreversible investments under high uncertainty regarding electricity and CO_2 price. The options to invest in CCS and CCU are investigated individually first, and then sequentially, leading to a hybrid CCUS plant that enables both CO_2 storage and methanol production. The prices of electricity, carbon and methanol are considered as stochastic and correlated with each other. Managerial flexibility exists regarding a postponement of the investment decision and the real-time optimization between selling methanol to the market or storing CO_2 for earning carbon credits after establishing the CCUS plant. We find that at CO_2 prices of around 40 €/t, CCS investment is economically rational, whereas CCU for methanol is not. Combining CCS with CCU increases the overall investment probability and potential for larger profits. Since methanol is more valuable than CO_2, CCU can be expected to dominate the value of the compound option for the case of favorable market conditions (i.e., sufficiently high methanol and CO_2 prices).

Keywords CCS · CCU · Real options · Methanol · Zero emissions · Sequential investments

Q. Yu · R. Madlener (✉)
Institute for Future Energy Consumer Needs and Behavior (FCN), School of Business and Economics/E.ON Energy Research Center, RWTH Aachen University, 52074 Aachen, Germany
e-mail: RMadlener@eonerc.rwth-aachen.de

Q. Yu
e-mail: qinghan.yu@rwth-aachen.de

R. Madlener
Department of Industrial Economics and Technology Management, Norwegian University of Science and Technology (NTNU), 7491 Trondheim, Norway

O. Grothe et al. (eds.), *Operations Research Proceedings 2022*, Lecture Notes in Operations Research, https://doi.org/10.1007/978-3-031-24907-5_28

Introduction

In addition to fuel switching and energy efficiency improvements, carbon capture and storage (CCS) is expected to play an important role in achieving deep cuts in global CO_2 emissions. This efficient technology captures CO_2 from large emitters and stores it permanently in suitable geological reservoirs (e.g., saline aquifers, salt caverns, and depleted gas fields). It is worth considering adding CCS equipment to existing power plants as well, a procedure known as CCS retrofitting, given the enormous amount of CO_2 emitted by power plants already in operation during their long lifespan. Although the savings in CO_2 emissions from stored CO_2 provide incentives for CCS retrofitting in countries with CO_2 certificate trading schemes in place, the heavy investment burdens, geographical restrictions, and often lack of social acceptance challenge its widespread implementation.

Captured CO_2 can have many uses in various industries, ranging from sparkling beverages to urea production, and thus does not necessarily have to be permanently stored. In light of recent breakthroughs in catalysts, methanol (MeOH) synthesis using hydrogen (H_2) and captured CO_2 as feedstock is attracting growing attention as a carbon capture and utilization (CCU) case [1, 2]. On the one hand, as a versatile chemical of high economic value, MeOH has the potential to generate profits that offset the investment costs of CCS retrofitting. On the other hand, MeOH synthesis provides a possible intersection for CCU and power to gas (PtG), if the H_2 used is produced from renewable energy via water electrolysis. Moreover, liquid MeOH serves as a more efficient medium for energy storage and fuel for transportation than gaseous H_2 does. Several pilot plants have been put into place to demonstrate the economic feasibility of CCU MeOH production. The largest of these is the $MefCO_2$ project in Niederaussem (Germany), which produces 1 ton of methanol per day. As an EU-funded project, $MefCO_2$ is jointly operated by various partners across Europe and aims to pave the way for industrial-scale deployment [3].

Since neither CCS retrofits nor CCU MeOH production has been deployed on a large scale to date, an interesting question is how these would interact with each other once they both became commercially available. Real options analysis (ROA; see [4]), which considers price uncertainties and the value of managerial flexibility, is an appropriate method to investigate this question. Several authors have conducted ROA to study the profitability of investments in CCS retrofits, but the existing literature has not determined whether the possibility of using captured CO_2 for CCU MeOH production accelerates CCS retrofits and thus serves as a stepping stone to permanent storage.

The original contribution of this paper is twofold: On the one hand, we model a complete CCUS MeOH production system based on ROA, which allows us to solve a path-dependent sequential investment problem. On the other hand, applying the proposed model, we identify the main factors that influence CCUS investments, and we determine how these affect option values and optimal operating policies.

Methodology

In answering our research questions, we consider an opportunity for a power plant owner to sequentially invest in both CCS and CCU MeOH production facilities. The investor has managerial flexibility in (i) how to proceed with the investment decisions and (ii) the subsequent real-time optimization, either by selling MeOH at the market price or storing CO_2 for carbon certificates after acquiring such a hybrid system of this kind. Figure 28.1 illustrates the relevant boundaries used in our study, including two possible subsystems centered and overlapped at the power plant itself, and a simplified mass and energy balance.

As the investor opts for profit-maximization by taking advantage of fluctuating market conditions, the investment opportunity can be viewed as a stochastic optimal control problem that is stated as follows:

$$J_0(s_0) = \max_{u_1,\ldots,u_T} E\left\{\sum_{t=0}^{T-1}(1+r)^{-t}g(s_t,u_t) + (1+r)^{-T}g(s_T)\right\} \tag{28.1}$$

$$g(s_t,u_t) = \pi(s_t) - I(u_t) \tag{28.2}$$

$$g(s_T) = \pi(s_T) \tag{28.3}$$

$$s_{t+1} = f(s_t,u_t) \tag{28.4}$$

$$s_t \in \Omega_{u,t} u_t \in \Omega_{u,t} \tag{28.5}$$

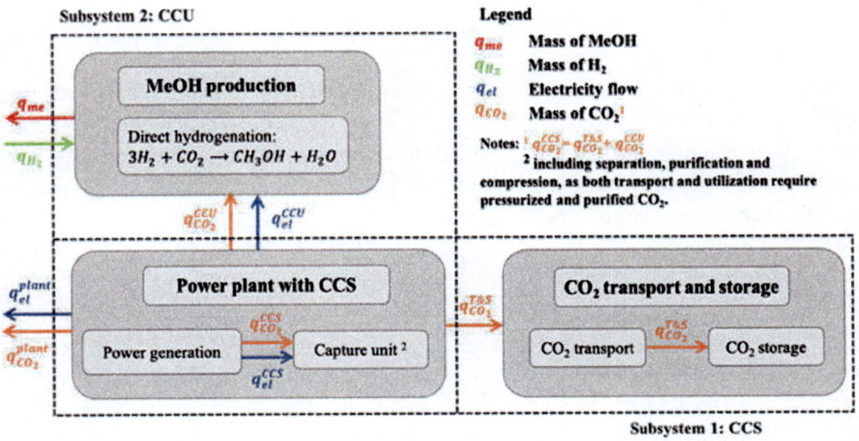

Fig. 28.1 Flow diagram illustrating the system boundaries and the mass and energy balance. Superscripts indicate the associated components (CCS, T&S, CCU and plant)

where $J_0(s_0)$ is the max. expected total profits over the planning horizon in the initial states, $g(s_t, u_t)$ is the net profit (the immediate payoff) at time step t, $\pi_t(s_t)$ is profit at time step t, $g(s_T)$ is the net profit (terminal payoff) at time step T, $\pi_T(s_T)$ is the profit at time step T, $I_t(u_t)$ is the investment costs incurred at time step t, s_t is state variable, u_t is the decision variable, r is the risk-adjusted discount rate and $\Omega_{s,u}$ are the discrete feasible sets for s and u. Hence, we formulate a discrete stochastic dynamic programming (SDP) model and apply Bertsekas' [5] framework in order to tackle the sequential investment problem.

Although we use a coal-fired power plant for illustrative purposes, the model can easily be modified and applied to other emission sources. For example, when used on a cement factory, electricity becomes an input to the system, and the opportunity cost of losses in electricity sales (a term in $\pi(\cdot)$, see [6] for details) turns into the energy cost of auxiliary equipment—the model can be thus simplified.

Data

Suppose that a modern baseload coal-fired power plant, which can later be retrofitted with a capture system, is built in 2021. For model applications, we first make state-of-the-art technology specifications and collect various cost parameter values from previous studies [7–9]. Then the prices of electricity $P_{el,t}$, emission certificates $P_{CO_2,t}$, and methanol $P_{me,t}$ are considered as uncertain and modeled as correlated geometric Brownian motion processes (GBMs):

$$\frac{dP_{i,t}}{P_{i,t}} = \mu_i t + \sigma_i dZ_{i,t} \tag{28.6}$$

$$dZ_{i,t} dZ_{j,t} = \rho_{i,j} \tag{28.7}$$

where $\{\mu_i; i = el, CO_2, me\}$ and $\{\sigma_i; i = el, CO_2, me\}$ are the drift and the volatility parameters, respectively, $\{dZ_{i,t}; i = el, CO_2, me\}$ represent the increments of a standard GBM. As reported in Table 28.1, the price processes are parametrized to approximate the price scenarios given in [10] and the offset prices are set based on current market conditions.

Table 28.1 Parametrization of the price processes

Parameter	$P_{i,2021}$ [1]	μ_i (%)	σ_i	$\rho_{i,el}$	ρ_{i,CO_2}	$\rho_{i,me}$
P_{el} [2]	56 €/MWh	3.00	0.05	1	0.579	0.405
P_{CO_2} [3]	43 €/t	5.00	0.20	0.579	1	0.315
P_{me} [4]	328 €/t	2.00	0.07	0.405	0.315	1

Notes: [1] reflecting the market conditions of March 12, 2021, based on data from Datastream; [2] base-load futures, F1BYc1; [3] EUA price, CFI2c1; [4] FOB Rotterdam, M2Tc1

Results and Discussion

In contrast to previous studies, such as [11], we find that CCS investments will be desirable by 2025, assuming there are low volatilities in the CO_2 price and given the current CO_2 price level of around 40 €/t (see Fig. 2a). Furthermore, we show that investments in CCU MeOH are economically not viable under the prevailing market conditions in Europe. The individual option to invest in CCU is, unsurprisingly, deep out-of-the-money. The prices of methanol and hydrogen are identified as the single most important factors that influence the profitability of CCU MeOH production. Investment costs and the opportunity costs of the losses in electricity sales caused by the retrofit are trivial compared to the huge profits that methanol sales are likely to generate. The base case study reveals that the critical threshold for investing in CCU is $P^*_{me} = 680$ €/t (see Fig. 2b). Otherwise, the price of hydrogen ought to drop below 1250 €/t to prompt investment in CCU (see Fig. 2c). However, we provide some evidence that CCU MeOH has the potential to fully outshine CCS in the case of high methanol prices (see Fig. 2d).

As of October 20, 2022, the carbon price increased to 68 €/t, while the methanol price remained relatively stable at 344 €/t. A high CO_2 price level dampens the

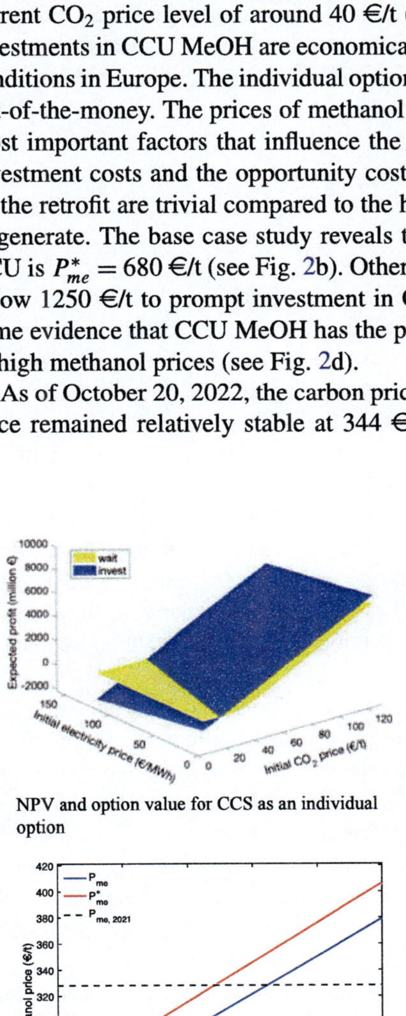

(a) NPV and option value for CCS as an individual option

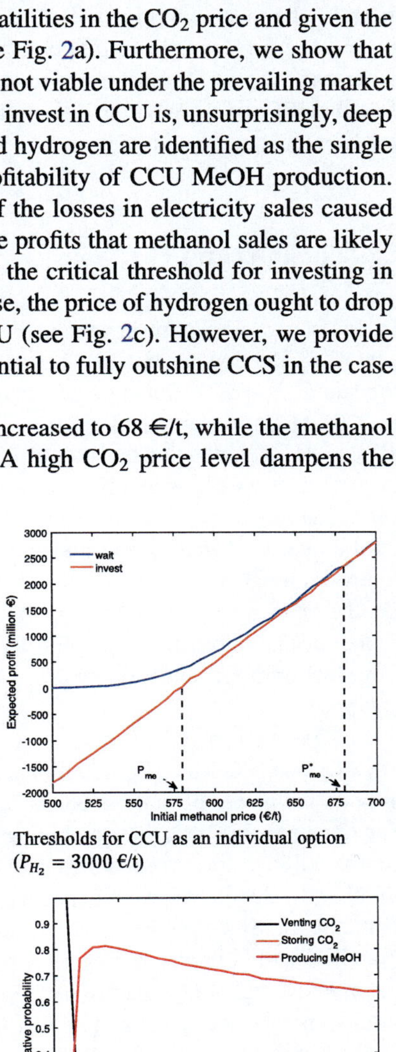

(b) Thresholds for CCU as an individual option ($P_{H_2} = 3000$ €/t)

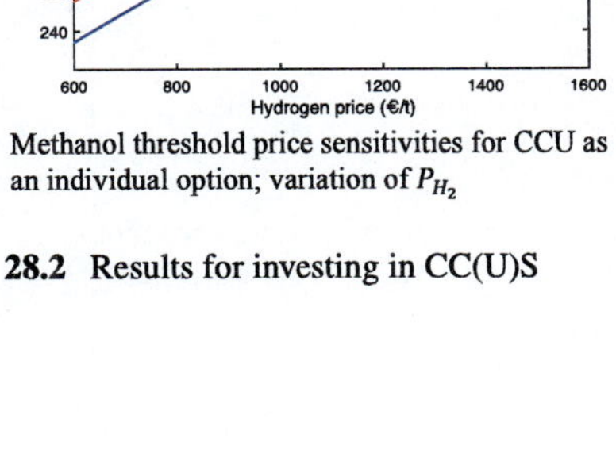

(c) Methanol threshold price sensitivities for CCU as an individual option; variation of P_{H_2}

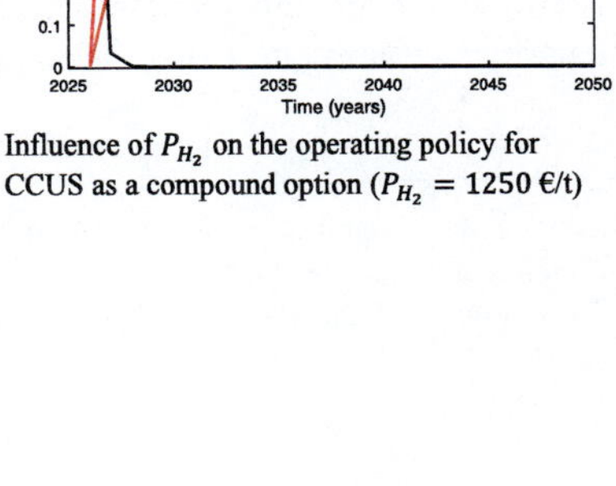

(d) Influence of P_{H_2} on the operating policy for CCUS as a compound option ($P_{H_2} = 1250$ €/t)

Fig. 28.2 Results for investing in CC(U)S

discouraging effects of high volatility and investment costs by reducing the waiting region, thus leading to earlier investment.[1] However, it is questionable how long the relatively high carbon price level will persist (given its volatile history) and whether the methanol price will remain low after the Russian-Ukraine war and the blow up of the North Stream Pipelines (as methanol is currently produced primarily from natural gas). A more detailed reporting of the results and discussion can be found in [6], and a similar analysis with a very different approach (fuzzy real options analysis) in a companion paper [12].

Conclusions and Outlook

In this paper, we first investigate the individual options to invest in CCS and CCU MeOH production separately. The option values, as well as the optimal investment thresholds, are determined using current market data and price scenarios provided by [10]. Under the assumed model parameters, we find that the critical threshold is $P^*_{CO_2} = 41$€/t, indicating that the currently rather high CO_2 price level already makes investments in CCS technology attractive. However, the investment opportunity is very sensitive to changes in volatility and investment costs. Policymakers should, on the one hand, reduce the uncertainty in the CO_2 price through consistent and rigorous climate policy if they aim to reduce CO_2 emissions through CCS retrofits. On the other hand, financial support should be provided. Otherwise, CCS is only economically viable for geographically well-positioned power plants, as they benefit from low transport and storage costs, thus leaving much of the technical potential untapped.

In the second part of the analysis, we assume that the power plant owner can invest in CCS and CCU MeOH production sequentially. And after the owner has invested in both, the power plant can operate in two modes, namely storage and utilization. The main conclusion is that combining CCS with CCU MeOH production increases the overall investment probability and the potential for larger profits. However, the additional value is somehow debatable as it stems from the uncertainty in the price of methanol. Since methanol is far more valuable than CO_2, it is thus very likely that CCU will dominate the value of the compound option once market conditions become favorable (e.g., the price of hydrogen falls dramatically).

In summary, we can conclude that CCU MeOH production will only accelerate CCS retrofits if the price of hydrogen decreases sharply. However, in that case, MeOH production, with its vast revenue potential, may outstrip carbon storage. Overall, it is therefore questionable whether CCU methanol production is a stepping stone to the permanent mitigation of CO_2 emissions.

[1] The same logic applies to the methanol price level as well. As the methanol price increases, the dashed line in Fig. 2c begins to rise and intersects the threshold lines at higher hydrogen prices, i.e., CCU becomes profitable at higher hydrogen prices.

In future work, some of the restrictive assumptions made in this paper could be relaxed. Decommissioning of the facilities, for instance, including the power plant itself, is an important aspect but is disregarded in our analysis. Another caveat is that we only considered the price uncertainty (P_{el}, P_{CO_2} and P_{me}) and used a fixed and constant hydrogen price throughout the planning horizon. The technical specifications and the cost parameters assumed reflect the current state of knowledge, and we did not consider learning factors for emergent technologies. Finally, the proposed model can be applied to similar investment problems faced by other CO_2 producers, or it can be extended to include market competition.

References

1. Bos, M. J., Kersten, S. R. A., & Brilman, D. W. F. (2020). Wind power to methanol: Renewable methanol production using electricity, electrolysis of water and CO_2 air capture. *Applied Energy, 264*, 114672.
2. Ravikumar, D., Keoleian, G., & Miller, S. (2020). The environmental opportunity cost of using renewable energy for carbon capture and utilization for methanol production. *Applied Energy, 279*, 115770.
3. MefCO_2 Homepage http://www.mefco2.eu/mefco2.php. Accessed Dec 31, 2021.
4. Dixit, A., & Pindyck, R. S. (1994). *Investment under uncertainty* (1st ed.). Princeton University Press.
5. Bertsekas, D. P. (1978). *Stochastic optimal control: The discrete time case* (1st ed.). Athena Scientific.
6. Yu Q., Madlener R. (2021). Industrial utilization or storage of CO_2? A compound real options valuation for the retrofitting of coal-fired power plants. FCN Working Paper No. 13/2021, Institute for Future Energy Consumer Needs and Behavior, RWTH Aachen University.
7. Turner, M., Iyengar, A., Woods, M. (2020). *Cost and performance baseline for fossil energy plants supplement: Sensitivity to CO_2 capture rate in coal-fired power plants* (No. NETL-PUB-22695). National Energy Technology Laboratory (NETL)
8. Leonzio, G., Foscolo, P. U., Zondervan, E., & Bogle, I. D. L. (2020). Scenario analysis of carbon capture, utilization (particularly producing methane and methanol), and storage (CCUS) systems. *Industrial and Engineering Chemistry Research, 59*(15), 6961–6976.
9. Pérez-Fortes, M., Schöneberger, J. C., Boulamanti, A., & Tzimas, E. (2016). Methanol synthesis using captured CO_2 as raw material: Techno-economic and environmental assessment. *Applied Energy, 161*, 718–732.
10. Öko-Institut. (2014). Klimaschutzszenario 2050. Studie im Auftrag des Bundesministeriums für Umwelt, Naturschutz, Bau und Reaktorsicherheit (2014). http://www.isi.fraunhofer.de/content/dam/isi/dokumente/ccx/2014/Bericht_Runde_1.pdf. Accessed Dec 31, 2021.
11. Rohlfs, W., & Madlener, R. (2011). Valuation of CCS-ready coal-fired power plants: A multi-dimensional real options approach. *Energy Systems, 2*(3), 243–261.
12. Yu Q., Madlener R. (2021). A fuzzy real options analysis of investments in coal-fired power plants retrofitted with CCS or CCU, FCN Working Paper No. 14/2021, Institute for Future Energy Consumer Needs and Behavior, RWTH Aachen University, December 2021

Chapter 29
Integration of Data Centers as Active Entities into Energy Systems Modeling

Juan Jerez Monsalves, Juan Gea-Bermúdez, Claire Bergaentzlé, and Dogan Keles

Abstract Energy system models allow exploring new interactions between energy supply and demand that new actors within the system enable. Data centers (DC) are such actors, becoming significant energy consumers in some regions while pressuring local energy systems, but also having the potential to contribute to the energy transition by providing demand response and waste-heat recovery. Yet, no past studies have investigated how these attributes can support energy system decarbonization. This article describes the integration of large DCs as an extension to the Balmorel energy system model, jointly optimizing their cooling portfolio along with the electricity and the district heating systems. The model allows DCs to invest in flexible-cooling and waste-heat technologies while competing freely with other traditional generators within the system. This article showcases a simplified example of DCs' potential in the Danish energy system through 2035. The results show that waste heat accounts for 9% of the national heating supply, leading to system-wide emission reductions of 0.8% in 2025 and cost savings of up to 3.0% in 2035. In the future, this model will be used for policy research to align the incentives of DC operators and society.

Keywords Energy system model · Data centers · Demand response · Waste heat

Introduction

The data center (DC) industry is expanding rapidly. However, this expansion has led to geographic clusters of large DCs in locations with favorable energy access, climate, and regulatory conditions, such as Dublin, the Netherlands, and Denmark [1]. These clusters put pressure on local energy systems through increased energy demand and grid congestion, potentially slowing down their energy transition [2]. Still, DCs also offer untapped opportunities to support the transition through their integration with the rest of the energy system. Integrated DCs can provide flexibility to electricity markets through demand response, while also recovering the waste heat

J. Jerez Monsalves (✉) · J. Gea-Bermúdez · C. Bergaentzlé · D. Keles
Technical University of Denmark, Kongens Lyngby, Copenhagen, Denmark
e-mail: jujmo@dtu.dk

O. Grothe et al. (eds.), *Operations Research Proceedings 2022*, Lecture Notes in Operations Research, https://doi.org/10.1007/978-3-031-24907-5_29

produced by their servers for further use. Demand response allows higher penetration of non-dispatchable sources (such as wind power), thus reducing the carbon footprint associated with DC electricity use. Waste-heat recovery substitutes other heat generators, reducing fuel consumption, costs, and emissions, especially if that waste heat comes from servers consuming already decarbonized electricity thanks to demand response.

Yet, policymakers face great uncertainty in assessing the industry's future energy use due to potential and unforeseen project cancelations or long-term technological developments in hardware. This uncertainty underscores the importance of energy system models as tools for exploring the range of possible outcomes on energy supply derived from different industry development paths.

The literature on energy use in DCs mainly focuses on individual facilities from the perspective of technical energy efficiency [3] or economic optimization of energy use [4]. In Petrović et al. [5], the authors use an energy-system perspective, using the Times-DK model, to optimize the Danish energy system along with DCs recovering waste heat to district heating (DH) systems through heat pumps coupled to their existing cooling systems. It is shown that waste heat from DCs can provide from 4 to 27% of DH after 2040. A similar analysis uses the Times-Ireland model to evaluate the impacts of DCs in Irish emissions [6]. To our knowledge, no study covers the energy value chain from flexible electricity consumption, to cooling generation, to waste-heat recovery.

This paper extends the scope of [5] by developing a tool for the joint optimization of cooling generation in DCs and the rest of the energy system in terms of investment decisions and operational planning. This tool has been implemented as an extension of the Balmorel energy system model and is tailored to consider competing cooling, thermal storage, and waste-heat recovery technologies with potential use in DCs.

This article is structured as follows: "Balmorel Energy System Model" introduces the Balmorel model. "Modeling of Integrated Data Centers" describes integrated DCs, the technologies considered, and the mathematical formulation. "Example Case" showcases an application example. "Concluding Remarks" concludes with a summary of its main contributions, limitations, and possible applications.

Balmorel Energy System Model

Balmorel is an open-source [7], deterministic, bottom-up energy system model that minimizes the socio-economic cost of energy supply. Investment decisions and operational scheduling of generation and transmission assets are optimized subject to resource-availability, technical, and policy constraints. Balmorel is tailored to represent closely connected DH and electricity systems with high detail. It also has a high degree of flexibility in spatiotemporal representation and extensibility. Although first developed for the Baltic Sea countries, Balmorel has been used worldwide by academia, authorities, and companies thanks to its modelling versatility: from hourly city-level dispatch to international long-term investment analyses. For a deeper

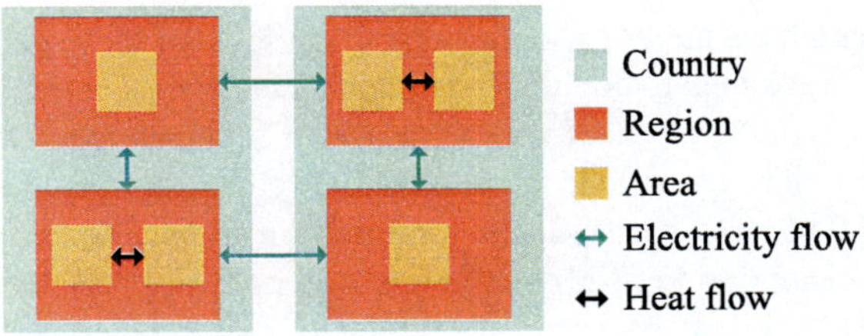

Fig. 29.1 Diagram of spatial hierarchy and energy flows in Balmorel [9]

description of the model and its use cases, the reader is referred to the work of Wiese et al. [8].

The spatial and temporal dimensions are each represented by three hierarchical levels. Space is represented through a nested Country-Region-Area structure. Balmorel defines the balance and transmission of electricity at the regional level, whereas those of heat are defined in and between areas, as illustrated in Fig. 29.1. Similarly, time is represented through a Year-Season-Term structure, where Years are composed of Seasons, and Seasons are composed of Terms. Investment decisions are made each Year, while operational scheduling is determined at each Term. Seasons serve to represent intra-seasonal and inter-seasonal storages with different cycling patterns.

Moreover, Balmorel can be extended through add-ons to enhance its modeling capabilities [9]. Add-ons allow for an improved representation of specific energy sectors, which this paper utilizes to optimize the cooling portfolio of DCs, instead of considering them as only an exogenous electricity demand.

Modeling of Integrated Data Centers

Demand Response and Waste-Heat Recovery in Data Centers

Most of the electricity consumed by DCs is used for IT operation or temperature/humidity control, any of which can be modulated to provide demand response. Yet they have different characteristics. On the one hand, DCs can shut down servers or shift non-critical workloads within seconds, but these actions interfere with their core business. Batteries can provide flexibility, too; however, their capacity and degradation limit their potential. On the other hand, cooling-based demand response is limited due to tightly controlled DC temperatures. Reduction of cooling output is unacceptable for business-critical facilities where reliability is paramount, and impractical for facilities with high power densities. Moreover, thermal systems' start-up costs and response times further limit this type of flexibility. However, cold storage units can overcome these issues, increasing operational reliability, and providing thermal inertia to accommodate flexibility from slow-reacting chillers. They are also cheaper and have larger capacities than batteries. For these reasons, the scope of this article is restricted to cooling-based demand response employing cold storage.

Another benefit of integrated DCs comes from recovering the heat produced by their operation. Air-cooled DCs generate waste heat at a temperature of about 30 °C, which is sufficient for direct use in space heating, but distribution losses limit this option to nearby buildings. Alternatively, DH networks can accept waste heat. However, a heat pump is required to bridge the temperature gap between them. Heat pumps can either be attached to existing cooling systems or replace them entirely, thereby providing cooling themselves. In this work, the latter approach is considered.

Technology Portfolio

The portfolio consists of several cooling technologies described by techno-economic parameters such as costs and efficiencies that are briefly described below as cooling generation capacities (G):

- Cold storage (G_{STO}): Different types of thermal storage can be defined by their seasonal pattern, charge/discharge rates, and other techno-economic parameters.
- Free cooling (G_{FC}): Cooling systems that operate when outdoor temperatures are lower than those of the server room, allowing heat to flow naturally outwards. Auxiliary equipment such as fans can be accounted by their efficiencies and costs.
- Electric chillers (G_{EL}): Electricity-driven mechanical cooling systems discharging waste heat outdoors without further use. These systems are the most commonly used in DCs, along with free cooling when climate conditions allow it.
- Absorption chillers (G_{ABS}): Heat-driven mechanical cooling systems discharging waste heat outdoors without further use. They use high-temperature heat and lower electricity consumption than electric chillers.
- Heat pumps (G_{HP}): Electricity-driven mechanical cooling systems providing waste heat for use in other Balmorel heat areas.

Mathematical Formulation

This work follows the modeling and data processing framework of Balmorel and uses features already present in the core model as much as possible. Therefore, only the changes and new additions made to the core model are highlighted in the following mathematical formulation, whose nomenclature is summarized in Table 29.1.

Figure 29.2 illustrates a DC within Balmorel, its energy flows, and relationship to the broader model. The cooling demand and allowed technologies are defined in areas specific to each type of DC where no other external technologies or demands are permitted. This approach separates cooling (inside the DC) from external DH flows, as both are modeled internally as heat. The core model handles the electrical aspect of DCs.

Table 29.1 Data center model nomenclature

Acronym		**Sets**		**Parameters**	
DC	Data center	G	Technology	x	Generation
STO	Storage	A	Area	x'	Storage consumption
FC	Free cooling	T	Term	d	Demand
EL	Electric chiller	**Indices**		β	Free-cooling availability
ABS	Absorption Chiller	c	Cooling	α	Heat-to-cold ratio
HP	Heat pump	h	Heat	v	Heat transmission

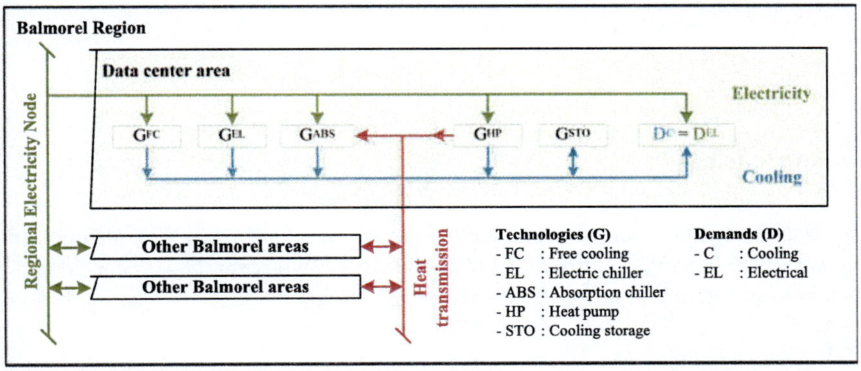

Fig. 29.2 Schematic representation of an integrated DC and its energy flows in Balmorel

The balance between cooling generation x_g from each technology g, storage loading x'_g, and demand d is described by Eq. (29.1), where t and a stand for time steps and areas, respectively, and A_{DC} is the set of all areas representing DCs.

$$\sum_{g \in G} x^c_{g,a,t} - \sum_{g \in G_{STO}} x'^c_{g,a,t} = d^c_{a,t} \quad \forall t, a \in A_{DC} \tag{29.1}$$

The ability of free cooling to cover demand may be partial or non-existent at specific time steps, as it depends on external weather conditions. Therefore, the total production of free cooling technologies in each area is thus limited by the availability parameter $\beta \in \mathbb{R} \geq 0$, as shown in Eq. (29.2).

$$\sum_{g \in G_{FC}} x^c_{g,a,t} \leq \beta_{a,t} d^c_{a,t} \quad \forall t, a \in A_{DC} \tag{29.2}$$

Heat production from heat pumps or consumption by absorption chillers is calculated through their heat-to-cold ratio, $\alpha_g \in \mathbb{R}$, as shown in Eq. (29.3). Positive values denote heat production, i.e., waste-heat recovery, while negative ones denote consumption.

$$x^h_{g,a,t} = \alpha_{g,a,t} x^c_{g,a,t} \forall t, a \in A_{DC}, g \in G_{HP \vee ABS} \tag{29.3}$$

Finally, the heat production/consumption determined in the above equation is connected to other heat areas in Balmorel, a', through Eq. (29.4). The left side of this equation is the total heat available for exchange from DC facilities, while the right side denotes heat transmission, v, to/from other areas.

$$\sum_{g \in G_{HP-\mathrm{ABS}}} x^h_{g,a,t} = \sum_{c'} (v^h_{a,c',t} - v^h_{c',a,t}) \forall t, a \in A_{DC} \tag{29.4}$$

Example Case

The Danish energy system is optimized to analyze how DCs with flexible cooling and waste-heat recovery impact cost and emissions. Two scenarios are considered, each with different technologies allowed in DCs: "BAU" (Business as Usual) and "IDC" (Integrated DCs). The former allows free cooling and electric chillers only, while the latter allows all the technologies described in "Technology Portfolio". All other parameters are kept constant so that "IDC" highlights the effects of integrated DCs.

The energy system consists of the electricity and DH sectors, both being part of the core Balmorel model; as well as hydrogen demand and generation technologies, industrial heat, and individual heating users as indicated in [10] (all included as add-ons). Finally, the DC cooling systems are included but not their investments in IT equipment. The optimal set of investment and dispatch decisions that minimize energy supply costs is found from the following exogenously defined parameters: spatiotemporally defined energy demand (including DCs) and renewable potentials, existing generation capacity, technology costs and efficiencies, fuel costs, and carbon taxes [9]. No curtailment costs have been assumed to incentivize investments in non-dispatchable renewable sources.

The optimization is run from 2025–2035, to account for all planned DC projects currently in the pipeline [10]. Each year consists of 192 time steps. For this example, a Balmorel version parametrized exclusively for Denmark is used, thereby excluding neighboring countries and cross-border electricity exchange. Denmark is then divided into several areas classified by their DH sizes or industrial heat temperature. DCs are assumed to be located next to the three largest DH areas.

Figure 29.3 shows how cooling is produced in a DC area in 2030. It is observed that heat pumps operate consistently throughout the year, being complemented by free

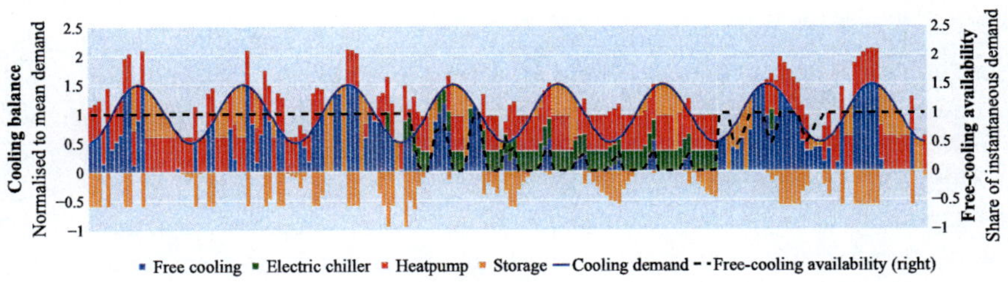

Fig. 29.3 Cooling balance of a data center in 2030

Table 29.2 Summary of main results from IDC scenario, shown as differences to BAU

Year	2025	2030	2035
DC demand—TWh_{el} [10]	3.39	7.49	9.63
System costs—million €	– 46.5 (– 1.7%)	– 76.5 (– 2.7%)	– 94.5 (– 3.0%)
System emissions—ktons CO_2	– 69 (– 0.8%)	+ 72 (+ 2.7%)	+ 78 (+ 5.8%)
DC waste-heat share in heat supply	5.07%	9.13%	9.13%

cooling during the winter and electric chillers during the summer when free cooling is less available. Intra-seasonal storage allows heat pumps to remain in operation even when other technologies meet all instantaneous demand. This way, DCs can provide waste heat when it is most needed in the system, thereby increasing its overall usage.

Table 29.2 summarizes the system-wide results, including all sectors considered in the model. Waste heat from integrated DCs covers up to 9% of the national heat supply in 2035, mostly displacing investments in heat pumps and, thus, in solar PV that powers them. This is reflected in lower system-wide costs, with savings from 1.7% in 2025 up to 3.0% in 2035, mostly in capital expenses.

Carbon emissions are reduced in the short term but increase in the future. Closer inspection shows waste heat displacing biomass cogeneration in some DH networks, thereby increasing the use of existing natural gas peak units in other areas to compensate for the lost electricity generation. Further analysis is required to verify if adding the effect of interconnectors reduces natural gas consumption, e.g., by importing cheaper electricity from hydropower-based Norway, or if it allows higher investments in non-dispatchable renewable electricity, made cost-competitive through the possibility of exporting electricity surplus and reducing curtailment.

Concluding Remarks

In the face of uncertain but disruptive developments, energy system models allow exploring the range of their potential impacts on energy supply. This article integrates

DCs providing demand response and waste-heat recovery, supporting decarbonization, into the Balmorel energy system model. Investment decisions and operational planning of DC's cooling systems are optimized from a socio-economic perspective.

The main limitation of this work is the lack of electrical flexibility from batteries or workload management. Another drawback of this work is that the internal implementation of cooling as heat complicates the final user's appropriate definition and use of the model. New energy carriers, however, cannot be implemented without significant modifications to the core Balmorel model.

The illustrative example reveals gains of up to 3.0% in system-wide cost reductions that integrated DCs would reach in Denmark. Even though carbon emissions are reduced in 2025, there is a net increase in 2030–2035 due to higher natural gas consumption. This can be a counterargument to the integration of DCs; however, further analysis is required to better represent the Danish electricity system, including electricity exchange with neighboring countries. This modeling tool will serve as a platform for future scenario analysis and policy testing directed at DC's integration and is adaptable to represent industries with similar energy uses as DCs.

References

1. Banet, C., Pollitt, M., Covatariu, A., Duma, D. (2021). Data Centres & the Grid. Centre on Regulation in Europe (CERRE), Brussels, Belgium
2. Kamiya, G., & Kvarnström, O. (2019). Data centres and energy—from global headlines to local headaches? International Energy Agency. https://www.iea.org/commentaries/data-centres-and-energy-from-global-headlines-to-local-headaches
3. Cioara, T., et al. (2018). Optimized flexibility management enacting data centres participation in smart demand response programs. *Future Generation Computer Systems, 78*, 330–342. https://doi.org/10.1016/j.future.2016.05.010
4. Díaz, A. J., et al. (2020). Effect of climate conditions on the thermodynamic performance of a data center cooling system under water-side economization. *Energy and Buildings, 208*, 109634. https://doi.org/10.1016/j.enbuild.2019.109634
5. Petrović, S., et al. (2020). The role of data centres in the future Danish energy system. *Energy, 194*, 116928. https://doi.org/10.1016/j.energy.2020.116928
6. Collina, L. (2022). Data centers in the future Irish energy system: scenario analysis using TIMES-Ireland Model. University College Cork
7. Balmorel Community. https://github.com/balmorelcommunity. Accessed July 27, 2022
8. Wiese, F., et al. (2018). Balmorel open source energy system model. *Energy Strategy Rev, 20*, 26–34. https://doi.org/10.1016/j.esr.2018.01.003
9. Gea-Bermúdez, J., et al. (2021). The role of sector coupling in the green transition: A least-cost energy system development in Northern-central Europe towards 2050. *Applied Energy, 289*, 116685. https://doi.org/10.1016/j.apenergy.2021.116685
10. Danish Energy Agency. (2022). Klimastatus og –fremskrivning 2022: 6A datacentre. Danish Energy Agency, Copenhagen, Denmark

Chapter 30
Inventing and Assessing Simple Heuristics for Bidding in Wholesale Electricity Markets

Jake Atkinson, Richard Allmendinger, and Joshua Knowles

Abstract In areas of commerce, fast and frugal heuristics have a demonstrated history of forecasting as or more accurately than complex strategies in situations of uncertainty. Financial participation in the deregulated electricity market requires accurate forecasting and presents an opportunity for the implementation of heuristics by traders seeking to arbitrage. Developmental research was undertaken to invent and assess simple heuristics for bidding in wholesale electricity markets. Two heuristic price-adjustment bidding strategies were developed, providing modest returns relative to existing strategies, greater risk aversion and reduced computational complexity.

Keywords Heuristics · Optimisation · Bidding · Arbitrage

Introduction

In deregulated markets, the trade of electricity between generators and retailers is facilitated through a wholesale dual-market, operated by independent system operators (ISO) [1]. Market mechanisms allow participants to submit bids to supply (INC bid) and distribute (DEC bid) energy across regions through a day-ahead (DAM) and real-time market (RTM). The dual market facilitates economic dispatch, generation scheduling and supply and demand forecasting.

Whilst the RTM represents a physical auction, the DAM is virtual and allows for participation by wholly financial entities. Entities may buy (sell) energy in the DAM with the obligation to sell (buy) it back in the RTM. INC bids are charged in the RTM and credited in the DAM and DEC bids charged in the DAM and

J. Atkinson (✉)
Spring Financial Group, Guildford, UK
e-mail: jake.atkinson@mqube.com

J. Atkinson · R. Allmendinger · J. Knowles
The University of Manchester, Manchester, UK
e-mail: richard.allmendinger@manchester.ac.uk

O. Grothe et al. (eds.), *Operations Research Proceedings 2022*, Lecture Notes in Operations Research, https://doi.org/10.1007/978-3-031-24907-5_30

credited in the RTM [2, 3]. Prices in both markets reflect locational marginal pricing (LMP), which accounts for the cost of electricity at specific times and locations [4].

Methodology

Data: Pricing data from the Pennsylvania-New Jersey-Maryland Interconnection (PJM) regional transmission organisation were used in the development and evaluation of bidding strategies. Data contained DAM and RTM LMPs (USD/unit) for 346 unique nodes, disaggregated by date and time throughout 2017. In forecasting, data were randomly divided, with 70% of forecasts reflecting values $\pm$ 5% of the LMP (within three standard deviations) and 30% of the forecasts reflecting historical mean prices, unique to the considered node.

The Mathematics of Virtual Bidding: Consider the trade of options (node-time pair), for each hour in the forward market. For each option, market participants forecast the DAM and RTM prices. In this notation, for virtual bids submitted on day d for option i, let $\gamma_{d,i}$ and $\pi_{d,i}$ represent the DAM and RTM clearing prices respectively. Also, let $\gamma'_{d,i}$ and $\pi'_{d,i}$ represent the respective forecasts of such prices. Further, let $\upsilon_{d,i}$ be the virtual bid submitted, as some function of $\gamma'_{d,i}$ or $\pi'_{d,i}$.

DEC bids are cleared by the ISO iff the virtual supply bid is greater than or equal to the DAM clearing price, i.e. $\upsilon_{d,i} \geq \gamma_{d,i}$. Conversely, INC bids are cleared by the ISO iff the virtual supply bid is less than or equal to the DAM clearing price, i.e $\upsilon_{d,i} \leq \gamma_{d,i}$.

As the distribution of prices is unknown, forecasts direct the value of virtual bids by market participants, such that submitted bids are representative of some function of the forecast. To maximise delta, where $\gamma'_{d,i} > \pi'_{d,i}$, a DEC bid is submitted, otherwise where $\gamma'_{d,i} < \pi'_{d,i}$, an INC bid is submitted.

Profit is realised when cleared bids are confluent with a positive δ. Such that for DEC bids:

$$\delta_{d,i} = \pi_{d,i} - \gamma_{d,i}, \text{ given that } \{\upsilon_{d,i} \geq \gamma_{d,i}\} \tag{30.1}$$

Conversely for INC bids:

$$\delta_{d,i} = \gamma_{d,i} - \pi_{d,i}, \text{ given that } \{\upsilon_{d,i} \leq \gamma_{d,i}\} \tag{30.2}$$

Hence, the cumulative δ over period s for cleared bids $\upsilon_{i,d}$ is denoted:

$$\sum_{d=1}^{s} (\pi_{d,i} - \gamma_{d,i}), \text{ given that } \{\upsilon_{d,i} \geq \gamma_{d,i}\} \tag{30.3}$$

In simulations, the required DAM and RTM prices as well as the submitted bids, over time period s, are passed as vectors $\gamma_d = [\gamma_{d,1}, \ldots, \gamma_{d,i}]^s$, $\pi_d = [\pi_{d,1}, \ldots, \pi_{d,i}]^s$ and $\upsilon_d = [\upsilon_{d,1}, \ldots, \upsilon_{d,i}]^s$ respectively, producing a vector of returns: $\delta_d = [\delta_{d,1}, \ldots, \delta_{d,i}]^s$.

Price-Setting Adjustment Mechanism: It follows that an evaluation of original DAM and RTM LMP forecasts determine whether a DEC or INC bid is initialised. Once initialised, a price adjustment on the forecast DAM LMP sets the submitted bid price, such that:

$$\upsilon_{d,i} = \gamma'_{d,i}(1 \pm \theta) \tag{30.4}$$

where, $(1 \pm \theta)$ is the price adjustment and θ is a tuneable parameter. The operator is conditional on the nature of the bid, i.e., DEC(+) and INC(−). Bid acceptance was contingent on the market clearing price as detailed in Eqs. (30.1) and (30.2). Strategies were compared on the number of bids cleared and the total market return achieved on a cumulative and per bid basis, following similar assessments in the literature [5]. Trading fees required by ISOs were not considered but may prevent the adoption of strategies benefiting from large bid volumes.

Benchmarking

Three distinct price-setting strategies used by dominant participants in the CAISO market were identified and replicated to establish benchmarks [5]. Strategies take advantage of the uncertainty of forecasts and volatility of the RTM. Following the replication of strategies, price adjustments were set in accordance with market participant behaviour [6]. Bids submitted through Strategy 1 set the price such that for INC bids $\upsilon_{d,i} < \gamma'_{d,i}$ and for DEC bids $\upsilon_{d,i} > \gamma'_{d,i}$. Strategy 1 represents the smallest price adjustment out of all considered strategies at 5%. In practice, this results in an INC bid of $0.95\gamma'_{d,i}$ and a DEC bid of $1.05\gamma'_{d,i}$. Price adjustments such as those implemented through this strategy result in the award of almost all bids, given accurate forecasts. Strategy 2 is ignorant of market volatility and results in most submitted bids being awarded, given that DAM clearance prices are not negative or extremely high for INC and DEC bids respectively. Bid prices are set such that $\upsilon_{d,i} = 0$ for INC bids and $\upsilon_{d,i} = 2\gamma'_{d,i}$ for DEC bids; this Strategy exposes bidders to entering the market when DAM prices are extreme. Strategy 3 is risk-averse and operates to exploit market volatility. As price distributions are unknown, this strategy offers an exceedingly lower bids clearance rate compared to other strategies, indeed, arbitrage is only realised on cleared bids. For INC bids: $\upsilon_{d,i} = 3\gamma'_{d,i}$, in contrast for DEC bids $\upsilon_{d,i} = -\gamma'_{d,i}$. Strategy 3 sets prices such that bids are almost always not awarded, securing bid clearance only in a volatile market. Strategy 1 and 2 produce a bid clearance greater than 80% and a return greater than 2.3m USD in simulation.

Table 30.1 Performance in bidding of benchmark strategies in the PJM (2017)

Strategy	Clearance (%)	Total delta ($)	$/Unit
(1)	89.60	2,333,090	0.95
(2)	99.40	2,398,269	0.80
(3)	1.60	33,036	18.66

Despite a relatively lower bid clearance of 1.6%, Strategy 3 realises an average 20-fold and 23-fold gain per accepted bid compared to Strategy 1 and 2 respectively (Table 30.1).

Simple Heuristics for Bidding in Energy Markets

As distributions are unknown and in lieu of complex forecasting, we present two heuristics with bid prices set as a function of price anchors (historical mean clearing prices on a per option basis). The use of historical means reduces time complexity and resource demand, inherent to existing bidding strategies [7, 8]. As the magnitude of the price adjustment can be adapted based on foreseen or known market conditions, heuristics may provide tools for use in the adaptive toolbox of market participants.

Price anchors consisted of historical means discretised by each option. Let $\bar{\gamma}_{d,i}$ and $\bar{\pi}_{d,i}$ be the historical mean clearing prices in the DAM and RTM respectively and let N represent the number of historical observations of cleared node-time pairs. Historical means were used to determine the type of bid submitted in simulation to maximise delta, such that where $\bar{\gamma}_{d,i} > \bar{\pi}_{d,i}$, a DEC bid was submitted and where $\bar{\gamma}_{d,i} < \bar{\pi}_{d,i}$, an INC bid was submitted.

Mean historical forward clearing prices were calculated (30.5), similarly, mean historical real-time clearing prices were set (30.6):

$$\bar{\gamma}_{d,i} = \frac{\sum_{d=1}^{s} \gamma_{d,i}}{N_{d,i}} \tag{30.5}$$

$$\bar{\pi}_{d,i} = \frac{\sum_{d=1}^{s} \pi_{d,i}}{N_{d,i}} \tag{30.6}$$

Zero-Adjustment Historical Average: The Zero-Adjustment Historical Average (ZAHA) heuristic omits the integration of a price adjustment, submitting bid prices exclusively reflecting the associated price anchor. Consequently, ZAHA requires relatively minimal understanding of its learning environment, similarly, bids are not made based on predictions and are only accepted given that DAM clearance prices are congruent with their historical averages, it is the most frugal heuristic developed in this research.

The value of submitted bids is equal to the historical mean, such that:

$$v_{d,i} = \bar{\gamma}_{d,i} \tag{30.7}$$

Some strategies reduce forecasting risk through the integration of a risk aversion parameter [7]. ZAHA may be considered risk averse given that submitted bids are reflective of historical means. The heuristic offers a compromise between high bid clearance, at risk of entering a volatile market, and higher cumulative returns compared to the most risk averse strategies in the literature [5].

High Clearance Historical Average: For each option, the High-Clearance Historical Average (HCHA) heuristic determines the bid type and sets the bid price, based on historical means. Bid price is established through a price adjustment function. For INC bids, prices are set significantly lower than the mean, such that all INC bids are set at 0 ($v_{d,i} = 0$). As DEC bids are most likely to be cleared when a bid is offered relatively higher than other bids, bid prices are submitted such that they represent double the historical average for the considered option: $v_{d,i} = 2\bar{\gamma}_{d,i}$. The HCHA heuristic is not risk averse and only fails to achieve bid clearance if the relationship between the DAM and RTM historical means is inconsistent with the actual DAM price. Further, bids may not be cleared if the actual DAM price is extremely high ($\gamma_{d,i} > 2\bar{\gamma}_{d,i}$), or negative ($\gamma_{d,i} < 0$).

Discussion

Table 30.2 details the clearance rate, total delta and average delta per unit cleared in simulation for developed heuristics. Whilst distinct in their mechanisms, both heuristics were successful in generating total returns in line with those of dominant participants in the CAISO market [5].

Whilst Strategies 1 and 2 generated the second highest cumulative delta amongst those considered, they are dependent on computationally intensive predictive algorithms. Novel Heuristics ZAHA (4) and HCHA (5) are fast-and-frugal and whilst obtaining lower than average ($\mu_{\delta,b} = 1.50 \times 10^6$) returns compared to benchmarks, they required minimal computation in setting price anchors and for two of three benchmark strategies, provided greater risk-aversion.

Figure 30.1 describes the cumulative delta versus bid clearance for all strategies. Given the implementation of frugal price anchors, the HCHA heuristic achieved

Table 30.2 Performance in bidding of novel heuristics in the PJM (2017)

Strategy	Clearance (%)	Total delta ($)	$/Unit
ZAHA (4)	73.00	520,833	0.33
HCHA (5)	99.40	1,925,032	0.64

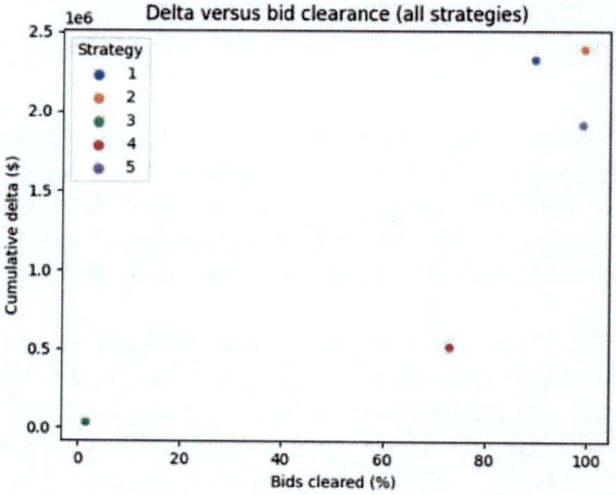

Fig. 30.1 Delta versus bid clearance amongst benchmark and novel strategies

cumulative returns of up to 80% that of Strategy 2 (the benchmark strategy with the highest clearance rate), without the associated forecasting expense. Similarly, the ZAHA heuristic, setting bid prices simply as historical means, achieved cumulative returns relative to 64% of that achieved through the deployment of Strategy 1. In the context of participation in the wholesale electricity market, price adjustment bidding strategies have the potential to provide traders with various methods to set bid prices dependent on the environment to which they are subjected.

Limitations on the extent of market simulation in the replication of the function of ISOs constrained the effective representation of market behaviour. Similarly, forecasting methods applied in research to facilitate bidding processes, may not have accurately represented those used in industry, impacting the generalisability of the results obtained.

Conclusion

Fast and frugal heuristics have shown to lead to more robust and ecologically rational decisions than complex decision-making methods in uncertain environments. Bidding (price-setting) in energy markets to maximize profits takes place in such an environment. We developed simple heuristics and validated them on pricing data from the Pennsylvania-New Jersey-Maryland Interconnection regional transmission organisation. Our experimental study concluded that simple heuristics are able to provide modest returns relative to existing strategies, greater risk aversion and reduced computational complexity. Future work can investigate the suitability of the proposed heuristics for other energy markets, consider the design of more advanced heuristics, and attempt simulating additional aspects of the energy market (such as trading fees).

References

1. PJM. (no date). *Market for electricity* [Online]. Available at: https://learn.pjm.com/electricity-basics/market-for-electricity.aspx. Last accessed: September 5, 2021.
2. Wang, W., & Yu, N. (2019). A machine learning framework for algorithmic trading with virtual bids in electricity markets. In 2019 IEEE Power and Energy Society General Meeting (PESGM) (pp. 1–5). IEEE.
3. Tang, W., Rajagopal, R., Poolla, K., & Varaiya, P. (2016). Model and data analysis of two-settlement electricity market with virtual bidding. In *2016 IEEE 55th Conference on Decision and Control (CDC)* (pp. 6645–6650). IEEE.
4. Fan, Z., Horger, T., Bastian, J., & Ott, A. (2008). An overview of PJM energy market design and development. In *2008 Third International Conference on Electric Utility Deregulation and Restructuring and Power Technologies* (pp. 12–17). IEEE.
5. Samani, E., & Mohsenian-Rad, H. (2021). A data-driven study to discover, characterize, and classify convergence bidding strategies in California ISO Energy Market. In *2021 IEEE Power and Energy Society Innovative Smart Grid Technologies Conference (ISGT)* (pp. 1–5). IEEE.
6. Albahli, S., Shiraz, M., & Ayub, N. (2020). Electricity price forecasting for cloud computing using an enhanced machine learning model. *IEEE Access, 8,* 200971–200981.
7. Xiao, D., Qiao, W., & Qu, L. (2018). Risk-constrained stochastic virtual bidding in two-settlement electricity markets. In 2018 IEEE Power and Energy Society General Meeting (PESGM) (pp. 1–5). IEEE.
8. Baltaoglu, S., Tong, L., & Zhao, Q. (2018). Algorithmic bidding for virtual trading in electricity markets. *IEEE Transactions on Power Systems, 34*(1), 535–543.

Chapter 31
Mathematical Optimization for Analyzing and Forecasting Nonlinear Network Time Series

Milena Petkovic and Nazgul Zakiyeva

Abstract This work presents an innovative short to mid-term forecasting model that analyzes nonlinear complex spatial and temporal dynamics in energy networks under demand and supply balance constraints using Network Nonlinear Time Series (TS) and Mathematical Programming (MP) approach. We address three challenges simultaneously, namely, the adjacency matrix is unknown; the total amount in the network has to be balanced; dependence is unnecessarily linear. We use a nonparametric approach to handle the nonlinearity and estimate the adjacency matrix under the sparsity assumption. The estimation is conducted with the Mathematical Optimisation method. We illustrate the accuracy and effectiveness of the model on the example of the natural gas transmission network of one of the largest transmission system operators (TSOs) in Germany, Open Grid Europe. The obtained results show that, especially for shorter forecasting horizons, proposed method outperforms all considered benchmark models, improving the average nMAPE for 5.1% and average RMSE for 79.6% compared to the second-best model. The model is capable to capture the nonlinear dependencies in the complex spatial-temporal network dynamics and benefits from both sparsity assumption and the demand and supply balance constraint.

Keywords Nonlinear time series · Mathematical optimization · Energy networks

Introduction

Since the EU introduced market regulations in 2005, the natural gas market is becoming increasingly competitive, moving towards short-term planning, e.g., day-ahead contracts, making the control of natural gas transmission networks even more challenging. The main task of TSOs is to fulfill all transport demands, ensuring the

M. Petkovic (✉) · N. Zakiyeva
Applied Algoritmic Intelligence Methods Department, Zuse Institute Berlin,Takustraße 7, 14195 Berlin, Germany
e-mail: petkovic@zib.de

N. Zakiyeva
e-mail: zakiyeva@zib.de

O. Grothe et al. (eds.), *Operations Research Proceedings 2022*, Lecture Notes in Operations Research, https://doi.org/10.1007/978-3-031-24907-5_31

security of supply safely and efficiently. Since gas in the pipes travels relatively slow with an average velocity of approximately 25 km/h [5], a high-precision short and mid-term forecast of supplies and demands is essential for the efficient and safe operation of the complex natural gas transmission networks and distribution systems.

This work is part of a joint project within the Energy Lab of a research campus MODAL [6] with one of Germany's largest transmission system operators, Open Grid Europe (OGE) [1]. Together with our industry partner, we develop a Network AutoRegressive Nonlinear model with a Balance constraint (NAR-NLB) model. The primary purpose of the proposed model is to provide a comprehensive understanding of the network dynamic and compute high-precision, multi-step, hourly forecasts for supply and demand nodes in the gas network. The focus is on forecasting shorter horizons (up to 8 h is the most relevant horizon in practice) to support the daily operations of the gas network. The results are used for optimizing gas transport, for example, routing the gas with compressors or valves and finding the optimal settings for these elements [9].

Methodology

Let N denote the number of nodes in a large-scale complex gas transmission network, and $y_{t,i}$ is the continuous response collected from node i at time point t with $0 \leq t \leq T$ and $1 \leq i \leq N$. In the network, N nodes are connected with pipelines but the flow connection is unknown. At any time point t, the total sum of gas in-flow and out-flow in the network equals zero. To capture the network effect of the N different nodes, we propose the NAR-NLB model, where the total gas in-flows and out-flows need to be balanced. Without loss of generality, we assume the demeaned process for the gas network and write the model without the intercept term. The NAR-NLB model with lag 1 is defined as:

$$\begin{aligned} y_{t,i} &= g_i\left(\sum_{j=1}^{N} y_{t-1,j} b_{j,i}\right) + \epsilon_{t,i}, \quad i, j = 1, \dots, N, \\ s.t. \ &\sum_{i=1}^{N} g_i\left(\sum_{j=1}^{N} y_{t-1,j} b_{j,i}\right) = 0 \text{ for all } t = 1, \dots, T, \end{aligned} \tag{31.1}$$

where $g(\cdot)$ is an unknown link function which is assumed to be smooth. $\epsilon_{t,i}$ is a strong white noise with zero mean and finite second moment $E||\epsilon_{t,i}||^2 < \infty$. When $j = i$, $b_{j,i}$ controls the autoregressive dependence. When $j \neq i$, the parameter $b_{j,i}$ tells us how the j-th node influences the i-th node; that is, the network influence of the past value of the j-th node on the current value of the i-th node. If $b_{j,i} = 0$ for all the $i = 1, \cdots, N$ and $i \neq j$, then the j-th node has no effect in the network. In (31.1), the constraint is imposed to the forecast of gas flows denoted as $\sum_{j=1}^{N} g(y_{t-1,j} b_{j,i})$ for a balanced demand and supply.

The model can be represented in a matrix form:

$$\begin{aligned} Y &= g(ZB) + E, \\ s.t. \quad g(ZB)1_N &= 0, \end{aligned} \tag{31.2}$$

where Y is a $T \times N$ matrix of the observed gas flow values. Let $y_t = (y_{t,1}, \ldots, y_{t,N})$ denote the gas values for all nods at time point t, we then have $Y = (y_1, \ldots, y_T)^\tau$. Z is a $T \times N$ matrix containing the past values of Y. Similarly, we have $Z = (z_1, \ldots, z_T)^\tau$. The parameter matrix B is a $N \times N$ matrix with unknown parameters $b_{j,i}$. Given the autoregressive dependence reflected by $b_{i,i}$ for $i = 1, \ldots, N$ in the diagonal elements of B, let the non-diagonal elements of matrix B define *the weighted adjacency matrix*. The column vector $B_i = (b_{1,i}, \ldots, b_{N,i})^\top$ of the weighted adjacency matrix represents the influence of other nodes in the network on the future value of the i-th node. The weighted adjacency matrix is assumed to be sparse. There is, however, no prior knowledge of the sparse structure in terms of location and number of significant elements. Finally, 1_N is a unit vector and E is a $T \times N$ matrix of white noise errors.

Next, we show the estimation of the unknown nonlinear function and parametric coefficients with mathematical programming (MP). In semiparametric models, as in (31.1), it is popular to approximate the unknown nonlinear functions using the spline smoothing approach, see [2]. We apply this technique for estimating the nonlinear link function $g(\cdot)$ for a given parameter value B using B-splines, or so-called basis splines [3]. We estimate the unknown coefficient matrix by applying the feature selection technique developed by [4] for the weighted adjacency matrix, with additional balance constraints for the demand and supply as follows.

$$\begin{aligned} &\min_B \textstyle\sum_{t=1}^T \sum_{i=1}^N (y_{t,i} - Z_t B_i)^2 \\ &\text{s.t.} \quad ||B_i||_0 \le L \quad \text{for } i = 1, \ldots, N \\ &\textstyle\sum_{i=1}^N Z_t B_i = 0, \quad \text{for } t = 1, \ldots, T \end{aligned} \tag{31.3}$$

where the upper bound L for l_0-norm of a column vector B_i given by

$$\sum_{j=1}^N 1(b_{j,i} \neq 0)$$

ensures the number of nonzeros in B_i to be less than integer L, where $j \neq i$ and $1(\cdot)$ denotes the indicator function. We use the estimated coefficient matrix $\hat{B}_i$ for $i = 1, \ldots, N$ to approximate the unknown nonlinear function $g(\cdot)$ using the B-spline interpolation [2]. To estimate the function, the B-spline requires the hyperparameters such as knots, spline coefficients, and degree of a spline.

$$B_{k,m}(\theta) = \frac{\theta - \theta_k}{\theta_{k+m} - \theta_k} B_{k,m-1}(\theta) + \frac{\theta_{k+m+1} - \theta}{\theta_{k+m+1} - \theta_{k+1}} B_{k+1,m-1}(\theta), \tag{31.4}$$

$$B_{k,0}(\theta) = \begin{cases} 1, & \text{if } \theta_k \leq \theta < \theta_{k+1} \\ 0, & \text{otherwise} \end{cases}.$$

Finally, we estimate the B-spline coefficients $\alpha_{k,i}$ with MP.

$$\min_{\alpha_{k,i}} \sum_{t=1}^{T} \sum_{i=1}^{N} (y_{t,i} - \sum_{j=1}^{N} \sum_{k=0}^{P_k} \alpha_{k,i} B_{k,m}(Z_t \hat{B}_i))^2, \quad i = 1, \ldots, N, \tag{31.5}$$

Experimental Setup

In this paper, we study the nonlinear dependencies and dynamic patterns of natural gas flows in the high-pressure gas pipeline network of OGE [1]. The dataset consists of demand and supply nodes with an hourly time resolution for a period of one year. To demonstrate the effectiveness of the proposed model, we consider a small network of one supply (S1) and four demand nodes (D1-D4). Figure 31.1 illustrates the temporal dependence among the five observed nodes. As it can be seen in the diagonal, there is a strong positive autocorrelation of each node with its own past values, while off the diagonal, the cross-correlations represent the dynamic temporal dependency among different nodes.

We calculate an out-of-sample forecast in real time starting from 05:00 and predict 1 to 24 h ahead forecast. Multi-step forecasts are made for the next day, for a total of 3 months. We use the training-validation technique to choose optimal hyperparameters in the NAR-NLB model for the sparsity estimation, spline order and number of knots. With the chosen parameters, we estimate the weighted adjacency matrix B at each point by training the model on the past seven days of balanced network data. With a rolling window size of 168 h (i.e., seven days), we move forward one period at a time to update the sparse adjacency matrix and then forecast until we reach the end of the sample.

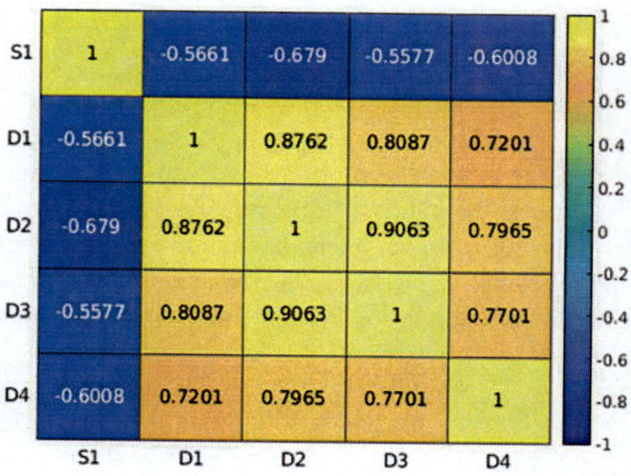

Fig. 31.1 Sample cross-correlation heatmap for one supply and four demand nodes in gas network

In order to evaluate the quality of the obtained results, we compare the performance of NAR-NLB model with well-known benchmarks: Baseline forecast (repeating value for the same hour of the previous day) and ARIMA as well as with Network Autoregressive Linear model with Balance constraint (NAR-LB) proposed by Zakiyeva and Petkovic in [8]. We determine the best ARIMA models for a univariate time series of five considered nodes according to an Akaike information criterion (AIC) using 28 days of the rolling window. The setup for NAR-LB model is identical to the proposed model. The performance of NAR-NLB model is measured and quantified by calculating the forecast accuracy for individual nodes, as well as the mean for the entire network. We use mean daily root mean squared error (RMSE) and mean daily normalized mean absolute percentage error (nMAPE) defined as:

$$\text{RMSE} = \frac{\sum_{d\in D_{test}} \sqrt{\frac{1}{H}\sum_{h=0}^{H-1}(q_{d,h} - \hat{q}_{d,h})^2}}{|D_{test}|},$$

$$\text{nMAPE} = \frac{\sum_{d\in D_{test}} (\frac{100\%}{H}\sum_{h=0}^{H-1} |\frac{q_{d,h}-\hat{q}_{d,h}}{max(q)}|)}{|D_{test}|},$$

where $q_{d,h}$ and $\hat{q}_{d,h}$ are the real and forecasted values of the natural gas flows on day d and hour h while H is a forecasting horizon.

Results

We demonstrate the multistep-ahead out-of-sample forecasting results in a balanced network. Table 31.1 shows an average $RMSE$ and $nMAPE$ for three different forecasting horizons (1h, 12h and 24h ahead) comparing to the alternative models for five gas nodes of the balanced network. The results show that NAR-NLB consistently outperforms all benchmark models. It can be noted that the difference is the smallest between NAR-LB and NAR-NLB models, which strongly indicates that using the network dynamic information as well as balancing constraint benefits the forecasting accuracy.

For shorter horizon, the NAR-NLB performs as the most accurate forecast model with the smallest forecast errors. This illustrates that modeling the nonlinear network dynamics improves the average forecast errors of the NAR-LB model from RMSE 6.294 and nMAPE 13.4% to RMSE 4.293 and nMAPE 1%. The difference in prediction performance with NAR-NLB is most significant for the shorter horizons, where nMAPE is improved for 5.1% and RMSE for 79.6% compared to the second best alternative model (ARIMA). As for the longer horizons (12h), NAR-NLB performs similar to NAR-LB model with the improvement of nMAPE by 1.4%. Similarly, for 24 h ahead forecast, NAR-NLB provides similar second-best accurate prediction as

Table 31.1 Comparison of the NAR-NLB model and the alternative time series models on multi step-ahead gas flow forecasts at five nodes in a balanced network.

	RMSE				nMAPE			
H	NAR-NLB	BAS	ARIMA	NAR-LB	NAR-NLB (%)	BAS (%)	ARIMA (%)	NAR-LB (%)
1	4.293	46.178	21.049	6.294	1	14.3	6.1	13.4
12	3.804	48.002	50.585	42.671	8.5	13.1	12.1	9.9
24	60.268	48.494	69.978	68.420	13.2	13.1	15.77	14.2

the BAS with a difference of nMAPE around 0.01%. Note that for calculating muti-step ahead forecast we are using recursive strategy, which can lead to accumulation of errors for longer horizons. The obtained results clearly show that proposed model benefits from modeling nonlinear temporal dependencies in the network.

By taking into account both nonlinear dynamics and sparse dependent structure, the NAR-NLB model provides superior performance compared to the three alternative predictive models. The NAR-NLB is able to capture the nonlinear dependencies in the complex spatial-temporal network dynamics. It improves out-of-sample forecast accuracy of individual nodes and consequently, there are fewer balancing errors in the network. Furthermore, the estimated adjacency matrix in NAR-NLB provides additional information on the cross-dependencies between the nodes, which shows the influential nodes that drive the network dynamics. In summary, it is useful to introduce both the nonlinearity and sparsity assumption together with the demand and supply balance constraint for accurate and stable forecasts in energy networks.

Conclusion

In this paper, we propose a network autoregression nonlinear model with balance constraint for robust short to mid-term forecasting and analyzing nonlinear complex spatial and temporal dynamics in energy networks under demand and supply constraints. The results show that the proposed model consistently outperforms the alternative models, improving the nMAPE by up to 5.1% compared to the second-best benchmark model, benefiting from modeling nonlinear dependencies between different nodes in the network and from implying balancing constraints on demand and supply.

Acknowledgements The work for this article has been conducted within the Research Campus Modal funded by the German Federal Ministry of Education and Research (fund numbers 05M14ZAM, 05M20ZBM).

References

1. Open Grid Europe GmbH, www.oge.net.
2. De Boor, C. (1978). *A practical guide to splines* vol. 27, p. 325. New York: Springer-Verlag.
3. Eubank, R. L. (1999). *Nonparametric Regression and Spline Smoothing*. CRC Press.
4. Bertsimas, D., King, A., & Mazumder, R. (2016). Best subset selection via a modern optimization lens. *The Annals of Statistics, 44*(2), 813–852.
5. Chen, Y., Koch, T., Zakiyeva, N., & Zhu, B. (2020). Modeling and forecasting the dynamics of the natural gas transmission network in Germany with the demand and supply balance constraint.*Applied Energy* vol. 278, ISSN 0306-2619.
6. Research Campus MODAL, EnergyLab. https://forschungscampus-modal.de
7. Petkovic, M., Chen, Y., Gamrath, I., et al. (2022). A hybrid approach for high precision prediction of gas flows. *Energy System* 13, 383–408. https://doi.org/10.1007/s12667-021-00466-4
8. Zakiyeva, N., & Petkovic, M. (2021). Modeling and forecasting gas network flows with multivariate time series and mathematical programming approach. In *Operations Research Proceedings 2021, GOR*.
9. Hoppmann-Baum, K., Hennings, F., Lenz, R., et al. (2021). Optimal Operation of Transient Gas Transport Networks. *Optimization and Engineering, 22*, 735–781.

Chapter 32
Maximization of the Smart Readiness Indicator of Buildings Under Budget Constraints

Tristan Emich, Shiva Faeghi, and Kunibert Lennerts

Abstract The Smart Readiness Indicator (SRI) is a method proposed by the European Commission which calls for better use of the potential of smart technologies in the building sector. The introduction of the SRI is intended to raise awareness of smart building technologies and make the added value more available for building users, owners, and providers of smart services. The technological smart readiness of buildings can be determined with the SRI assessment method. The method has 54 questions, which are divided into nine domains: heating, domestic hot water, cooling, ventilation, lighting, electricity, electric vehicles, dynamic envelope and monitoring & control. Each question is assessed with up to five different levels, representing incremental levels of technological equipment. These questions form the basis for the calculation of the SRI score. When improving the SRI score of a building, to choose the technologies that will provide the maximum SRI score improvement with a limited budget can be challenging. Therefore, the aim of this paper is to help the decision makers to come up with the correct choices that have the highest impact on the SRI score. The chosen solution method here is a specific non-dominated sorting genetic algorithm (NSGA II) algorithm. The proposed method is then applied to a hypothetical building to demonstrate its applicability and capability. The results show which SRI domains and questions to focus on. This gives future directions regarding choosing technologies to be implemented.

Keywords Combinatorial optimization · Energy policy and planning · Decision support systems

T. Emich (✉) · S. Faeghi · K. Lennerts
Institute of Tecnology and Management in Construction, Karlsruhe Institute of Technology, Karlsruhe, Germany
e-mail: tristan.emich@kit.edu

S. Faeghi
e-mail: shiva.faeghinezhad@kit.edu

K. Lennerts
e-mail: kunibert.lennerts@kit.edu

O. Grothe et al. (eds.), *Operations Research Proceedings 2022*, Lecture Notes in Operations Research, https://doi.org/10.1007/978-3-031-24907-5_32

Introduction

The European Union is the 3rd largest energy consumer – with 11% – after China (22%) and the United States (14%) [2], and it is very reliant on energy imports. There are environmental concerns and the urge to keep the global temperature increase to 1.5°. Therefore, large efforts are being made to cut down on imported energy and to reduce the CO_2 emissions.

Buildings are responsible for 40% of energy consumption and 36% of total CO_2 emissions in the EU [3], which presently is about 1100 Mt CO_2e [6]. Hence, there is an urge to look for cost-efficient strategies like smart technologies to reduce energy consumption and accordingly CO_2 emissions. The potential savings through digitization can be estimated at around 30 Mt CO_2e per year [6]. To reach this goal, the EU Energy Performance of Buildings Directive (EPBD) [4] recommends to make the buildings smart-ready by adopting controlling and automation systems. In this regard, the EPBD defines cost-effective measures to contribute to the above-mentioned goals, facilitate using renewable energy sources and provide a healthy and comfortable environment for the occupants. This method is called Smart Readiness Indicators (SRI). SRI is a useful method to measure the smartness level of a building.

With the EPBD, the European Commission calls for a better use of the potential of smart technologies in the building sector. Based on this directive, SRI was developed [5]. The introduction of such a uniform EU evaluation system is intended to assess the technological maturity of buildings, the extent to which interaction with the user as well as the energy network is possible and whether this enables more efficient management of a building. The intelligence capability indicator also aims to raise awareness of smarter building technologies and make their added value more tangible for building users, owners, tenants and smart service providers. It is intended to support technological innovation in the real estate industry and promote an incentive system for the integration of modern, intelligent and innovative technology in buildings.

The Smart Readiness Methodology categorises the building services, known as "Smart-Ready-Services" into nine domains: (1) heating, (2) cooling, (3) domestic hot water, (4) ventilation, (5) lighting, (6) dynamic building envelope, (7) electricity, (8) electric vehicle charging and (9) monitoring and control. The assessment of Smart-Ready-Services are done based on seven impact criteria: (1) energy efficiency, (2) maintenance and fault prediction, (3) comfort, (4) convenience, (5) health, well-being and accessibility, (6) information to occupants and (7) energy flexibility and storage. The impact criteria are taken into account by an individual weight, depending on the domain, the location (northern, western, southern, north-eastern and south-eastern Europe) and the type of building (residential or non-residential)[5]. Each domain has a variable number of questions, which can have up to 5 different levels. There are two method options: the simplified method with 27 questions and the detailed method with 54 questions. Not all of the questions are mandatory. For example, a building that has no air conditioning can still achieve the highest SRI score (100%). Each level of each question has a different impact on the SRI score of the building and imposes different costs. This means that focusing on some measures is more cost-effective than others.

At this point, the following question arises: Having calculated the smartness level of a building, on which domains and questions should be focused to maximise the smartness of a building considering budget limitations? In this paper, we propose a method to maximise the SRI score considering budget constraints. In this regard, we adopted the solution approach of genetic algorithms to search the decision scope. For the sake of simplicity, this paper assumes that all 54 questions have to be considered.

The paper is outlined as follows: Section "Materials and Methods" discusses the material and methods, describes the calculation of the SRI score, defines the problem, the objective function and the solution approach. Section "Numerical Example and Results" provides the numerical example and analyses the results and finally Section "Conclusions" concludes the paper and suggests ideas for future investigations.

Materials and Methods

Calculation of the SRI Score

Each level of each question has different impact value of each impact criteria. In addition, each domain and each impact criteria has an individual weighting that was defined by the European Commission: A distinction is made between a non-residential building and a residential building, and the location of the building determines the Climate zones (Northern Europe, Western Europe, Southern Europe, Northeastern Europe, Southeastern Europe), which has also an influence[5]. After each individual question is evaluated and taken into account with the individual weighting, the SRI score [%] can be calculated by using the following equation:

$$SRI = \sum_{q=1}^{54} \sum_{i=1}^{7} \frac{I_{i(level)}}{I_{i(max)}} * \omega_{i,q} \tag{32.1}$$

where q is the question number, i the impact criteria, $I_{i(level/max)}$ the highest Impact or the Impact of level, $\omega_{i,q}$ is the weight for the individual impact criteria.

Problem Definition

This paper aims at maximising the SRI score constrained by budget limitations using the genetic algorithm. If the SRI should be improved for a given building, this can be done by increasing the level of each individual question. Since each of the 54 questions of the SRI method can be evaluated independently, there is a wide range of possible combinations: 11 of the questions have 3 levels, 20 of the questions have 4 levels and 23 of the questions can reach 5 independent levels. This means that there are $5^{23} \cdot 4^{20} \cdot 3^{11} \approx 2.3 \cdot 10^{33}$ number of possible combinations in total. Therefore, the problem is a large combinatorial problem.

In addition, the improvement of each level is associated with costs. Since normally only a limited budget is available for a project, it makes sense to use the budget in such a way that the highest SRI score is achieved. Here we assume that each and every level improvement of the SRI score is associated with a cost of 1 monetary unit. Therefore, the optimization problem, considered as a single objective constrained optimization one, can be described as followed:

$$Maximize \quad SRI \tag{32.2}$$

$$Subject \quad To \quad \sum_{q=1}^{54} C_{q,\Delta level} \leq B \tag{32.3}$$

where $\Delta level$ is the difference between the current level and the improved level through implementation of SRI measures, $C_{q,\Delta level}$ is the cost of $\Delta level$ in questions q, and B is the budget limit.

Solution Approach

Since the decision space of the described problem is very large, the chosen solution approach is a genetic algorithm. Here a modified version of non-dominated sorting genetic algorithm II (NSGA-II) [1] is presented. In this modified version there are two mutation functions and one crossover function. The first mutation function mutates the best 100 combinations of each population and the second one applies random mutation to the whole population. These two mutations are supposed to increase diversity in the population and avoid local optima. Every mutation part is iterated and after each iteration the population is sorted to simulate tournament selection and multiple-mutation. Further, the crossover function is also iterated on the best 10 combinations. Here, there is a sorting after each iteration to simulate tournament selection and multi-crossover. We also added non-dominated sorting to each function to preserve elitism. The proposed algorithm is summarised in Fig. 32.1 and in the following steps:

Fig. 32.1 Code flowchart

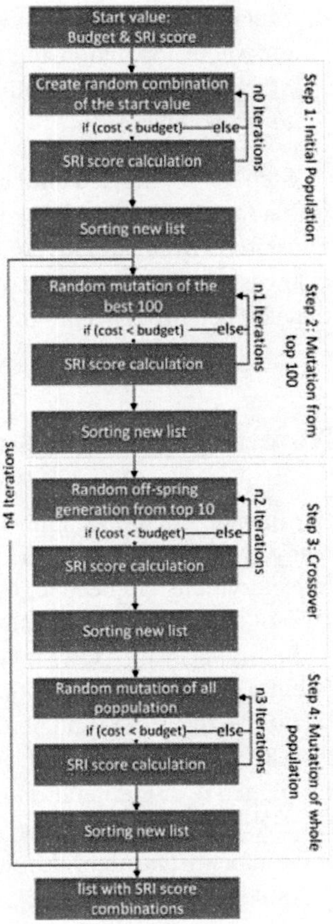

1. Initial population: At the beginning, random combinations are created, where the number of repetitions of the iterations is n0. Each of the 54 questions is assigned a value between the existing value and the maximum value to be achieved for each question. A subsequent review of the budget provides information on whether the randomly created combination can be used further on. If the combination meets the budget constraint (Eq. 32.3), it is transferred to a new list. Then, the surviving combinations is sorted and passed to the next step.
2. Create mutated population from top 100: A random combination of the best 100 is chosen and proceeded analogously as in the first step, where the number of repetitions of the iterations is n1. The generated population form Step 2 is added to the population.
3. Crossover: Two random combinations are chosen from the top 10 existing combinations to generate off-springs. The crossover function here is a k-point crossover, where k is chosen randomly. The number of repetitions of the iteration describes n2. The generated population form Step 3 is added to the population.

4. Mutation of the whole population: 100 combinations are chosen randomly from the whole population, mutated and proceeded as in the first step, where the number of repetitions of the iterations is n3. The generated population form Step 4 is added to the population.

Step 2-4 are iterated until the convergence criteria is met. The convergence criteria is that the changes in SRI score in the last 5 iterations are less than 1%. As it can be understood from the above-mentioned steps, that the populations size is dynamic and changes after each iteration. To prevent an uncontrolled growth of the population, they are penalized according to Equation (32.3).

Numerical Example and Results

For the numerical example, a hypothetical non-residential building located in Germany with the current SRI of 24.7% is chosen. To achieve a SRI of 100%, 130 monetary units are needed. However, the budget available in the simulation is only 10% of the 130 monetary units.

In this paper we examine four scenarios, which are defined in Table 32.1. The different scenarios investigate the influence of the repetitions (n0-n3) of each algorithm section on the results and performance. As the results show, the individual graphs in Figs. 32.2 and 32.3 reflect the number of iterations of four steps of the algorithm and are plotted over the iteration n4. All experiments are run on a station with an Intel® Core™ i7-8550U processor at 1.80 GHz and 16.0 GB RAM under Windows® 10 environment. The algorithm was scripted in Python®.

Table 32.1 shows the results of different iteration scenarios, the obtained SRI score, and gives the iteration number (n4) at which a meaningful SRI score is output and iterations can be theoretically terminated. In addition, Fig. 32.3 shows the average SRI score after each iteration step of all scenarios. It can be seen that all scenarios converge to approximately the same level of SRI Score.

All graphs in Fig. 32.4 showing that all *Random mutation (n3)* has little influence on the 10 best combinations. In the future, it can be refrained from simulating this part or, if only, at the beginning.

The biggest influence on the changes of the top 10 is always the *Crossover (n2)* combination of the best 10. However, this algorithm is also very time-intensive, because more often the costs are smaller than the budget. In contrast, one can see that the *Mutation from top 100* gets less important with each successive iteration. This is the result of the budget check: if there are already combinations in the top 100 with the marginal cost, the SRI score is not calculated for this combination, because each increase would lead to costs going over the budget. The scenario a gives a very good result after about 14 minutes and 9 iterations. Even after a high number of iteration steps, the SRI score is not changing, which means that this can also be used as a convergence criteria.

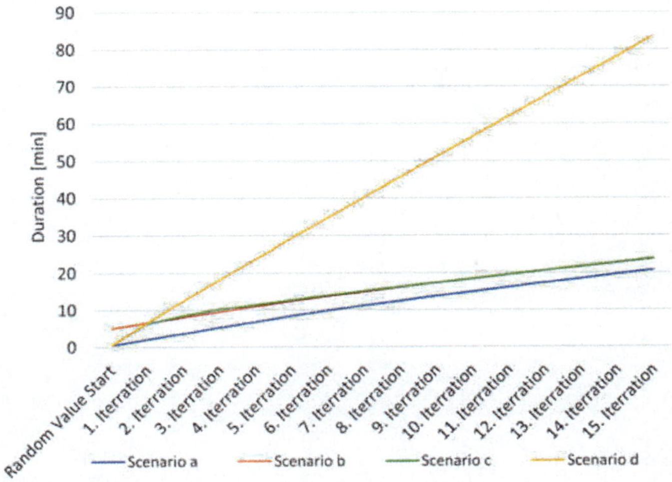

Fig. 32.2 Run duration for different scenarios

Analysis of the duration of the individual steps as a function of the number of repetitions has shown that there is a proportional relationship no significant differences

Table 32.1 Iteration scenarios, their characteristics and results

Scenario	a	b	c	d
Step 1: Initial population (n0)	100	1000	100	100
Step 2: Mutation from top 100 (n1)	100	100	1000	100
Step 3: Crossover (n2)	100	100	100	1000
Step 4: Mutation of whole population (n3)	100	100	100	100
SRI score [%]	43.28	43.27	43.23	43.29
No. iteration [-]	9	8	9	4
Duration [min]	13.6	16.0	17.3	23.6

Table 32.2 Probability in % of level distribution of all scenario results, where **bold** is the initial level

level // question	5	7	9	10	25	35	45	47	49	51	52
0	**20**	**0**	**34**	**14**	0	0	**95**	**0**	**0**	0	0
1	0	41	66	0	0	**75**	4,5	0	100	**25**	**0**
2	0	0	0	0	**98**	25	0	91	0	75	100
3	80	59	0	84	2,3	0	0	9,1	0	0	0
4	0	0	0	2,3	0	0	0	0	0	0	0

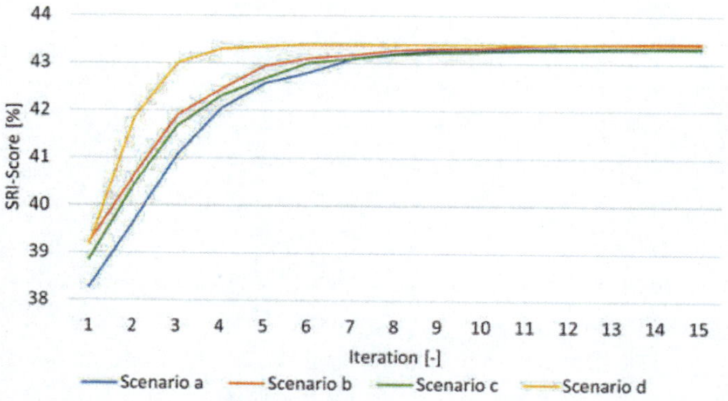

Fig. 32.3 SRI score changes in different scenarios

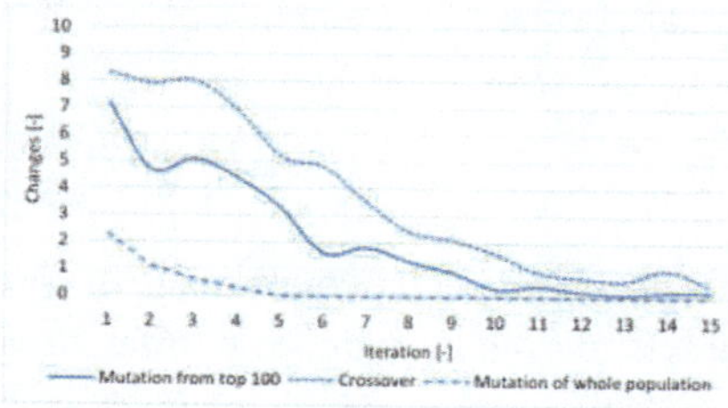

(a) Scenario a

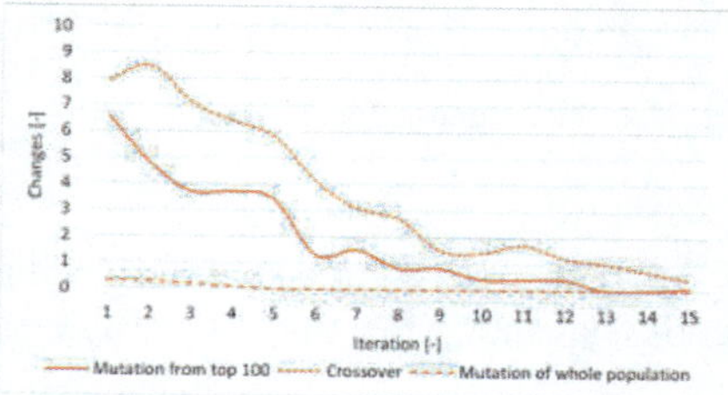

(b) Scenario b

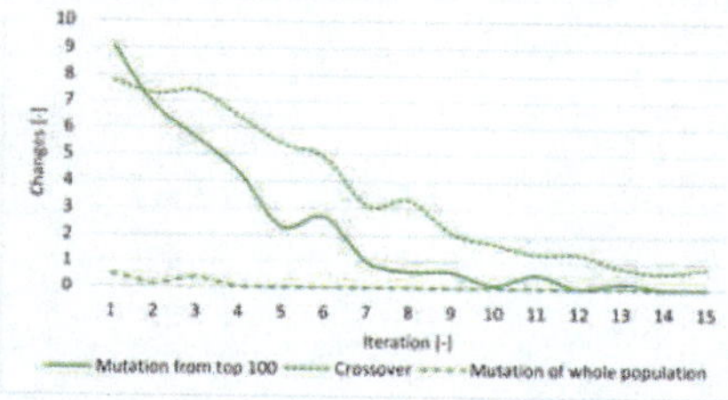

(c) Scenario c

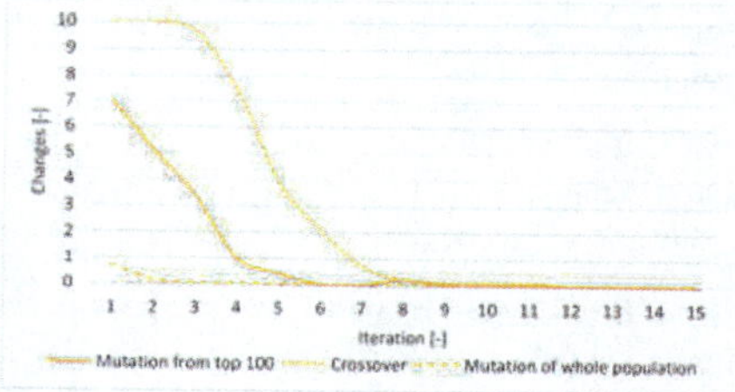

(d) Scenario d

Fig. 32.4 Number of newly added combination among the top 10

as can be seen in Fig. 32.2. The duration per 100 steps is about 0.5 minutes, but the duration of the *Mutation from top 100* decreases with each iteration step, which is due to the fact described above.

Table 32.2 shows the probability of which question was changed and to what degree. The results were compiled from all simulation scenarios. The result presented here reflects the weighting since all costs were set to 1 monetary unit. It can be seen that some questions, for example 47, 49 and 52, are more effective than others. These three questions are in the domain of "Monitoring and Control", which implies that for the case building we should focus on services like HVAC systems management

(47), occupancy detection (49), and reporting information regarding demand side (52).

Conclusions

In this paper we proposed a modified version of NSGA-II algorithm to maximize the SRI score for an existing building considering budget constraints. A hypothetical numerical example is solved using the proposes algorithm to demonstrate its performance and efficiency.

There are limitations in this work which are elaborated here and will be investigated in future work. The proposed algorithm is designed for the described single-objective problem, and its behavior for multi-objective problems must be studied. Particularly, the behavior of the proposed tournament selection in dense decision areas should be studied. For the sake of simplicity, the cost changes according to level changes are considered to be uniform and linear, which does not correspond to the reality and should be investigated further. Moreover, in future works, practical indications regarding technologies that would lead to improved SRI scores will be investigated. Another important point which should be investigated regarding SRI improvements, is the contribution of such improvements on CO_2 emissions. A research is planned to scrutinize this relationship.

References

1. Deb, K., Pratap, A., Agarwal, S., & Meyarivan, T. (2002). A fast and elitist multiobjective genetic algorithm: NSGA-II. *IEEE Transactions on Evolutionary Computation, 6*(2), 182–197.
2. European Commission. Statistical Office of the European Union. (2020). The EU in the world: 2020 edition. Publications Office. 10.2785/932944.
3. European Commission. (2020). A Renovation Wave for Europe-Greening Our Buildings, Creating Jobs, Improving Lives. Communication from the Commission to the European Parliament, the Council, the European Economic and Social Committee and the Committee of the Regions.
4. Union, European. (2018). Directive (EU) 2018/844 of the European Parliament and of the Council of 30 May 2018 amending Directive 2010/31/EU on the energy performance of buildings and Directive 2012/27/EU on energy efficiency. *Official Journal of the European Union, 156,* 75–91.
5. Verbeke, S., Aerts, D., Reynders, G., Ma, Y., & Waide, P. (2020). Final Report on the Technical Support to the Development of a smart Readiness Indicator for Buildings. European Commission: Brussels, Belgium.
6. Wolf, S., Teitge, J., Mielke, J., Schütze, F., & Jaeger, C. (2021). The European Green Deal-more than climate neutrality. *Intereconomics, 56*(2), 99–107.

Chapter 33
Optimal Design and Operation of Community Hydrogen Generation and Storage Applications

Manuel Katholnigg, Armin Golla, Frederik vom Scheidt, Sarah Henni, and Christof Weinhardt

Abstract The European energy crisis and the global climate crisis call for a strong reduction of fossil fuel usage in the residential heating sector. Given the rising role of energy communities, we address this challenge by optimizing the design and operation of different energy community systems with a linear program and a genetic algorithm for rolling horizon control, respectively. In particular, we compare status quo systems that are based on natural gas, with purely electricity-based systems, and systems based on electricity as well as the local production, storage, and usage of green hydrogen in the respective community. Applying our method to a case study of a community with 19 households, across various regulatory scenarios, and two different objective scenarios, we find that including hydrogen can achieve considerable CO_2 emission reduction, higher self-sufficiency, and lower costs than systems using natural gas for heating. Set-ups without hydrogen, but with larger electric heat pumps, achieve similar emission reductions at lower costs but enable less self-sufficiency in the community.

Keywords Hydrogen economy · Hydrogen storage · Power-to-gas · Linear programming · Sustainable development

Introduction

With the introduction of a national hydrogen strategy, Germany aims to decrease its dependency on fossil natural gas and reach its climate targets. A key part of this strategy is the generation of green hydrogen from renewable electricity through electrolysis. In energy communities, electrolysis can be combined with hydrogen storage and fuel cells. The combination offers the potential to reduce seasonal imbalances between electricity generation from solar photovoltaic plants, which occurs mainly in summer, and residential electricity and heat consumption, which is highest in winter. With fuel cells, hydrogen systems can provide both heat and electricity, which

M. Katholnigg (✉) · A. Golla · F. Scheidt · S. Henni · C. Weinhardt
Karlsruhe Institute of Technology, Kaiserstraße 12, 76131 Karlsruhe, Germany
e-mail: manuel.katholnigg@student.kit.edu

O. Grothe et al. (eds.), *Operations Research Proceedings 2022*, Lecture Notes in Operations Research, https://doi.org/10.1007/978-3-031-24907-5_33

makes them attractive for meeting residential energy needs. While economic analyses regarding hydrogen are increasingly performed on a national level [1], few studies exist on hydrogen viability in energy communities. For these, seasonal hydrogen storage systems are discussed, which can be planned with linear optimization problems [2, 3]. In a linear optimization problem over one year, the full seasonal cycle is covered, which creates economic incentives for charging in summer and moderate discharging in winter. These incentives are absent in an operation strategy based on short-term rolling horizon forecasts, which is not addressed by previous studies. Furthermore, there is a lack of comparisons with other technology options for residential sector coupling under different market and policy scenarios.

Mathematical Energy Community Model

The proposed energy community is mathematically modeled using Calliope, a multi-level energy systems framework. To present the mathematical model in a compact way, we use binary variables and a mixed-integer linear program notation. Due to Calliope framework specifications, it is implemented as a linear program without binary variables. Furthermore, our mathematical representation includes only the components of the framework used in this study. The full framework-based mathematical formulation can be found in the Calliope documentation [4].

A set of technologies Θ is available that includes technologies for resource conversion Θ^{ρ}, carrier conversion Θ^{ζ}, storage Θ^{s}, transmission Θ^{t} and demand Θ^{d}. A technology θ is placed at a location $\lambda \in \Lambda$. The location set Λ consists of locations in the household set H and the community hub $(\Lambda \setminus H)$. A technology can interact with energy carriers $\zeta \in Z$ produced in the community and with energy resources $\rho \in P$ given externally. The model operates with hourly time steps $t \in T$. Cost variables have an index $k \in K$, indicating monetary costs (k^m) in € and emissions (k^e) in kg CO_2e. The community aims to minimize total costs or emissions, with feed-in revenue considered as negative costs. Depending on the scenario, the weight $w_k \in \{0, 1\}\ \forall k \in K$ is set to favor cost ($w_{k^m} = 1$, $w_{k^e} = 0$) or emissions ($w_{k^m} = 0$, $w_{k^e} = 1$). In the objective function (Eq. (33.1)), c_k corresponds to the technology-related cost of cost class k, u is unsatisfied demand and M is a factor by which unsatisfied demand is penalized.

$$\min \sum_{\theta \in \Theta} \sum_{\lambda \in \Lambda} \sum_{k \in K} c_k(\lambda, \theta) \cdot w_k + \sum_{\zeta \in Z} \sum_{t \in T} u(\zeta, t) \cdot M \tag{33.1}$$

The overall system is balanced, if the consumed carrier quantity ζ^c equals the sum of the produced and exported carrier quantity (ζ^p and ζ^x) of a carrier in every time step and at every location (Eq. (33.2)). Carrier input is modeled by values < 0, output by values > 0. The energy capacity e^c limits the sum of all produced carriers by a technology (Eq. (33.3)). For demand, storage, and transmission technologies, the energy capacity also limits the consumed carrier quantity (Eq. (33.4)).

$$-\sum_{\theta\in\Theta}\zeta^c(\lambda,\theta,\zeta,t)=\sum_{\theta\in\Theta}(\zeta^p(\lambda,\theta,\zeta,t)+\zeta^x(\lambda,\theta,\zeta,t)) \quad \forall\zeta\in Z,\lambda\in\Lambda,t\in T \tag{33.2}$$

$$\sum_{\zeta\in Z}(\zeta^p(\lambda,\theta,\zeta,t)+\zeta^x(\lambda,\theta,\zeta,t))\leq e^c(\lambda,\theta) \quad \forall\zeta\in Z,\lambda\in\Lambda,\theta\in\Theta,t\in T \tag{33.3}$$

$$-\zeta^c(\lambda,\theta,\zeta,t)\leq e^c(\lambda,\theta) \quad \forall\zeta\in Z,\lambda\in\Lambda,\theta\in\{\Theta^d,\Theta^s,\Theta^t\},t\in T \tag{33.4}$$

The maximum carrier quantity that can be consumed per household is limited by the household demand d (modeled by values ≤ 0).

$$\zeta^c(\lambda,\theta,\zeta,t)\geq d(\lambda,\theta,\zeta,t) \quad \forall\zeta\in Z,\lambda\in H,\theta\in\Theta^d,t\in T \tag{33.5}$$

The available resource quantity ρ^{av} limits how much carrier quantity can be produced and exported using resource efficiency η^ρ (Eq. (33.6)). When a resource conversion technology is located at a household, the available resource quantity is determined by the time-dependent resource quantity per kW, energy capacity ρ^t and the used resource area ρ^{ar} (Eq. (33.7)). The roof area r limits the resource area per household (Eq. (33.8)).

$$\frac{\zeta^p(\lambda,\theta,\zeta,t)+\zeta^x(\lambda,\theta,\zeta,t)}{\eta^\rho(\lambda,\theta,\zeta)}\leq\rho^{av}(\lambda,\theta,t) \quad \forall\zeta\in Z,\lambda\in\Lambda,\theta\in\Theta^\rho,t\in T \tag{33.6}$$

$$\rho^{av}(\lambda,\theta,t)=\begin{cases}\rho^t(\lambda,\theta,t)\cdot\rho^{ar}(\lambda,\theta), & \text{if } \lambda\in H \text{ and } \theta\in\Theta^\rho\\ \infty, & \text{else}\end{cases}$$

$$\forall\lambda\in\Lambda,\theta\in\Theta,t\in T \tag{33.7}$$

$$\sum_{\theta\in\Theta^\rho}\rho^{ar}(\lambda,\theta)\leq r(\lambda) \quad \lambda\in H \tag{33.8}$$

A carrier can be transported between two locations λ and λ^t through transmission technologies with an energy efficiency η per distance d (Eq. (33.9)).

$$\frac{\zeta^p(\lambda,\lambda^t,\theta,\zeta,t)}{\eta^d(\lambda,\theta,t)}=-\zeta^c(\lambda,\lambda^t,\theta,\zeta,t) \quad \forall\lambda,\lambda^t\in\Lambda,\zeta\in Z,\theta\in\Theta^t,t\in T \tag{33.9}$$

Carrier conversion technologies can have multiple inputs and outputs. The relationship between the first input carrier ζ^{fi} and the first output carrier ζ^{fo} under consideration of the conversion efficiency η^ζ is shown in Eq. (33.10). The input carrier ratio ζ^{ri} and the output carrier ratio ζ^{ro} describe relations between first and second input and output carriers ζ^{si} and ζ^{so} in Equations (33.11) and (33.12).

$$\frac{\zeta^{p}(\lambda,\theta,\zeta^{fo},t)}{\eta^{\zeta}(\lambda,\theta,\zeta^{fi},\zeta^{fo})} = -\zeta^{c}(\lambda,\theta,\zeta^{fi},t) \qquad \forall \zeta^{fi},\zeta^{fo} \in Z, \lambda \in \Lambda, \theta \in \Theta^{\zeta}, t \in T \tag{33.10}$$

$$\frac{\zeta^{c}(\lambda,\theta,\zeta^{si},t)}{\zeta^{ri}(\lambda,\theta,\zeta^{fi},\zeta^{si})} = \zeta^{c}(\lambda,\theta,\zeta^{fi},t) \qquad \forall \zeta^{fi},\zeta^{si} \in Z, \lambda \in \Lambda, \theta \in \Theta^{\zeta}, t \in T \tag{33.11}$$

$$\frac{\zeta^{p}(\lambda,\theta,\zeta^{so},t)}{\zeta^{ro}(\lambda,\theta,\zeta^{fo},\zeta^{so})} = \zeta^{p}(\lambda,\theta,\zeta^{fo},t) \qquad \forall \zeta^{fo},\zeta^{so} \in Z, \lambda \in \Lambda, \theta \in \Theta^{\zeta}, t \in T \tag{33.12}$$

The storage level s at time t is determined by the storage level at the previous time step $t-1$, the storage loss rate l between $t-1$ and t, and the charged and discharged energy (Eq. (33.13)). The charge and discharge efficiency are represented by η^{ζ}. The storage capacity s^c limits the storage level, as described through $s(\lambda,\theta,t) \leq s^c(\lambda,\theta), \forall \lambda \in \Lambda, \theta \in \Theta^s, t \in T$. A minimum state of charge s^d as a fraction of s^c is ensured by $s(\lambda,\theta,t) \geq s^d(\lambda,\theta) \cdot s^c(\lambda,\theta), \forall \lambda \in \Lambda, \theta \in \Theta^s, t \in T$. Charging and discharging speed is limited depending on the storage capacity through $e^c(\lambda,\theta) \leq s^c(\lambda,\theta) \cdot e^s(\lambda,\theta)$, where e^s is the maximum e^c per s^c.

$$\begin{aligned} s(\lambda,\theta,\zeta,t) = {} & s(\lambda,\theta,\zeta,t-1)\cdot(1-l(\lambda,\theta,\zeta)) + \zeta^{c}(\lambda,\theta,\zeta,t)\cdot\eta^{\zeta}(\lambda,\theta,\zeta) \\ & - \frac{\zeta^{p}(\lambda,\theta,\zeta,t)}{\eta^{\zeta}(\lambda,\theta,\zeta)} \qquad \forall \lambda \in \Lambda, \theta \in \Theta^s, \zeta \in Z, t \in T \end{aligned} \tag{33.13}$$

For all technologies besides $\theta \in \Theta^d$, technology-related costs c_k consist of investment costs c_k^i and variable costs c_k^v (Eq. (33.14)). Investment costs include annual capacity-related costs c_k^c and annual fixed operation and maintenance costs c_k^f (Eq. (33.15)). Capacity-related costs depend on capacity-related costs per energy or storage capacity (c_k^e or c_k^s) and the yearly depreciation rate d_k (Eq. (33.16)). Depending on the technology, fixed operation and maintenance costs per energy capacity c_{km}^{ef} or per storage capacity c_{km}^{sf} are assumed (Eq. (33.17)).[1]

$$c_k(\lambda,\theta) = \begin{cases} c_k^i(\lambda,\theta) + \sum_{t\in T} c_k^v(\lambda,\theta,t), & \text{if } \theta \in \{\Theta^p,\Theta^{\zeta},\Theta^t,\Theta^s\} \\ 0, \text{if} & \theta \in \Theta^d \end{cases}$$
$$\forall k \in K, \lambda \in \Lambda, \theta \in \Theta \tag{33.14}$$

[1] Further mathematical formulations, such as equations for building energy demands and variable costs, are used to make the model executable.

$$c_k^i(\lambda, \theta) = c_k^c(\lambda, \theta) + c_k^f(\lambda, \theta) \qquad \forall k \in K, \lambda \in \Lambda, \theta \in \Theta \setminus \Theta^d \quad (33.15)$$

$$c_k^c(\lambda, \theta) = \begin{cases} d_k(\lambda, \theta) \cdot c_k^e(\lambda, \theta) \cdot e^c(\lambda, \theta), & \text{if } \theta \in \{\Theta^\rho, \Theta^\zeta, \Theta^t\} \\ d_k(\lambda, \theta) \cdot c_k^s(\lambda, \theta) \cdot s^c(\lambda, \theta), & \text{if } \theta \in \Theta^s \end{cases}$$
$$\forall k \in K, \lambda \in \Lambda, \theta \in \Theta \setminus \Theta^d \quad (33.16)$$

$$c_k^f(\theta) = \begin{cases} c_k^{ef}(\theta) \cdot e^c(\theta), & \text{if } \theta \in \{\Theta^\rho, \Theta^\zeta, \Theta^t\} \\ c_k^{sf}(\theta) \cdot s^c(\theta), & \text{if } \theta \in \Theta^s \end{cases} \qquad \forall k \in K, \theta \in \Theta \setminus \Theta^d \quad (33.17)$$

Case Study

Based on the mathematical model, we investigate an energy community of 19 households as displayed in Fig. 33.1. Energy consumption data is derived from Edinburgh [5], according input data for photovoltaic and solar thermal systems is retrieved from [6]. We simulate an average year between 2020 and 2050 under German policy conditions using price and emission forecasts.[2] To reflect uncertainty about future policies, we solve and compare six scenarios. In scenario A, we assume German subsidies, including a state-guaranteed feed-in tariff and a reduction of capacity-related monetary costs for solar thermal (30%) and fuel cell technology (40%). Scenario B assumes a gas-covered heat supply. In response to the current energy crisis in Germany, Scenario C analyzes a 50% higher electricity price compared to the average projected price until 2050. To investigate stronger independence from fossil energy, we allow 25% less grid electricity consumption in scenarios D (planning optimization) and E (operation) compared to scenario A. To investigate the emission reduction potential of seasonal hydrogen storage, scenario F investigates purely ecological optimization.

All scenarios, except for scenario E, are solved for a full-year horizon using Gurobi. In scenario E we develop an operation strategy with an 48-hour rolling horizon to demonstrate the real-world application of the energy system resulting from scenario D. The rolling horizon splits the year-round optimization problem into subproblems. Seasonal patterns of renewable energy production are not reflected within these subproblems. Therefore, there is a lack of economic incentives to charge hydrogen storage in the summer months or not to discharge it too quickly in the winter months.

To evaluate the theoretical potential of an operating strategy that integrates such incentives, we first divide the analyzed year into charging and discharging periods of the seasonal storage based on the results from scenario D. Subsequently, two decision variables are added to the optimization problem: For the summer months, we add

[2] The Calliope model as well as references for the parameter values are available at github.com/manuelkat/seasonal-storage.

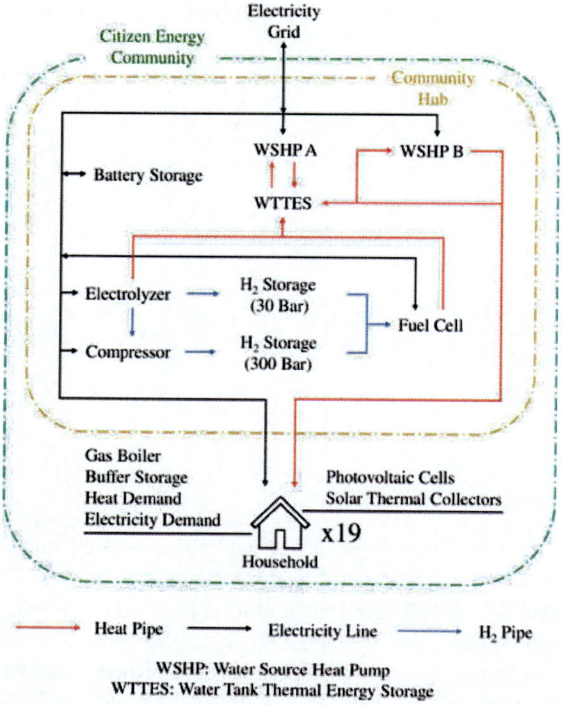

Fig. 33.1 Residential energy community model

variable monetary cost ≤ 0 when using the electrolyzer, and in the winter months, we add variable monetary cost ≥ 0 for the fuel cell to the actual variable cost. To integrate these decision variables, an iterative two-stage optimization is necessary due to framework limitations. In the first stage, a genetic algorithm is used to set the two added decision variables to discrete values, which are passed to Calliope in the second stage to initiate the solution with Gurobi. The economic implications of the artificial incentives are removed from the economic result. The result serves the fitness function of the genetic algorithm, which minimizes monetary cost and iteratively repeats the two-stage process. The genetic algorithm originates from the open-source Python library PyGAD, and uses steady-state parent selection, random mutation, and one-point crossover [7]. To minimize computational cost, a grid search is then performed to systematically analyze promising value ranges of the decision variables determined by the genetic algorithm for new optima and robustness.

We determine the addition of variable costs -0.06€/kWh for the electrolyzer and 0.20€/kWh for the fuel cell for an optimal operation of the hydrogen storage. The economic result generated with the operation strategy for scenario D differs 4.2% from scenario E, which optimizes over 365 d without artificial incentives and embodies the theoretical optimum. The resulting storage levels of the hydrogen storage tanks from the optimization of both scenarios are shown in Fig. 33.2 over one year. The

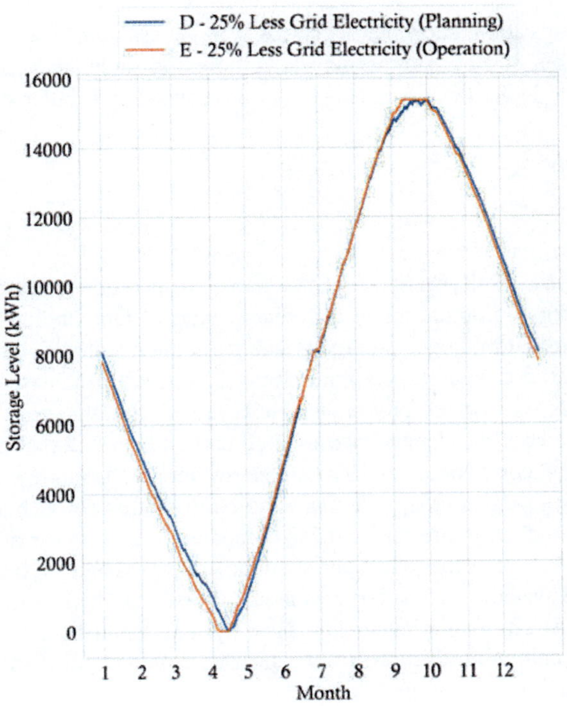

Fig. 33.2 Hydrogen storage level curves

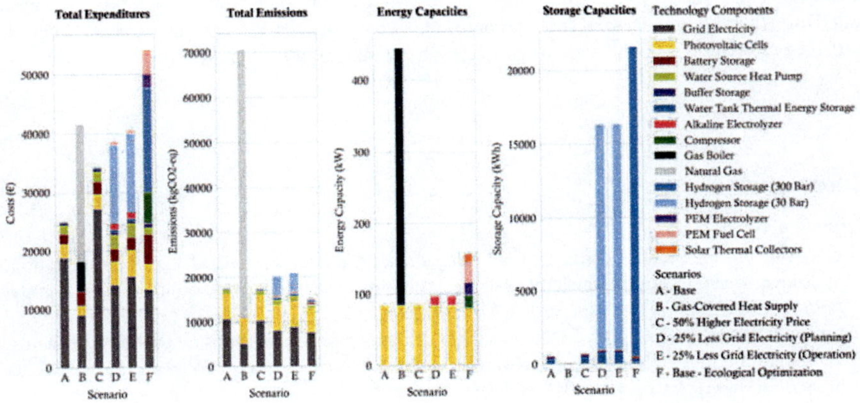

Fig. 33.3 Scenario results

economic and ecologic results, including the capacities of the components resulting from the optimization, are shown in Fig. 33.3.

Further investigations show that scenarios D and E, replacing the 30 bar hydrogen storage through a 300 bar storage with a compressor, undercut scenario A's total emissions due to lower manufacturing emissions caused by smaller steel tanks.

Conclusion

This study investigates the optimized installation and operation of seasonal hydrogen storage for energy communities in different scenarios. Our findings indicate that community seasonal hydrogen storage is not yet economically viable compared to renewable generation without seasonal storage, even with increased subsidies or strongly increased grid electricity prices. However, we find that seasonal hydrogen storage can decrease the dependence of a community on external energy supply without increased costs compared to fossil energy supply. Assuming a perfect price forecast for one year and energy demand and consumption for 48 h, our operating strategy is economically inferior by 4.2% compared to a one-year perfect forecast on all data. We also show that when photovoltaic systems, a battery, heat pumps, and thermal storage systems are used, lifecycle emissions from hydrogen storage tanks are crucial in determining whether hydrogen systems contribute to an additional reduction of the community's carbon footprint. This should be considered in potential subsidy policies.

Acknowledgements The research of Armin Golla is funded by the German Federal Ministry of Education and Research (BMBF) within the Innovations for Tomorrow's Production, Services and Work (funding number 02K18D000) and implemented by the Project Management Agency Karlsruhe (PTKA). The authors are responsible for the content of this publication.

References

1. vom Scheidt, F., Qu, J., Staudt, P., Mallapragada, D. S., & Weinhardt, C. (2022). Integrating hydrogen in single-price electricity systems: The effects of spatial economic signals. *Energy Policy.* https://doi.org/10.1016/j.enpol.2021.112727
2. Gabrielli, Paolo, et al. (2020). Seasonal energy storage for zero-emissions multi-energy systems via underground hydrogen storage. *Renewable and Sustainable Energy Reviews, 121*, 109629. https://doi.org/10.1016/j.rser.2019.109629
3. Petkov, I., Gabrielli, P. & Spokaite, M. (2021) The impact of urban district composition on storage technology reliance: trade-offs between thermal storage, batteries, and power-to-hydrogen. Energy *224*. https://doi.org/10.1016/j.energy.2021.120102.
4. Pfenninger, S., & Pickering, B. (2018). Calliope: a multi-scale energy systems modelling framework. *Journal of Open Source Software, 3*, 825. https://doi.org/10.21105/joss.00825
5. Goddard, N. et al. (2021). IDEAL household energy dataset. https://datashare.ed.ac.uk/handle/10283/3647 (University of Edinburgh, 2021).

6. Pfenninger, S., & Staffell, I. (2016). Long-term patterns of European PV output using 30 years of validated hourly reanalysis and satellite data. *Energy, 114*, 1251–1265. https://doi.org/10.1016/j.energy.2016.08.060
7. Fawzy Gad, A. (2021). PyGAD: An Intuitive Genetic Algorithm Python Library. arXiv e-prints, arXiv-2106.

Chapter 34
Optimal Design of Building Energy Supply—A Case Study

Elisabeth Halser, Elisabeth Finhold, Neele Leithäuser, and Karl-Heinz Küfer

Abstract In this paper, we optimize the design of a new office building's energy supply with respect to costs and carbon emissions for one example year. The aim is to gain an overview over the impact of the supply design decision. For that, we combine known concepts to one decision support workflow. We simulate the expected heating loads of the building with a thermal network and use them to set up a mixed-integer linear program for the problem, which we can solve with one or both objectives.

Keywords Thermal network · Mixed-integer linear programming · Multi-objective decision making

Introduction

Energy supply of buildings is a significant driver of human caused carbon emissions [6] and hence of climate change. Therefore, significant emission reductions in this field are inevitable. As the Fraunhofer society wants to become carbon neutral by 2030, its Institute for Industrial Mathematics (ITWM) wants to purchase devices for its new building which deal with uncertain future heating and cooling needs (called loads) in a money and carbon saving way. This case study answers the question what range of energy related costs and carbon emissions is achievable for the new building in an example year. In this, we are particularly interested in a thorough analysis of the trade-offs between the two objectives. In subsequent studies we will deal with uncertainty of prices and loads to definitely select one device configuration.

The planned office building is located in Kaiserslautern as an annex to an already existing office complex, where heating and cooling loads of the past years are known in hourly resolution. Cooling loads in this case are almost exclusively caused by computing centers in the building.

In our case study, we focus on the design and operation of the new office building's energy supply. In order to find a suitable combination of heating and cooling

E. Halser (✉) · E. Finhold · N. Leithäuser · K.-H. Küfer
Fraunhofer ITWM, Fraunhofer-Platz 1, Kaiserslautern 67663, Germany
e-mail: elisabeth.halser@itwm.fraunhofer.de

O. Grothe et al. (eds.), *Operations Research Proceedings 2022*, Lecture Notes in Operations Research, https://doi.org/10.1007/978-3-031-24907-5_34

generators for the new building, we first need reasonable heating and cooling loads for at least one year. We start with a draft of the building cubature and its insulation properties and use the approach of a thermal network to simulate hourly heating loads for different climate scenarios. A network model of the already existing building is used for calibration. The generated hourly heating loads together with extrapolated cooling loads from the existing complex are then the data basis to find the optimal selection of generators (heating, cooling, electrical power), thermal storages and their dimensions for the new building. For that purpose, we model and solve the problem as a bicriteria mixed-integer linear program (MILP), where costs and carbon footprint are minimized simultaneously.

The thermal network is a well-known approach and for example used in [8] and [4]. It exploits an electrical analogy [10] and is therefore also called resistance capacitance (RC) network. It can be derived as a finite difference approximation to the mono-dimensional heat conduction equation [3].

MILP formulations for building energy supply are for example used in [1, 7, 9]. Ashouri et al. [1] also use a thermal network and consider the converter's individual constraints to minimize costs. In contrast to this work, we simultaneously minimize CO_2-emissions and solve the problem with the well-known ϵ-constraint method [5]. A bicriteria optimization is for example also done in [9].

Thermal Network

To obtain the building's heating loads, we simulate the ambient-temperature dependent thermal behavior of the building and add the energy that is required to heat the building to a time-dependent target temperature in each time step. The building is modeled as a network of 306 thermal capacitors, which are connected by 592 thermal resistances. Every room corresponds to a capacitor. Walls, as the connection between two rooms, are modeled as a combination of one capacitor and two connecting resistances. In every time step there is temperature exchange between the capacitors through the resistances. We consider radiation through windows with a solar heat gain coefficient of 0.6 (60% of radiation energy passing a window stays in the building).

For calibration, we visually compare the actual loads of the existing building from 2016 to 2021 with the simulated loads of this building over the same period of time. To compensate for the closing of jalousies and the varying heat over the day, the target temperature of the rooms and the radiation area are slightly adjusted. Moreover, a factor is multiplied to the resistances, which compensates for the fact that in practice the air in the rooms is changed over time, which is not represented in the model. Results for one typical week can be seen in Fig. 34.1. As the reference data is very noisy in the warmer months, we were only able to archive a relative error of 32%, but the balance fits quite well with 7% error.

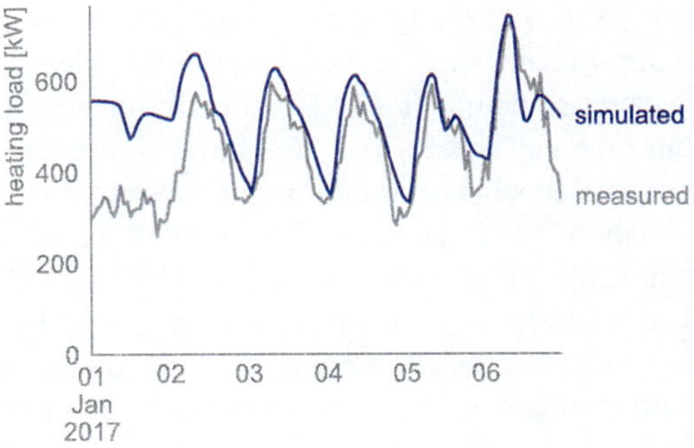

Fig. 34.1 Heating loads of the existing building in the first week of 2017

Table 34.1 Notation for the MILP

$\mathcal{M}$	Set of all possible devices
$\mathcal{T}$	Set of all hours of one year
$\mathbf{b}_i$	1 if converter i is bought, else 0 (binary variable)
$\mathbf{d}_i$	Dimension of converter i
$\mathbf{s}_{it}$	Dimension-scaled load of converter i in time step t
$\mathbf{d}^{hs}/\mathbf{d}^{cs}$	Dimension of heat/cold storage [kWh storage potential]
$\mathbf{s}_t^h/\mathbf{s}_t^c$	Heat/cold storage charge in time step t [kWh]
H_t/C_t	Heating/cooling loads in time step t [kWh]
$e_{it}/k_{it}/w_{it}/g_{it}$	Maximum electric/heat/wood/gas consumption of converter i in time step t [kWh]
$el_{it}/h_{it}/c_{it}$	Maximum electric/heat/cold production of converter i in time step t [kWh]
$p^e/p^h/p^w/p^g$	Price of electricity/district heat/wood/gas [€/kWh]
$c^e/c^h/c^w/c^g$	Carbon equivalent emissions of electricity/district heat/wood/gas [kg/kWh]
p_i^f/c_i^f	Base depreciation costs/carbon emission of converter i [€]/[kg]
p_i^d/c_i^d	Size dependent depreciation costs/carbon emission of converter i [€]/[kg]
p^{hs}/p^{cs}	Depreciated price factor of heat/cold storage [€]
c^{hs}/c^{cs}	Depreciated carbon emission factor of heat/cold storage [kg]
l^{hs}/l^{cs}	Loss rates of the heat and cold storage

MILP Model for the Energy Supply Design

The loads obtained in the last section can now be used as the basis of the device selection model. The problem is modeled as a MILP, where costs or carbon emissions shall be minimized. Not only operational costs/emissions are considered, but also the depreciated device costs, respectively, the device generation carbon emissions. For an overview over the notation of the model see (Table 34.1). For each converter type we have a representative of a certain size. The dimension can be understood as a scaling factor w.r.t. the representative device. If for example the representative district heat has 200 kW capacity and the model computes a dimension of 1.5, the district heat index d in the building should have 300 kW capacity. If this upscaled district heating is operated with a load of 5/6, it therefore effectively receives $5/6 \cdot 1.5 \cdot 200$ kW in this time step with $s_{dt} = 1.25$. The maximum energy in- and outputs of the converters are time-dependent, as they might depend for example on the ambient temperature or radiation. The depreciated costs and carbon emissions of the converters are modeled affine linear with respect to the converter size and linear over the converters' lifetime. The affine linearity assumption is suggested by a dataset of real world converters. In contrast to the converters, we model the storages and their depreciation linearly without offset, as real world data suggested negligible offset in this case.

Objective Functions The two objective functions are defined as follows, where d is the technical index for district heat. For the cost minimization we have for the optimization variables $\mathbf{s}_{it}, \mathbf{s}_t^c, \mathbf{s}_t^h, \mathbf{d}^{hs}, \mathbf{d}^{cs}, \mathbf{b}_i, \mathbf{d}_i \; \forall i \in \mathcal{M}, t \in \mathcal{T}$

$$\begin{aligned} min \sum_{\substack{i \in \mathcal{M} \\ i \neq d}} \sum_{t \in \mathcal{T}} \mathbf{s}_{it} \left(e_{it} p^e - el_{it} p^{el} + w_{it} p^w + g_{it} p^g\right) + \sum_{t \in \mathcal{T}} \mathbf{s}_{dt} k_{dt} p^d \\ + \mathbf{d}^{hs} p^{hs} + \mathbf{d}^{cs} p^{cs} + \sum_{i \in \mathcal{M}} \left(\mathbf{b}_i p_i^f + \mathbf{d}_i p_i^d\right). \end{aligned} \tag{34.1}$$

For greenhouse gas minimization we have

$$\begin{aligned} min \sum_{\substack{i \in \mathcal{M} \\ i \neq d}} \sum_{t \in \mathcal{T}} \mathbf{s}_{it} \left(e_{it} c^e + w_{it} c^w + g_{it} c^g\right) + \sum_{t \in \mathcal{T}} \mathbf{s}_{dt} k_{dt} c^d \\ + \mathbf{d}^{hs} c^{hs} + \mathbf{d}^{cs} c^{cs} + \sum_{i \in \mathcal{M}} \left(\mathbf{b}_i c_i^f + \mathbf{d}_i c_i^d\right). \end{aligned} \tag{34.2}$$

General Constraints There are constraints that are relevant for all devices. Only bought converters can have non-zero size (34.3). Storages are empty in the beginning (34.4). Neither converters (34.5) nor storages ((34.6) and (34.7)) can work beyond their technical limits. Finally, (34.8) and (34.9) ensure that enough heating and cooling is available in every time step. This links the time steps via the storage states. Let M be a sufficiently large number.

$$0 \leq \mathbf{d}_i \leq M \cdot \mathbf{b}_i \quad \forall i \in \mathcal{M} \tag{34.3}$$

$$\mathbf{s}_0^c = \mathbf{s}_0^h = 0 \tag{34.4}$$

$$0 \leq \mathbf{s}_{it} \leq \mathbf{d}_i \quad \forall i \in \mathcal{M}, t \in \mathcal{T} \tag{34.5}$$

$$0 \leq \mathbf{s}_t^c \leq \mathbf{d}^{cs} \quad \forall t \in \mathcal{T} \tag{34.6}$$

$$0 \leq \mathbf{s}_t^h \leq \mathbf{d}^{hs} \quad \forall t \in \mathcal{T} \tag{34.7}$$

$$\mathbf{s}_{t+1}^c = l^{cs} \mathbf{s}_t^c + \sum_i \mathbf{s}_{it} c_{it} - C_t \quad \forall t \in \mathcal{T} \tag{34.8}$$

$$\mathbf{s}^h_{t+1} = l^{hs}\mathbf{s}^h_t + \sum_i \mathbf{s}_{it}(h_{it} - k_{it}) - H_t \quad \forall t \in \mathcal{T} \tag{34.9}$$

Almost all converters come with individual technical constraints, such as the fact that a cogeneration unit can only work between 40 and 100% if it is turned on or a heat pump can work in different modes. Listing all of them would go beyond the scope of this article, but they are implemented in the model.

Single Objective Optimization Example

If we solve the problem with climate data from 2017 with the single objectives (34.1) and (34.2), respectively, we obtain the device selections from Table 34.2. The proposed dimensions therein are realistic in comparison with the devices in the existing building. Determining all the model parameters was a challenging task. For carbon data we mainly used the ÖKOBAUDAT [2] and extrapolated this information to get estimates for devices missing in the data base. For the depreciated device costs, we applied linear regression to a data set of about 150 reference devices. These devices are also used to derive generic performance data such as efficiency factors which might depend on the outside temperature. The energy prices are from recent years.

Table 34.2 Optimization results of either cost or carbon emission minimization—only devices selected by the optimization are included in the table. Further converter types are: gas boiler, reversible and non-reversible air-water and brine-water heat-pump, non-reversible water-water heat pump, adsorption chiller, compression chiller and district heating

Device	Cost optimization	CO_2 optimization
Cogeneration unit (heating/electricity)	334 kW/220 kW	–
Water-water heat pump (cooling/heating)	243 kW/195 kW	305 kW/244 kW
Absorption chiller	274 kW	262 kW
Pellet boiler	–	164 kW
Photovoltaic	201 m^2	–
Solar thermal	–	1050 m^2
Free cooling	–	116 kW
Cold storage	11822 l	33,067 l
Heat storage	6467 l	100,587 l

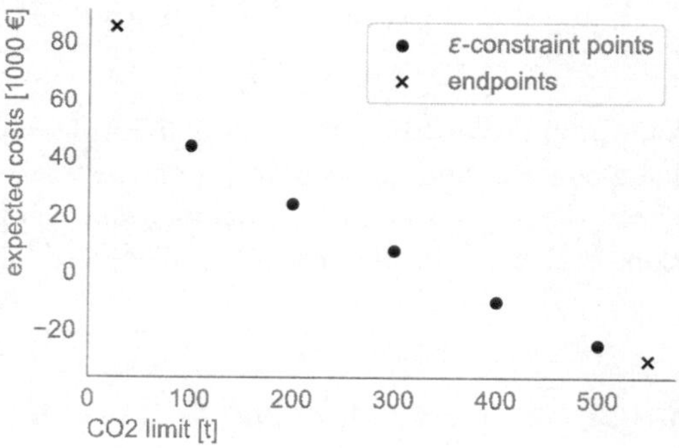

Fig. 34.2 ϵ-Constraint method results 2017

Multiobjective Optimization Example

For the multiobjective optimization we use the ϵ-constraint method. For that, various strict upper limits are set for carbon emissions and then a cost optimization is executed. To find a reasonable range for the emission values on the Pareto front, we run two lexicographic optimizations (i.e., we first minimize costs and then the emissions among all solutions with minimal costs and vice versa). Note that the solutions in Table 34.2 are chosen to be in fact lexicographically optimal and correspond to the endpoints of the Pareto front. We solve the problem with Gurobi. We allow for a 5% optimality gap, as computation time of one point even in this case exceeds 12 h on an intel CORE i7 vPro 8th Gen. Figure 34.2 shows the results.

Figure 34.2 shows that the yearly costs differ by more than 100.000 € depending on the CO_2-budget. The other way around, CO_2 emissions can be reduced by up to 95% when accepting higher costs. The relationship between costs and carbon emissions seems almost linear for carbon values above 100 t. The negative cost values in the graph for CO_2 higher than 300 t are due to electricity generated with a cogeneration unit and photovoltaic and sold on the market.

Conclusion and Outlook

In this paper, we gained an overview over Pareto-optimal energy supply configurations for one sample year and observed that there is a large potential for trade-offs. However, we do not know whether these solutions are Pareto-optimal or even feasible in other years. Therefore, we have to find a way of making the solution more robust and at the same time not too conservative. We are currently working on an inverse robust approach, where we introduce a third dimension to the objective space, which measures the degree of undersupply of device selections over different years.

Another possible field of research is to use the problem formulation with fixed design variables in a rolling horizon approach for controlling the converters once they are installed.

References

1. Ashouri, A., Fux, S. S., Benz, M. J., & Guzzella, L. (2013). Optimal design and operation of building services using mixed-integer linear programming techniques. *Energy, 59*, 365–376. https://doi.org/10.1016/j.energy.2013.06.053
2. BBSR: Ökobaudat (31.05.2022). https://www.oekobaudat.de/no_cache/datenbank/suche.html
3. Fraisse, G., Viardot, C., Lafabrie, O., & Achard, G. (2002). Development of a simplified and accurate building model based on electrical analogy. *Energy and Buildings, 34*(10), 1017–1031. https://doi.org/10.1016/S0378-7788(02)00019-1
4. Hatanaka, T., Ikawa, T., & Okamoto, D. (2019). Remodeling of rc circuit building thermodynamics model with solar radiation based on a regularization-like technique. In *2019 12th Asian Control Conference (ASCC)*, pp. 7–12
5. Hillier, F.S., & Miettinen, K. (1998). Nonlinear multiobjective optimization, vol. 12. Springer US. https://doi.org/10.1007/978-1-4615-5563-6
6. IEA: Global energy use and energy-related co_2 emissions by sector. (2020). (21.06.2022). https://www.iea.org/data-and-statistics/charts/global-energy-use-and-energy-related-co2-emissions-by-sector-2020
7. Iturriaga, E., Aldasoro, U., Terés-Zubiaga, J., & Campos-Celador, A. (2018). Optimal renovation of buildings towards the nearly zero energy building standard. *Energy, 160*, 1101–1114. https://doi.org/10.1016/j.energy.2018.07.023
8. Mukherjee, S., Mishra, S., & Wen, J. T. (2012). Building temperature control: A passivity-based approach. In *2012 IEEE 51st IEEE Conference on Decision and Control (CDC)*, pp. 6902–6907. IEEE (2012). https://doi.org/10.1109/CDC.2012.6426676
9. Petkov, I., Mavromatidis, G., Knoeri, C., Allan, J., & Hoffmann, V. H. (2022) Mangoret: An optimization framework for the long-term investment planning of building multi-energy system and envelope retrofits. *Applied Energy* **314**, 118–901 (2022). https://doi.org/10.1016/j.apenergy.2022.118901
10. Robertson, A.F., & Gross, D. (1958). An electrical-analog method for transient heat-flow analysis. *Journal of research of the National Bureau of Standards* **61**

Chapter 35
Optimal Trading of a Hybrid Electric, Hydrogen and Gas Fueling Station in Day-Ahead and Intra-day Markets: Modeling Aspect

Farnaz Sohrabi, Mohammad Rohaninejad, Mohammad Reza Hesamzadeh, and Július Bemš

Abstract Energy crisis and environmental concerns encourage the adoptions of substitute transportation options instead of the conventional internal combustion engine vehicles. The electric, hydrogen and natural gas vehicles are promising alternatives, so more attention should be paid to the economic and operational features of their charging stations. This paper proposes a multifunction charging station to refill electric, hydrogen and natural gas vehicles which takes part in the day-ahead and intra-day markets. The objective of this station is to maximize its profit by attaining the optimal operation of the devices and bidding curves. Coordinated bidding is considered since this charging station participates in the sequential markets with different price scenarios. The clearing prices and dispatched amounts in both markets are unknown at the time of bidding. This problem is formulated as a two-stage stochastic program since the markets are cleared sequentially and the prices are revealed gradually. Finally, the economic effectiveness of the proposed multifunction charging station is analyzed in different scenarios.

Keywords Charging station · Electricity markets · Stochastic programming · Coordinated bidding

Nomenclature

Indices

h, ω, i Index for hours, scenarios and bidding prices.

Constants

$\eta_{B,Ch}$, $\eta_{B,Dis}$	Charging and discharging efficiency of battery.
η_E, η_F, η_M	Efficiency of electrolyzer, fuel cell and methanation.
$\eta_{T,Ch}$, $\eta_{T,Dis}$	Charging and discharging efficiency of tank.

F. Sohrabi (✉) · M. Rohaninejad · J. Bemš
Czech Technical University in Prague, Prague, Czech Republic
e-mail: sohrafar@fel.cvut.cz

M. R. Hesamzadeh
KTH Royal Institute of Technology, Stockholm, Sweden

O. Grothe et al. (eds.), *Operations Research Proceedings 2022*, Lecture Notes in Operations Research, https://doi.org/10.1007/978-3-031-24907-5_35

$\lambda_{DA}, \lambda_{ID}$	Day-ahead and intra-day electricity market prices.
$\lambda_E, \lambda_H, \lambda_M$	Electricity, hydrogen and methane selling prices.
π	Probabilities associated with scenarios.
$H_{Ch,Max}, H_{Ch,Min}$	Maximum and minimum charging capacity of tank.
$H_{Dis,Max}, H_{Dis,Min}$	Maximum and minimum discharging capacity of tank.
$H_{T,Max}, H_{T,Min}$	Maximum and minimum capacity of tank.
K	Coefficient for converting hydrogen to power.
$P_{B,Max}, P_{B,Min}$	Maximum and minimum capacity of battery.
$P_{Ch,Max}, P_{Ch,Min}$	Maximum and minimum charging capacity of battery.
$P_{Dis,Max}, P_{Dis,Min}$	Maximum and minimum discharging capacity of battery.
$P_{E,Max}, P_{E,Min}$	Maximum and minimum capacity of electrolyzer.
$P_{F,Max}, P_{F,Min}$	Maximum and minimum capacity of fuel cell.
P_L, H_L, M_L	Power, hydrogen and methane demand of station.
y	Fixed bid prices for electricity market.

Variables

H_{Ch}, H_V, H_M	Produced hydrogen for tank, vehicles and methanation.
$H_{F,Dis}, H_{V,Dis}, H_{M,Dis}$	Discharged hydrogen for fuel cell, vehicles and methanation.
P_B, H_T	Inventory level of the battery and tank.
P_{DA}, P_{ID}	Dispatched level in day-ahead and intra-day markets.
P_F	Produced power of fuel cell.
$P_{V,Dis}, P_{H,Dis}$	Discharged battery power for vehicles and electrolyzer.
P_V, P_{Ch}, P_H	Purchased power for vehicles, battery and electrolyzer.
u_B, u_H	Selection of charging mode for the battery and tank.
u_E	Selection of working mode for the electrolyzer.
x_{DA}, x_{ID}	Bidding volume to day-ahead and intra-day markets.

Introdution

The transportation sector has the largest share of total energy consumption growth in the world caused excessive pollution and energy crisis. One solution is using electric, hydrogen and natural gas vehicles as alternative transportation options. The demand of these environment friendly vehicles can be satisfied from the power grid. The hydrogen demand of the vehicles can meet through power to gas process which converts electricity to hydrogen through electrolyzer. The generated hydrogen can react with carbon dioxide to produce synthetic natural gas and provide the demand of vehicles [1].

There have been many studies about single refueling station; however, only few papers investigate the integration of electric and hydrogen refueling station. The multifunction charging station can also provide the flexibility as stated in [2] An autonomous hybrid hydrogen electricity charging station powered by photovoltaic in which the objective function maximizes the profit is modeled for a remote off-grid area in [3]. The flexible operation of a multi-product charging station to provide hydrogen and electricity for public transportation systems and private vehicles with the goal of profit maximizing is studied in [4]. A robust model for optimal design and planning of a standalone charging station for electric and hydrogen vehicles powered by the solar system and diesel generator is presented in [5]. Mehrjerdi addresses an optimal algorithm to design and set the off-grid electric and hydrogen charging station powered by the solar system and diesel generator with the aim of cost minimization [6]. A risk-oriented design and planning of a standalone hydrogen and electric refueling station based on information-gap decision theory is obtained in [7]. Beside previous studies, the combined electric, hydrogen and gas refueling station has been investigated recently by Xu et al. [8] for the first time in which a risk-based scheduling of an off-grid station powered by renewable energy based on information gap decision theory is proposed.

By far, a grid-connected solution has not been addressed in literature. Therefore, this study proposes a price-taker multifunction charging station to support electric, hydrogen and gas vehicles which is powered by the grid. A price-taking hybrid fueling station cannot affect the market price by changing its own actions [9]. Optimal volume allocation to the different markets is a complex task due to the sequential and stochastic nature of the power markets which has been not taken into account by previous works. The coordinated bidding process is considered based on [10], which employs a stochastic linear programming model for constructing piecewise-linear bidding curves. This study tries to obtain bidding curves of the charging station and optimal operation of devices by considering the uncertainty of market prices with the purpose of profit maximization.

Problem Formulation

The proposed multifunction charging station consists of a battery, electrolyzer, tank, fuel cell and methanation device as shown in Fig. 35.1. The electric vehicles demand is satisfied by the grid, battery and fuel cell. The hydrogen demand of the vehicles is met by the electrolyzer and tank. The demand of the natural gas vehicles is provided through methanation process.

Equation (35.1) maximizes the profit of the charging station owner including the revenue of selling the power, hydrogen and methane to the vehicles minus the cost of purchasing power from the day-ahead and intra-day markets where the set of variables Ξ includes P_{DA}, P_{ID}, P_V, P_{Ch}, P_H, $P_{V,Dis}$, $P_{H,Dis}$, P_B, H_{Ch}, H_V, H_M, H_T, $H_{F,Dis}$, $H_{V,Dis}$, $H_{M,Dis}$, P_F, u_B, u_H, u_E, x_{DA} and x_{ID}. Equation (35.2) specifies that the purchased power from markets should be equal to the devoted power for electric vehicles, battery and electrolyzer. Equation (35.3) indicates that the power level of the battery should be equal to the sum of its level in an hour ago and the bought power for charging the battery minus the discharged power for the electric vehicles and electrolyzer. Constraint (35.4) limits the inventory level of the battery by its capacity. Constraints (35.5) and (35.6) permit the battery to be in

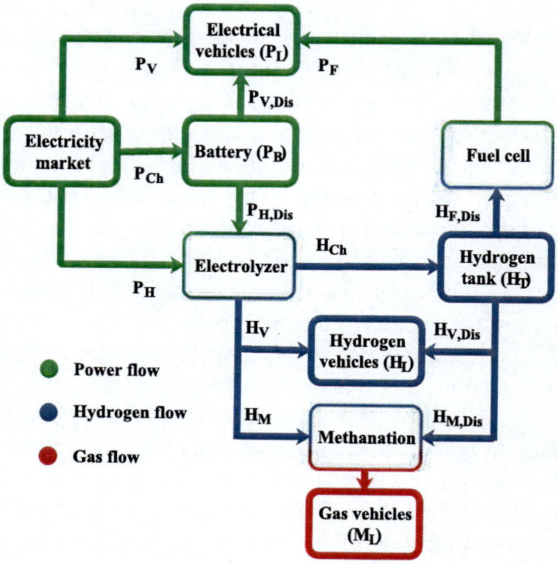

Fig. 35.1 Electric, hydrogen and gas charging station model

charge, discharge and shut-down mode. Equation (35.7) affirms that the summation of purchased and discharged battery power for running the electrolyzer that is multiplied by the efficiency and divided to a constant should be equal to the devoted hydrogen for the tank, hydrogen vehicles and methanation device. Constraint (35.8) bounds the used power of the electrolyzer by its capacity where the binary variable avoids the simultaneous operation of the fuel cell and electrolyzer. Equation (35.9) states that the hydrogen level of the tank should be equal to its level in an hour ago and the charged volume minus the discharged amount for fuel cell, hydrogen vehicles and methanation device. Constraint (35.10) restricts the hydrogen level of the tank by its capacity. Constraints (35.11) and (35.12) let the hydrogen tank to be in charge, discharge or shut-down mode. Equation (35.13) asserts that the discharged hydrogen from the tank for the fuel cell that is multiplied by the constant and efficiency should be equal to its generated power. Constraint (35.14) bounds the generated power of the fuel cell by its capacity. Equation (35.15) states that the summation of the power from the market, battery and fuel cell for charging electric vehicles should be equal to the demand. Equation (35.16) expresses that the summation of the electrolyzer and tank hydrogen for refueling vehicles should be equal to the load. Equation (35.17) declares that the summation of the electrolyzer and tank hydrogen for the methanation process that is multiplied by the efficiency should be equal to the needed methane. Constraints (35.18) and (35.19) impose the required market rules that the bidding curves should be non-increasing. Equations (35.20) and (35.21) formulate the bidding curves in the day-ahead and intra-day markets, respectively.

$$\underset{\Xi}{\textbf{Maximize}} \sum_{h\omega} \pi(\omega) \Big(P_L(h)\lambda_E + H_L(h)\lambda_H + M_L(h)\lambda_M$$
$$- P_{DA}(h,\omega)\lambda_{DA}(h,\omega) - P_{ID}(h,\omega)\lambda_{ID}(h,\omega) \Big) \quad (35.1)$$

$$P_{DA}(h,\omega) + P_{ID}(h,\omega) = P_V(h,\omega) + P_{Ch}(h,\omega) + P_H(h,\omega) \quad \forall h, \forall \omega \quad (35.2)$$

$$P_B(h,\omega) = P_B(h-1,\omega) + \eta_{B,Ch} P_{Ch}(h,\omega)$$
$$- \left(P_{V,Dis}(h,\omega) + P_{H,Dis}(h,\omega)\right)/\eta_{B,Dis} \quad \forall h, \forall \omega \quad (35.3)$$

$$P_{B,Min} \le P_B(h,\omega) \le P_{B,Max} \quad \forall h, \forall \omega \quad (35.4)$$

$$P_{Ch,Min} u_B(h,\omega) \le P_{Ch}(h,\omega) \le P_{Ch,Max} u_B(h,\omega) \quad \forall h, \forall \omega \quad (35.5)$$

$$P_{Dis,Min}(1 - u_B(h,\omega)) \le P_{V,Dis}(h,\omega) + P_{H,Dis}(h,\omega)$$
$$\le P_{Dis,Max}(1 - u_B(h,\omega)) \quad \forall h, \forall \omega \quad (35.6)$$

$$H_{Ch}(h,\omega) + H_V(h,\omega) + H_M(h,\omega) = \eta_E \left(P_H(h,\omega) + P_{H,Dis}(h,\omega)\right)/K$$
$$\forall h, \forall \omega \quad (35.7)$$

$$P_{E,Min} u_E(h,\omega) \le P_H(h,\omega) + P_{H,Dis}(h,\omega) \le P_{E,Max} u_E(h,\omega) \quad \forall h, \forall \omega \quad (35.8)$$

$$H_T(h,\omega) = H_T(h-1,\omega) + \eta_{T,Ch} H_{Ch}(h,\omega)$$
$$- \left(H_{F,Dis}(h,\omega) + H_{V,Dis}(h,\omega) + H_{M,Dis}(h,\omega)\right)/\eta_{T,Dis} \quad \forall h, \forall \omega \quad (35.9)$$

$$H_{T,Min} \le H_T(h,\omega) \le H_{T,Max} \quad \forall h, \forall \omega \quad (35.10)$$

$$H_{Ch,Min} u_H(h,\omega) \le H_{Ch}(h,\omega) \le H_{Ch,Max} u_H(h,\omega) \quad \forall h, \forall \omega \quad (35.11)$$

$$H_{Dis,Min}(1 - u_H(h,\omega)) \le H_{F,Dis}(h,\omega) + H_{V,Dis}(h,\omega) + H_{M,Dis}(h,\omega)$$
$$\le H_{Dis,Max}(1 - u_H(h,\omega)) \quad \forall h, \forall \omega \quad (35.12)$$

$$P_F(h,\omega) = \eta_F H_{F,Dis}(h,\omega) K \quad \forall h, \forall \omega \quad (35.13)$$

$$P_{F,Min}(1 - u_E(h,\omega)) \le P_F(h,\omega) \le P_{F,Max}(1 - u_E(h,\omega)) \quad \forall h, \forall \omega \quad (35.14)$$

$$P_L(h) = P_V(h,\omega) + P_{V,Dis}(h,\omega) + P_F(h,\omega) \quad \forall h, \forall \omega \quad (35.15)$$

$$H_L(h) = H_V(h,\omega) + H_{V,Dis}(h,\omega) \quad \forall h, \forall \omega \quad (35.16)$$

$$M_L(h) = \eta_M \left(H_M(h,\omega) + H_{M,Dis}(h,\omega)\right) \quad \forall h, \forall \omega \quad (35.17)$$

$$x_{DA}(i+1,h) \le x_{DA}(i,h) \quad \forall h, i \in \{1, ..., I-1\} \quad (35.18)$$

$$x_{ID}(i+1,h,\omega) \le x_{ID}(i,h,\omega) \quad \forall h, \forall \omega \in \Omega_{DA}, i \in \{1, ..., I-1\} \quad (35.19)$$

$$P_{DA}(h,\omega) = \frac{\lambda_{DA}(h,\omega) - y(i)}{y(i+1) - y(i)} x_{DA}(i+1,h)$$
$$+ \frac{y(i+1) - \lambda_{DA}(h,\omega)}{y(i+1) - y(i)} x_{DA}(i,h) \quad if \quad y(i) \le \lambda_{DA}(h,\omega) \le y(i+1)$$
$$\forall h, i \in \{1, \ldots, I-1\} \quad (35.20)$$

$$P_{ID}(h,\omega) = \frac{\lambda_{ID}(h,\omega) - y(i)}{y(i+1) - y(i)} x_{ID}(i+1,h,\omega)$$
$$+ \frac{y(i+1) - \lambda_{ID}(h,\omega)}{y(i+1) - y(i)} x_{ID}(i,h,\omega) \quad if \quad y(i) \le \lambda_{ID}(h,\omega) \le y(i+1)$$
$$\forall h, \forall \omega \in \Omega_{DA}, i \in \{1, \ldots, I-1\} \quad (35.21)$$

Case Study

The detailed parameters of the devices are given in Table 35.1. The coefficient of converting hydrogen to methane and power is considered 0.27 and 3.5 kWh/m^3, respectively. The electricity, hydrogen and gas selling prices are assumed to be 0.4 EUR/kWh, 1.2 and 1.6 EUR/m^3, respectively. The bidding curves for selected hours in markets are shown in Fig. 35.2. The bidding curves for the day-ahead market are first stage variables and scenario independent while the bidding curves for intra-day market are second stage variables and scenario dependent. Consequently, different bidding curves are submitted to the second market based on the realized scenario in the first market which reveals the sensitivity of bidding curves in the intra-day market to the deviations in the day-ahead market prices. The profit of the charging station equals to 6610 EUR. The revenue of selling electricity, hydrogen and gas equals to 5000, 4552 and 610 EUR, respectively. This illustrates that refueling electric vehicles and gas vehicles has the biggest and smallest share in gaining profit.

Table 35.1 Detail specification of the devices.

Parameter	Value	Parameter	Value
Battery		**Tank**	
$\eta_{B,Ch}$, $\eta_{B,Dis}$	85%, 85%	$\eta_{T,Ch}$, $\eta_{T,Dis}$	85%, 85%
$P_{B,\text{Min}}$, $P_{B,\text{Max}}$	0, 400 (kWh)	$H_{T,\text{Min}}$, $H_{T,\text{Max}}$	0, 300 (m^3)
$P_{Ch,\text{Min}}$, $P_{Ch,\text{Max}}$	0, 80 (kW)	$H_{Ch,\text{Min}}$, $H_{Ch,\text{Max}}$	0, 60 (m^3/h)
$P_{Dis,\text{Min}}$, $P_{Dis,\text{Max}}$	0, 80 (kW)	$H_{Dis,\text{Min}}$, $H_{Dis,Max}$	0, 60 (m^3/h)
Electrolyzer		**Fuel cell**	
η_E	85%	η_F	50%
$P_{E,\text{Min}}$, $P_{E,\text{Max}}$	0, 2000 (kWh)	$P_{F,\text{Min}}$, $P_{F,\text{Max}}$	0, 100 (kWh)

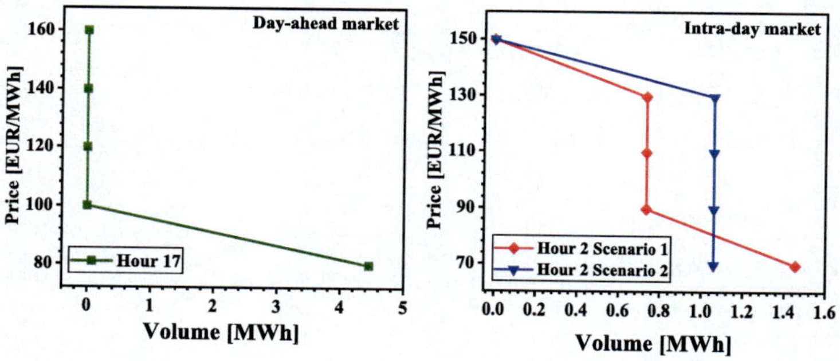

Fig. 35.2 Bidding curves in the day-ahead and intra-day markets

Conclusion

A two-stage stochastic program is formulated for a grid-connected multifunction charging station participated in day-ahead and intra-day markets. The objective function maximizes profit by deciding on optimal operation of the devices and bidding curves under market price uncertainties. The outcomes prove the feasibility and states that different plans have to be taken to moderate impacts of uncertainties. The extended version of this paper with more contributions is currently under preparation to be published in near future.

Acknowledgements The research has been supported by the Czech Technical University in Prague, grant number SGS22/109/OHK5/2T/13 and the Ministry of Education, Youth and Sports within the dedicated programme ERC CZ under the project POSTMAN with reference LL1902.

References

1. Zhang, X., Bauer, C., Mutel, C. L., & Volkart, K. (2017). Life cycle assessment of power-to-gas: Approaches, system variations and their environmental implications. *Applied Energy, 190*, 326–338.
2. Moiseeva, E., Wogrin, S., & Hesamzadeh, M. R. (2017). Generation flexibility in ramp rates: Strategic behavior and lessons for electricity market design. *European Journal of Operational Research, 261*(2), 755–771.
3. Xu, X., Hu, W., Cao, D., Huang, Q., Liu, W., Jacobson, M. Z., & Chen, Z. (2020). Optimal operational strategy for an offgrid hybrid hydrogen/electricity refueling station powered by solar photovoltaics. *Journal of Power Sources, 451*, 227810
4. Panah, P. G., Bornapour, M., Hemmati, R., & Guerrero, J. M. (2021). Charging station stochastic programming for hydrogen/battery electric buses using multi-criteria crow search algorithm. *Renewable and Sustainable Energy Reviews, 144*, 111046
5. Wang, Y., Kazemi, M., Nojavan, S., & Jermsittiparsert, K. (2020). Robust design of off-grid solar-powered charging station for hydrogen and electric vehicles via robust optimization approach. *International Journal of Hydrogen Energy, 45*(38), 18995–19006.
6. Mehrjerdi, H. (2019). Off-grid solar powered charging station for electric and hydrogen vehicles including fuel cell and hydrogen storage. *International Journal of hydrogen Energy, 44*(23), 11574–11583.
7. Sriyakul, T., & Jermsittiparsert, K. (2021). Risk-constrained design of autonomous hybrid refueling station for hydrogen and electric vehicles using information gap decision theory. *International Journal of Hydrogen Energy, 46*(2), 1682–1693.
8. Xu, X., Hu, W., Liu, W., Wang, D., Huang, Q., Huang, R., & Chen, Z. (2021). Risk-based scheduling of an off-grid hybrid electricity/hydrogen/gas/refueling station powered by renewable energy. *Journal of Cleaner Production, 315*, 128155
9. Biggar, D. R., & Hesamzadeh, M. R. (2014). The economics of electricity markets. Wiley Blackwell
10. Bringedal, A. S., Søvikhagen, A. M. L., Aasgård, E. K., & Fleten, S. E. (2021). Backtesting coordinated hydropower bidding using neural network forecasting. *Energy Systems*, 1–21

Chapter 36
Optimized Congestion Management in Balancing Markets for Electricity Transmission System Operator

Sinan Eren and Ali Nezih Güven

Abstract A core function of each Electricity Transmission System Operator (TSO) is the procurement of ancillary services in real time balancing markets, necessary for a stable and reliable operation of the system. TSO controls the active power to manage congestions in transmission network based on security criteria. Congestion management is achieved by rescheduling generation using mainly experience of operators. However, TSO has legal obligations to procure ancillary services in accordance with economic, transparent and non-discriminatory procedures. This work aims to develop an algorithm/software system for a TSO to decide on the optimum redispatch in the sense of economy and security. The redispatch problem is formulated as an optimal power flow (OPF) problem. The objective function of the optimization is the minimization of the total cost of the generation shifts necessary. The security constraints will be satisfied by power flow equations which are non-linear and non-convex by nature. In order to obtain a robust solution due to the large scale nature of the problem, linearization techniques are applied to power flow equations. Compared with the commonly used DC OPF methods, reactive power and bus voltages are taken into account in the formulation. The system developed is deployed as a real time running application for real large-scale networks.

Keywords Power system security · Optimum power flow · Redispatch in balancing market

S. Eren (✉)
TÜBİTAK Marmara Research Center, Power Systems Group, Gebze Kocaeli 41470, Turkey
e-mail: sinan.eren@tubitak.gov.tr

A. N. Güven
Electrical and Electronics Engineering Department, Middle East Technical University, Çankaya Ankara 06800, Turkey
e-mail: guven@metu.edu.tr

O. Grothe et al. (eds.), *Operations Research Proceedings 2022*, Lecture Notes in Operations Research, https://doi.org/10.1007/978-3-031-24907-5_36

Introduction

Diversified number of actors affect energy transfer in the electricity network such as generation companies, consumers, regulators and operators. Impacts of some of these participants are predictable but not determinable, such as consumers and renewable energy sources, and there are those that are manageable, such as conventional power plants. In this business, Electricity Transmission System Operator (TSO) is responsible for the secure, reliable and economic operation of the electricity grid.

Market Operator (MO) is responsible for managing and operating electricity market, where members send their bids to buy or sell energy in determined delivery platforms. Its task is to match all buy or sell orders in a transparent manner, according to the regulations and to establish a reference price. However, majority of market designs aim for the economic operation of the system (i.e., price based unit commitment) rather than security constrained unit commitment.

This market design may result in security violations in the grid such as loading of one or multiple elements, bus voltage deviations and loss of stability. In real time operation, TSO manages transmission congestions via generation rescheduling or load shedding in a balancing market. In this work, an operational assistant is developed to determine the redispatch in accordance to economic, transparent, non-discriminatory procedures for transmission system operation.

Congestion Management

In a deregulated environment, different methods of accomplishing the congestion management have been implemented [1]. Price area congestion control and transaction-based control can be given as example models. For the market models considering transmission conditions in unit commitment (UC) processes, less number of violations of security is expected in real time operation. However, transmission bottlenecks can still be encountered in operation due to the dynamic nature of consumption and renewable generation, any failure in an equipment or the complex security conditions such as stability, which may not be covered in UC.

Moreover, generation scheduling procedures may be free of security constraints since they require complex business processes to integrate between the market model and the network model, robustness issues and non-deterministic designs. In this case, TSO has to deal with more corrective control actions in real time system operation. In most cases, congestion management is based on operator experience. Although their decisions are not based on analytical calculations and do not yield optimal solutions, this approach can be successful in solving congestion problems. In more systematic approaches, power transfer distribution factors (PTDF) are used to deal with the problem [2]. PTDF approach presents the relation between bus injections and branch flows so that the operator can decide on the generation reschedule. However, the solution obtained by this approach may not be the optimum solution, and is limited to cases that can be solved by single action. In most professional cases, the problem is defined as optimal power flow (OPF) and redispatch requirements are obtained from optimization results.

Optimal Power Flow

After the first formulation of OPF problem in 1970s,s, numerous solution techniques have been proposed [3]. In OPF methods with AC network model, original power flow equations are considered in the optimization problem. Nonlinear optimization methods (i.e., Successive Linear Programming, Interior Point Method) are used to solve the problem due to the non-convex and non-linear nature of the problem. Some methods linearize network model to approximate ones, also called DC-OPF models. They assume voltage angle differences of neighboring buses are close to zero, and voltage magnitudes are close to nominal values. This approach reduces the computational complexity, and guarantees the convergence; however, accuracy of the solution is sacrificed. There are also metaheuristic methods for OPF based on machine learning techniques (i.e., Genetic Algorithm, Particle Swarm Optimization) to overcome this complex problem [4].

There is always a trade off between speed of solution, risk of convergence and accuracy of results. In this study, AC linearized OPF model is preferred to meet the time and convergence requirements of the system operator in real time operation.

Problem Formulation

Problem Statement

The main objective function of the problem is defined as minimization of total redispatch cost (1). Redispatch cost is calculated based on piecewise linear function of generation up/down bids, this formulation introduces integer variables into the problem. There exist penalty terms in the objective function such as bus voltage magnitude deviation from base case results, since voltage adaptations require lots of control actions which are not easily applicable in real time operation.

$$\min \sum_{p \subset P} f(P_R^p) + \sum_i \varepsilon \left| v_i^2 - v_{i,0}^2 \right| \tag{1}$$

where p is power plant index of plant set P, f(.) is piecewise linear function of redispatch cost, ε is penalty term for bus voltage magnitude deviation from base case results. Redispatch magnitude of a plant (P_R^p) is summation of the redispatch of generators (P_R^g) related to that power plant (2). There exists a minimum value of plant redispatch ($P_R^{\min}$) in both up/down directions. After the redispatch, final active power generation of a generator should comply with the physical power limits ($P_P^{g,\min}$ and $P_P^{g,\max}$) unless the generator is shut-down.

$$P_R^p = \sum_{g \subset G^p} P_R^g, \; where \; |P_R^p| > P_R^{\min} \; and \; P_P^{g,\min} \leq P_R^g + P_P^{g,0} \leq P_P^{g,max} \quad (2)$$

The constraints of this problem are power flow and injection equalities, voltage and angle limits, generator power limits, supply - demand balance equality. And also there exist branch power flow limits which are quadratic inequalities based on active and reactive power flows (3).

$$P_{ij}^2 + Q_{ij}^2 \leq (S_{ij}^{\max})^2 \quad (3)$$

Linearization of Power Flow Equations

OPF problem suggested in this work consists of network branch flow constraints as in (4) and (5), and nodal power balance equations as in (6) and (7). P_{ij}/Q_{ij} represent active/reactive power flows from bus i to bus j; P_i/Q_i represent active/reactive power injections at bus i; G_{ij}/B_{ij} are real/imaginary parts of Y_{ij} in the bus admittance matrix; g_{ij}/b_{ij} are conductance/susceptance of branch (i, j); finally v_i/θ_i are voltage magnitude/angle at bus i.

$$P_{ij} = g_{ij}\left(v_i^2 - v_i v_j \cos\theta_{ij}\right) - b_{ij} v_i v_j \sin\theta_{ij} \quad (4)$$

$$Q_{ij} = -b_{ij}\left(v_i^2 - v_i v_j \cos\theta_{ij}\right) - g_{ij} v_i v_j \sin\theta_{ij} \quad (5)$$

$$P_i = \sum_{(i,j)\in\mathcal{K}} P_{ij} + \left(\sum_{j=1}^{N} G_{ij}\right) v_i^2 \quad (6)$$

$$Q_i = \sum_{(i,j)\in\mathcal{K}} Q_{ij} + \left(\sum_{j=1}^{N} -B_{ij}\right) v_i^2 \quad (7)$$

These equations constitute a non-linear and non-convex problem. In order to transform this problem into a linear problem, linearization techniques are utilized. Second order Taylor series expansions of sine and cosine functions are applied (8), assuming θ_{ij} is generally low. To decouple v and θ, voltage magnitudes are assumed to close to 1 p.u. (9).

$$\sin\theta_{ij} \approx \theta_{ij}, \;\; \cos\theta_{ij} \approx 1 - \frac{\theta_{ij}^2}{2} \quad (8)$$

$$v_i v_j \theta_{ij} \approx \theta_{ij},\ \ v_i v_j \theta_{ij}^2 \approx \theta_{ij}^2 \tag{9}$$

Problem still constitutes non-convex behaviour due to θ_{ij}^2 and v_{ij}^2 terms. These terms are linearized using first order Taylor series expansion (10) around operating point of base case load flow conditions $(\theta_{i,0}, v_{i,0})$.

$$\theta_{ij}^2 \approx 2\theta_{ij,0}\theta_{ij} - \theta_{ij,0}^2,\ \ v_{ij}^2 \approx 2\frac{v_{i,0} - v_{j,0}}{v_{i,0} + v_{j,0}}\left(v_i^2 - v_j^2\right) - \left(v_{i,0} - v_{j,0}\right)^2 \tag{10}$$

When θ_i and v_i^2 are chosen as independent variables, the problem becomes a linear optimisation problem, as given in (11) and (12). In the implementation, original approach stated in [5] is adhered.

$$P_{ij}^L \approx g_{ij}\theta_{ij,0}\theta_{ij} - \frac{1}{2}g_{ij}\theta_{ij,0}^2 + g_{ij}\frac{v_{i,0} - v_{j,0}}{v_{i,0} + v_{j,0}}(v_i^2 - v_j^2) - \frac{1}{2}g_{ij}(v_{i,0} - v_{j,0})^2 \tag{11}$$

$$Q_{ij}^L \approx -b_{ij}\theta_{ij,0}\theta_{ij} + \frac{1}{2}b_{ij}\theta_{ij,0}^2 - b_{ij}\frac{v_{i,0} - v_{j,0}}{v_{i,0} + v_{j,0}}(v_i^2 - v_j^2) + \frac{1}{2}b_{ij}(v_{i,0} - v_{j,0})^2 \tag{12}$$

Solution Technique

The stated problem formulation constitutes a Mixed Integer Quadratically Constrained Linear Programming problem. In order to solve the problem efficiently, IBM ILO CPLEX software package is utilized. CPLEX Optimizer has a modeling layer called Concert that provides interfaces to Java language. The problem is built in Analysis Server environment, then it is sent to the developed Optimization Server application which has installed CPLEX software. This optimization application is designed using features such as parallel execution and kernel management. Implementation experiences will be mentioned in the next section.

Architecture of the Economic Redispatch Module

Economic Redispatch Module operates periodically and automatically in every hour (it will be rescheduled as in every 15 min). Initially, this module obtains the network model data, and performs pre-processing. Then it calculates network states by performing a load flow (LF) analysis. LF algorithm uses fully coupled Newton-Raphson technique. Network states predicate voltage violations and overloadings in branches. If there is no congestion observed, system presents LF analysis results. If there is, redispatch candidates are obtained from Balancing Market. Generation

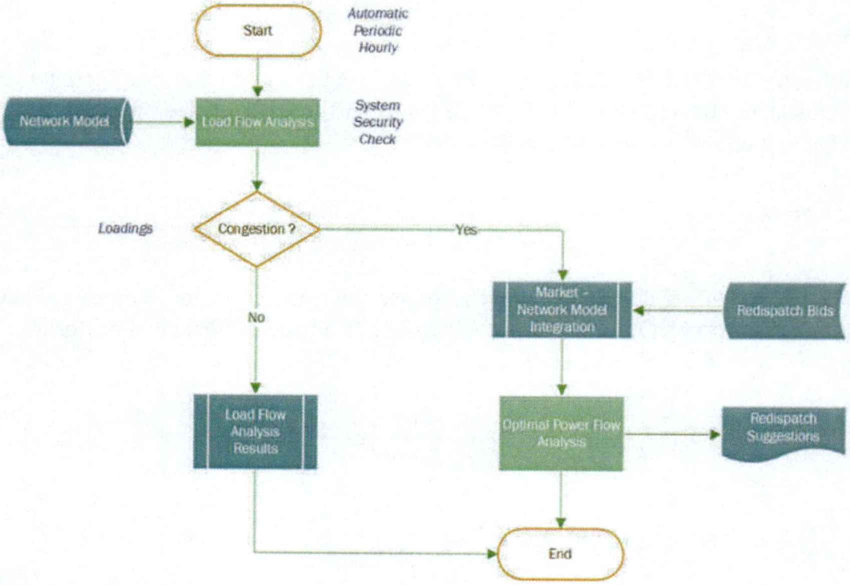

Fig. 36.1 Workflow of the economic redispatch module

up/down costs are based on the bids of generation companies. OPF problem finally is carried out to obtain the most economical solution for eliminating congestion. Workflow diagram is summarized in Fig. 36.1.

Application of the Algorithm

Turkish Electricity Market Structure

Organized and physical Wholesale Power Market of Turkey is operated under two entities: TSO (TEİAŞ) and MO (EPİAŞ). EPİAŞ is responsible for the day-ahead and intra-day markets, and TEİAŞ is responsible for balancing and ancillary markets. Actions in the balancing market are conducted under two objectives: to restore frequency reserve and to eliminate congestion management.

In the balancing market, producers can bid their redispatch offers up to maximum 15 price levels in up/down direction. The offers are based on bidding units, which are converted and distributed to physical generators using market model - network model integration tool. The rules of balancing market are stated in Balancing and Settlement Regulation of Energy Market Regulatory Authority. The module considers all the related constraints of the Regulation.

Implementation Results

Dispatcher Information System (YTBS) is an Energy Management System (EMS) developed for the TSO of Turkey in order to meet the requirement of system monitoring and analysis facilities [6]. The system has a network modelling component to construct a mathematical model of the grid for the EMS applications. Network information consists of more than 1400 substations, 2200 transmission lines, 2700 transformers, 1800 power plants of 100 GW total installed capacity. The algorithm is developed as a part of this system.

The application is designed to operate in real-time for real large scale electricity systems. For this purpose, it is deployed in National Load Dispatch Control Center of Turkey. In the optimization model, only high voltage ($\geq$ 36 kV) network is considered. After parallel bus aggregation, there has been an average of 1400 active buses. The number of integer variables changes between 750 and 1000.

In test trials, the optimization process, which includes constructing the optimization problem in Concert technology, solution and the communication between servers takes around 9 s. The computer processor is an Intel(R) Xeon(R) CPU E5-2699 v4 @ 2.20 GHz. In addition, other processes such as querying the network model, acquiring measurements, topology processing, load flow solution, market data correlation are observed to take about 30 s.

Results of the analysis are presented via interfaces under the Dispatch Information System web application (Fig. 36.2). Operators monitor congestions in the base case condition and the final operating condition, and then report the suggested redispatch orders. The system has been in a process of monitoring and evaluating the results since May 2022. The system operator has been giving the operational feedbacks and based on these improvements have been done. The overall performance and the results are considered to be successful by the TSO. Tests on the historical congestion cases indicate that the algorithm can offer 30% more economical redispatch suggestions compared to the experience-based preferences.

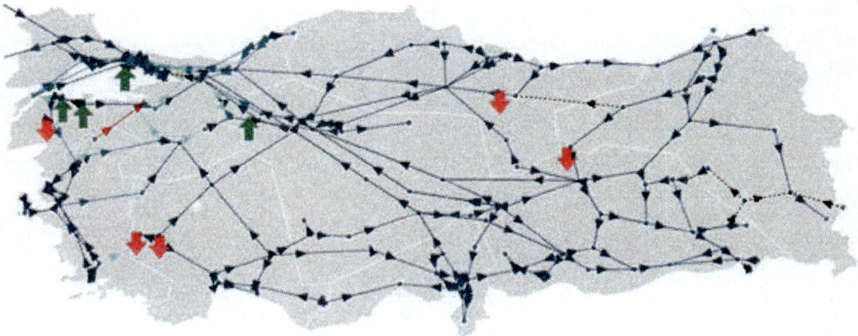

Fig. 36.2 Presentation of congestion and suggested redispatch orders

Conclusion

The congestion management is one of the critical tasks of the electricity system operation and requires a complicated business process. This work presents a methodological approach to determine generation up/down orders within a balancing market in order to eliminate transmission bottlenecks. Redispatch is considered as an OPF problem, and enriched with features to obtain practicable results such as the minimization of the number of control actions. The redispatch calculations conducted for historical congested cases indicate that developed approach makes possible to obtain more economical operational results.

Acknowledgements This research and technology development work is carried out within the scope of the Load Dispatch Center Technical Consultancy Project, developed for TEİAŞ, by TÜBİTAK MRC Energy Technologies. The project is funded by Turkish Electricity Transmission Company (TEİAŞ).

References

1. Christie, R. D., Wollenberg, B. F., & Wangensteen, I. (2000). Transmission management in the deregulated environment. *Proceedings of the IEEE, 88*(2), 170–195.
2. López, P. C., Sadikovic, R., Pinto, H., & Magnago, F. (2015). Swiss TSO experience with an ac security-constrained optimal power flow application for real-time security management. *IEEE Eindhoven PowerTech, 2015*, 1–6.
3. Falvo, M., Panella, S., Caprabianca, M., & Quaglia, F. (2021). A Review on unit commitment algorithms for the Italian electricity market. *Energies, 15*, 18.
4. Huang, W., Pan, X., Chen, M., & Low, S. H. (2022). DeepOPF-V: Solving AC-OPF problems efficiently. *IEEE Transactions on Power Systems, 37*(1), 800–803.
5. Yang, Z., Zhong, H., Bose, A., Zheng, T., Xia, Q., & Kang, C. (2018). A Linearized OPF model with reactive power and voltage magnitude: A pathway to improve the MW-Only DC OPF. *IEEE Transactions on Power Systems, 33*(2), 1734–1745.
6. Eren, S., Küçük, D., Ünlüer, C., Demircioğlu, M., Arslan, Y., & Sönmez, S. (2015). A Web-based dispatcher information system for electricity transmission grid monitoring and analysis. In *2015 9th International Conference on Electrical and Electronics Engineering (ELECO)*, pp. 986-990.

Chapter 37
Quantifying Capacity Adequacy in Energy System Modelling Through Stochastic Optimization

Shima Sasanpour and Karl-Kiên Cao

Abstract Energy system optimization models (ESOMs) can be helpful tools to determine the optimal structure of future energy systems. They usually optimize the expansion and dispatch of the energy system's components through a minimization of total system costs. The obtained energy systems are designed to cover the energy demand for the specific assumptions made within the underlying scenarios. However, if such energy systems are exposed to slight deviations, such as a lower availability of wind energy, situations of uncovered demand may occur. The uncertainties in the scenario assumptions can be indirectly captured via excess generation capacities. However, the required amount of these excess capacities is unclear. This study analyzes capacity adequacy by considering uncertainties in a decarbonized German power system through stochastic optimization within an ESOM. Different uncertainties, such as technology investment costs, total annual demand and different weather conditions are considered and their influence on the power system is compared. Therefore, a variety of different assumptions for these uncertainties are extracted from literature and included in the stochastic optimization. As a result, the impact of the uncertainties on the structure of the energy system are identified and the excess capacity needed is estimated.

Keywords Stochastic programming · Energy system optimization model · Decarbonized energy system

Introduction

The structure of our current energy system has to alter significantly in order to be decarbonized. Due to the usually applied cost-minimization approach, energy systems determined by ESOMs represent "minimum designs", which can ensure security of supply only if real developments are congruent with the considered model inputs. However, these input parameters are subject to uncertainties. While these

S. Sasanpour (✉) · K.-K. Cao
German Aerospace Center (DLR), Institute of Networked Energy Systems, Stuttgart, Germany
e-mail: Shima.Sasanpour@dlr.de

O. Grothe et al. (eds.), *Operations Research Proceedings 2022*, Lecture Notes in Operations Research, https://doi.org/10.1007/978-3-031-24907-5_37

uncertainties are often not considered within the optimization, they can have a strong influence on the structure of the energy system. There are different possibilities to consider uncertainties in energy system modeling [1]. One method is stochastic optimization, which has the advantage that it provides one hedging strategy to cope with the possible risks. Usually, time varying inputs, such as demand and weather profiles are considered with stochastic optimization. Furthermore, Nadal et al. found that techno-economic parameter uncertainties can have a significant influence on the energy system as well [2].

This paper compares the influence of a variety of uncertainties on the structure of the energy system in a decarbonized energy system of Germany and gives an estimate on the excess capacities needed to achieve capacity adequacy. Additionally, two probability distributions and their impact on the energy system are compared.

Method

Model

For our study, we use the ESOM REMix [3]. It optimizes the expansion and dispatch of technologies of the energy system by minimizing the total system costs. The input data consists of historic weather and demand profiles and techno-economic parameters, such as investment costs, fuel costs and efficiencies. REMix is written in the mathematical optimization language GAMS. For each random variable the stochastic optimization method is included with the Extended Mathematical Programming (EMP) feature which allows an implementation as add-on to the deterministic model [4].

Model Setup

The considered model is based on [5] and consists of ten German regions and nine neighboring countries. The electricity imports and exports of the neighboring countries are fixed. The model considers the power sector with several power plant, storage and grid technologies and a green-field approach in terms of capacity expansion planning. We assume that no CO_2 emissions are allowed within the power sector. Therefore, only a limited number of power plant technologies can be expanded, such as renewable energy converters, biomass power plants and hydrogen-fueled gas power plants. Green hydrogen can be imported for a specific price.

Uncertainties

The analyzed uncertainties include weather, demand and parameter uncertainties. The demand is split up between the normalized demand profile and the annual demand, which allows the separate adjustment of each. The techno-economic parameter uncertainties include fuel and investment costs. The corresponding parameter values are collected through literature research, e.g. [6, 7]. Table 37.1 shows the standard deviation and the minimum, maximum, and median values of the parameter uncertainties. The weather and demand profile uncertainties are represented by several historic weather (2006–2012) and demand (2006–2015) years.

For the weather and demand profile uncertainty the same probability is assigned to each historic year. For each parameter uncertainty the interval between the minimum and maximum value is divided into ten sections with an equal range. The mean of each section is used as value of the random variable, resulting in ten scenarios per random variable.

Our approach to consider the probability of the parameter uncertainties is to use a truncated normal distribution. Due to the limited number of values from literature for each random variable, the median instead of the mean value is used as expected value μ for the normal distribution. The standard deviation σ is calculated by using a standard normal distribution. The minimum or maximum value x, that has a further distance to the expected value μ, is assigned to a z-value of -3 or 3, which corresponds to a probability of 99.87%. The standard deviation can then be calculated as

$$\sigma = (x - \mu)/z \tag{37.1}$$

Table 37.1 Parameter uncertainties

Random variable	Min	Max	Median	Standard deviation	Unit
Biomass cost	0.03	0.10	0.06	0.018	€/kWh
Hydrogen cost	0.10	0.21	0.12	0.03	€/kWh
BioPower invest	2589	7412	4161	1084	€/kW
CCGT invest	574	1195	757	146	€/kW
OCGT invest	300	568	453	51	€/kW
HydroRoR invest	1040	3107	1665	481	€/kW
LiIonBattery invest	17	50	26	8	€/kW
PumpedStor invest	34	663	402	123	€/kW
PV invest	216	1300	347	318	€/kW
WindOff invest	1605	5202	2527	892	€/kW
WindOn invest	870	2400	1160	413	€/kW
Annual demand	459	841	535	101	TWh

The normal distribution is then cut off at the minimum and maximum value and normalized to receive a truncated normal distribution. Furthermore, as there exists no evidence about the real distribution of uncertain parameters, we benchmark this distribution function using an alternative uniform distribution, which assigns the same probability to each scenario. Therefore, the extreme values receive a higher weighting compared to the truncated normal distribution.

For each uncertainty and probability distribution, a stochastic optimization model run is executed. The Base case serves as reference of a future German power system and is solved by deterministic optimization. Therefore, the median values of the parameter uncertainties and the weather and demand profile of 2006 are included.

Indicators

To analyze the impact of uncertainties on the structure of the energy system, two indicators are used. The relative root mean square error (RRMSE)

$$\mathrm{RRMSE} = \frac{\sqrt{\frac{1}{n}\sum_{i=1}^{n}(S_i - B_i)^2}}{\overline{B}} \tag{37.2}$$

compares the expanded capacities of the stochastic optimization S_i with the capacities of the Base case B_i. It provides an indication of the impact of the uncertainty on the model results.

For the weighted average of the deterministic scenarios (*WADS*), each stochastic scenario of the random variables is calculated with perfect foresight. The resulting capacities are then weighted with the same probabilities as the stochastic scenarios to receive the *WADS*. This can be compared to the results of the stochastic optimization.

Results

The results of all stochastic model runs are compared to the Base case with the RRMSE. The model runs with fuel cost and most investment cost uncertainties deviate from the Base case by less than 10%. Only the biomass power plant (BioPower), photovoltaic (PV) and wind offshore (WindOff) investment cost uncertainties, as well as the annual demand (AnnDemand) and weather profile uncertainties show a higher deviation.

Figure 37.1a shows the capacity differences of PV and WindOff investment cost uncertainties compared to the Base case. The capacities of the Base case are shown in Table 37.2 as reference. As PV investment cost uncertainties with a truncated normal

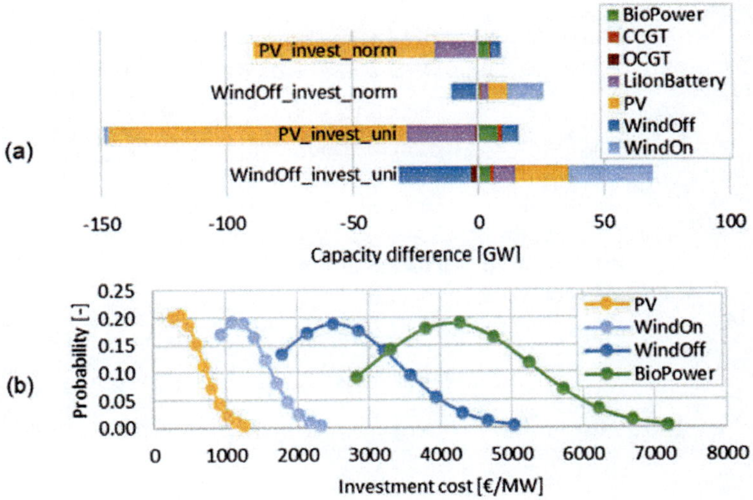

Fig. 37.1 **a** Capacity differences of investment costs uncertainties compared to base case. **b** Probability distribution of investment costs

Table 37.2 Capacities of Base case in GW

BioPower	CCGT	OCGT	Hydro	LiIonBatt	PV	PumpedStor	WindOff	WindOn
6.7	22.4	17.1	6.2	43.8	236.8	8.8	60.9	3.4

distribution are included, significantly less PV and lithium-ion batteries (LiIonBatteries) are expanded. Similarly, with a WindOff investment cost uncertainty less WindOff capacities are available.

The explanation for this effect can be found in Fig. 37.1b. In this figure, the truncated normal distributions of the investment costs are shown. For both PV and WindOff the median is close to the minimum value. As higher investment costs are considered in the stochastic optimization compared to the deterministic Base case, both technologies are expanded less. The uniform distribution amplifies the effect observed for the normal distribution, since high investment costs get an even higher probability.

In Fig. 37.2a, capacity differences of the time-dependent profile uncertainties compared to the Base case are shown. The consideration of different demand profiles results in a stronger expansion of PV and LiIonBatteries while less WindOff is expanded. The availabilities of PV seem to better align with the different demand profiles. With different weather years considered, PV and wind onshore (WindOn) become partially replaced by WindOff capacities, which have a higher availability and seem to be less dependent on the weather years. For both the weather and demand profile uncertainties, the hydrogen-fueled combined cycle gas turbine (CCGT) and open cycle gas turbine (OCGT) power plants are built to a higher extend, since they provide flexibility and are independent of time-dependent uncertainties.

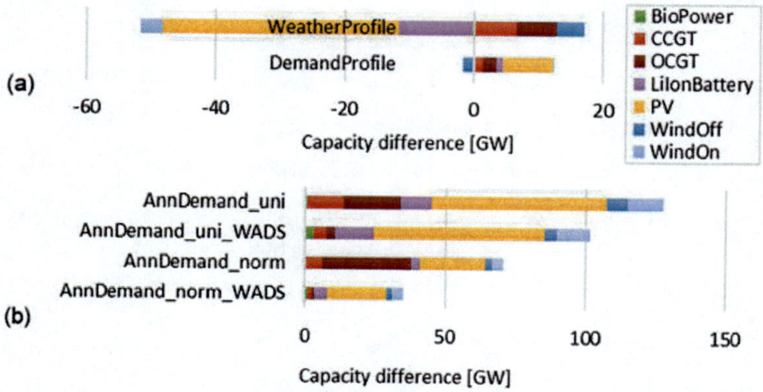

Fig. 37.2 **a** Capacity difference of profile uncertainty compared to Base case. **b** Capacity difference of annual demand uncertainty compared to Base case

Figure 37.2b illustrates the impact of annual demand uncertainties compared to the Base case. With a truncated normal distribution and a stochastic optimization approach, especially OCGT power plants with low investment costs are expanded, since the probability of extremely high annual demands is relatively low. As greater annual demands have a higher probability with a uniform distribution, technologies with higher investment costs but lower operational costs, such as CCGT, PV, and WindOn, are expanded to a higher extent. At the same time, the amount of excess capacities increases. In comparison, when the deterministic *WADS* approach is used, the additional CCGT and OCGT capacities are clearly smaller. Therefore, with the stochastic approach the hedging against the risk of higher demands is higher, since more secured capacities are available.

Discussion and Outlook

This paper analyzes the impact of different uncertainties on the structure of a decarbonized German power system with a stochastic optimization approach. While most parameter uncertainties have little impact on the results, others alter the structure of the energy system significantly. Especially for uncertainties concerning the future weather and annual demand, hydrogen-fueled power plants can provide risk hedging, almost independent from costs of green hydrogen. The consideration of the annual demand uncertainty can indicate the excess capacities needed to achieve capacity adequacy [8]. Depending on the assumed probability distribution, and therefore on the hedging strategy, 70 to 130 GW of additional capacities are required. A uniform distribution needs higher hedging than a normal distribution, since extreme scenarios have a higher probability. For each parameter and time-dependent profile uncertainty other technologies become more attractive. This indicates that by including several random variables within the stochastic optimization, the energy system might become more diverse. The combined consideration of the uncertainties with the highest impact on the structure of the energy system could be analyzed in future research.

Acknowledgements We thank our UNSEEN colleagues, a project funded by the German Federal Ministry for Economic Affairs and Climate Action, grant FKZ 03EI1004. We gratefully acknowledge the Gauss Centre for Supercomputing e.V. for providing computing time on the GCS Supercomputer JUWELS at Jülich Supercomputing Centre (JSC).

References

1. Yue, X., et al. (2018). A review of approaches to uncertainty assessment in energy system optimization models. *Energy strategy reviews, 21*, 204–2017.
2. Nadal, A., et al. (2020). Accounting for techno-economic parameters uncertainties for robust design of remote microgrid. *International Journal of Electrical Power & Energy Systems, 116*, 105531.
3. Gils, H. C., et al. (2017). Integrated modelling of variable renewable energy-based power supply in Europe. *Energy, 123*, 173–188.
4. GAMS https://www.gams.com/39/docs/UG_EMP.html. Accessed 18 July, 2022
5. Cao, K. K., et al. (2018). Incorporating power transmission bottlenecks into aggregated energy system models. *Sustainability, 10*(6), 1916.
6. Tsiropoulos, I., et al. (2018). Cost development of low carbon energy technologies. *Scenario-based cost trajectories to 2050*. Publications Office of the EU
7. IRENA. (2017). Electricity storage and renewables: Costs and markets to 2030. International Renewable Energy Agency
8. Boffino, L., et al. (2019). A two-stage stochastic optimization planning framework to decarbonize deeply electric power systems. *Energy Economics, 84*, 104457.

Chapter 38
Soft-Coupling Energy and Power System Models to Analyze Pathways Toward a De-fossilized German Transport Sector

Danial Esmaeili Aliabadi, Niklas Wulff, Matthias Jordan, Karl-Friedrich Cyffka, and Markus Millinger

Abstract The transport sector is a major consumer of energy worldwide. Unfortunately, there is no silver bullet to de-fossilize the transport sector due to its intricacy; therefore, many concepts and technologies should be combined to have a noteworthy impact on this hard-to-abate sector. As such, the required diverse set of expertise for making correct decisions cannot be achieved by merely utilizing one model. In this study, we connect multiple datasets and models that employ various methodologies with different purposes to exhibit a pathway to a green transport sector. The extended bioenergy optimization (BENOPTex) and renewable energy mix (REMix) models are coupled iteratively to produce coherent results while considering different sets of constraints. The combined effects of bioenergy and synthetic fuel—using renewable electricity—on the German transport sector are investigated via a scenario. Two demand models are also used to capture the specificities of the energy demands of the mainly behavior-driven road transportation as well as technology-driven aviation sector. The outcome of the resulted soft-coupled model respects biomass availability, regulatory circumstances, techno-economic properties, and power sector expansion for the production of synthetic fuels.

D. Esmaeili Aliabadi (✉) · M. Jordan
Department of Bioenergy, Helmholtz Centre for Environmental Research, Leipzig, Germany
e-mail: danial.esmaeili@ufz.de

M. Jordan
e-mail: matthias.jordan@ufz.de

N. Wulff
Department of Energy Systems Analysis, Institute of Engineering Thermodynamics, German Aerospace Center, Stuttgart, Germany
e-mail: niklas.wulff@dlr.de

K.-F. Cyffka
Department of Bioenergy Systems, Deutsches Biomasseforschungszentrum gGmbH, Leipzig, Germany
e-mail: karl-friedrich.cyffka@dbfz.de

M. Millinger
Department of Space, Earth and Environment, Chalmers University of Technology, Gothenburg, Sweden
e-mail: markus.millinger@chalmers.se

O. Grothe et al. (eds.), *Operations Research Proceedings 2022*, Lecture Notes in Operations Research, https://doi.org/10.1007/978-3-031-24907-5_38

Keywords Soft-coupling · Bioenergy · Renewable energy · Transportation · Optimization models

Introduction and Background

The transport sector is considered as a major consumer of energy worldwide, and accounts for over 28% of the total final energy consumption in the EU [9]. Unfortunately, this rapidly growing sector has the highest reliance level on fossil fuels [11]. These ratios also apply to Germany, where the renewable share in the transport sector is currently at only 6.8% [1].

In order to de-fossilize transport, experts from different disciplines should unite to capture the technological, social, and economical aspects. These disciplines may benefit from various modeling techniques in their studies. For instance, researchers may use agent-based models to simulate consumer behavior [14] or exploit optimization models to minimize total system costs considering different transportation modes [8, 18].

The two main modeling perspectives are top-down macroeconomic and bottom-up engineering approaches [2]. Energy system optimization and simulation models are members of the latter, while e.g. input-output, general equilibrium, and system dynamics models are part of the former. There are pros and cons associated to these approaches. While top-down models account for macroeconomic feedback and microeconomic realism, they ignore technical aspects for the sake of tractability, which makes them unreliable for making long-term future projections. On the other hand, bottom-up models can introduce new technologies and evaluate their impact on future energy market fundamentals. Although the technology explicitness of bottom-up models can assist researchers and policymakers to predict future trends, these models lack behavioral realism: they assume a perfect substitution of competing technologies. Figure 38.1 compares the contrasting approaches with respect to technology explicitness, behavioral realism, and macroeconomic feedback.

To develop a comprehensive energy system model with the advantage of both approaches, a growing number of researchers began investigating hybrid models [5]. There are two well-known methods for combining models from different disciplines: soft-linking and hard-linking [2]. The hard-linking method attempts to embed one model into another and solve both models simultaneously, whereas soft-linked models are solved iteratively to reach an equilibrium (i.e., a consistent solution) across all models. Employing a soft-linking approach is suggested when the involved models exploit different techniques.

The developed hybrid models can enrich results and discussions in sector- or technology-specific models via exchange. Also, the results of hybrid models often cover a broader set of stakeholders. Finally, the set of assumptions in different models can be reviewed, verified and harmonized while coupling. However, one should note that reaching an equilibrium solution is not guaranteed when soft-coupling models.

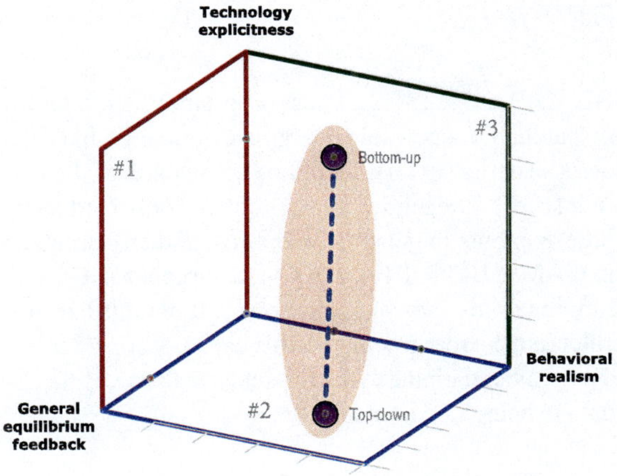

Fig. 38.1 Top-down, bottom-up, and all the combinations in between. The bottom-up models are often on the green surface (i.e., #3) and top-down models on the blue surface (i.e., #2)

Moreover, harmonizing models is a complex and time-consuming process; therefore, a subset of models might need to be streamlined.

In this study, we generate a hybrid model by soft-linking the extended bioenergy optimization (BENOPTex) and the Renewable Energy Mix (REMix) models to find a sustainable pathway to de-fossilize the German transport sector.

Methodology

In this section, we first introduce the linked models. Then, we elaborate on our strategy to couple these models considering the data flow from external datasets. The termination condition is also defined at the end of this section.

REMix

The Renewable Energy Mix (REMix) [10] model is a deterministic linear optimization model that optimizes capacity development and electricity dispatch, storage, and transmission in an hourly resolution. REMix consists of two components: the Energy Data Analysis Tool (EnDAT) and an optimization model, which is implemented in GAMS. REMix benefits from a high spatial and temporal resolution; nonetheless, the technological dimension of the REMix model can be improved [15].

BENOPTex

BENOPTex [4, 13] is a linear programming model which optimizes the allocation of dispatchable renewable energy carriers (e.g., hydrogen, energy crops, and residues) across energy sectors, including some parts of the wider bioeconomy (e.g., chemical industries). The model consists of two connected sections: a MATLAB module which digests inputs to generate scenarios, and an optimization model that is implemented in GAMS. BENOPTex enjoys a comprehensive set of technologies for transforming raw materials into fuel and power. It also utilizes the database provided in [12] to calculate the maximum possible capacity considering available biomass in Germany. The provided inputs to the model have been verified by experts to ensure high quality data is being used.

Coupling Models

Figure 38.2 illustrates the suggested methodology for soft-coupling the REMix and BENOPTex models. There are shared inputs for both models such as the energy used by the freight transport and the number of road passenger vehicles by fuel type, which are extracted from Vector21 [14]. We also use the forecasted consumed energy by the aviation sector as input from another tool developed by DLR (i.e., 4D-RACE) [17].

First, the REMix model optimizes the capacity expansion considering the electricity and energy demand in each sector. Accordingly, the optimal solution provides electricity spot price projections, the technology mix, and the excess renewable electricity given regional load. These pieces of information are sent to the BENOPTex model. Based on the spot price, we calculate the final electricity price considering announced taxes and levies (e.g., from the German Renewable Energy Act – EEG) in four categories: private households, trade and commerce, industries, and privileged industries. Also, the carbon intensity of electricity is calculated using the technology mix provided in the previous step.

BENOPTex generates the scenario in MATLAB and calls the GAMS solver to solve the underlying model. The model delivers a Pareto optimality curve, by which the modeler can observe the GHG abatement level against the total system cost. The optimal solution indicates the amount of synthetic fuel that should be produced considering the available excess renewable electricity, the amount of produced biofuels (e.g., LNG, bioethanol, and biodiesel) considering regulations (e.g., EEG and RED II[16]) and available capacities, and the price of final products. To adjust inputs for the next round, the optimal values of decision variables (such as the installed capacity of each technology, the production from each technology at different time slices for each technology, and the consumed feed by each technology [13]) that maximize the GHG abatement level are sent to REMix.

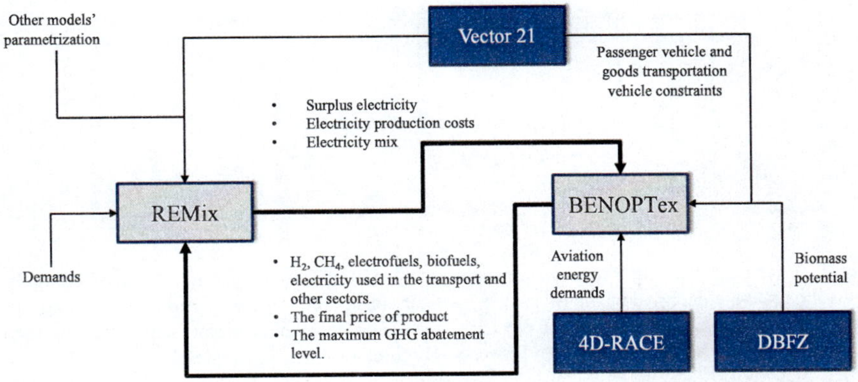

Fig. 38.2 The proposed methodology to couple REMix and BENOPTex in an iterative manner. DBFZ: German biomass research center

This process is repeated iteratively to reach an equilibrium, when the solution of the BENOPTex model does not change the REMix's optimal solution considerably and vice versa. The termination condition is determined to be a maximum 10% change of the sectoral fuel production from 2020 until 2050 between two consecutive iterations. The termination condition is fulfilled in the fourth cycle with a 6.61% difference. Please note that, in general, there is no theoretical proof for the convergence of soft-coupled models, as the involved models optimize different objective functions considering different sets of constraints [6].

Results and Discussions

Figure 38.3 depicts the production of bio- and e-fuels for the transport sector considering RED II constraints. As one can see, the available excess renewable electricity and domestic biomass are not sufficient to de-fossilize the transport sector completely; hence, e-fuels need to be imported (the top most area in the total sub-plot, which is shown in light blue). The domestic production of e-fuels (i.e., PtL) also is not viable since Germany needs to utilize the excess renewable electricity in BEVs in order to satisfy the predetermined GHG quota. The reported results for the reference scenario are achieved after the fourth solving round between REMix and BENOPTex. Passenger road transport can be thoroughly decarbonized thanks to the decarbonization of electricity and the higher efficiency of BEVs. However, decarbonization of other sectors require investments in technologies that do not exist yet. In maritime and freight transport by road vehicles (e.g., trucks), liquefied natural gas plays a vital role, while sustainable aviation fuel and e-fuels are major options in aviation.

While integrating models for the BEniVer project,[1] we noticed that underlying parameters that depend on socio-political factors are changing frequently. We expect

[1] https://www.ufz.de/index.php?en=46330.

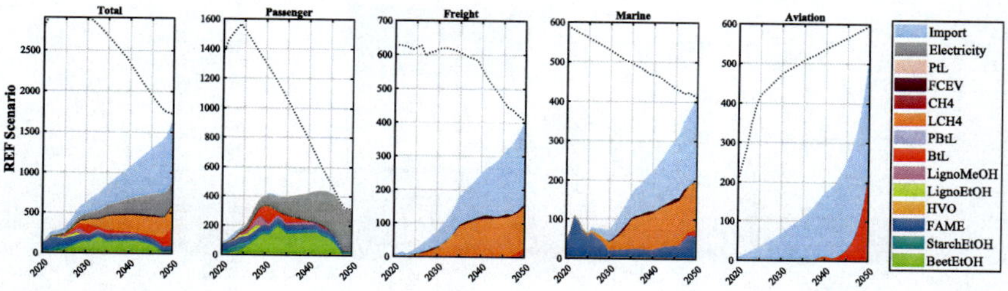

Fig. 38.3 The produced fuels for each transport sector (in PJ) under the reference scenario. The dash lines illustrate the energy demand by each sub-sector considering energy efficiency improvement. PtL: Power-to-Liquid, FCEV: fuel cell electric vehicle, LCH4: liquefied biomethane, BtL: Biomass to liquids via Fischer-Tropsch, PBtL: Power-to-Hydrogen + BtL, LignoMeOH: Lignocellulose-based methanol, LignoEtOH: Lignocellulose-based ethanol, HVO: Hydrotreated vegetable oil, FAME: Fatty-acid methyl ester, StarchEtOH: Starch-based ethanol, and BeetEtOH: Sugar beet-based ethanol. (For interpretation of the references to color in this figure legend, the reader is referred to the web version of this article.)

to witness more rapid and frequent updates in declared regulations and directives to respond to socio-political shocks such as the pandemic and conflicts between countries. Consequently, the available time for scientists is expected to decrease to assist policymakers in designing suitable response strategies. The changing environment was a serious impediment to our harmonization process between the REMix and BENOPTex models; hence, we suggest creating an autonomous super-model in order to minimize the role of humans in interpreting, transferring, and transforming outputs and inputs between various models.

Concluding Remarks

In this paper, we explain the significance and the challenges of linking energy system models to support policymakers in making correct decisions. A new framework has been established by which the outputs of REMix and BENOPTex models converge in a scenario. Although the methodology is fixed, the presented results in this study are subject to change as the underlying models are constantly expanding to embrace higher spatial, technological, temporal, and institutional resolutions. Finally, we suggest developing an autonomous interlinked model that acts as the first responder to the constantly changing environment.

One future research direction can be to link climate and energy system models to incorporate changing meteorological parameters imposed by climate change [7]. Another avenue for future research is to analyze the strategic behavior of generating companies considering their techno-economic characteristics determined by BENOPTex and REMix [3].

Acknowledgements This work has been supported by the German Federal Ministry for Economic Affairs and Energy (BMWi), Grant number 03EIV116F, through the research project Begleitforschung Energiewende im Verkehr - BEniVer.

References

1. Erneuerbare Energien im Verkehr. (2022). https://www.umweltbundesamt.de/daten/energie/erneuerbare-energie-im-verkehr
2. Aliabadi, D. E., Çelebi, E., Elhüseyni, M., & Şahin, G. (2021). Modeling, simulation, and decision support. In *Local electricity markets* (pp. 177-197). Elsevier. https://doi.org/10.1016/B978-0-12-820074-2.00017-4
3. Aliabadi, D. E., & Chan, K. (2022). The emerging threat of artificial intelligence on competition in liberalized electricity markets: A deep Q-network approach. *Applied Energy, 325*, 119813
4. Aliabadi, D. E., Chan, K., Jordan, M., Millinger, M., & Thrän, D. (2022). Abandoning the residual load duration curve and overcoming the computational challenge. In *2022 Open Source Modelling and Simulation of Energy Systems (OSMSES)* (pp. 1–6). IEEE.
5. Bataille, C., Jaccard, M., Nyboer, J., & Rivers, N. (2006). Towards general equilibrium in a technology-rich model with empirically estimated behavioral parameters. *The Energy Journal* (Special Issue# 2).
6. Cao, K. K., Haas, J., Sperber, E., Sasanpour, S., Sarfarazi, S., Pregger, T., Alaya, O., Lens, H., Drauz, S. R., & Kneiske, T. M. (2021). Bridging granularity gaps to decarbonize large-scale energy systems-the case of power system planning. *Energy Science & Engineering, 9*(8), 1052–1060.
7. Craig, M. T., Wohland, J., Stoop, L. P., Kies, A., Pickering, B., Bloomfield, H. C., Browell, J., De Felice, M., Dent, C. J., Deroubaix, A., et al. (2022). Overcoming the disconnect between energy system and climate modeling. *Joule*.
8. Esmaeili Aliabadi, D., Thrän, D., Bezama, A., & Avşar, B. (2022). A systematic analysis of bioenergy potentials for fuels and electricity in Turkey: A bottom-up modeling. In *Transitioning to Affordable and Clean Energy* (pp. 295–314). MDPI. https://doi.org/10.3390/books978-3-03897-777-3-10
9. Eurostat. Energy statistics-an overview https://ec.europa.eu/eurostat/statistics-explained/index.php?title=Energy_statistics_-_an_overview
10. Gils, H. C., Scholz, Y., Pregger, T., de Tena, D. L., & Heide, D. (2017). Integrated modelling of variable renewable energy-based power supply in Europe. *Energy, 123*, 173–188. https://doi.org/10.1016/j.energy.2017.01.115
11. IEA. (2021). World Energy Outlook. OECD Publishing. https://doi.org/10.1787/14fcb638-en. Available at https://www.iea.org/reports/world-energy-outlook-2021
12. Kalcher, J., Naegeli de Torres, F., Gareis, E., Cyffka, K. F., & Brosowski, A. (2021). Dashboard biogene rohstoffe in deutschland. https://doi.org/10.48480/95ct-gn40. https://www.openagrar.de/receive/openagrar_mods_00074316
13. Millinger, M., Tafarte, P., Jordan, M., Musonda, F., Chan, K., Meisel, K., & Aliabadi, D. E. (2022). A model for cost-and greenhouse gas optimal material and energy allocation of biomass and hydrogen. *SoftwareX, 20,* 101264.
14. Mock, P. (2010). Development of a scenario model for the simulation of future market shares and CO2 emissions from vehicles (vector21).
15. Prina, M.G., Manzolini, G., Moser, D., Nastasi, B., & Sparber, W. (2020). Classification and challenges of bottom-up energy system models—A review. *Renewable and Sustainable Energy Reviews, 129*, 109917. https://doi.org/10.1016/j.rser.2020.109917
16. RED II. (2018). Directive (EU) 2018/2001 of the European Parliament and of the Council of 11 December 2018 on the promotion of the use of energy from renewable source. Tech. rep., Publications office of the European Union.

17. Scheelhaase, J. D., Dahlmann, K., Jung, M., Keimel, H., Nieße, H., Sausen, R., Schaefer, M., & Wolters, F. (2016). How to best address aviation's full climate impact from an economic policy point of view? Main results from AviClim research project. *Transportation Research Part D: Transport and Environment, 45*, 112–125.
18. Yue, X., Pye, S., DeCarolis, J., Li, F. G., Rogan, F., & Gallachóir, B. Ó. (2018). A review of approaches to uncertainty assessment in energy system optimization models. *Energy Strategy Reviews, 21*, 204–217. https://doi.org/10.1016/j.esr.2018.06.003

Chapter 39
Towards Decentralized Models for Day-Ahead Scheduling of Energy Resources in Renewable Energy Communities

Louise Sadoine, Martin Hupez, Zacharie De Grève, and Thomas Brihaye

Abstract We address electricity consumption scheduling on a day-ahead basis within a community of prosumers that own renewable generation. We establish two market designs that enable coordination between members and where excess production can be valued outside or inside the community. For each, we propose two formulations: centralized schemes where the common objective is optimized, while in decentralized schemes, each member optimizes its own objective. The natural interdependence between members sharing a common network leads to the formulation of non-cooperative games. We solve some proposed models on a use-case by using distributed algorithms that ensure confidentiality.

Keywords Energy communities · Optimization · Game theory

Introduction

Context. Introduced by the EU Commission in its Directive 2018/2001 [1], Renewable Energy Communities (RECs) are gaining an increasing momentum in the energy sector, as an important factor for accelerating the energy transition [2]. These consist in organised entities gathering electricity consumers and prosumers (i.e., consumers and producers of renewable electricity) who can exchange energy locally, without resorting to the classical market structure. They thus aim at fostering local investment in renewable generation and local flexibility.

Related work. The energy exchanges rule inside the community may, however, vary depending on the authors: peer-to-peer exchanges communities [11], local markets communities [6] or cooperative communities [3, 4, 9, 10]. A review of models and solution concepts, together with an extensive bibliography can be found in [15].

L. Sadoine (✉) · T. Brihaye
Effective Mathematics, UMONS, Mons, Belgium
e-mail: louise.sadoine@umons.ac.be

M. Hupez · Z. De Grève
Electrical Power Engineering Unit, UMONS, Mons, Belgium
e-mail: zacharie.degreve@umons.ac.be

O. Grothe et al. (eds.), *Operations Research Proceedings 2022*, Lecture Notes in Operations Research, https://doi.org/10.1007/978-3-031-24907-5_39

Contributions. This article focuses on the day-ahead scheduling of the demand in RECs based on a cooperative framework. We propose two market designs, inspired by [9, 10], which dictate the exchanges inside the RECs. The first design implements a cooperative demand management scheme inside a community, whereas the second model allows virtual mutualization of excess resources among community members. The contributions of this paper are threefold:

1. we extend the formalism of [9] and [10] by modeling the selling of local excess renewable generation to the classical retail market, and assigning a non zero value to the energy exchanged locally,
2. we develop a centralized and decentralized version of the proposed models, relying respectively on convex optimization and game theory,
3. we solve the proposed models on a use-case, and highlight limitations that require further research effort.

Community Framework

We assume cooperative communities of prosumers connected to the same LV distribution feeder. Each prosumer is equipped with a bi-directional metering device, or smart meter.

Prosumer Load Profile

Model 1. Let $\mathcal{N} = \{1, \ldots, N\}$ be the set of community prosumers, and $\mathcal{T} = \{1, \ldots, T\}$ the set of time steps of duration Δt for a given day. The consumption profile of member $i \in \mathcal{N}$ divides into different load components.

Let A_i be the set of flexible appliances (i.e. for which electricity consumption can be shifted in time, such as a washing-machine) of member i. For each device $a \in A_i$, we define the flexible load scheduling vector $x_{i,a} = (x_{i,a}^1, \ldots, x_{i,a}^T)$.

The non-flexible load (e.g. fridge) of user i is modeled by $d_i = (d_i^1, \ldots, d_i^T)$. A user i might also be equipped with non-dispatchable energy generation (e.g. photovoltaic panels), represented by $g_i = (g_i^1, \ldots, g_i^T)$. Note that, in the context of a deterministic approach, the non-flexible load and local generation can be seen as state variables.

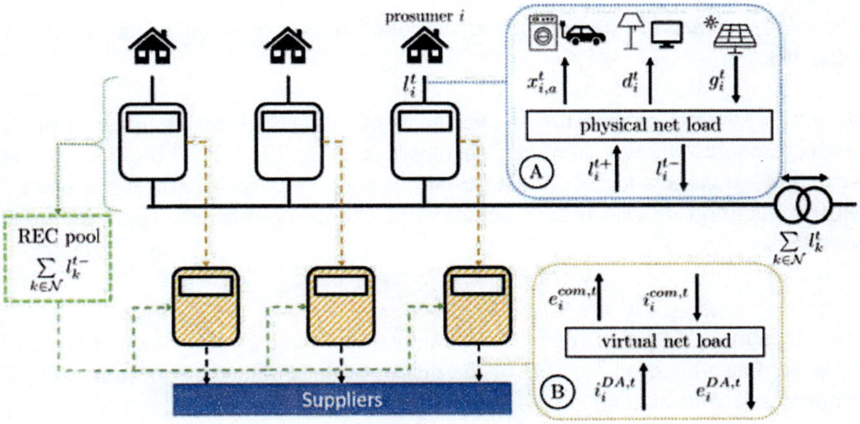

Fig. 39.1 Cooperative demand-side models

Then, the net load of prosumer i at time $t \in \mathcal{T}$ can be modeled:

$$l_i^t = \sum_{a \in A_i} x_{i,a}^t + d_i^t - g_i^t. \tag{39.1}$$

Net load is negative if the prosumer's production exceeds his electrical needs, so he must inject this surplus into the network, and positive if the local generation doesn't cover all the user's consumption, so he will import energy to fill the gap. We define two distinct variables $l_i^{t+} = \max(0, l_i^t)$ and $l_i^{t-} = \max(0, -l_i^t)$ respectively the positive and negative net load, such as $l_i^t = l_i^{t+} - l_i^{t-}$.

Furthermore, flexible appliances are subject to different individual needs and physical constraints. The temporal flexibility consented to device a by individual i, is defined by a daily binary vector $\delta_{i,a} = (\delta_{i,a}^1, \ldots, \delta_{i,a}^T)$. A value of 1 indicates that member i agrees to schedule a over time slot $t \in \mathcal{T}$, otherwise it will be set to 0. Beside, the predetermined total amount of energy that application a must consume for the day is denoted $E_{i,a}$. Finally, for the sake of simplicity, we consider flexible devices with fully modular consumption cycles, i.e., each of them is limited only by maximum power $M_{i,a}$. These constraints are formalized as follows:

$$\delta_{i,a}.x_{i,a}^\top = E_{i,a} \tag{39.2}$$

$$0 \leqslant x_{i,a}^t \leqslant M_{i,a} \,.\, \delta_{i,a}^t \,.\, \Delta t, \quad \forall t \in \mathcal{T}. \tag{39.3}$$

This article adapts a simplified grid model considering maximum injection and withdrawal connection power for each house:

$$l_i^{t+} \leqslant l_i^{\max}, \quad \forall t \in \mathcal{T} \tag{39.4}$$

$$l_i^{t-} \leqslant l_i^{\min}, \quad \forall t \in \mathcal{T} \tag{39.5}$$

where $l_i^{\min} \geqslant 0$ and $l_i^{\max} > 0$ are the lower bound and upper bound of the member's capacities.

Model 2. This setup extends the first design by allowing the virtual mutualization of excess resources among community members (see Fig. 39.1). For billing purposes, we define virtual flow variables that deviate from grid physical flows. A prosumer i with production surplus can sell a quantity $e_i^{com,t}$ to the community. This variable is bounded as:

$$e_i^{com,t} \leqslant l_i^{t-}, \quad \forall t \in \mathcal{T}. \tag{39.6}$$

The allocation $i_i^{com,t}$ of surplus locally produced renewable energy to member i is considered as a decision variable of the optimization problem. In addition, excess production is allocated only to members with an energy deficiency:

$$i_i^{com,t} \leqslant l_i^{t+}, \quad \forall t \in \mathcal{T}. \tag{39.7}$$

Moreover, the total excess production allocated to the community must equal the total quantity imported by members:

$$\sum_{i \in \mathcal{N}} i_i^{com,t} = \sum_{i \in \mathcal{N}} e_i^{com,t}, \quad \forall t \in \mathcal{T}. \tag{39.8}$$

Finally, the energy imported $i_i^{DA,t}$ and exported $e_i^{DA,t}$ on the retail market by individual i are obtained by:

$$i_i^{DA,t} = l_i^{t+} - i_i^{com,t}, \quad \forall t \in \mathcal{T} \tag{39.9}$$

$$e_i^{DA,t} = l_i^{t-} - e_i^{com,t}, \quad \forall t \in \mathcal{T}. \tag{39.10}$$

Cost Structure

We suppose that members aim to minimize their electricity bill, which comprises different components:

Gray energy costs. These are the costs charged by the electricity supplier. Each community member is free to choose an external supplier from the set $\mathcal{M} = \{1, \ldots, M\}$ for the portion of consumption not covered by local energy. Every supplier $j \in \mathcal{M}$ has a set of customer $\mathcal{N}_j \subseteq \mathcal{N}$. Recall that only the positive net load l_i^{t+} of an individual i is charged. Thus, the supplier j applies for each customer $i \in \mathcal{N}_j$: $C_j(l_i^{t+}) = \lambda_{imp,j}^t . l_i^{t+}$, with $\lambda_{imp,j}^t$ in €/kWh.

Local energy costs (second model only). The electricity withdrawn from the REC pool is subject to a specific tariff λ_{iloc}^t €/kWh. For each user $i \in \mathcal{N}$, we have $C_l(i_i^{com,t}) = \lambda_{iloc}^t . i_i^{com,t}$.

Upstream grid costs. These are the costs of using the transmission grid and the distribution network. Accounting for an accurate image of incurred costs implies to aggregated net load of all members transiting the MV/LV transformer denoted as $L^t = \sum_{i \in \mathcal{N}} l_i^t$. The network costs therefore amount to $C_{gr}(L^t) = \alpha.(L^t)^2 = \alpha.(\sum_{i \in \mathcal{N}} l_i^t)^2$ with α €/kWh^2 the price for the upstream network use.

Revenue from exported energy. Community members may receive income through the sale of excess local production directly on the day-ahead retail market at a fixed price $\lambda_{exp,j}^t$ €/kWh such that the electricity export is a benefit for the user i: $R_j(l_i^{t-}) = \lambda_{exp,j}^t.l_i^{t-}$. In the second model, a prosumer could export his production surplus in the REC pool at a tariff λ_{eloc}^t €/kWh. For each user i, we have $R_l(e_i^{com,t}) = \lambda_{eloc}^t.e_i^{com,t}$.

Day-Ahead Power Exchange Scheduling Problems

The day-ahead power exchange scheduling problem minimizes the total cost of electricity consumption by a REC. In Section "Optiproblem", we present an optimal but idealistic resolution while Section "GT" formulates a more realistic method.

Centralized Optimization Formulations

Assuming a central operator with some control over the REC's flexible appliances, we formulate the centralized optimization problem as:

$$F_1 := \begin{cases} \min\limits_{\Theta} & \sum\limits_{t \in \mathcal{T}} \Big[\sum\limits_{j \in \mathcal{M}} \sum\limits_{i \in \mathcal{N}_j} (C_j(l_i^{t+}) - R_j(l_i^{t-})) + C_{gr}(L^t) \Big] \\ \text{s.t.} & \Theta \in \Omega^1 := \{(x_i, l_i^+, l_i^-)_{i=1}^N \in \mathbb{R}^n : (39.1) - (39.5)\}. \end{cases} \tag{39.11}$$

We also adopt a second design, in which a prosumer can share his excess energy production with other end-users. We have in that case:

$$F_2 := \begin{cases} \min\limits_{\Theta} \sum\limits_{t \in \mathcal{T}} \Big[\sum\limits_{j \in \mathcal{M}} \sum\limits_{i \in \mathcal{N}_j} (C_j(i_i^{DA,t}) - R_j(e_i^{DA,t})) + C_l(i_i^{com,t}) \\ \qquad -R_l(e_i^{com,t}) + C_{gr}(L^t) \Big] \\ \text{s.t. } \Theta \in \Omega^2 := \{(x_i, i_i^{com}, e_i^{com}, i_i^{DA}, e_i^{DA}, l_i^+, l_i^-)_{i=1}^N \in \mathbb{R}^n : (39.1) - (39.10)\}. \end{cases} \tag{39.12}$$

These models are convex optimization problems with quadratic objective functions, convex inequality constraints and affine equality constraints. They can be solved in a centralized way by standard algorithms such as interior-point [5].

Noncooperative Game Formulations

Centralized models (39.11)–(39.12) do not take into account the fact that each member has control over his load profile that impact his own aims, which may conflict with those of the other users. To model interactions between strategic prosumers competing for a common resource, we resort to non-cooperative game theory.

Costs distributions. In this paper, the total cost is distributed among community members at each time slot $t \in \mathcal{T}$ [9]. The daily bill of user i for F_1 and F_2 is respectively:

$$b_i^1(\Theta) = \sum_{t\in\mathcal{T}} C_j(l_i^{t+}) - R_j(l_i^{t-}) + l_i^t.\alpha L^t; \tag{39.13}$$

$$b_i^2(\Theta) = \sum_{t\in\mathcal{T}} C_j(i_i^{DA,t}) - R_j(e_i^{DA,t}) + C_l(i_i^{com,t}) - R_l(e_i^{com,t}) + l_i^t.\alpha L^t \tag{39.14}$$

There are many billing methods, each impacting the problem differently [9, 10].

Model 1. Each prosumer $i \in \mathcal{N}$ is a selfish player who chooses his individual strategy Θ_i to minimize his own daily bill $b_i(\Theta_i, \Theta_{-i})$ defined as in (39.13). We model the Problem (39.11) as a Nash equilibrium problem (NEP) where $\Omega := \prod_{i=1}^N \Omega_i$ and $b := (b_i^1)_{i=1}^N$ is the billing vector. Each user i aims at solving the following optimization problem, given other players strategies Θ_{-i}

$$G := \begin{cases} \min\limits_{\Theta_i} \; b_i^1(\Theta_i, \Theta_{-i}) & \forall i \in \mathcal{N} \\ \text{s.t.} \quad \Theta_i \in \Omega_i. \end{cases} \tag{39.15}$$

A Nash equilibrium (NE) of the game G is a strategy profile Θ^* such that for all $i \in \mathcal{N}$ and $\Theta_i \in \Omega_i$, $b_i^1(\Theta_i^*, \Theta_{-i}^*) \leq b_i^1(\Theta_i, \Theta_{-i}^*)$.

Model 2. We add the sharing of excess local energy including global constraint (39.8). Hence, each player's strategy set can depend on the rival player's strategies Θ_{-i}: $\Omega_i(\Theta_{-i}) \subseteq \mathbb{R}^{n_i}$, unlike a NEP. This is a generalized Nash equilibrium problem (GNEP) [7]. Each prosumer i minimize his individual function b_i as in (39.14), depending of Θ_{-i}:

$$\mathcal{G} := \begin{cases} \min\limits_{\Theta_i} \; b_i^2(\Theta_i, \Theta_{-i}) & \forall i \in \mathcal{N} \\ \text{s.t.} \quad \Theta_i \in \Omega_i(\Theta_{-i}). \end{cases} \tag{39.16}$$

A generalized Nash equilibrium (GNE) of the game $\mathcal{G}$ is a strategy profile Θ^* such that for all $i \in \mathcal{N}$ and $\Theta_i \in \Omega_i(\Theta_{-i}^*)$, $b_i^2(\Theta_i^*, \Theta_{-i}^*) \leq b_i^2(\Theta_i, \Theta_{-i}^*)$. The variability of the strategy sets makes GNEPs more complicated to solve than NEPs. Solution analysis is built on the variational inequalities theory [7, 8].

Nash Equilibrium Computation. We attest the existence of a NE for game (39.15) and GNE for (39.16), where each user updates his strategy by minimizing bill function (39.13) or (39.14). For both games, we rely upon distributed algorithms such as proximal decomposition (PDA) [8, 14] to efficiently compute the equilibrium, while considering prosumer privacy.

Case Study

We illustrate the performance of proposed REC mechanisms, on a use-case made by $N = 3$ houses from New-York, US. We use hourly electricity consumption and production patterns extracted from the Pecan Street Project [12], and simulate for a whole day $T = 24$. The parameters considered can be found in [13].

The left graph of Fig. 39.2 compares the total costs without mutualization (39.11) and with mutualization (39.12), and depicts the total costs achieved by individual optimizations on commodity costs only [10]. The right side of Fig. 39.2 shows the split of the total costs among the REC members of model 1 (39.15). By jointly planning their flexible load, members can find the cost-optimal trade-off between arbitrage in the commodity (dynamic pricing) and the upstream grid costs, which are minimized by smoothing the total load over time. In comparison, users who minimize commodity costs individually concentrate their consumption at time slots with the lowest prices, leading to higher total costs. Model 2 allows members to share excess energy from non-dispatchable sources, hence decreasing the total costs. We see that total costs of decentralized model 1 are similar to the centralized case. We note a slight inefficiency for continuous billing, compensated by an allocation more representative of the real actions of individuals.

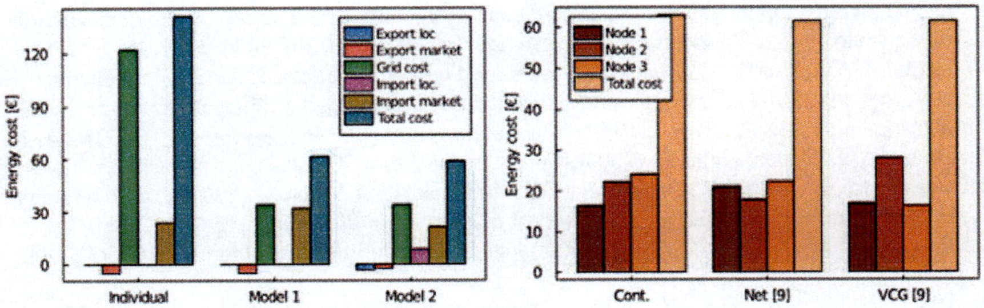

Fig. 39.2 Comparison of the total costs division for different models (left) and individual costs for three considered billings of model 1 (right)

Conclusion and Outlook

We proposed two cooperating models for optimally scheduling day-ahead energy exchanges within communities of prosumers. Model 1 implements a coordinated demand-side management scheme, whereas model 2 extends model 1 by considering mutualization of local generation excess. Compared to the case where everyone acts individually, models 1 and 2 save 56.12% and 58.3% resp. on the whole REC bill, through an increased resort to local available flexibility for the same amount of mobilized resources. We showed that community mechanisms appear also as promising solutions to better manage the local grid via significantly lowered grid costs. We illustrated that the choice of the billing scheme can incentivize different prosumers' behaviors, although the ideal scheme depends on political choices.

Adequate parameters and a suitable termination criterion for the PDA algorithm of Problem (39.16) still need to be determined. The theoretical characterization of generalized Nash equilibria shall be addressed. Including storage systems may help to further reduce import and upstream grid costs.

Acknowledgements This work was partly supported by the Fonds de la Recherche Scientifique - FNRS under grant n°T.0027.21.

References

1. Directive (EU) 2018/2001 of the European Parliament and of the Council of 11 December 2018 on the promotion of the use of energy from renewable sources. https://eur-lex.europa.eu/legal-content/EN/TXT/?uri=uriserv:OJ.L_.2018.328.01.0082.01.ENG&toc=OJ:L:2018:328:TOC
2. Clean energy package for all Europeans. European Commission (2019).
3. Atzeni, I., Ordóñez, L. G., Scutari, G., Palomar, D. P., & Fonollosa, J. R. (2013). Noncooperative and cooperative optimization of distributed energy generation and storage in the demand-side of the smart grid. *IEEE Transactions on Signal Processing, 61*(10), 2454–2472.
4. Atzeni, I., Ordóñez, L. G., Scutari, G., Palomar, D. P., & Fonollosa, J. R. (2014). Noncooperative day-ahead bidding strategies for demand-side expected cost minimization with real-time adjustments: A GNEP approach. *IEEE Transactions on Signal Processing, 62*(9), 2397–2412.
5. Boyd, S., & Vandenberghe, L. (2004). *Convex optimization*. Cambridge University Press.
6. Cornélusse, B., Savelli, I., Paoletti, S., Giannitrapani, A., & Vicino, A. (2019). A community microgrid architecture with an internal local market. *In Applied Energy, 242*, 547–560.
7. Facchinei, F., & Kanzow, C. (2007). Generalized Nash equilibrium problems. 4OR, *5*(3), 173–210.
8. Facchinei, F. & Pang, J.-S. (2007). Finite-dimensional variational inequalities and complementarity problems. Springer.
9. Hupez, M., Toubeau, J.-F., Atzeni, I., De Grève, Z., & Vallée, F. (2022). Pricing electricity in residential communities using game-theoretical billings. *IEEE Transactions on Smart Grid.*
10. Hupez, M., Toubeau, J.-F., De Grève, Z, & Vallée, F. (2021). A new cooperative framework for a fair and cost-optimal allocation of resources within a low voltage electricity community. *IEEE Transactions on Smart Grid, 12*(3), 2201–2211.
11. Le Cadre, H., Jacquot, P., Wan, C., & Alasseur, C. (2019). Peer-to-peer electricity market analysis: From variational to generalized Nash equilibrium. *European Journal of Operational Research.*

12. Pecan Street Inc. (2020). Residential data New York 15 min. [Online]. https://dataport.pecanstreet.org/
13. Sadoine, L., Hupez, M., De Grève, Z., & Brihaye, T. (2022). Online companion—Towards decentralized models for day-ahead scheduling of energy resources in renewable energy communities. [Online]. https://doi.org/10.5281/zenodo.6847243
14. Scutari, G., Facchinei, F., Pang, J.-S., & Palomar, D. P. (2014). Real and complex monotone communication games. *IEEE Transactions on Information Theory, 60*(7), 4197–4231.
15. Tsaousoglou, G., Giraldo, J. S., & Paterakis, N. G. (2022). Market mechanisms for local electricity markets: A review of models, solution concepts and algorithmic techniques. *Renewable and Sustainable Energy Reviews, 156*, 111890.

Part VII
Finance

Chapter 40
Alternative Prize Money Distributions for Higher Gender Equity in Sports

Maren Martens and Verena Starflinger

Abstract In many sports, women receive less prize money than men. This issue has been discussed extensively over decades. While many people consider different prize structures for men and women as unfair, others argue, e.g., that men attract greater public interest or that the competition is harder among men than among women (in particular, when there participate more men than women). In this paper, we focus on the discussion of fairness in the distribution of prize money in endurance sports concerning the severity of the competition. We present two methods to distribute prize money across gender based on the individual performances w.r.t. gender specific records. We suppose these "across gender distributions" to be fair, as they suitably respect that women generally are slower than men. Furthermore, we compare commonly used prize distributions to our across gender distributions, introducing a statistical fairness measure. For our investigations, we focus on triathlon, but the results can easily be adapted to any other endurance sports.

Keywords OR in sports · Statistics · Gender equity · Ethics

Introduction

Problem Definition. In sports it often happens that women receive less prize money than men. More and more disciplines are deciding to join the gender-equal prize distribution, but differences in the prize structure are still common, e.g., in soccer or golf. On first sight, it seems unfair to reward women less than men. Nevertheless, economic motives due to the number of spectators or physical differences between men and women in terms of absolute performance are often decisive for the amount of prize money. In this paper, we concentrate on triathlon, which by the majority of people is considered to be generally fair w.r.t. the prize structure, such as other endurance sports are as well. Triathlon is often selected as a positive example for not disadvantaging women, see, e.g., [11]. Though, it needs to be examined carefully

M. Martens (✉) · V. Starflinger
Landshut University, Landshut 84036, Germany
e-mail: martens@haw-landshut.de

O. Grothe et al. (eds.), *Operations Research Proceedings 2022*, Lecture Notes in Operations Research, https://doi.org/10.1007/978-3-031-24907-5_40

what can be called “fair”. For the reason that often there are competing way more male than female athletes, it still happens—even in triathlon—that there are more prize ranks for men than for women. So our core questions will be: What would be fair distributions of prize ranks across men and women? How fair are existing distributions of prize ranks? We will not discuss, however, what shall be fair amounts of prize money among the prize ranks. This leaves room for further research. Though, we assume that the prize money is decreasing for increasing ranks.

Related Results from the Literature. To the best of our knowledge, there exist no scientific investigations on fair distributions of prize ranks across men and women. However, there exists literature presenting results on how athletes’ performances are related to the total amount of prize money and its spread over the prize money ranks. O’Toole [8] gives a good overview on those results and additionally investigates gender disparities in the incentive effects of different total amounts and structures of prizes in marathon races, but his samples show only insignificant differences between genders. However, O’Toole does not compare the spread of prize ranks between men and women w.r.t. fairness, which we will do in this paper. Brown [2] suggests ten ways to best distribute prize money w.r.t. the principles of stakes fairness in different situations and mentions performance-based proportionality, but also he does not elaborate the gender criterion.

Contribution of this Paper. In this paper, we will first give an overview on existing methods of prize rank distributions and introduce prize rank distributions that are supposed to be fair w.r.t. gender equity (Section “Existing Versus Fair Prize Rank Distributions”). We further introduce a fairness measure for prize rank distributions using mathematical-statistical methods (Section “A Fairness Measure”) and finally use that function to assess which existing methods of prize rank distributions can be considered as fair (Section “Evaluation of Existing Prize Rank Distributions”). For the assessment we evaluate different competitions in triathlon, differentiating between amateurs (in triathlon usually called *age groupers*) and professionals. We restrict our approaches to the distribution of prize money ranks in triathlon, but the results can easily be adapted to other endurance sports. We are quite aware of the fact that in many countries there exists a third agency recognized gender, the diverse gender. However, since this gender is not considered in a separate category in sports yet, we omit it in our investigations.

Existing Versus Fair Prize Rank Distributions

Method 1. A method frequently used for prize rank distributions is to award the first w places in the men’s ranking as well as in the women’s ranking. E.g., the international organization “World Triathlon” and the “World Triathlon Corporation” (organizer of the “Ironman” event series) set up their competitions according to this guideline [5, 11].

Method 2. Another method of rewarding athletes' performances is to distinguish between the number of prizes for men and women, thus allocating prize money for the first m men and the first w women, with $m \neq w$. Generally this is done with $m > w$. Examples for this method are the Müritz-Triathlon [7] with $m = 6, w = 3$ or the Austria-Triathlon [1], which had $m = 10, w = 5$ for many years (e.g., in 2019).

In addition to Method 1, some organizers apply time limits for athletes to receive prize money (or bonuses). Per default none of these methods, in particular the most commonly applied Methods 1 and 2, can be considered as being fair. In most competitions the number of male athletes is significantly higher than that of female athletes, e.g., in the "Ironman" series the average female participation only adds up to 21% of all entrants [6]. Therefore, in Method 1, men have a much harder time getting hold of the prize money, whereas in Method 2 high performing women seem to have a disadvantage from having fewer prize money ranks.

To evolve alternative, supposedly fair models for gender equitable distributions of prize money, the idea is to take the world records of the considered discipline as an orientation to generate a gender-neutral ranking across men and women, in which the sequence of places is determined in ascending order w.r.t. the male world record for men or, respectively, the female world record for women. Here, we consider two options:

Quotient Model (QM). The sequence of places is determined in ascending order according to the quotient of the finish time and the respective world record (male/female). Starting with place number one with the smallest quotient, everyone until a predefined number gets prize money. The proportion of women and men who receive prizes is not predetermined.

Difference Model (DM). The approach is the same as in the QM, but instead of the quotient of the finish time and the world record, the difference is taken.

While the DM is easier to understand and to mentally calculate, the QM seems to be fairer w.r.t. different finish times for men and women. It is to be expected that higher finish times directly lead to higher differences to the world record and therefore disadvantage slower finishers (generally women). Variants of these models, e.g., with time limits for being rewarded or variable prize money w.r.t. finish time, are imaginable, but will not be discussed in this paper.

A Fairness Measure

In order to compare typical methods for prize rank distribution like the ones in Section "Existing Versus Fair Prize Rank Distributions" to the fair QM and DM, we introduce an *unfairness factor*. For all prize ranks (across gender) given in the method that is to be evaluated, we compute the "error" in ranking w.r.t. the fair ranking taken from the QM or DM. I.e., in a competition with m prize money ranks for men and w for women, we compare the ranks of the first m men and the first w women to the ranks that they would reach in the across gender ranking (AGR). Adapting the standard error from statistics, we define, e.g., for a woman placed on rank $i \leq w$ in the split ranking (SR) and on rank $W_i > i$ in the AGR, the error value as $(i - W_i)^2$. Though we do not consider the actual spread of prize money, we surely assume that the prize money is strictly decreasing for increasing ranks (up to w and m, resp.) and therefore define the ranking error to be higher for higher differences between i and W_i. For the unfairness factor, this "standard error" is scaled such that its minimum is 1. In the following we introduce the unfairness factor w.r.t. the QM, but all definitions and proofs apply the same way w.r.t. other models, in particular the DM.

Throughout the rest of this paper, we assume, w.l.o.g., that $w \leq m$. For a given competition with given finish times, let $1, \ldots, m$ be the prize money ranks for men and $1, \ldots, w$ those for women (in their gender specific rankings). We define the AGR (M, W) as the pair of the two functions $M : \{1, \ldots, m\} \to \mathbb{N}, i \mapsto M_i$, and $W : \{1, \ldots, w\} \to \mathbb{N}, i \mapsto W_i$, such that $M_1 < \ldots < M_m$ and $W_1 < \ldots < W_w$ are the ranks of the first m men and the first w women w.r.t. the QM. We assume, also w.l.o.g., that there participate at least m men and at least w women in the competitions under consideration. (Otherwise m (or w) can be reduced to the number of participating men (or women).)

Definition 1 *For $w > 0$, the* unfairness factor $F(M, W)$ *of an across gender ranking (M, W) is defined as* $F(M, W) := \sqrt{\frac{\sum_{i=1}^{m}(i-M_i)^2+\sum_{i=1}^{w}(i-W_i)^2}{w^2m+\frac{1}{3}w-\frac{1}{3}w^3}}$.

One issue might be irritating at first sight: Assuming we have $m + w$ prize ranks in the AGR, it seems like $F(M, W)$ accounts for those athletes $i \leq m$ (or w, resp.) who receive prize money in the SR, but not in the AGR, whereas it does not account for those who receive prize money in the AGR, but not in the SR (they are not considered in $F(M, W)$). However, this is not true as the ones mentioned secondly push back others (from the other gender), thus causing an increment in their ranking errors. Note that $F(M, W)$ is not defined for $w = 0$. For now we assume $w \geq 1$, but we will discuss this again in Section "Evaluation of Existing Prize Rank Distributions".

Theorem 1 *The minimum of $F(M, W)$ is 1. This is attained exactly for those across gender rankings (M, W) with $M_i = 2i$ and $W_i = 2i - 1$ or vice versa, for all $i = 1, \ldots, w$, and $M_i = w + i$, for all $i = w + 1, \ldots, m$.*

The mentioned ranking property of minimum AGRs can be proven switching ranks for any (M, W) that does not meet the property, thus constructing a ranking of smaller $F(M, W)$. It is then easy to compute $F(M, W)$ for a minimum (M, W) to show this to be 1. We omit the detailed proof for brevity.

Evaluation of Existing Prize Rank Distributions

We evaluate Methods 1 and 2 from Section "Existing Versus Fair Prize Rank Distributions" w.r.t. fairness compared to the QM and the DM, using $F(M, W)$. For this purpose, we consider the rankings from 26 long-distance triathlon competitions between 2011 and 2020. We distinguish between rankings of professional athletes (16 races) and amateurs (26 races), as it is also done in most major competitions. After an extended search for races, which provide both good access to complete ranking lists and a good mixture of professionalism and amateurism, we decided for: Ironman World Championship Professionals 2011–2019 [4], Ironman World Championship Amateurs 2011–2019 [4], Ironman Texas Professionals 2011–2015, 2017, 2019 [3], Ironman Texas Amateurs 2011–2015, 2017, 2019 [3], Austria-Triathlon Ironman Distance 2011–2020 [10]. (For Ironman Texas, we cleaned the data from the competitions in 2016 and 2018 to account for their shortened bike tracks.) In 2020 and 2021 competitions were very limited due to the corona pandemic.

To have a standard evaluation method, all comparisons were done with $m = 10$. Thus, for Method 1, also $w = 10$. For Method 2, we ran our analyses for $w = 1, \ldots, 9$. We chose 10 as standard m, as on the one hand 10 seems to be widely-used for the number of prize ranks in triathlon and on the other hand there participated at least 10 women in all competitions we considered (which is not the case for numbers bigger than 10). Actual proportions between prize ranks for men and women have already been discussed in Section "Existing Versus Fair Prize Rank Distributions", but note that for us primarily the finish times are relevant rather than actual proportions, as we aim for a general consideration of what distribution would be fairest.

Figure 40.1 shows the average unfairness factors over all considered competitions. It can be seen that, for amateurs, Method 1 on average provides the highest unfairness. For the QM the unfairness is minimal at $w = 2$, while for the DM the unfairness is minimal at $w = 1$. Having a smaller w at minimum and generally higher unfairness factors for the DM is not surprising. We already discussed in Section "Existing Versus Fair Prize Rank Distributions" that the higher value of the women's world record—compared to men—would let you expect to generally have higher finish times with a bigger bandwidth and therefore also have higher differences to the world record. Therefore, the differences between the women's ranks in the SRs and in the AGRs are bigger.

In Section "A Fairness Measure" we mentioned the difficulty to define a standardized unfairness factor for $w = 0$. Nevertheless, the question arises whether it would be even fairer to have no prize ranks for women at all. One could say that if and only if no woman was among the first 11 across gender ranks, it would be fairer to

Amateur competitions

	w	1	2	3	4	5	6	7	8	9	10
QM	#	20	24	25	25	25	26	26	26	26	–
	F	2.30	2.06	2.09	2.22	2.40	2.62	2.86	3.12	3.41	3.71
DM	#	22	26	26	26	26	26	26	26	26	–
	F	2.62	2.90	3.12	3.45	3.72	4.04	4.40	4.74	5.11	5.47

Professional competitions

	w	1	2	3	4	5	6	7	8	9	10
QM	#	2	4	9	8	10	12	13	14	14	–
	F	2.85	1.84	1.52	1.39	1.33	1.33	1.34	1.35	1.39	1.44
DM	#	2	7	12	13	13	14	15	15	15	–
	F	2.52	1.80	1.60	1.51	1.47	1.51	1.54	1.58	1.63	1.68

Fig. 40.1 Number of competitions where Method 2 is fairer (#) and approximate average unfairness factors (F), for $w = 1, \ldots, 9$. The unfairness factors for $w = 10$ are those of Method 1

have $w = 0$. However, in the 26 competitions, it happens only in 1 for the QM and 4 for the DM that there is no woman among the first 11 across gender ranks. So we conclude that on average it is not fairer to have $w = 0$.

For professionals, the highest unfairness factors are reached at $w = 1$, while the minimum is reached at $w = 5$. Generally, the unfairness factors are lower than those of amateur races. This and the higher value of w at minimum indicate that in professional competitions the difference between highly performing men and highly performing women is not as big as in amateur races.

Figure 40.1 also shows the numbers of competitions where Method 2 is fairer than Method 1. Those prove that the average values of F are not a product of single peaks, but that for the big majority of the competitions it is fairer to have $w < m$. Interestingly, for professionals, the number of competitions, in which Method 2 is fairer, is not constantly growing for the QM. The troublemaker is Ironman World Championship 2015, where Method 2 is fairer for $w = 1, 2, 3$, but not for $w = 4, \ldots, 7$. This can be explained by its specific AGR, but we omit a detailed analysis for brevity. However, note this setting to be unique.

Conclusion and Future Work

We introduced a fair measure evaluating athletes' performances across gender. The analyses of numerous triathlon competitions revealed that, w.r.t. this measure, it would be fairer to have less prize ranks for women than for men. As we analyzed quite a high number of competitions, we claim that our results can be extended to generality, with few exceptional competitions. However, the situation differs a bit between amateurs and professionals. Whereas for amateurs it would be fairest to award only the best one or two women, while ten men are awarded, for professionals, the fairest would be to award the best five women. As already discussed above, we consider the QM to be the fairer choice for AGRs, but the DM has the advantage that it is easier to compute the ranking relevant times. Therefore, we would see the DM as a considerable option, if AGRs were introduced in reality. However, we want to mention that there are still good reasons not to use AGRs, e.g., athletes might feel being robbed of medals (or trophies in general), as the number of awarded medals would drop from six to three. (Check [9] for discussions on this issue w.r.t. horse-riding competitions.) Also there are strong arguments to stick to Method 1 (instead of applying Method 2). E.g., it might discourage (high potential) women from pushing their limits or, even worse, from participating at all. In summary, it stays open what shall be considered best. An idea to combine the advantages of the different schemes could be to choose the classical way with the same number of prize ranks for men and women, but then add rewards based on the fictitious AGR.

References

1. Austrian Triathlon Verein. (2020). Austria Triathlon Podersdorf. https://www.austria-triathlon.at. Accessed: 2021-06-10
2. Brown, A. (2015). Principles of stakes fairness in sport. *Politics, Philosophy & Economics, 14*(2), 152–186.
3. Cox, R. (2022). Ironman Texas: Summary. https://www.coachcox.co.uk/imstats/series/25. Accessed: 2022-01-17
4. Cox, R. (2022). Ironman World Championship: Summary. https://www.coachcox.co.uk/imstats/series/12. Accessed: 2022-01-17
5. SportsEngine Inc. (2022). Competition rules. https://www.ironman.com/competition-rules. Accessed: 2022-04-19
6. Lacke, S. (2020). Women in Triathlon: Where are we now? https://www.triathlete.com/culture/news/women-in-triathlon-where-are-we-now. Accessed: 2022-02-04
7. MSC Müritz-Sportclub Waren e.V. Ausschreibung. (2012). 27. Müritz-Triathlon. Preisgelder. https://docplayer.org/81991559-22-duathlon-27-triathlon.html. Accessed: 2022-02-04
8. O'Toole, D. (2009). Prize structures, gender, and performance in competitive marathon running. https://scholarship.tricolib.brynmawr.edu/bitstream/handle/10066/3639/2009O'TooleD.pdf. Accessed: 2022-02-04
9. Peiffer, A., & Lesser, E. (2021). Das Biathlon Doppelzimmer. Folge 36. *Podcast*.
10. Pentek-timing GmbH. (2022). Ergebnisübersicht. Erfolge auf einen Blick. https://balancer.pentek-timing.at/resultsoverview.html. Accessed: 2022-01-17
11. World Triathlon. (2018). Prize money breakdown. https://www.triathlon.org/uploads/docs/World_Triathlon_Prize_Money_Breakdown.pdf. Accessed: 2022-02-04

Chapter 41
Explainable Machine Learning and Economic Panel Data

Theo Berger

Abstract We apply boosted trees and Shapley values to analyze economic spillover effects within a customer-supplier network and assess economic interpretability. We translate conditional volatility into a Value-at-Risk universe and generate innovative economic features based on Natural Language Processing. Our results provide evidence for the economic relevance of spillover within a customer-supplier network for applied risk measurement. Furthermore, we demonstrate that the application of machine learning to panel data leads to innovative insights.

Keywords Boosted trees · Interpretable machine learning · Economic data

Introduction

In this paper, we focus on the interpretability of black box machine learning approaches. We apply large panel data set as discussed in [1], fit boosted regression trees and assess interpretability via Shapley values and discuss the results in comparison to interpretable econometric benchmark, namely pooled OLS.

Specifically, we translate conditional volatility into a Value-at-Risk (VaR) universe, generate innovative features (i.e. we identify economic links via SFAS 131) and analyze economic spillover effects within the built customer-supplier network. We examine whether the VaRs of companies' customer help to explain their own VaR.

Instead of looking at aggregate equity market volatility, we investigate conditional stock return volatility from the perspective of individual firms. We also focus on one special channel of volatility—the transferring of volatility along the supply chains.

T. Berger (✉)
Department of Business and Administration, University of Applied Sciences Harz, Friedrichstr. 57-59, 38855 Wernigerode, Germany
e-mail: tberger@hs-harz.de
URL: https://www.hs-harz.de/tberger/zur-person

Department of Business and Administration, University of Bremen, Enrique-Schmidt-Str. 2, 28334 Bremen, Germany

O. Grothe et al. (eds.), *Operations Research Proceedings 2022*, Lecture Notes in Operations Research, https://doi.org/10.1007/978-3-031-24907-5_41

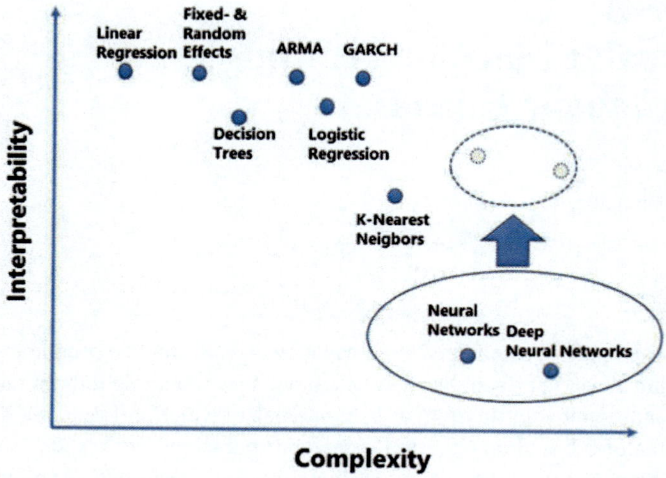

Fig. 41.1 The goal of our study is to assess interpretability of complex machine learning algorithms in the context of economic data

The contribution of our study is the application of tree based ML techniques to economic panel data. Moreover we apply SHAP values to assess the local interpretability of ML based predictions and compare the results with an econometric benchmark as published in [1].

Figure 41.1 provides an overview of the applied approaches. The study is part of our research that deals with competing machine learning algorithms and the assessment interpretability within a Big Data framework. As presented in [3], boosted trees describe a sensible choice in the context of financial data. Therefore, we assess the precision of boosted trees and discuss explainability via Shapley values as described in [6]. Up to now, Shapley values are the only concept of explainable machine learning which draws on solid theoretical foundation from game theory. Therefore, predictions are fairly distributed among the feature values. For an in-depth discussion on Shapley values, we refer to [7].

The remainder of this paper is structured as follows. Section "Methodology" provides a brief overview about the methodology and Section "Empirical Findings" presents the investigated data and preliminary results of the empirical assessment. Conclusion and Outlook are in Section "Conclusion".

Methodology

In this section, we introduce the notation used throughout the paper, we define the desirable properties of the linear model and tree ensembles, specify boosting for trees and present SHAP as our interpretable machine learning setting.

Boosted Trees

We apply an ensemble approach, which sums the prediction of multiple trees together:

$$\hat{y}_i = \sum_{k=1}^{K} f_k(x_i), \quad f_j \in F, \tag{1}$$

where K is the number of trees, f is a function in the functional space F, and F is the set of all possible CARTs.

We follow [2]: Let the regularized learning objective function be defined as follows:

$$obj = \underbrace{\sum_{i=1}^{n} l(y_i, \hat{y}_i^{(t)})}_{training\ loss} + \underbrace{\sum_{k=1}^{K} \Omega(f_k)}_{regularization}, \tag{2}$$

with

$$\Omega(f) = \gamma T + \frac{1}{2}\lambda\omega^2. \tag{3}$$

Here l is a differentiable convex loss function that measures the difference between prediction $\hat{y}_i$ and target variable y_i. Ω penalizes the complexity of the model. Each f_k corresponds to an independent tree structure and leaf weights ω and T is the number of leaves in tree.

Then, we perform gradient tree boosting. The model consists of functions instead of parameters, therefore it is trained in an additive manner: Let $\hat{y}_i^{(t)}$ be the prediction of the ith instance at the tth iteration, we need to add f_t in order to minimize the following objective:

$$obj^t = \sum_{i=1}^{n} l(y_i, \hat{y}_i^{(t-1)} + f_t(x_i)) + \Omega(f_t). \tag{4}$$

Chen and Guestrin [2] provide a scalable algorithm to tackle this issue: namely, XG-Boost. For a thorough introduction to boosting we refer to [4, 5].

Model Interpretability: Shapley Values

As described in [8], the Shapley value is the average marginal contribution of a feature value across all possible coalitions. The SHAP (SHapley Additive exPlanation) framework describes a model agnostic approach, which allows to estimate Shapley values expressing predictions as linear combinations [6].

Shapley value ϕ_j of jth feature is defined as follows:

$$\phi_j = \sum_{S \subseteq \{1,\ldots,p\} \backslash \{j\}} \frac{|S|!(p - |S| - 1)!}{p!} (f_{S \cup \{j\}}(x_{S \cup \{j\}}) - f_S(x_S)) \tag{5}$$

With x_S as a subset of the input features in the set S, $f_{S \cup \{j\}}$ is the trained model with feature j and f_S without. Hence, the Shapley value is the feature contribution to the prediction.

Empirical Findings

We apply the data set as discussed in [1]. That is, we study daily data ranging from January 2010 to December 2019, which is 2608 days. Our sample comprises 594 companies and 6216 economic links. Once a link is identified, we assume that the link is valid for one year prior to the reporting data. We weight each customer link by its sales normalized by the total amount of sales (as reported by the supplier). We assess 95% Value-at-Risk forecasts. Then, we apply competing XGBoost and SHAP-Values and discuss the results in comparison to pooled OLS, the interpretable benchmark, which is given as follows:

$$\Delta VaR_{i,t} = \alpha + \beta \ customer_{i,t-1} + \gamma \ control_{t-1} + \epsilon_{i,t-1} \tag{6}$$

with

$$customer_{i,t-1} = \{(G_{t-1} \cdot \Delta VaR_{t-1})_i, (G_{t-1} \cdot Hit_{t-1})_i\} \tag{7}$$

and

$$\begin{aligned} control_{t-1} =& \{Hit_{i,t-1}, RV1m_{i,t-1}, RV3m_{i,t-1}, RV6m_{i,t-1}, RV12m_{i,t-1}, \\ & SP500_{t-1}, VIX_{t-1}, Baa - Aaa_{t-1}, YC_{t-1}, YCSlope_{t-1}\}. \end{aligned} \tag{8}$$

The results of the applied XG-boost algorithm in combination with Shapley values are illustrated in Fig. 41.2.

The applied XG-boost approach provides a better in-sample fit than the pooled OLS approach as described in [1]. Via Shapley values, we are able to identify relevant determinants for the investigated target variable. In contradiction to the finding provided by the interpretable benchmark, Shapley values identify the realized variance for 3 months and 1 month (i.e. RV3m and RV1m) as the most important variables to describe a companies' VaR. In line with the interpretable benchmark, volatility spillover, gvar, are also characterized as a relevant determinant. Hence, the applied black box algorithm in combination with Shapley values allows to unbox black box ML approach and provides economic insights.

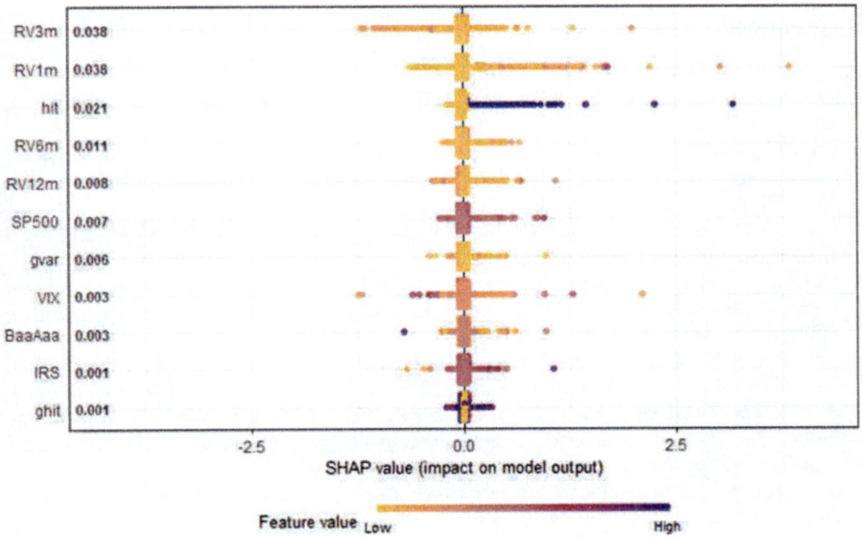

Fig. 41.2 Shapley values. This figure presents a ranked list of explanatory variables on the y-axis and the Shapley value for each observation on the x-axis. The values on the y-axis are the averaged Shapley values for each variable and describe the average marginal contribution of each feature to the target variable

Conclusion

We applied competing machine learning algorithms to a large panel data set and discussed interpretability via Shapley values. We compared results with pooled OLS and find that XGBoost in combination with SHAP describes a sensible choice. We have modeled economic links as described in [1] and can confirm the findings, there is empirical evidence that changes in customers' VaR spill over to suppliers' VaR.

In this vein, this study presents an initial step towards a deeper understanding of explainable ML in the context of financial data. Based on our preliminary methodological framework, our results indicate that explainable ML allows for the identification of relevant economic explanatory variables. However, this study draws on data as presented in [1], and therefore an in-depth discussion on theoretical and empirical robustness as well as a comparison of competing ML techniques present fruitful next steps.

References

1. Berger, T., & Gençay, R. (2020). Volatility spillover along the supply chains. *Journal of Risk, 22*(5), 83–113.
2. Chen, T., & Guestrin, C. (2016). XGBoost: A scalable tree boosting system. In: *KDD 16: Proceedings of the 22nd ACM SIGKDD International Conference on Knowledge Discovery and Data Mining* (Vol. 1, No. 1, pp. 785–794).
3. Gu, S., Kelly, B., & Xiu, D. (2020). Empirical asset pricing via machine learning. *The Review of Financial Studies, 33*(5), 2223–2273.
4. Hastie, T., Tibshirani, R., & Friedman, J. (2017). *The elements of statistical learning*. Springer.
5. James, G., Witten, D., Hastie, T., & Tibshirani, R. (2021). *An introduction to statistical learning*. Springer.
6. Lundberg, S. M., & Lee, S.-I. (2017). A unified approach to interpreting model predictions. In I. Guyon, U. V. Luxburg, S. Bengio, H. Wallach, R. Fergus, S. Vishwanathan, & R. Garnett (Eds.), *Advances in neural information processing systems* (Vol. 30, No. 1, pp. 4765–4774).
7. Molnar, C. (2022). Interpretable machine learning—A guide for making black box models explainable. https://christophm.github.io/interpretable-ml-book.
8. Shapley, L. S. (1953). A value for n-person games. *Contributions to the Theory of Games, 2*(25), 307–317.

Part VIII
Game Theory and Behavioral Management Science

Chapter 42
Considering Short and Long Term Fairness in Recurrent Auctions with an Application to Collaborative Rostering

T. Heller and S. Velten

Abstract Collaborative duty rostering can increase the satisfaction of employees in healthcare. For the acceptance of a final rostering, a fair selection of included wishes is essential. As in rostering problems various constraints must be respected, it is generally not possible to accept all bids (wishes for a single free time slot) in a planning period and fairness is important. In this paper we present a weighted Vickrey-Clarke-Groves (VCG) mechanism approach where past auction results are incorporated by a fairness factor and where the underlying winner determination problem is given by a hitting set problem (HSP). We present numerical results of simulation runs over several planning periods.

Keywords Auctions/competitive bidding · Game theory · Health care

Motivation

The shortage of skilled nursing staff is enormous in many regions of Germany and throughout Europe. The lack of attractiveness of the profession, which is limited by various factors, is often cited as a reason for this. On the one hand, there are the physical and psychological demands of the job, which often must be performed under high time pressure and is relatively poorly paid. On the other hand, shift work significantly impairs the compatibility of work and private life.

Collaborative duty rostering can increase the satisfaction of employees in healthcare and, in particular, improve the compatibility of work and private life. This was demonstrated, for example, in the research project GamOR (see [3]), in which, among other things, Constraint Programming based models and algorithms were developed for decision support collaborative duty rostering (see [2]).

A key aspect in [2] is the determination of conflicting employee requests for shifts or free time before the roster is actually created. In the GamOR approach,

T. Heller · S. Velten (✉)
Fraunhofer Institute for Industrial Mathematics ITWM, Kaiserslautern 67663, Germany
e-mail: sebastian.velten@itwm.fraunhofer.de

O. Grothe et al. (eds.), *Operations Research Proceedings 2022*, Lecture Notes in Operations Research, https://doi.org/10.1007/978-3-031-24907-5_42

these conflicts are communicated to the employees concerned, e.g., via app, so that they can be resolved within the team. This approach is evaluated positively by both employees and managers (see [3]), but it also has several drawbacks:

- In large planning groups and with many requests, resolving conflicts can be very complex.
- It cannot be expected that employees will always be able or willing to agree on conflict resolution.
- When conflicts are resolved in a team, some employees are better at asserting themselves than others.

In these cases, automatically generated conflict resolutions based on objective criteria can help the planner, or the manager who ultimately must make the decision, resolve the conflicts.

A Recurrent Auction Framework

To generate proposals for conflict resolutions that take decisions of previous planning periods into account, we propose a recurrent auction framework in which bids are wishes for a single free time slot and weights as well as fairness factors are considered. We assume that the time span of a wish is the minimum free time that the employee needs so that granting the wish is beneficial for her or him (e.g., to keep an appointment). Therefore, a partial fulfillment of bids is not considered.

In general, auction models are widely used for distributing units among a group of participating players (see e.g. [5]). Each of the players can place *points* on each of the goods, reflecting their valuation. In a classical auction setting, one is interested in solving the *winner determination problem*, i.e. finding the subset of all submitted bids such that the sum of individual valuations is maximized (see [5]). In *recurrent auctions*, the same group of players participates in several auctions where the same or equal units are distributed (see e.g. [4]).

In some sense we deviate from this setting. Instead of focusing solely on finding an optimal auction solution, we are interested in subsets of bids which cannot be fulfilled at the same time, i.e. a set of bids of which at least one has to be a losing bid. Since we are considering recurrent auctions, we want to incorporate past auction results into the decision on the winning bids.

We start with introducing some definitions. A set of bids that cannot be fulfilled together is called *conflict set*. A conflict set is called *minimal* if no contained bid can be removed and the resulting set is still a conflict set. A bid b_i contained in a conflict set is called *conflict bid*. The *fairness factor* f_j of a player is a measure on past auction results and lies in the interval [0, 1]. The *weight* w_i of a bid (or wish) is defined as the product of the points π_i placed on the bid and the fairness factor f_j of the submitting player: $w_i = \pi_i * f_j$.

Computing fairness factors, budgets and updating step The *fairness factor* f_j of a player is a measure on the individual past auction results. We define the *number-based fairness factor* as 1 minus the quotient of the number of accepted bids divided by the number of submitted bids, and the *point-based fairness factor* as 1 minus the quotient of the sum of points placed on accepted bids divided by the sum of points over all submitted bids. For both of these definitions it holds: the more bids accepted in the past, the smaller the fairness factor.

Each player starts with a budget B of points which can be used to weight the bids. Depending on the result of the auction, invested points must be paid or not (see below). After each period, the players gain a fixed number of points, regardless of the auction results. This allows players to place bids in the upcoming planning period even if they spent all their points in the last planning periods.

Computing winning bids Given the set of bids and the set of minimal conflicts, the task is to find a subset of bids with minimal weight such that at least one bid from each minimal conflict is chosen. This can be formulated as a weighted hitting set problem (weighted HSP) (see [1]). In the weighted HSP, one is given a set of elements U with a corresponding weight and a collection $\mathcal{C}$ of subsets $C_i \subseteq U$. The task is to find a minimal weighted selection of elements such that in each of the subsets C_i at least one is chosen.

Let U denote the set of all submitted bids and let $\mathcal{C}$ denote the set of minimal conflicts C. Recall that the weight of a bid is defined as the product of the fairness factor of the submitting player and the spent points on said bid. We introduce a binary decision variable x_i for every bid i which is equal to 1 if the bid is not accepted. The weighted HSP is then defined by:

$$\min \sum_{i \in U} w_i \cdot x_i \tag{1}$$

$$\text{s.t.} \sum_{i \in C} x_i \geq 1 \qquad \forall C \in \mathcal{C} \tag{2}$$

$$x_i \in \{0, 1\} \tag{3}$$

Remark 1 The formulation as a weighted HSP is used because by the time the conflicts need to be resolved in the collaborative duty rostering process developed in GamOR, the set of minimum conflicts is already known (see [2]). However, the size of $\mathcal{C}$, and therefore the number of constraints in (2) may be exponentially large. If this is the case, (2) and (3) may be replaced by a formulation of the feasibility constraints of the duty rostering problem (like the CP formulation given in [2]) and some additional constraints connecting the variables x_i $(i \in U)$.

A solution of this problem is denoted by x and the set of possible outcomes by X. The valuation of $x \subseteq X$ for each bid i is given by a *bid valuation function* from X to $\mathbb{R}_+$ denoted by $v_i^b(x)$. Note that we evaluate a solution by the bid valuation function v_i^b for each bid and not by the selection of winning bids, that yields the optimal solution.

In some sense, the valuation function is "reversed"—being a selected bid in the hitting set problem means that the bid is in fact a losing bid and vice versa. The bid valuation function v_i^b is therefore defined by

$$v_i^b(x) := w_i(1 - x_i). \tag{4}$$

For bid i the valuation of a solution is equal to zero, if it is contained in the solution of the weighted HSP. Conversely, it is equal to its weight, if it is not part of the optimal solution. With this, the hitting set problem above can be reformulated as an equivalent welfare maximization problem:

$$\max \sum_{i \in U} v_i^b(x) \tag{5}$$

$$\text{s.t. (2) and (3)} \tag{6}$$

We assume here that the *player valuation function* v_j^p of a player j does not depend on the fairness factor and is defined as

$$v_j^p(x) := \frac{1}{f_j} \sum_{i \text{ is a bid of } j} v_i^b(x), \tag{7}$$

i.e. the sum over all evaluations $v_i^b(x)$ of the submitted bids divided by the fairness factor f_j. Note that additional constraints can be incorporated into the optimization problem, e.g. a bounded number of winning bids for each player.

Computing costs After solving the weighted HSP, the mechanism has to update the available points and the fairness factor of each player. We distinguish the following cases after a solution to the weighted HSP is computed.

- If a bid is not a conflict bid, then no second bid on the same good exists. The bet is returned and the bid is a winning bid.
- If a bid is a conflict on its own, it violates the feasibility conditions. The submitted points will be returned and the bid cannot be granted.
- If a bid is a conflict bid and another bid on the same good exists, the hitting set computation from above has to determine the winning bids and, thus, their costs have to be computed.

The set X of computed solutions of the HSP is called *considerable solution set*, a solution $x \in X$ is called *(considerable) solution* and the set of all bids is denoted by N. Since every bid i is associated with a player j, the fairness factor of a bid is given by the fairness factor of the associated player.

The *Vickrey payments* is a famous payoff measure for auction solutions which is based on results from Vickrey, Clarke and Groves (see e.g. [6]). For an optimal solution x^* of the weighted HSP, we define the payment p_i for a bid i by

$$p_i := \max_{x \in X} \sum_{j \in N \setminus i} v_j^b(x) - \sum_{j \in N \setminus i} v_j^b(x^*), \tag{8}$$

i.e. the maximum over the sum of valuations of others when bid i is considered minus the sum of valuations of bids other than i when bid i is missing. Note that if bid i is not chosen in the optimal solution x^*, the payment p_i is zero. Given a solution x, the utility u_i of a bid is defined by

$$u_i := v_i^b(x) - p_i. \tag{9}$$

Note that in general the computation of a single payment requires to solve

$$\max_{x \in X} \sum_{j \in N \setminus i} v_j^b(x), \tag{10}$$

which again requires the solution of a (smaller) weighted HSP. However, we can construct a feasible solution as follows. Set variable x_i to 1. When bid i is not considered in the objective function, the corresponding decision variable x_i can be set to 1 without changing the solution value. All decision variables x_j that are contained only in conflict sets with x_i can be set to 0, since all conflict sets containing x_i are resolved. All other decision variables are set accordingly to their value from x^*. Thus, we obtain a feasible solution for (10) and the payment as in (8) is always non-negative.

Furthermore, it holds that there are *no positive transfers*, i.e. the payment to a bid is at least zero, and *individual rationality for bids* holds as well, i.e. the utility of a winning bid is non-negative. The following two results are direct consequences of VCG mechanisms (see e.g. [5]).

Lemma 1 (Truthfulness on a bid basis) *Bidding the true value of a bid is an optimal strategy under the consideration of the utility function in* (9).

We consider the utility of a player as follows:

$$u_j := v_j^p(x) - \frac{1}{f_j} \sum_{i \text{ is a bid of } j} p_i. \tag{11}$$

Lemma 2 (Truthfulness on player basis) *Assume each player only submits one bid. Then, bidding the true value of a bid is an optimal strategy under the consideration of the utility function* (11).

Simulation Results

For our simulation, we run for both fairness factor variants 10 randomly generated planning processes which consist of 100 periods each. In each planning process, 20

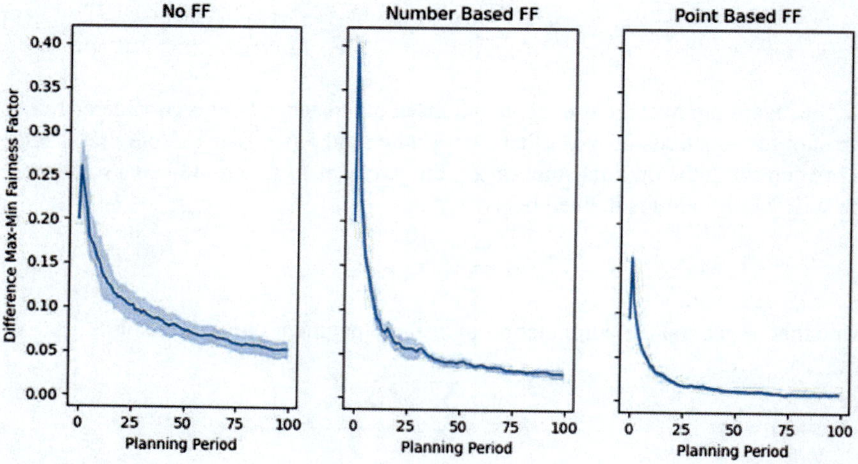

Fig. 42.1 Evaluation on the difference between minimal and maximal FF for the different FF cases

participants are present. Their initial budget of points to spend is 100. In each period, each participant places a random number of wishes and spends a random percentage of its available points in an arbitrary way on all of its wishes. After each period, each participant gets 20 additional points.

In Fig. 42.1 the evaluation of the difference between the maximal and the minimal fairness factor among all participants is shown. Besides the mean value, also the 95% confidence interval is shown in the charts. We distinguish different fairness factors (FF), i.e. no FF, the number-based FF and the point-based FF. For the case of no FF, the number-based FF is used for the evaluation. We see that the difference shrinks quickly after only a few planning periods. Due to the random behaviour of the agents, we see a spike in the difference for the number-based FF in the first few periods. Overall we see that the difference gets smaller for all cases. Also due to the randomness in the bidding behaviour of the different participants, these fairness factors always differ by a small margin. However, in the case of no FF, the 95% confidence interval is considerably larger than in the other cases. In total we see that incorporating a FF into the underlying conflict resolving process, leads faster to a more 'fair' distribution of winning bids (with respect to the considered FF).

Conclusion and Outlook

In this paper we describe a mechanism for finding fair solutions in an auction within a group of players over the course of many auction runs. This allows a decision support for duty rostering where past planning results are included in the optimization. We experimentally show that with this approach a convergence to a fair distribution of wishes can be achieved. Moreover, applying the defined fairness factor leads to a faster and more stable convergence. Our further research includes the extension to a bi-criteria model, in order to find trade-offs between short term and long term fairness aspects.

References

1. Garey, M. R., & Johnson, D. S. (1979). *Computers and intractability* (Vol. 174). Freeman.
2. Heydrich, S., Schroeder, R., & Velten, S. (2020). Collaborative duty rostering in health care professions. *Operations Research for Health Care, 27.*
3. Kubek, V., Velten, S., Eierdanz, F., & Blaudszun-Lahm, A. (Eds.). (2020). *Digitalisierung in der Pflege zur Unterstützung einer besseren Arbeitsorganisation.* Springer.
4. Murillo, J., Muñoz, V., López, B., & Busquets, D. (2008). A fair mechanism for recurrent multi-unit auctions. In *German Conference on Multiagent System Technologies* (pp. 147–158). Springer.
5. Roughgarden, T. (2010). Algorithmic game theory. *Communications of the ACM, 53*(7), 78–86.
6. Vickrey, W. (1961). Counterspeculation, auctions, and competitive sealed tenders. *The Journal of Finance, 16*(1), 8–37.

Chapter 43
Coopetition and Knowledge Sharing in Dynamic Business Environments

Ayesha Alhosani, Richard Allmendinger, and Mercedes Bleda

Abstract Strategic alliances among competitors are increasingly gaining relevance as business organizations recognize the benefits of learning and sharing resources to succeed in dynamic environments. In an increasingly interconnected business world, it is crucial for companies to reap the benefits that arise from this type of collaboration. We use multi-agent simulation coupled with principles from operations research to study how learning and knowledge sharing within alliances can impact the performance of competing firms in an endogenously changing environment. We consider two types of competing firms: ones that search for business solutions in the business landscape (*searchers*), and ones that search and also have the power to endogenously change or reshape the landscape to their advantage (*shapers*). Our model allows us to analyze under which circumstances strategic alliances can help firms to successfully adapt to these endogenously generated changes by allowing them to learn and build up an enhanced knowledge base of the business landscape.

Keywords Coopetition · NK model · Endogenously-changing landscape · Shaping · Searching · Adaptation

Introduction

A strategic alliance is a collaboration between two or more business organizations to share resources, exchange products and co-develop or co-provision services in an effort to pursue a commonly beneficial goal for a defined period [5]. We focus on a specific type of alliance where the collaborating companies are also competitors, in a phenomenon known as *coopetition* [1]. A recent example of coopetition is the

A. Alhosani (✉) · R. Allmendinger · M. Bleda
The University of Manchester, Manchester, UK
e-mail: ayesha.alhosani@postgrad.manchester.ac.uk

R. Allmendinger
e-mail: richard.allmendinger@manchester.ac.uk

M. Bleda
e-mail: mercedes.bleda@manchester.ac.uk

O. Grothe et al. (eds.), *Operations Research Proceedings 2022*, Lecture Notes in Operations Research, https://doi.org/10.1007/978-3-031-24907-5_43

collaboration and knowledge sharing between competitors, Pfizer and BionTech to produce a COVID-19 vaccine which allowed the two companies to release their vaccine first and outperform other pharmaceuticals.

A critical factor that influences an organization's decision to form or join an alliance is the ability to access resources, information, and knowledge otherwise inaccessible. Alliances grant companies with resources and knowledge that can be utilized to improve their competitive position [5]. In highly dynamic environments, forming alliances with competitors might be crucial for survival, particularly if these environments change in an endogenous way. Changes in an industry's environment are endogenously generated by companies that have the ability to change the business landscape. These companies are known in the management literature as *shapers* [2]. Shapers are companies that are not only able to search for business solutions in the landscape to improve their performance but also to change the landscape to their advantage endogenously. Existing work has shown that these companies tend to outperform competitors which are only able to search (searchers). Strategic alliances in this setting can, however, influence these dynamics, as they may provide firms and searchers in particular, with the necessary information and knowledge to improve their performance and successfully adapt to these changes.

In this work, we build a simulation model where competing companies searching for solutions in an endogenously changing landscape also collaborate within alliances. The model allows us to analyze how knowledge sharing among alliance members affects their performance and under which circumstances searchers can successfully adapt to changes that shapers generate in the business landscape. The following section presents the description of the simulation model. Next, the experimental study is shown in Section "Experimental study", followed by conclusions in Section "Conclusions".

Model Description

The model extends an NK co-opetition simulation model developed by the authors [4] to incorporate the influence of alliances and explicit memory approaches on the performance of firms. As in the standard NK model [3], the complexity (ruggedness) of the landscape is represented by the parameter K. The business solutions in the landscape are represented by strings of 0s and 1s of length $N + Z$, i.e. $\boldsymbol{x} = (x_1, \ldots, x_N, x_{N+1}, \ldots, x_{N+Z})$. Searchers companies search the landscape by modifying one of the N dimensions of the solutions at each time step (one-step hill climbing), while shapers can choose to search (in the same manner) or to shape, that is, change one of the Z solutions' dimensions; we explain below how a searcher/shaper decides which option to commit to. An important to note is: that the values of Z solutions' dimensions are the same for all solutions (i.e., we can think of them as global variables). Hence, shaping will impact the fitness of the solutions for all companies in the landscape, i.e., as a result of a shaping action, some companies will be better or worse off in terms of their performance. The parameter E defines the interdependency between searching and shaping solutions dimensions and the degree of the malleability of the environment.

To analyze the influence of alliances on these search and shape dynamics, we group agents of the same type into alliances, i.e., alliances are formed by searchers or shapers only. For the sake of simplicity, we do not include heterogeneity in group composition. Companies within an alliance share knowledge through a central unbounded memory, accessible to all alliance members at all times. Companies are randomly grouped into alliances, and throughout the simulation, each company takes a turn to improve its current solution. Once each agent has had its turn, each alliance adopts the solution found by its members with the highest fitness or performance and stores it in the alliance knowledge base (memory) for later use.

At each turn, searchers can: (a) search the landscape (by flipping a randomly selected bit out of the N bits) and find a new solution; (b) Adopt the decision variables values of a solution from the alliance's memory that has the same values of the Z shaping bits; or (c) not take action. Searchers will choose the new business solution (a) or (b) that has the higher fitness or will not take action (c) if the found or shared solutions are of inferior fitness to their current one. At each turn, shapers have the same options as a searcher, plus they can (d) change a shaping dimension (one of the Z variables) in their current solution (shape the environment). Shapers then will choose the new business solution (a), (b) or (d) that has the highest fitness, or will not take action (c) if the found, shared, and "shaped" solutions are of inferior fitness to their current one. If shaping (d) is selected, then the fitness of all other agents is recomputed as a change in the (global) Z variables affects the bit strings and hence the fitness of all solutions.

To analyse how alliances influence the dynamics and performance of searchers and shapers, we compare our simulation results with the results of a simulation model in which agents do not form alliances (our baseline model).

Experimental Study

This section covers the experimental setup, followed by an analysis of the experimental results.

Experimental Setup

We run simulations for $N = 12$, $Z = 12$, and low, medium, and high levels of complexity ($K = \{1, 2, 3, 4, 5\}$) in high malleability environments ($E = 5$). As the focus of this initial study is to isolate and understand the impact of alliances on agents (companies), we have decided to fix the alliance group size and group composition as follows: The agent population is composed of 16 firms—8 searchers and 8 shapers—with firms being grouped into two alliances of 4 per group—one alliance with 4 searchers, and another alliance with 4 shapers; the remaining firms do not join alliances and thus operate individually in the same manner as in the baseline model. Future work will look at varying alliance group sizes and group compositions. Simulations were run for 200 generations (one generation involves iterating over all agents) across 100 runs to compare the performance between searchers, shapers within alliances, and those operating outside alliances (on the same landscape). We then compare the results to the data collected from the baseline model; in which the experiment is ran separately with the same parameters and for the same number of generations and runs.

Experimental Analysis

The first finding we observe from Fig. 43.1 is that alliances boosts the performance for all types of firms, specifically in environments characterized by low complexity. In a business context, alliances enable companies to share and access knowledge they otherwise would not have access to. This phenomenon resembles innovative companies (alliances) developing new technologies that improves operations if known and adapted by them. An example in the telecom industry is when google develops a new version of its Android operating system (OS) for mobile devices. The new version will give all partnering and competing telecom companies that use the same OS, a kick in terms of their devices' performance and potential sales. This is visible in Fig. 43.1, where shapers in an alliance are always the highest performers in a setting of low complexity. However, as confirmed in [2], shaping is a powerful tool, but the agents lose the advantage of shaping as K increases. As shown in Fig. 43.1, when $K = 3$, shapers lose the advantage of shaping and knowledge sharing. The figure shows that shapers from the baseline model catch up to the shapers alliance, and both groups perform at the same level. This finding aligns with the observation made in [2] for environments with high complexity and high malleability.

In comparison to less complex (rugged) landscapes, $K = 1$, a shapers alliance has a short-lived advantage over its competition, which they lose as everyone converges to the optima mid-way through the optimization run, as shown in Fig. 43.1. The justification for the phenomenon of this short-lived advantage is explained by Fig. 43.2: shapers are able to identify the best strategy (shaping policy) early on, and they only need to tune the searching parameters to identify the best location in the landscape. As they have identified the shaping policy ahead of everyone else, and since they are sharing the knowledge with each other, they are able to reach the highest fitness possible. Furthermore, since the shaping policy is a global strategy that will impact everyone in the landscape (even companies from other alliances), the other companies will eventually catch up as the landscape is simple. Other agents/alliances will try different search solutions until they find the optimum solution for this shaping policy. This is also apparent from Fig. 43.2, where one can see that the total changes to the shaping policy drastically subsides at about $T = 36$.

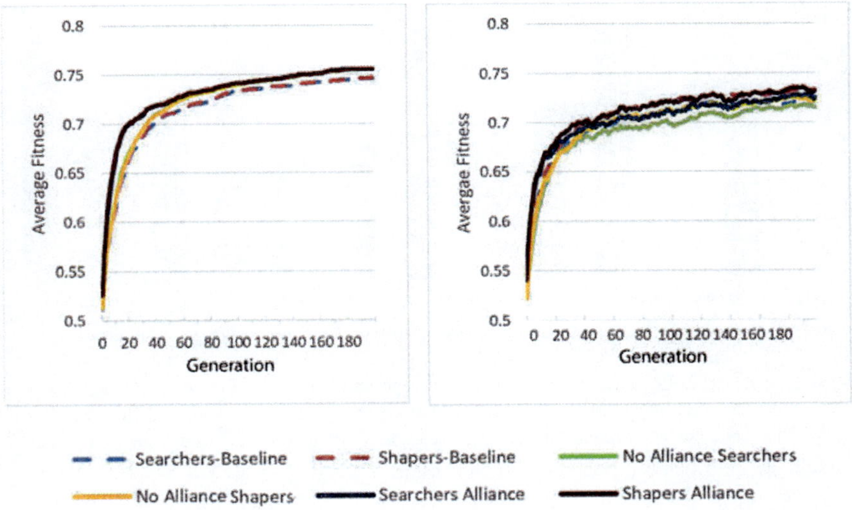

Fig. 43.1 Comparison of average fitness values for agents in a population that are in alliances (Searcher and Shaper Alliance) and not in an alliance (No Alliance Shapers and Searchers). Searchers and Shapers-Baseline are average fitness values for searchers and shapers in a population without alliances. Results are shown for $K = 1$ (left) and $K = 3$ (right)

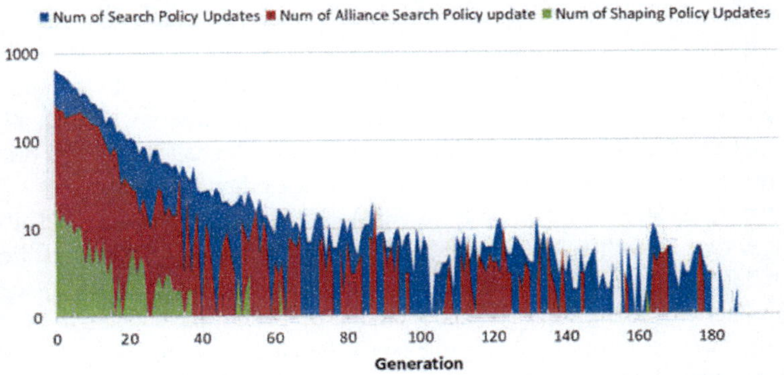

Fig. 43.2 Number of times when agents mutated search policy (num of search policy updates), shaping policy (num of shaping policy update), or learnt from group knowledge (num of alliance search policy updates) over 100 generations when $K = 1$

Furthermore, as shown in Fig. 43.1, in the case of $K = 3$, the shapers are still able to maintain their competitive advantage temporarily as compared to the Shapers from the baseline model who catches up to the same level of performance later on. The justification for this behavior is: in addition to the above reasoning, a shapers alliance can evaluate up to three solutions (corresponding to options (a), (b) and (d)) in one step and pick the solution with the highest performance giving them more options to scan the landscape more widely and thus encounter better solutions faster

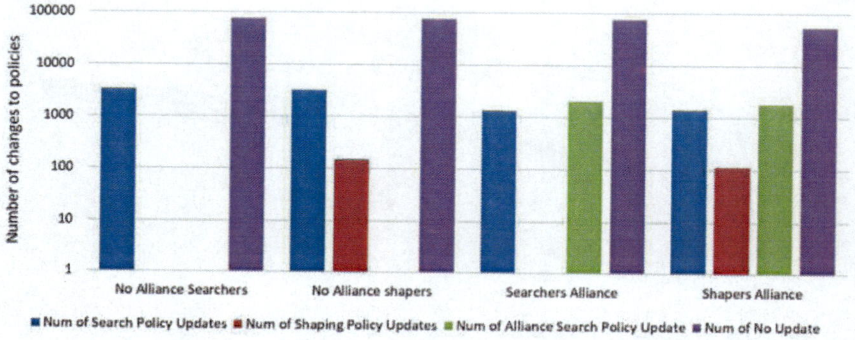

Fig. 43.3 average policy updates per alliance; average number of times of policy (search/shaping policy) were mutated by agents in a population that are in alliances (searcher and shaper alliance) and not in an alliance (no alliance shapers and searchers); compared to "no update". i.e. when an agent decides to remain in the same location, when $K = 1$

compared to their competition. Figure 43.3 supports this justification as it shows the number of changes to search and shaping policy by agents in a population that are in alliances (Searcher and Shaper Alliance) and not in an alliance (No Alliance Shapers and Searchers). The figure shows the "no Update" data referring to the instances in which an agent decides that their current solution yields higher fitness than if it were to change search/shaping policies or learn from the alliance's experience. For the shapers alliance, the number of times the alliance members chose not to update any policy is lower than for the searchers' alliance and the agents acting individually (no alliance). This indicates that members of the shapers alliance are the least prone to stay in one place and not take action compared to the other alliances.

Another interesting observation from Fig. 43.1 is that the searchers in an alliance will outperform searchers and shapers in the baseline model. This is justified by the searchers' alliance sharing the knowledge, which provides accessibility to more superior solutions through the alliance memory. So even if the shapers in the baseline model have the power to change the landscape, the searchers in an alliance can exceed an individual shaper's performance if searchers collaborate and share knowledge. As shown in Fig. 43.3, searchers in an alliance are less likely to remain idle than individual shapers, allowing them to explore more and reach areas with higher performance in the landscape.

Conclusions

We proposed a simulation model to investigate the influence of alliances on the performance of firms competing in an endogenously changing environment. In the model, firms within an alliance share knowledge and information (via explicit memory) they acquire about superior solutions with their peers in the alliance. We found that the presence of alliances in dynamic environments improves the performance of all firms in the landscape, regardless of whether they are in an alliance or not. The average performance of companies in industries with alliances is higher than in industries without alliances. This performance improvement, however, is less significant in more complex environments. Searchers (in principle, less powerful organizations) can benefit significantly from being in an alliance in low-complexity environments, allowing them to exceed the performance of shapers that are not in an alliance. Future research can consider environments where agents are part of multiple alliances and can exit/enter alliances during search. Studying different alliance sizes and compositions is also important.

References

1. Brandenburger, A. M., & Nalebuff, B. J. (2011). *Co-opetition. Currency*
2. Gavetti, G., Helfat, C. E., & Marengo, L. (2017). Searching, shaping, and the quest for superior performance. *Strategy Science, 2*(3), 194–209.
3. Kauffman, S. A. (1993). *The origins of order: Self-organization and selection in evolution.* Oxford University Press.
4. Lim, C. W., Allmendinger, R., Knowles, J., Alhosani, A., & Bleda, M. (2022). Cooperative multi-agent search on endogenously-changing fitness landscapes. In *PPSN* (pp. 265–278). Springer.
5. Lin, H., & Darnall, N. (2015). Strategic alliance formation and structural configuration. *Journal of Business Ethics, 127*(3), 549–564.

Chapter 44
Decreasing Viability of Tychastic Controlled Systems

Sigifredo Laengle and Tomás Laengle-Aliaga

Abstract The viability kernel in Viability Theory depends on control variables and usually also on uncontrolled ones. Control variables try to increase viability, and uncontrolled ones instead destroy it. Tyches are uncertainties without statistical regularity that diminish viability. We progress in the study of both effects. We use a necessary condition of the system viability and apply it to the linear case by introducing the Minkowski difference between sets. We also find such a difference interprets the problem adequately.

Keywords Viability theory · Viability kernel · Decreasing viability

Introduction

The Viability Theory (VT) has aroused enormous practical interest in areas such as electronics [7] and natural resources [10]. Also, the applications to Economic Theory represent another essential development [1]. However, few systematic studies of the viability kernel of controlled tychastic systems. Exceptionally we find isolated developments that are theoretical [3] or applied to ecological problems [4] and finance [5]. However, much other management and economic problems are tychastic systems. Moreover, many of them are also linear. This article is part of a research project on the economic and managerial applications of VTs with the presence of *tychastic variables*, that is, disturbers with an unknown probabilistic description. In a previous article [9], we formulated the classic distributive bargaining problem as a

S. Laengle (✉)
Faculty of Economics and Business, University of Chile, Diagonal Paraguay 257, 8320000 Santiago, Chile
e-mail: slaengle@fen.uchile.cl

T. Laengle-Aliaga
Faculty of Physical and Mathematical Sciences, University of Chile, Beauchef 850, 8320000 Santiago, Chile
e-mail: tlaengle@dim.uchile.cl

O. Grothe et al. (eds.), *Operations Research Proceedings 2022*, Lecture Notes in Operations Research, https://doi.org/10.1007/978-3-031-24907-5_44

controlled linear tychastic system. The main result of this article generalizes such a previous work.

The following section introduces the Viability Theorem and a lemma we need to solve our problem. Section "The Viable Fund Management" solves an example of viable fund management that the Viability Theorem can only solve. Section "General Solution of the Controlled Linear Tychastic System" contains the main result, and Section Conclusion" presents the conclusion with an interpretation of such result.

Viability Theorem and a Previous Result

Let $T \in \mathbb{R}_{++}$, the *time-space* interval be $[0, T]$, and $(\mathcal{H}, \langle \cdot \mid \cdot \rangle)$ be the *state space*, a Euclidean space with induced norm $\| \cdot \|$. Now consider the set of absolutely continuous functions[1] on $[0, T]$. Let $\varphi : \mathcal{H} \to 2^{\mathcal{H}}$ be a set-valued operator that defines the following differential inclusion with the initial conditions $\dot{x} \in \varphi x$ and $x(0) = x_0$. We then say that x is an *evolution* for φ starting at x_0 if x is a solution of the differential inclusion $\dot{x} \in \varphi x$ and $x(0) = x_0$.

An *environment* is a nonempty subset C of $\mathcal{H}$. The *support function* of C is $\sigma_C : \mathcal{H} \to [-\infty, +\infty] : u \mapsto \sup \langle C \mid u \rangle$. Letting $x \in \mathcal{H}$ and C be a non-empty convex subset, we define the *normal cone* of C at x as $N_C x \doteq \{u \in \mathcal{H} : \sup \langle C - x \mid u \rangle \leq 0\}$ in case that $x \in C$ and $N_C x \doteq \emptyset$ otherwise. Now let $x \in C$ and $u \in \mathcal{H} \setminus \{0\}$. If $\sup \langle C \mid u \rangle = \langle x \mid u \rangle$, then $\{y \in \mathcal{H} : \langle y \mid u \rangle = \langle x \mid u \rangle\}$ is a *supporting hyperplane* of C at x, and x is a *support point* of C with *normal vector* u. The set of support points of C is denoted $\operatorname{spts} C$ and its closure $\overline{\operatorname{spts}}\, C$. Given a normal vector u, the set of support points wrt that vector is $\operatorname{spts}_u C$. Also let $\operatorname{bdry} C$ be the *boundary* of C.

The following lemma is used in the proof of Proposition 1:

Lemma 1 *Let C be a nonempty convex set of $\mathcal{H}$. Then the following equality holds:*

$$\{(x, u) : x \in \operatorname{bdry} C \text{ and } u \in N_C x\} = \left\{(x, u) : u \in \mathcal{H} \setminus \{0\} \text{ and } x \in \operatorname{spts}_u C\right\}.$$

Proof First, let $(x, u) \in \operatorname{bdry} C$ and $u \in N_C x$. Then $u \in \mathcal{H} \setminus \{0\}$ and $u \in N_C x \leftrightarrow \sup \langle C - x \mid u \rangle \leq 0 \leftrightarrow \sup \langle C \mid u \rangle \leq \langle x \mid u \rangle$, that is, $x \in \operatorname{spts}_u C$. Second, let $u \in \mathcal{H} \setminus \{0\}$ and $x \in \operatorname{spts}_u C$. Then $u \in N_C x$ and $\operatorname{spts}_u C \subset \operatorname{spts} C \subset \overline{\operatorname{spts}}\, C$, and by the Bishop-Phelps Theorem [6, p. 134] we obtain $x \in \operatorname{bdry} C$. □

We say that the environment C is *viable* under the set-valued map φ if for each $x_0 \in C$, x is the trajectory for φ starting at x_0 that lies within C (that is, $x(t) \in C$ for each $t \in [0, T]$). We now state the Viability Theorem, which gives the sufficient and necessary conditions for the viability of C under φ in the case where C is convex [8, p. 8] and [2, p. 152].

[1] Functions $x : [0, T] \to \mathcal{H}$ such that $\|x\|$ and $\|\dot{x}\|$ are integrable.

Theorem 1 (Viability Theorem) *Let C be a non-empty closed convex subset of $\mathcal{H}$ and φ an upper semi-continuous set-valued operator with non-empty compact convex values. The following assertions are then equivalent: (a) C is viable under φ; (b) $(\forall x \in C)(\forall u \in N_C x)$ $\sup \langle \varphi x \mid -u \rangle \geq 0$; and (c) $(\forall x \in C)(\exists y \in C)$ $y - x \in \varphi y$.*

Reference [2, p. 152] proves that (c) is a necessary condition for the viability of C under φ. There also exists a proof, not given here, that (c) is also sufficient.

The Viable Fund Management

The fund management example evolves in time-space, and is explained in what follows. Assume first of all that an agent controls the *fund flow* x through the *rate of cash flow* p, such that $\dot{x} = p$ where $\dot{x} = \frac{d}{dt}x$. Also assume that the reserve fund is subject to a *demand rate* defined by an uncontrollable variable q that determines the *fund reserve* y according to the differential equation $\dot{y} = x - q$. Thus, the system's dynamic is described by the equations $\dot{x} = p$ and $\dot{y} = x - q$.

Now assume further that the fund y is greater than 0 but not greater than a value $a \in \mathbb{R}_{++}$. Thus, the *environment* C is the set of admissible values of variables x and y, that is, $(x, y) \in C \doteq \mathbb{R} \times [0, a]$. The uncontrollable variable q satisfies $q \in [-\nu, \nu]$ where $\nu \in \mathbb{R}_{++}$, while the control variable $p \in [-\mu, \mu]$, where $\mu \in \mathbb{R}_{++}$.

However, solutions can lead to collapse. Consider, for example, the case where the external situation imposes a maximum demand for funds, say, $q \doteq \nu$. At the same time, the controller estimates that the capacity to generate funds will be insufficient to avoid shortages, and therefore injects a maximum flow of $p \doteq \mu$. Suppose also that the state of the system has an initial flow given by[2] $x(0) \doteq \nu - \sqrt{2a\mu}$ and an initial fund of $y(0) \doteq \frac{1}{2}a$. In this situation, the regulator will calculate whether there is enough time to avoid collapse. This is done by solving the following system of differential equations: $\dot{x} = \mu$, $\dot{y} = x - \nu$, with initial states $x(0)$ and $y(0)$. Thus, he integrates the first equation and then the second one, both with respect to t obtaining

$$x(t) = \mu t + \nu - \sqrt{2a\mu} \text{ and } y(t) = \frac{1}{2}\mu t^2 - \sqrt{2a\mu}t + \frac{1}{2}a.$$

Next, the controller calculates the time to collapse. After a number of algebraic manipulations it can be seen that at any time t, if $0 \leq t \leq \sqrt{a\mu}(\sqrt{2} - 1)/\mu$ the fund reserve y will be non-negative, and if $t > \sqrt{a\mu}(\sqrt{2} - 1)/\mu$ it will be negative (see the blue flow in Fig. 44.1).

In what follows, we solve analytically the problem. According to the Viability Theorem 1 (equivalent assertion (b)), the computation of the fund management example's viability kernel proceeds as shown in the following proposition:

---

[2] We will see later why this quantity was chosen.

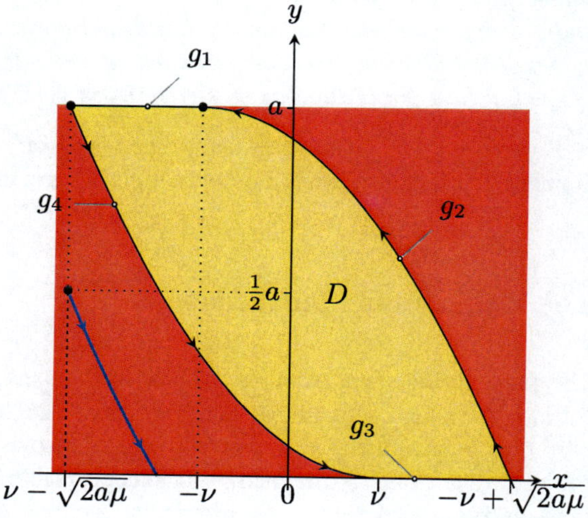

Fig. 44.1 The tychastic viability kernel is given by $D = \bigcap_{i \in I} \text{lev}_{\leq 0}\, g_i$ with $I \doteq \{1, \ldots, 4\}$, where each g_i is given in Table 44.1. The blue arrow is the unique solution derived in Section "The Viable Fund Management" that inevitably reaches a non-viable state

Table 44.1 The tychastic viability kernel is $D = \bigcap_{i \in I} \text{lev}_{\leq 0}\, g_i$ with $I \doteq \{1, \ldots, 4\}$

i	$g_i(x, y)$
1	$y - a$
2	$\frac{1}{2}(x + \nu)^2 + \mu y - a\mu$
3	$-y$
4	$\frac{1}{2}(x - \nu)^2 - \mu y$

These functions are obtained by Proposition 1, which uses the Viability Theorem 1 condition (b). Thus, we obtain Fig. 44.1

Proposition 1 *Let* $C \doteq \mathbb{R} \times [0, a]$ *be the environment, where* $a \in \mathbb{R}_{++}$, $P \doteq [-\mu, \mu] \times \{0\}$ *and* $Q \doteq \{0\} \times [-\nu, \nu]$ *where* $\mu, \nu \in \mathbb{R}_{++}$. *The set-valued operators for every* $(0, q) \in Q$, *are given by* $(\dot{x}, \dot{y}) \in \varphi_q(x, y) \doteq (0, x) + P - (0, q)$. *Then the tychastic viability kernel of C for the family* $\{\varphi_q\}_{q \in Q}$ *is given by*[3] $\bigcap_{i \in I} \text{lev}_{\leq 0}\, g_i$, *where* $I = \{1, \ldots, 4\}$ *and each continuous convex function* g_i *is given by* $g_i : \mathbb{R}^2 \to \mathbb{R}$ *as shown in Table 44.1.*

Proof In this case, $\mathcal{H} \doteq \mathbb{R}^2$. The proof consists in determining the largest non-empty subset $D \subset C$ that satisfies Condition (b) in Theorem 1. We proceed in three steps.

(1) Suppose a priori that D is a closed convex non-empty subset of C. Since $(x, y) \in \text{int}\, D \leftrightarrow N_D(x, y) = \{0\}$, Condition (b) of Theorem 1 is equivalent for all $q \in Q$ to

$$(\forall (x, y) \in \text{bdry}\, D)(\forall (u, v) \in N_D(x, y)) \ \sup \ \langle \varphi_q(x, y) \mid -(u, v) \geq 0.$$

[3] Let $\mathcal{X}$ be a set, $g : \mathcal{X} \to [-\infty, +\infty]$ be a function, and $\xi \in \mathbb{R}$. The *lower level set* of g at height ξ is the set $\text{lev}_{\leq \xi}\, g \doteq \{x \in \mathcal{X} : g(x) \leq \xi\}$.

Then, by Lemma 1, the above inequality is equivalent to

$$(\forall(u, v) \in \mathcal{H} \setminus \{0\})(\forall(x, y) \in \text{spts}_{(u,v)} D) \ \sup \ \langle \varphi_q(x, y) \mid -(u, v) \geq 0.$$

Therefore, if we take $(u, v) \in \mathcal{H} \setminus \{0\}$ and $(x, y) \in \text{spts}_{(u,v)} D$, then for all $(0, q) \in Q$ we have

$$\begin{aligned} 0 \leq \sup \ \langle \varphi_q(x, y) \mid -(u, v)\rangle &= \sup \ \langle (0, x) + P - (0, q) \mid -(u, v)\rangle \\ &= -xv + qv - \inf_p \ pu. \end{aligned}$$

or equivalently, $xv + \inf_p \ pu - \inf_q \ qv \leq 0$. Considering this inequality, the problem is to find, for all $(u, v) \in \mathcal{H} \setminus \{0\}$, all points $(x, y) \in D$ such that (x, y) is an optimum of $\sup_{x,y} \ (xu + yv)$ subject to (a) $xv + \inf_p \ pu - \inf_q \ qv \leq 0$; (b) $\dot{x} = p$ and $\dot{y} = x - q$; (c) $(x, y) \in \mathbb{R} \times [0, a]$; and finally, (d) $(p, q) \in [-\mu, \mu] \times [-\nu, \nu]$.

(2) To solve the foregoing problem, we consider separately the eight possible combinations of values of (u, v), numbered **(2.1)** through **(2.8)**, depending on whether the values of each are positive, negative or zero.

(2.1) $u > 0$ and $v \doteq 0$. In this case, $\inf_p \ pu = -\mu u$ and Restriction (a) is then equivalent to $-\mu u \leq 0$ or $\mu \geq 0$, which is a tautology.

(2.2) $u > 0$ and $v > 0$. In this case, $\inf_p \ pu = -\mu u$ and $\inf_q \ qv = -\nu v$. The problem is then to find $(x, y) \in D$ such that (x, y) is an optimum of $\sup_{x,y} \ (xu + yv)$ subject to (a) $xv - \mu u + \nu v \leq 0$; (b) $\dot{x} = -\mu$ and $\dot{y} = x + \nu$; and (c) $(x, y) \in \mathbb{R} \times [0, a]$. Upon solving the system of differential equations (b) by the first integral method, we obtain $\mu y + \frac{1}{2}(x + \nu)^2 = k$, where k is a constant to be determined.

(2.3) $u \doteq 0$ and $v > 0$. Here, $\inf_p pu = 0$ and $\inf_q qv = -\nu v$. The objective function is $\sup yv = av$. Observe that for any value of (x, y), Restriction (b) is $\dot{x} = p$ and $\dot{y} = x + \nu$ so $\dot{y} \leq 0$ if $x \leq -\nu$, which is satisfied if $(x, y) \in]-\infty, -\nu] \times \{a\}$. At this point we return to (2.2) to determine the value of constant k subject to the condition that $y(-\nu) = a$, which gives a value of $k \doteq \mu a$.

(2.4) $u < 0$ and $v > 0$. In this case, $\inf_p pu = \mu u$ and $\inf_q qv = -\nu v$. We again observe that for any value of (x, y), Restriction (b) is $\dot{x} = \mu$ and $\dot{y} = x + \nu$, so $\dot{y} \leq 0$ if $x \leq -\nu$. This is satisfied if $(x, y) \in]-\infty, -\nu] \times [0, a]$. Therefore, considering cases (2.1)–(2.4) we can define $g_1(x, y) \doteq y - a$ and $g_2(x, y) \doteq \frac{1}{2}(x + \nu)^2 + \mu y - a\mu$.

(2.5)–(2.8). These cases are treated symmetrically with (2.1)–(2.4), obtaining g_3 and g_4. The various g_i functions are all set out in Table 44.1 and set D is shown in yellow in Fig. 44.1.

(3) The third step consists in proving that D is the largest viable subset of C. The proof itself involves a number of trivial algebraic manipulations that add nothing to the key arguments in steps (1) and (2) so is not given here. □

General Solution of the Controlled Linear Tychastic System

Now we will calculate the viability kernel assuming the presence of tychastic variables when φ has a *linear* form.

Proposition 2 *Let C be a nonempty compact convex subset of $\mathcal{H}$, $A \in \mathcal{B}(\mathcal{H})$, and both P and Q be nonempty compact convex subsets of $\mathcal{H}$. Let $\varphi_q \doteq A + P - q$ be a family of set-valued operators indexed by the tychastic variables $q \in Q$. Then $(\forall x \in C)(\exists y \in C)\, y - x \in Ay + (P \ominus Q)$ holds.*

Proof Since C and φ_q satisfy the conditions of the Viability Theorem 1, then C is viable for φ_q, i.e., according to the equivalent assertion (c): $(\forall x \in C)(\exists y \in C)(\forall q \in Q)\, y - x \in \varphi_q y$. Let us take any point $x \in C$, then there exists a point $y \in C$ such that

$$(\forall q \in Q)\, y - x \in \varphi_q y = Ay + P - q \text{ or } (\forall q \in Q)\, (y - Ay - x) \in P - q,$$

and, since $P - q$ is a closed convex subset, we have for all $u \in \mathcal{H}$

$$(\forall q \in Q)\, \langle y - Ay - x \mid u \rangle \leq \sup \langle P - q \mid u \rangle = \sup \langle P \mid u \rangle - \langle q \mid u \rangle,$$

or

$$\langle y - Ay - x \mid u \rangle \leq \sup \langle P \mid u \rangle + \inf \langle -Q \mid u \rangle = \sigma_P(u) - \sigma_Q(u),$$

which it is equivalent to

$$(\forall x \in C)(\exists y \in C)\, y - x \in Ay + (P \ominus Q),$$

the wished expression. □

Conclusions

Consider the following interpretation of Proposition 2, where Q represents the sets of values of the tychastic variables and P of the controlled variables ones. Thus, the larger the volume of the set P or the smaller the volume of Q, the velocity $y - x$ belongs to a larger set $P \ominus Q$ (note that Ay only displaces the set $P \ominus Q$). Therefore, the volume of the viability kernel increases as the volume of Q decreases or the volume P increases.

References

1. Aubin, J. P. (1997). Dynamic economic theory: A viability approach. *Studies in Economic Theory*. Springer. https://doi.org/10.2307/41794699
2. Aubin, J. P. (1998). *Optima and equilibria: An introduction to nonlinear analysis*. Springer
3. Aubin, J. P. (2013). Tychastic viability. *Acta Biotheoretica, 61*(3), 329–340. https://doi.org/10.1007/s10441-013-9194-4. 32nd Seminar of the French-Speaking Society for Theoretical Biology, Saint-Flour, France, June 10-13, 2012.
4. Aubin, J. P., Chen, L., & Durand, M. H. (2012). Dynamical allocation method of emission rights of pollutants by viability constraints under tychastic uncertainty. *Environmental Modeling & Assessment, 17*(1–2), 7–18. https://doi.org/10.1007/s10666-011-9272-4
5. Aubin, J., & Haddad, G. (2002). History path dependent optimal control and portfolio valuation and management. *Positivity, 6*(3), 331–358. https://doi.org/10.1023/a:1020244921138
6. Bauschke, H. H., & Combettes, P. L. (2017). Convex analysis and monotone operator theory in Hilbert Spaces. *CMS books in mathematics*. Springer. https://doi.org/10.1007/978-3-319-48311-5
7. Blanchini, F., & Miani, S. (2015). Set-theoretic methods in control. *Systems & Control: Foundations & Applications* (2nd edn.). Birkhäuser. https://doi.org/10.1007/978-3-319-17933-9
8. Clarke, P. H., Ledyaev, Y. S., Stern, R. J., & Wolenski, P. R. (1995). Qualitative properties of trajectories of control systems: A survey. *Journal of Dynamical and Control Systems, 1*(1), 1–48. https://doi.org/10.1007/bf02254655
9. Laengle, S. (2021). Articulating bargaining theories: Movement, chance, and necessity as descriptive principles. *Central European Journal of Operations Research, 29*(1), 49–71. https://doi.org/10.1007/s10100-020-00729-y
10. Oubraham, A., & Zaccour, G. (2018). A survey of applications of Viability Theory to the sustainable exploitation of renewable resources. *Ecological Economics, 145*, 346–367. https://doi.org/10.1016/j.ecolecon.2017.11.008

Part IX
Health Care Management

Chapter 45
Locating Relief Trains for Patient Transports in Case of Mass-Casualty Incidents

Florentina Hager and Melanie Reuter-Oppermann

Abstract In case of a mass-casualty incident with several hundred or even thousands of patients, providing fast medical treatments is one of the main goals. If close-by hospitals cannot provide sufficient capacities to treat all victims, they need to be transported to more remote hospitals. In these cases, mass transportation modes such as trains or ships could assist to promote faster transportation. However, to be useful, they must arrive at the site of the incident as fast as possible. Therefore, we present a stochastic mathematical model that simultaneously determines the optimal fleet size of relief trains to be stationed as well as the optimal locations to prepare for patient transport after mass-casualty events. We test our model in a case study in the German state of Bavaria.

Keywords Mass-casualty incidents · Patient transportation · Relief trains

Introduction

Mass-casualty incidents with several hundred or even thousands of patients are often very complex and time-critical. Then, ensuring the survival of as many patients as possible is the main goal. In order to achieve that, patients need to receive the first treatment as soon as possible and if necessary, must be transported to a hospital. In these scenarios, close-by hospitals might not be able to treat all patients, but patients might need to be transported to hospitals in other parts of the country. Then, different modes of transport like trains or ships would be necessary for transporting them. In order to start transport fast, the means of transport should be available as soon as possible. Therefore, countries should prepare for those incidents by strategically locating appropriate modes of transport.

F. Hager · M. Reuter-Oppermann (✉)
Informations Systems, Technical University of Darmstadt, Darmstadt, Germany
e-mail: oppermann@is.tu-darmstadt.de

O. Grothe et al. (eds.), *Operations Research Proceedings 2022*, Lecture Notes in Operations Research, https://doi.org/10.1007/978-3-031-24907-5_45

In this work, we present a stochastic programming model to determine the optimal fleet size as well as the optimal locations for relief trains to be stationed to prepare for patient transport after mass-casualty incidents. We use the German state of Bavaria as a case study in this work and build a set of scenarios with varying demands.

The remainder of this paper is structured as follows. In the next section, we present the relevant literature. We present our model in Section "Problem Description and Mathematical Formulation" followed by the case study in Section "Case Study". We conclude this paper in Section "Conclusion" with a short summary and outlook.

Foundations

Over the past years, literature has extensively addressed disasters management (e.g., [1]). While several authors illustrate the importance of considering relief trains, pointing out that "a significant portion of the population will need to be evacuated by mass-transportation vehicles" [2], relatively few authors address train location and allocation in their models. Bababeik et al. present a multi-objective model that maximises link exposure while minimising travel times [3]. Another model dealing with relief train location and allocation is presented by Tripathi et al. [4]. Similar to [3] they include link exposure (link importance) in their multi-objective model. Besides maximising link exposure, the objective aims to minimise travel times, while maximising overall coverage and node assignment. However, both models only consider locating facilities and do not incorporate upstream decisions. The literature for determining the minimal number of facilities in emergency cases (e.g., relief trains) dates back to 1971 when the set covering problem (SCP) was introduced by [5]. More recent models have been extended to also account for stochasticity as done by [6, 7], for example. Literature reviews on emergency facility locations have been provided by [8, 9], for example. This paper extends current literature by strategically locating relief trains while simultaneously considering the actual allocation of trains to potential mass casualty incidents.

Problem Description and Mathematical Formulation

For the problem description, we assume a specified region with potential demand points for relief trains. These demand points can be larger cities or other locations with a high number of injured people. The demand in each potential demand point depends on whether the location is hit by a disaster (which we assume with a certain probability), and if hit, by the number of injured patients to be transported out of the original location. Each location is weighted by a factor that should reflect a location's expected demand and medical resources. If a demand location shall be served by a relief train, the location must be reached within a specified threshold, otherwise, we assume that other transport modes are chosen. Both, the number of available relief trains as well as their location before the disaster occurs must be made before the demand is known (here-and-now). Once a disaster hits the region, the demand can be observed and the trains can be allocated to the demand points (wait-and-see).

To formulate the problem, we use the following notation. The set of potential locations for positioning relief trains is noted as I, while J describes the set of demand locations to be covered by relief trains. As described above, we assume the demand for relief trains to be stochastic. Following [7], we address this problem by a random variable ξ_j indicating the demand in each location j, and ζ_{ij} as the ride time between location i and j. We define a scenario as the full realisation of ξ as well as ζ, with the given probability for each scenario. Ω describes the set of potential scenarios, each scenario occurs with a probability of π_ω and results in a demand of relief trains $d_{j\omega}$ in location $j \in J$. The scenario depending ride time from $i \in I$ to $j \in J$ is noted as $t_{ij\omega}$ to account for potential track damages or closures as part of the incident. We use the following notation for the remaining parameters and the decision variables: T describes the maximum acceptable ride time, Tr the maximum number of trains located at a certain location. The weight of city $j \in J$ is denoted as w_j and M is an arbitrary high number. Decision variable $x_{ij\omega}$ takes the value of 1 if location $j \in J$ is covered by a relief train located at $i \in I$ in scenario $\omega \in \Omega$, else 0, while decision variable $y_{ij\omega}$ describes the number of relief trains located at $i \in I$ covering location $j \in J$ in scenario $\omega \in \Omega$. The number of relief trains located at $i \in I$ is described by z_i.

The model can be formulated as follows:

$$\text{maximise } f_1: \sum_{i\in I}\sum_{j\in J}\sum_{\omega\in\Omega} y_{ij\omega}\cdot w_j \cdot \pi_\omega \tag{1}$$

$$\text{minimise } f_2: \sum_{i\in I} z_i \tag{2}$$

$$\text{subject to } x_{ij\omega}\times t_{ij\omega} \leq T \qquad i\in I, j\in J, \omega\in\Omega \tag{3}$$

$$\sum_{i\in I} y_{ij\omega} \leq d_{j\omega} \qquad j\in J, \omega\in\Omega \tag{4}$$

$$x_{ij\omega} \leq y_{ij\omega} \qquad i\in I, j\in J, \omega\in\Omega \tag{5}$$

$$y_{ij\omega} \leq x_{ij\omega}\cdot M \qquad i\in I, j\in J, \omega\in\Omega \tag{6}$$

$$\sum_{j\in J} y_{ij\omega} \leq z_i \qquad i\in I, \omega\in\Omega \tag{7}$$

$$z_i \leq Tr \qquad i\in I \tag{8}$$

$$x_{ij\omega}\in\{0,1\},\ y_{ij\omega}, z_i \in \mathbb{N}_0 \qquad i\in I, j\in J, \omega\in\Omega \tag{9}$$

Thereby, the objective function f_1 maximises the weighted number of allocations, while the second objective function f_2 minimises the total number of relief trains. Constraints (3) ensure that each location is only covered if it can be reached within the specified threshold in each scenario, while by constraints (4) we restrict the number of allocated trains by the demand. A location is only considered as covered if at least one relief train is allocated (constraints (5)) and the location is admissible (constraints (6)), and the sum of allocated relief trains originating in location i must not exceed the number of relief trains positioned at location i (constraints (7)). Moreover, the maximum number of trains located at a potential location is restricted by constraints (8). Constraints (9) are domain constraints.

To deal with the multi-objective problem by applying the ϵ-constrained method, we solve the model iteratively only with objective f_1, start with one train to be located and add the following constraint, with f_2 representing the number of trains located in the previous run:

$$\sum_{i \in I} z_i \geq f_2 + \epsilon \tag{10}$$

Case Study

Data

To test the model in a case study, we assume all cities with a population of more than 100,000 inhabitants located in the state of Bavaria to be both, locations to be covered by a relief train as well as potential locations for positioning. According to the Statistisches Bundesamt (Federal Statistical Office), by the end of 2020, 8 cities fulfilled these conditions, thereof two cities with a population of more than 500,000 inhabitants [10]. We assume twelve different scenarios and approximate the travel time for relief trains by the travel time for freight trains according to [11], with additionally accounting for the required time to transport the casualties to the relief train. We estimate the demand for relief trains for each city based on the scenario (whether a city is hit by an emergency), the city's population and available hospital beds. For estimating the probability for each scenario, we consider a city's political exposure, logistical and industrial relevance and its population [12]. Except for the population, we approximate the four factors by the gross value added in the respective sectors. For the whole of Germany, a relief train should arrive within 24 hours from the beginning of the incident, including making the decision about using the trains and preparing them, leading to a maximum ride time of about 6 hours. In this work, the maximal acceptable ride time is set to 90 minutes due to the smaller area considered in this work. Depending on the number of wagons and the configuration, a relief train can transport between 42 critically injured and up to 1000 slightly injured patients. To calculate the weights for each city, we follow the equation for the risk index [13]:

$$RiskIndex = P(Likelihood) \times I(Impact) \tag{11}$$

Thereby, we approximate the impact by the number of inhabitants per hospital bed [14], while we estimate the likelihood by the city's population [10].

Results

The problem was implemented in Julia 1.7.3 and solved using GLPK Solver 5.0.0 on a Intel Core i7-1185G7 with 3 GHz and 16 GB RAM. The results were obtained after a maximal total computational time of 0.07 CPU seconds. The results are presented in Table 45.1. Note that Ingolstadt(2) means that 2 trains are located in Ingolstadt in that solution.

Sensitivity Analysis

To test the model's sensitivity to its parameters, we additionally ran the model with the following modifications: (1) Weighting each city equally (EqW), (2) weighting each city only based on its population (this modification equals the objective function of maximising the covered population, PopW), and (3) with equal probability for each scenario (EqProb).

In all sensitivity tests the same results are obtained for one to three relief trains, slight modifications can be observed when locating and allocating four to five relief trains (Table 45.2).

Table 45.1 Results with 1–5 relief trains

# of trains	Locations	Weighted pop. covered
1	{Ingolstadt}	1,365,705
2	{Ingolstadt(2)}	2,681,447
3	{München, Ingolstadt(2)}	3,993,742
4	{München, Augsburg, Ingolstadt(2)}	4,154,648
5	{München, Nürnberg, Augsburg, Ingolstadt(2)}	4,219,725

Table 45.2 Sensitivity analysis

4 relief trains		
Modification	Locations	Weighted pop. covered
EqW	{München, Nürnberg, Ingolstadt(2)}	3,687,499
PopW	{München, Augsburg, Ingolstadt(2)}	4,155,682
EqProb	{München, Augsburg, Ingolstadt(2)}	3,670,425
5 relief trains		
Modification	Locations	Weighted pop. covered
EqW	{München, Regensburg, Ingolstadt(2), Fürth}	3,783,700
PopW	{München, Regensburg, Ingolstadt(2), Augsburg}	4,224,779
EqProb	{München, Augsburg, Ingolstadt(2), Fürth}	3,728,037

Conclusion and Outlook

In this paper, we proposed a stochastic programming model in order to determine the optimal fleet size of relief trains and the optimal locations and allocations of relief trains to disaster sites. Thereby, we aimed to maximise the demand locations covered while keeping the fleet size moderate. To account for uncertainty, we proposed a scenario-based two-stage model, and tested the model for the German state of Bavaria. In the case study, we performed a sensitivity analysis with different weights and scenario probabilities.

Addressing the strategic planning level in this work, the presented model is based on the assumption of sufficient local transportation resources such that all casualties can be transported from the disaster site to the relief train. Also, the model does not account for any other transportation modes such as air crafts or busses. In reality, the complete casualty transport from the disaster site to the final destination (such as hospitals) must be considered, with efficient solutions incorporating multiple transport modes.

In the next step, we will build an instance for the whole of Germany and analyse the scenarios and results within a discrete-event simulation. Moreover, we plan to extend the model by considering the complete patient transport chain to address these limitations to better reflect reality. In addition, we aim to extend the model to include multiple transportation modes, e.g. airplanes, busses and ships, in order to provide decision support for decision makers in Germany where to locate which modes of transport in order to prepare for mass-casualty incidents.

References

1. Farahani, R. Z., Lotfi, M. M., Baghaian, A., Ruiz, R., & Rezapour, S. (2020). Mass casualty management in disaster scene: A systematic review of OR&MS research in humanitarian operations. *EJOR, 287*(3), 787–819.
2. Bayram, V. (2016). Optimization models for large scale network evacuation planning and management: A literature review. *Surveys in Operations Research and Management Science, 21*(2), 63–84.
3. Bababeik, M., Khademi, N., & Chen, A. (2018). Increasing the resilience level of a vulnerable rail network: The strategy of location and allocation of emergency relief trains. *Transportation Research Part E: Logistics and Transportation Review, 119*, 110–128.
4. Tripathi, G., Tanksale, A. N., & Verma, M. (2022). Optimal location of accident relief facilities in a railway network. *Safety Science, 146*, 105560.
5. Toregas, C., Swain, R., ReVelle, C., & Bergman, L. (1971). The location of emergency service facilities. *Operations Research, 19*(6), 1363–1373.
6. Beraldi, P., & Bruni, M. E. (2009). A probabilistic model applied to emergency service vehicle location. *EJOR, 196*(1), 323–331.
7. Nickel, S., Reuter-Oppermann, M., & Saldanha-da-Gama, F. (2016). Ambulance location under stochastic demand: A sampling approach. *Operations Research for Health Care, 8*, 24–32.
8. Reuter-Oppermann, M., van den Berg, P. L., & Vile, J. L. (2017). Logistics for emergency medical service systems. *Health Systems, 6*(3), 187–208.

9. Bélanger, V., Ruiz, A., & Soriano, P. (2019). Recent optimization models and trends in location, relocation, and dispatching of emergency medical vehicles. *EJOR, 272*(1), 1–23.
10. Statistisches Bundesamt: Städte (Alle Gemeinden mit Stadtrecht) nach Fläche, Bevölkerung und Bevölkerungsdichte am 31.12.2020. https://www.destatis.de/DE/Themen/Laender-Regionen/Regionales/Gemeindeverzeichnis/Administrativ/05-staedte.html. Accessed June 10, 2022
11. Railway Routing. https://signal.eu.org/osm/#locs=53.550341,10.000654;52.517037,13.388860. Accessed June 22, 2022
12. Bruttoinlandsprodukt, Bruttowertschöpfung in den kreisfreien Städten und Landkreisen der Bundesrepublik Deutschland. https://www.statistikportal.de/de/veroeffentlichungen/bruttoinlandsprodukt-bruttowertschoepfung-0. Accessed July 11, 2022
13. Nirupama, N. (2012). Risk and vulnerability assessment: A comprehensive approach. *International Journal of Disaster Resilience in the Built Environment, 3*(2), 103–114.
14. Verzeichnis der Krankenhäuser und Vorsorge- oder Rehabilitationseinrichtungen 2020. https://www.destatis.de/DE/Themen/Gesellschaft-Umwelt/Gesundheit/Krankenhaeuser/Publikationen/Downloads-Krankenhaeuser/krankenhausverzeichnis-leseprobe-3500100.html. Accessed June 23, 2022

Part X
Heuristics, Metaheuristics and Matheuristics

Chapter 46
A Hybrid Metaheuristic for the Clustered Travelling Salesman Problem

Abtin Nourmohammadzadeh and Stefan Voß

Abstract In this work, a special type of the travelling salesman problem (TSP), namely the clustered TSP (CTSP), is addressed. In the CTSP, the cities are already divided into clusters and the salesman seeks to find the shortest tour through all the cities which includes each city exactly once while being restricted to visit the cities of each cluster contiguously. Due to the NP-hardness of the focused problem, a hybrid metaheuristic consisting of the artificial bee colony (ABC) and the tabu search (TS) algorithm is proposed to deal with it. The results of our solution approach on two sets of benchmark instances are presented and compared with those of two other methods from the literature.

Keywords Clustered travelling salesman problem · Hybrid metaheuristic · Artificial bee colony algorithm · Tabu search

Introduction

The clustered travelling salesman problem (CTSP) is a very practical variant of the TSP, in which the cities are pre-clustered and presented in some groups. Like in the TSP, the salesman has to find the shortest tour which includes each city exactly once. However, the difference is that the salesman must visit the cities of each cluster directly after each other.

The CTSP has found applications in many fields such as automated warehouse routing, emergency vehicle dispatching, production planning, disk defragmentation, and commercial transactions with supermarkets, shops and grocery suppliers [13]. It means that finding efficient approaches to this problem provides precious solutions for the important analogous industrial problems.

A. Nourmohammadzadeh (✉) · S. Voß
Institute of Information Systems, University of Hamburg, Hamburg, Germany
e-mail: abtin.nourmohammadzadeh@uni-hamburg.de

S. Voß
e-mail: stefan.voss@uni-hamburg.de

O. Grothe et al. (eds.), *Operations Research Proceedings 2022*, Lecture Notes in Operations Research, https://doi.org/10.1007/978-3-031-24907-5_46

Although there has been a huge amount of research on the TSP leading to many powerful methods, available works on the CTSP are still limited. The CTSP is firstly introduced, modelled and solved by a branch and bound approach in [4]. Some other examples using exact solution approaches are [9, 16], while [1, 2] are two examples of applying approximation algorithms to the problem. A trend to use heuristics and metaheuristics for the CTSP is also evident. Some related works are: [7, 11, 15, 17]. Reference [13] is a recent work, which transforms the CTSP to the TSP by defining very high costs for edges with vertices belonging to different clusters, and then, applies a GA-EAX algorithm. By an overview of the related literature, one can notice that there is a considerable room for further research in the CTSP field, specially to develop fast efficient alternative metaheuristics. Therefore, in this work, we aim at investigating this problem and propose an efficient hybrid metaheuristic to provide good quality fast solutions.

The problem can be easily shown to be NP-hard based on the already proved NP-hardness of the TSP. This is due to the fact that the CTSP is equivalent to the TSP if it has only one cluster or if each of its clusters includes exactly one city (vertex).

We design a hybrid metaheuristic algorithm, which consists of an artificial bee colony algorithm (ABC) [10] as its framework and embedded local searches based on the tabu search (TS) [8] concept. The reason behind choosing this combination is the promising performance of the swarm intelligence algorithms on the efficient path finding problems in networks [14] and the verified searching ability of the TS for the CTSP [11]. Our method is examined on two sets of benchmark instances. The results of our hybrid approach are compared with a state-of-the-art metaheuristic called GA-EAX presented in [13], which has been proved to outperform its prior counterparts, and also with the exact Concorde solver [5], which is able to provide optimal solutions for the instances.

The organisation of the rest of this paper is as follows: Our solution methodology is explained in Section "Solution Methodology". Subsequently, Section "Computational Results" shows the numerical results and the related comparisons. Finally, the conclusions of this work and future directions are drawn in Section "Conclusions".

Solution Methodology

Our solution methodology consists of an ABC algorithm, in which the moves of bees are according to the TS principles. Therefore, it is denoted as ABC-TS.

The ABC algorithm is inspired by the behaviour of three types of bees namely employees, onlookers and scouts while searching for better food sources. The procedure of this algorithm is so that a number of food sources are randomly initialised at the beginning. Consequently, one employee bee is responsible for each source and searches around it. If it finds a better solution, it memorises the solution and forgets the previous one. The onlookers, whose number is equal to the number of employees, choose each a source (solution) from the best sources located by the employees probabilistically based on the quality of solutions. They also make some changes in the chosen solutions to improve them. If a source is not improved after a number of consecutive iterations, it is abandoned and replaced by a new random solution. This act is associated with scouts. The best found solution of the system is saved upon completing each iteration and the overall best is reported at the end of the algorithm as the final result.

The TS algorithm is a kind of guided local search to improve the solutions. The TS works based on two major rules at each step: 1. Worsening moves can be accepted in the case that no improving move is possible in order to avoid local optima, and 2. Some moves are prohibited or put in the tabu list in order to avoid returning to already visited solutions. Our hybridisation approach is to embed the TS concept in each iteration of the ABC algorithm for the neighbourhood search attempts of employee and onlooker bees. The termination condition of the overall hybrid algorithm is stagnation over a number of consecutive iterations.

The solution encoding structure is a string containing a permutation of the labels of all cities. This structure is translated to a feasible solution in a way that the salesman visits, firstly, the clusters, and then, their cities according to their order in the string. It means that the salesman begins from the first city of the string and its cluster, then completes visiting that cluster by passing through its other cities based on their order in the string. After this step, all the elements belonging to the visited cluster are removed from the structure and the previous procedure is repeated with the first city of the remaining structure and its cluster, which is the next cluster in the solution. This continues until all of the clusters and cities are visited. As a simple example, if there are six cities, the clusters are $\{2, 3, 5\}$, $\{1, 6\}$, $\{4\}$, and the encoding structure is 6-3-1-4-2-5, then it is converted to this order of visiting the cities (tour): 6-1—3-2-5—4 (— means cluster change).

A neighbourhood move of a bee in the ABC is exchanging the contents of two positions in the encoding structure and as it happens, the same pair of positions is not tried by that bee until a number of other moves are done. The main parameters of the whole hybrid algorithm are set by the response surface method (RSM) [3] specifically for each instance. As an example, the number of moves that a specific exchange remains in the tabu list (denoted as C) is an important parameter, which has an initial interval of [5, 50] in our parameter setting. The RSM determines a suitable integer value for this parameter within this interval once for each instance. The pseudocode of our hybrid ABC-TS approach is shown as Algorithm 1.

Computational Results

Two sets of test instances are used. Set 1 includes 15 CTSP instances which are built according to the data of benchmark TSP instances obtained from TSPLIB [18]. The number of clusters is $\lceil |V|/10 \rceil$ and the cities are clustered based on their coordinates by the K-means algorithm [12]. Other than by our ABC-TS algorithm, the instances are also processed by the Concorde solver [5] and GA-EAX [13], which has been already proved to be the best metaheuristic alternative for the CTSP. Concorde and

Algorithm 1: The proposed hybrid ABC-TS metaheuristic

Data: Problem inputs, parameters of the algorithm
Result: A good quality feasible solution
1 Generate F random candidate solutions as initial food sources.
2 Assign one employee bee e ($e \in EA$, EA: the set of employed bees, $|EA| = F$) to each food source.
3 $It = 0$, saves the number of consecutive iterations without improvement.
4 $TL = \emptyset$, tabu list.
5 **while** *$It \leq STI$ (STI:The maximum allowable number of stagnant iterations)* **do**
6 **for** $e \in EA$ **do**
7 **while** *$i \leq M$, M: the number of moves of the employee bees* **do**
8 Move e by exchanging the contents of two positions (genes) g_1 and g_2, $(g_1, g_2) \notin TL$ in the solution encoding structure, and then, include (g_1, g_2) in TL.
9 Evaluate the objetive value of the new solution.
10 **if** *the new solution is better than the previous one* **then**
11 The new solution is the new position of e.
12 Memorise the best solution of e so far.
13 Update TS by eliminating the pair which was executed C moves (rounds) ago.
14 $i = i + 1$
15 **else**
16 Go back to Line 8
17 **end**
18 **if** *There has not been any improving move* **then**
19 Choose one of the moves randomly
20 $i = i + 1$
21 Go to Line 8
22 **end**
23 **end**
24 Choose for each onlooker bee o ($o \in ON$, $|ON| = |F|$) a solution from the best results of all employee bees probabilistically according to the quality of solutions.
25 **for** $o \in ON$ **do**
26 **while** *$j \leq N$, N: the number of moves of the onlooker bees* **do**
27 Move o by exchanging the contents of two positions (genes) g_1 and g_2, $(g_1, g_2) \notin TL$ in the solution encoding structure, and then, include (g_1, g_2) in TL.
28 Evaluate the objetive value of the new solution.
29 **if** *the new solution is better than the previous one* **then**
30 The new solution is the new position of o
31 Memorise the best solution of o so far.
32 Update TL by eliminating the pair which was executed C moves (rounds) ago.
33 $j = j + 1$
34 **else**
35 Go back to Line 27
36 **end**
37 **if** *There has not been any improving move* **then**
38 Choose one of the moves randomly
39 $j = j + 1$
40 Go to Line 27
41 **end**
42 **end**
43 **end**
44 Update the food sources by eliminating those which are not improved over FSE consecutive iterations.
45 Replace the eliminated food sources by new random solutions (the act of scout bees).
46 **if** *No improvement is observed in the whole iteration* **then**
47 $It = It + 1$
48 **end**
49 **else**
50 It=0
51 **end**
52 **end**
53 **end**
54 Report the overall best solution.

Table 46.1 The results of Set 1 (best values regarding the comparison between GA-EAX and ABS-TS are shown in bold)

Instance	|V|	m	Concorde		GA-EAX			ABC-TS		
			Opt.	Time	LG	AG	Time	LG	AG	Time
bayg29	29	3	1610	0.8	**10T**	**0**	1.5	**10T**	**0**	**1.3**
att48	48	5	10,628	0.9	**10T**	**0**	1.5	**10T**	**0**	**1.3**
brazil58	58	6	25,395	1.0	**10T**	**0**	1.6	9T	0.0010	**1.4**
gr96	96	10	55,209	1.1	**10T**	**0**	1.7	8T	0.0017	**1.5**
gr120	120	12	6942	1.3	**9T**	0.0011	1.8	**9T**	**0.0008**	**1.6**
ch130	130	13	6110	1.4	8T	0.0018	1.9	**10T**	**0**	**1.7**
ch150	150	15	6528	1.6	7T	0.0032	2.1	**8T**	**0.0026**	**1.8**
d198	198	20	15,780	2.7	**8T**	**0.0037**	2.3	7T	0.0043	**1.9**
a280	280	28	2579	4.5	7T	0.0046	2.6	**9T**	**0.0012**	**2.2**
fl417	417	42	11,861	22.3	**10T**	**0**	3.0	**10T**	**0**	**2.5**
d493	493	50	35,002	28.9	**8T**	**0.0020**	3.2	7T	0.0028	**2.9**
d657	657	66	48,912	35.6	7T	0.0036	3.4	**8T**	**0.0030**	**3.2**
d1291	1291	130	50,801	72.8	**6T**	0.0061	5.5	**6T**	**0.0055**	**5.2**
fl1400	1400	140	20,127	126.7	**7T**	**0.0072**	5.7	6T	0.0084	**5.4**
fl1577	1577	158	22249	150.4	6T	0.0098	5.8	**7T**	**0.0090**	**5.6**
Average				30.1	8.2	0.0029	2.9	**8.3**	**0.0027**	**2.6**

GA-EAX are re-implemented by us for the instances of Set 1 through manipulating the edge weights and converting them to their equivalent TSP (see the method in [13]). All the algorithms are programmed in Python and some computers with a Core(TM) i7 processor, 3.10GHz CPU and 16GB of RAM are used.

Concorde is applied once to each instance, while GA-EAX and ABC-TS are run ten times for each instance due to the possibility of a different result in each run. Table 46.1 shows the obtained results. Its columns contain the instance name, number of cities, number of clusters, optimum result obtained by Concorde, its execution time in seconds, and the least and average optimality gap (LG and AG) as well as the average execution time (s) for the GA-EAX and ABC-TS, respectively. In case that the best result of the metaheuristic methods is equal to the optimal result (Opt.), the number of runs which reached the optimal result is shown in column "LG" and after that comes the letter "T". For example, 10T means that the optimal result is found by all of the 10 runs. Therefore, the related average gap (LG) is 0.

Set 2 contains the same instances solved in [13]. For this set, the available results of Concorde and GA-EAX reported in [13] are used and our ABC-TS is implemented. The results of this set are summarised in Table 46.2.

As it is evident from the results of the two sets, ABC-TS and GA-EAX can provide promising results. Except for one case with GA-EAX, both metaheuristics can find the optimal solutions obtained by the Concorde solver within their ten runs for all instances. The solution quality of GA-EAX and ABC-TS is quite the same, although the overall average gap of all instances is slightly lower for the ABC-TS. To better clarify this, the Wilcoxon signed-rank test [6] is conducted based on all runs of all

Table 46.2 The results of Set 2 (best values regarding the comparison between GA-EAX and ABS-TS are shown in bold)

Instance	$\|V\|$	*m*	Concorde		GA-EAX			ABC-TS		
			Opt.	Time	LG	AG	Time	LG	AG	Time
49-pcb1173	1173	49	61,600	5638.3	4T	0.0326	35.0	**5T**	**0.0274**	**32.7**
100-pcb1173	1173	100	63,382	588.3	**8T**	**0.0013**	32.5	6T	0.0041	**31.3**
144-pcb1173	1173	144	62,142	38.4	**10T**	**0**	18.6	9T	0.0011	**15.2**
10-nrw1379	1379	10	58,783	562.9	6T	0.0070	26.8	**8T**	**0.0051**	**23.9**
12-nrw1379	1379	12	59,129	58.5	9T	0.0007	27.6	**10T**	**0**	**25.8**
1500-10-503	1500	10	1116	65.5	**10T**	**0**	28.4	**10T**	**0**	**26.5**
1500-20-504	1500	20	15,698	40.7	5T	0.0172	34.5	**7T**	**0.0091**	**31.7**
1500-50-505	1500	50	22,900	67.0	5T	0.0044	35.1	**6T**	**0.0036**	**33.4**
1500-100-506	1500	100	29,799	108.7	**8T**	**0.0020**	39.5	7T	0.0021	**38.0**
1500-150-507	1500	150	34,068	114.7	**10T**	**0**	32.3	9T	0.0007	**30.1**
2000-10-a	2000	10	105,360	7214.3	0.0826	0.1167	45.3	**1T**	**0.1150**	**43.7**
2000-10-h	2000	10	33,708	812.7	**10T**	**0**	35.6	**10T**	**0**	**33.5**
2000-10-z	2000	10	33,509	200.9	9T	0.0003	37.3	**10T**	**0**	**35.3**
2000-10-x1	2000	10	33,792	1325.4	6T	0.0136	35.6	**8T**	**0.0061**	**32.6**
2000-10-x2	2000	10	33,509	170.9	**10T**	**0**	39.6	9T	0.0005	**37.1**
Average				1133.81	7.33	0.0131	33.58	**7.66**	**0.0117**	**31.4**

instances with the initial hypothesis that the average gaps of the two metaheuristics are equal. This hypotheses cannot be rejected at the significance level of 0.05 due to the obtained p-value of 0.56. Regarding the required computation time, we do not observe a sharp increase by larger instances with the metaheuristics. However, the execution time of the Concorde increases rapidly as it copes with instances containing more cities and clusters. Comparing the GA-EAX and ABC-TS in terms of the computation time, it can be stated that the latter is moderately faster in dealing with all of the instances. Here again the same statistical test is done, which results in the rejection of the hypothesis of the equal average execution times of the two methods due to a very low related p-value (= 0.003).

In general, it is deduced that our hybrid ABC-TS performs on the investigated CTSP instances at least as good as the best state-of-the-art metaheuristic. In addition, it is a little faster. The good performance of this hybrid algorithm is due to the compatibility of the ABC and TS algorithm with the CTSP and the hybridisation, which enables us to benefit from the advantages of both algorithms.

Conclusions

In this work, a very practical variant of the TSP called CTSP is investigated and a hybrid metaheuristic is devised for it. The results of numerous instances of different sizes indicate that our proposed solution methodology is capable of finding good quality solutions when they are compared with the known optimal results and the results of the previous best metaheuristic. Furthermore, its execution times are moderately shorter. This success owes to the agent-based searching mechanism, which is used in a swarm intelligence algorithm like ABC, and its suitability for the CTSP as well as the practicability of TS for efficient local search. For the continuation of this research direction, applying our method to larger CTSP benchmark instances and also other variants of the TSP can be considered.

References

1. Anily, S., Bramel, J., & Hertz, A. (1999). An approximation algorithm for the clustered traveling salesman tour and path problems. *Operations Research Letters, 24*(1–2), 29–35. https://doi.org/10.1016/S0167-6377(98)00046-7
2. Bao, X., & Liu, Z. (2012). An improved approximation algorithm for the clustered traveling salesman problem. *Information Processing Letters, 112*(23), 908–910. https://doi.org/10.1016/j.ipl.2012.08.020
3. Box, G. E. P., & Draper, N. R. (2007). *Response surfaces, mixtures, and ridge analyses* (2nd ed.). Wiley.
4. Chisman, J. A. (1975). The clustered traveling salesman problem. *Computers & Operations Research, 2*, 115–119. https://doi.org/10.1016/0305-0548(75)90015-5
5. Concorde Home. (2020, October 11). https://www.math.uwaterloo.ca/tsp/concorde.html
6. Conover, W. J. (1999). *Practical nonparametric statistics* (3rd ed.). Wiley.
7. Ding, C., Cheng, Y., & He, M. (2007). Two-level genetic algorithm for clustered traveling salesman problem with application in large-scale TSPs. *Tsinghua Science and Technology, 12*(4), 459–465. https://doi.org/10.1016/S1007-0214(07)70068-8
8. Glover, F. W., & Laguna, M. (2013). Tabu search. *Springer*. https://doi.org/10.1007/978-1-4615-6089-0
9. Jongens, K., & Volgenant, T. (1985). The symmetric clustered traveling salesman problem. *European Journal of Operational Research, 19*(1), 68–75. https://doi.org/10.1016/0377-2217(85)90309-1
10. Karaboga, D., & Basturk, B. (2007). A powerful and efficient algorithm for numerical function optimization: Artificial bee colony (ABC) algorithm. *Journal of Global Optimization, 39*(3), 459–471. https://doi.org/10.1007/s10898-007-9149-x

11. Laporte, G., Potvin, J. Y., & Quilleret, F. (1997). A Tabu search heuristic using genetic diversification for the clustered traveling salesman problem. *Journal of Heuristics, 2*(3), 187–200. https://doi.org/10.1007/BF00127356.
12. Lloyd, S. (1982). Least squares quantization in PCM. *IEEE Transactions on Information Theory, 28*(2), 129–137. https://doi.org/10.1109/TIT.1982.1056489
13. Lu, Y., Hao, J. K., & Wu, Q. (2022). Solving the clustered traveling salesman problem via traveling salesman problem methods. *PeerJ. Computer Science, 8*, e972. https://doi.org/10.7717/peerj-cs.972
14. Mathew, A. T., Paul, A., Rojan, A., & Thomas, A. (2021). Implementation of swarm intelligence algorithms for path planning. *Journal of Physics: Conference Series, 1831*(1), 012008. https://doi.org/10.1088/1742-6596/1831/1/012008
15. Mestria, M. (2018). New hybrid heuristic algorithm for the clustered traveling salesman problem. *Computers & Industrial Engineering, 116*, 1–12. https://doi.org/10.1016/j.cie.2017.12.018
16. Mestria, M., Satoru Ochi, L., & de Lima Martins, S. (2013). GRASP with path relinking for the symmetric Euclidean clustered traveling salesman problem. *Computers & Operations Research, 40*(12), 3218–3229. https://doi.org/10.1016/j.cor.2012.10.001
17. Potvin, J. Y., & Guertin, F. (1996). The clustered traveling salesman problem: A genetic approach. In *Meta-heuristics* (pp. 619–631). Springer. https://doi.org/10.1007/978-1-4613-1361-8_37
18. TSPLIB: MP-testdata—The TSPLIB symmetric traveling salesman problem instances (2021). http://elib.zib.de/pub/mp-testdata/tsp/tsplib/tsp/index.html

Chapter 47
A Study of Scalarisation Techniques for Multi-objective QUBO Solving

Mayowa Ayodele, Richard Allmendinger, Manuel López-Ibáñez, and Matthieu Parizy

Abstract In recent years, there has been significant research interest in solving Quadratic Unconstrained Binary Optimisation (QUBO) problems. Physics-inspired optimisation algorithms have been proposed for deriving optimal or sub-optimal solutions to QUBOs. These methods are particularly attractive within the context of using specialised hardware, such as quantum computers, application specific CMOS and other high performance computing resources for solving optimisation problems. Examples of such solvers are D-wave's Quantum Annealer and Fujitsu's Digital Annealer. These solvers are then applied to QUBO formulations of combinatorial optimisation problems. Quantum and quantum-inspired optimisation algorithms have shown promising performance when applied to academic benchmarks as well as real-world problems. However, QUBO solvers are single objective solvers. To make them more efficient at solving problems with multiple objectives, a decision on how to convert such multi-objective problems to single-objective problems need to be made. In this study, we compare methods of deriving scalarisation weights when combining two objectives of the cardinality constrained mean-variance portfolio optimisation problem into one. We show significant performance improvement (measured in terms of hypervolume) when using a method that iteratively fills the largest space in the Pareto front compared to a naïve approach using uniformly generated weights.

Keywords Digital Annealer · QUBO · Multi-objective optimisation

M. Ayodele (✉)
Fujitsu Research of Europe, Slough, UK
e-mail: mayowa.ayodele@fujitsu.com

R. Allmendinger · M. López-Ibáñez
The University of Manchester, Manchester, UK
e-mail: richard.allmendinger@manchester.ac.uk

M. López-Ibáñez
e-mail: manuel.lopez-ibanez@manchester.ac.uk

M. Parizy
Fujitsu Limited, Kawasaki, Japan
e-mail: parizy.matthieu@fujitsu.com

O. Grothe et al. (eds.), *Operations Research Proceedings 2022*, Lecture Notes in Operations Research, https://doi.org/10.1007/978-3-031-24907-5_47

Introduction

In recent years, there has been significant research interest in solving Quadratic Unconstrained Binary Optimisation (QUBO) formulations of optimisation problems. This is a common formulation used by hardware solvers classified as quantum or quantum-inspired machines. They have been shown to achieve a speed up compared to classical optimisation algorithms implemented on general purpose computers [2]. Ising machines such as Fujitsu's Digital Annealer (DA) [7] and D-wave's Quantum Annealer [9] are single objective solvers. Many optimisation problems however have more than one objective, e.g. the Cardinality Constrained Mean-Variance Portfolio Optimisation Problem (CCMVPOP) [5] considered in this study entails selecting assets that maximise returns while minimising the associated risks. Typically, multi-objective problems are converted to single objective problems before the Ising machines are applied to them. For example, the ϵ-constraint approach was used in the Quantum Annealer to solve a portfolio optimisation problem [11]. Scalarisation has also been used when solving multi-objective QUBO in previous work [4, 12]. One of the main challenges to using scalarisation is how to define a set of weights resulting in a diverse set of solutions on the Pareto front (PF). A common approach is to generate weights uniformly using, for example, the simplex lattice design [12]. However, a uniform choice of weights does not necessarily translate to a diverse set of Pareto-optimal solutions [8, 12]. Previous studies have therefore also considered iterative method which uses a dichotomic procedure to derive new weights perpendicular to two solutions that have the largest distance between them [6, 8]. In this study, we propose a method for deriving scalarisation weights which targets less explored regions of the PF. The proposed method utilises the weights used during previous scalarisations in addition to the relative position of the corresponding solutions in the PF and relies less on the weights and fitness being perfectly correlated.

The following section presents the problem description of the CCMVPOP. Methods of generating scalarisation weights used in this study are described in Section "Scalarisation Methods". Results and conclusions are presented in Sections "Results" and "Conclusion".

Cardinality Constrained Mean-Variance Portfolio Optimisation Problem

Portfolio Optimisation entails selecting assets that maximise returns while minimising the associated risks. In the CCMVPOP [5], cardinality constraints on the number of asset types to be considered are imposed. Given the number of asset types to consider (n), the fixed number of assets a portfolio must contain (K), the expected return of asset i (μ_i) and the covariance between assets i and j ($\sigma_{i,j}$), the minimum (ϵ_i) and maximum (δ_i) proportion of a chosen asset i, we aim to find the proportion of each asset i to hold ($w_i \in [0, 1]$). Binary variables z_i are used to indicate whether an asset i is selected or not. The CCMVPOP is formally defined as follows.

$$\text{minimise} \quad \lambda_1 \left(\sum_{i=1}^{n} \sum_{j=1}^{n} w_i w_j \sigma_{i,j} \right) + \lambda_2 \left(-\sum_{i=1}^{n} w_i \mu_i \right) \tag{1}$$

$$\text{subject to} \quad \sum_{i=1}^{n} w_i = 1, \quad \sum_{i=1}^{n} z_i = K \tag{2}$$

$$\epsilon_i z_i \le w_i \le \delta_i z_i, \quad z_i \in \{0, 1\}, \quad i = 1, \dots, n \tag{3}$$

The first objective is the first term in Eq. (1) and minimises the risk (sum of covariance between all pairs i, j of chosen assets) of the chosen assets of the portfolio. The second objective is the second term in Eq. (1) and maximises returns (sum of expected return of each asset i) of chosen assets. A negative sign is appended to the second objective to convert it to a minimisation problem. $\lambda = (\lambda_1, \lambda_2)$ is a set of scalarisation weights. The cardinality constraint (Eq. 2) forces the number of chosen assets to be equal to K, and Eq. (3) ensures the proportion of a chosen asset w_i to be within given bounds. The QUBO formulation of the CCMVPOP ($K = 10, \epsilon_i = 0.01$ $\delta_i = 1$) used is based on the binary representation presented in [10].

Scalarisation Methods

In [4], a scalarisation framework, Scalarisation Based DA (SB-DA), was proposed for obtaining multiple non-dominated solutions for the bi-objective quadratic assignment problem formulated as QUBO. A CPU implementation of the 1st generation DA algorithm [1] was used in that study. However, in this study, we use the 3rd generation DA [7] which is designed to be faster and more efficient than previous generations of the DA, it also benefits from hardware speedup [7]. For simplicity, we use DA to refer to 3rd generation DA in the rest of this work.

We propose two extensions of the SB-DA, which we call SB-DA*s* and SB-DA*i* (Algorithm 1). Parameters B, D and G are QUBO matrices representing the first objective, second objective and constraint functions, respectively. The number of scalarisation weights is denoted by k and $time$ is the total time allowed for all DA executions. To allow more solutions to be considered for non-dominance, n_top is a parameter used to define the number of top solutions (solutions with the lowest

Algorithm 1 SB-DA Algorithm

Require: $B, D, G, k, time, n\_top, s\_type$
1: $\Lambda \leftarrow \{(0, 1), (1, 0)\}$
2: **if** s_type in {*random*, *uniform*} **then** Mode ← static **else** Mode ← iterative
3: **if** s_type is *random* **then** add $k - 2$ sets of random weights to Λ
4: **if** s_type is *uniform* **then** $\Lambda \leftarrow \text{SLD}(H = k, m = 2)$
5: **if** Mode is static **then** $A \leftarrow$ execute SB-DA*s* **else** $A \leftarrow$ execute SB-DA*i*
6: **return** all non-dominated solutions from archive A

Algorithm 2 SB-DAi

Require: $B, D, G, k, time, n\_top, \Lambda$

$A \leftarrow \emptyset$, $W \leftarrow \{\}$ ▷ Initialise archive and mapping between weights and solutions
for each $i \in \{1, \ldots, k\}$ **do**
  if $i \leq 2$ **then**
    $\lambda = (\lambda_1, \lambda_2) \leftarrow \Lambda_i \quad R, S \leftarrow B, D$
  else
    $R, S \leftarrow rescale(B, D)$
    $\lambda \leftarrow \emptyset, max\_d \leftarrow 0$ ▷ Initialise weights and maximum distance
    for $j \in [1, i-2]$ **do**
      $d \leftarrow Distance(W_j, W_{j+1})$ ▷ Manhattan distance
      $[(\lambda_1^{sol1}, \lambda_2^{sol1}), sol1], [(\lambda_1^{sol2}, \lambda_2^{sol2}), sol2] \leftarrow W_j, W_{j+1}$
      $\lambda\_temp \leftarrow \left(avg(\lambda_1^{sol1}, \lambda_1^{sol2}), avg(\lambda_2^{sol1}, \lambda_2^{sol2})\right)$
      if $(d > max\_d)$ and $(\lambda\_temp \notin W)$ **then** $\lambda \leftarrow \lambda\_temp, max\_d \leftarrow d$
    end for
    if $\lambda == \emptyset$ **then** $\lambda \leftarrow$ Random weights ▷ each set of weights sums to 1
  end if
  $Q \leftarrow (\lambda_1 \cdot R + \lambda_2 \cdot S) + \alpha \cdot G$
  $Y \leftarrow$ ExecuteDA($Q, n\_top, time\_limit = \frac{T}{k}$), add all solutions in Y to A
  $W_i \leftarrow [\lambda, Y_0]$ ▷ save weight and best solution in Y
end for
return A

Algorithm 3 SB-DAs

Require: $B, D, G, k, time, n\_top, \Lambda$

$A \leftarrow \emptyset$ ▷ Initialise archive
for each $i \in \{1, \ldots, k\}$ **do**
  $\boldsymbol{\lambda} = (\lambda_1, \lambda_2) \leftarrow \Lambda_i$
  if $i > 2$ **then** $R, S \leftarrow rescale(B, D)$ **else** $R, S \leftarrow B, D$
  $Q \leftarrow (\lambda_1 \cdot R + \lambda_2 \cdot S) + \alpha \cdot G$
  $Y \leftarrow$ ExecuteDA($Q, n\_top, time\_limit = \frac{T}{k}$), add all solutions in Y to A
end for
return A

energies) to be returned during each DA execution. In this study, we compared three methods of deriving scalarisation weights, s_type set to *random*, *uniform* or *iterative*. Where s_type is set to *random*, k sets of randomly generated weights are pre-computed. For each set of weights $\lambda = (\lambda_1, \lambda_2)$, λ_1 is a random value between range [0, 1] while $\lambda_2 = 1 - \lambda_1$. For s_type set to *uniform* method, k sets of evenly distributed weights are pre-computed. In this study, we use the Simplex Lattice Design (SLD) (Line 4 of Algorithm 1) to generate evenly distributed weights. Where s_type is *iterative*, weights are derived with the aim of finding solutions that fall within the less crowded region of the Pareto front. To achieve this aim, a mapping between each set of weights and the best solution found by the DA using such set of weights are stored as W. W is sorted in ascending order of λ_1. For any two adjacent solutions in W, the Manhattan distance between the solutions are recorded. Solutions $sol1$ and $sol2$ that correspond to the largest gap in the Pareto front are saved. A set of scalarisation weights used to derive $sol1$ and $sol2$ are $\lambda^{sol1} = (\lambda_1^{sol1}, \lambda_2^{sol1})$

and $\lambda^{sol2} = (\lambda_1^{sol2}, \lambda_2^{sol2})$ respectively. An average of λ^{sol1} and λ^{sol2} becomes the scalarisation weights used in the new iteration (Lines 9–11). The new set of weights is however only used if this has not been used in previous iterations (Line 12). If there is no unique set of weights that can be derived using this procedure, randomly generated weights are used.

SB-DA can be executed in one of two modes. It is executed in static mode (SB-DAs: Algorithm 3), if s_type is *random* or *uniform* and in iterative mode (SB-DAi: Algorithm 2) if s_type is set to *iterative*. In SB-DAs, scalarisation weights are pre-computed while in SB-DAi, a set of scalarisation weights at a given iteration is influenced by the set of scalarisation weights used in previous iterations. An iteration of SB-DA refers to a run of the DA with a given set of scalarisation weights. In both modes of the SB-DA, each objective is optimised independently ($\lambda = (0, 1)$ and $\lambda = (1, 0)$) before other weights are used. This is because QUBO matrices B and D are rescaled using information about the Lower Bound (LB) and Upper Bound (UB). These bounds are achieved by minimising each objective independently. LB of B (or D) is derived by minimising B (or D) independently. Conversely, UB of B (or D) is derived by minimising D (or B) independently. The LB or UB are updated if smaller or larger energies are found for any individual objective at any iteration of the SB-DA. In Line 4 of Algorithm 3 and Line 6 of Algorithm 2, $rescale(B, D)$ is computed such that $R = \max_{1 \leq i \leq k}(\mathrm{UB}_i)/(\mathrm{UB}_1 - \mathrm{LB}_1) \cdot B$ and $S = \max_{1 \leq i \leq k}(\mathrm{UB}_i)/(\mathrm{UB}_2 - \mathrm{LB}_2) \cdot D$. This is done to reduce bias towards any of the objectives, allowing the algorithm to control the bias using the scalarisation weights only. QUBO matrix Q (Algorithm 3: Line 5, Algorithm 2: Line 16) is an aggregate of QUBO matrices representing the objectives and constraint, penalty weight (α) is set using Maximum change in Objective function divided by Minimum Constraint function of infeasible solutions (MOMC) originally proposed in [3]. ExecuteDA (Algorithm 3: Line 6, Algorithm 2: Line 17) runs DA on Q for $time\_limit$ seconds, returning the non-dominated solutions found amongst the best n_top solutions.

Results

To generate the results presented in this section, default parameters of the DA are used. Table 47.1 shows that the *uniform* method consistently found the highest number of non-dominated solutions while the *iterative* method consistently found the lowest number of non-dominated solutions across all problem instances. However, using the proposed *iterative* method consistently led to the highest hypervolume. Higher hypervolume values were reached because the iterative method was able to find weights that allowed the algorithm to focus on harder and more extreme regions of the search space.

Table 47.1 Average and standard deviation number of non-dominated solutions and hypervolume across 20 runs of the DA (stopping criteria: $0.05n$, number of weights: 10)

Problem instance (n)	Scalarisation method	Mean ± Stdev non-dominated solutions	Mean ± Stdev hypervolume	Problem instance (n)	Scalarisation method	Mean ± Stdev non-dominated solutions	Mean ± Stdev hypervolume
Port1 (248)	Random	324 ± 50	37.06 ± 14.38	Port3 (712)	Random	628 ± 132	121.64 ± 25.30
	Uniform	**354** ± 39	28.71 ± 9.50		Uniform	**866** ± 51	94.27 ± 1.90
	Iterative	250 ± 25	**60.37** ± 0.22		Iterative	417 ± 71	**136.46** ± 1.45
Port2 (680)	Random	514 ± 82	159.92 ± 7.79	Port4 (784)	Random	552 ± 112	117.79 ± 25.35
	Uniform	**676** ± 61	155.38 ± 0.86		Uniform	**798** ± 49	116.73 ± 1.23
	Iterative	275 ± 50	**164.44** ± 0.92		Iterative	398 ± 49	**136.42** ± 2.25

Hypervolume values have been divided by 10^{23}. Reference points used for computing the hypervolume values are the sum of all positive QUBO coefficients in B and D $\left(\sum_{i=1}^{m}\sum_{i=1}^{m}\max\{0, B_{ij}\}, \sum_{i=1}^{m}\sum_{i=1}^{m}\max\{0, D_{ij}\}\right)$. Best values presented in bold (Student t-test used for test of significance)

Conclusions

This study compared three simple methods of generating scalarisation weights within the context of bi-objective QUBO solving. The methods were applied to QUBO formulation of the CCMVPOP. We show that considering more than one best solution during each scalarisation can lead to finding more non-dominated solutions. We also show that for this problem, higher hypervolume can be reached by using adaptive methods of generating scalarisation weights when compared to random or evenly distributed weights.

References

1. Aramon, M., Rosenberg, G., Valiante, E., Miyazawa, T., Tamura, H., & Katzgraber, H. G. (2019). Physics-inspired optimization for quadratic unconstrained problems using a digital Annealer. *Frontiers in Physics, 7*, 48.
2. Ayodele, M. (2022). Comparing the digital Annealer with classical evolutionary algorithm. arXiv preprint arXiv:2205.13586
3. Ayodele, M. (2022). Penalty weights in QUBO formulations: Permutation problems. In *European Conference on Evolutionary Computation in Combinatorial Optimization (Part of EvoStar)* (pp. 159–174). Springer.
4. Ayodele, M., Allmendinger, R., López-Ibánez, M., & Parizy, M. (2022). Multi-objective QUBO solver: Bi-objective quadratic assignment problem. In *Proceedings of the Genetic and Evolutionary Computation Conference, GECCO '22*. ACM Press. https://doi.org/10.1145/3512290.3528698
5. Chang, T. J., Meade, N., Beasley, J., & Sharaiha, Y. (2000). Heuristics for cardinality constrained portfolio optimisation. *Computers and Operations Research, 27*(13), 1271–1302.
6. Dubois-Lacoste, J., López-Ibáñez, M., & Stützle, T. (2011). Improving the anytime behavior of two-phase local search. *Annals of Mathematics and Artificial Intelligence, 61*(2), 125–154.
7. Hiroshi, N., Junpei, K., Noboru, Y., & Toshiyuki, M. (2021). Third generation digital Annealer technology. https://www.fujitsu.com/jp/documents/digitalannealer/researcharticles/DA_WP_EN_20210922.pdf
8. Liefooghe, A., Verel, S., Paquete, L., & Hao, J. K. (2015). Experiments on local search for bi-objective unconstrained binary quadratic programming. In *International Conference on Evolutionary Multi-criterion Optimization* (pp. 171–186). Springer.
9. McGeoch, C., & Farré, P. (2020). *The d-wave advantage system: An overview* (Tech. Rep.). D-Wave Systems Inc.
10. Parizy, M., Sadowski, P., & Togawa, N. (2022). Cardinality constrained portfolio optimization on an ising machine. In *2022 IEEE 35th International System-on-Chip Conference (SOCC) (SOCC 2022)*, Belfast, United Kingdom (Great Britain).
11. Phillipson, F., & Bhatia, H. S. (2021). Portfolio optimisation using the D-wave quantum Annealer. In M. Paszynski, D. Kranzlmüller, V. V. Krzhizhanovskaya, J. J. Dongarra, & P. M. A. Sloot (Eds.), *Computational Science—ICCS 2021* (pp. 45–59). Springer.
12. Zhou, Y., Wang, J., Wu, Z., & Wu, K. (2018). A multi-objective Tabu search algorithm based on decomposition for multi-objective unconstrained binary quadratic programming problem. *Knowledge-Based Systems, 141*, 18–30.

Chapter 48
Low Budget Traveling: The Orienteering Problem with Hotel Selection and Budget Constraint

Paul Päprer and Benedikt Zipfel

Abstract In this paper, we consider the orienteering problem with hotel selection (OPHS) and introduce an additional cap on the available budget, frequently experienced in practice but neglected in the literature. We present a heuristic solution approach for the modified problem configuration, which comprises the construction of initial solutions and the improvement of those solutions using a multi-start VNS heuristic with adaptive adjustment. Within computational studies on adjusted benchmark data, we evaluate the impact of the considered budget constraint by comparing the results of the original OPHS with the results of its budget constraint extension. Further, we show the efficiency of the proposed metaheuristic.

Keywords Routing · Orienteering problem · Capacity restricted · Multi-trip

Introduction

The orienteering problem with hotel selection (OPHS) represents a variant of the classical orienteering problem (OP) that considers planning multi-day tours with consecutive trips. Given a complete graph $G = (\mathcal{V}, \mathcal{E})$, the set of vertices $\mathcal{V}$ can be split into two disjoint subsets. There are $H + 1$ hotels ($i = 0, \ldots, H$) and N nodes ($i = H + 1, \ldots, H + N$), which represent the attractions to be visited. While each of the latter vertices has a score s_i that reflects the preferences, hotels have no score. Attractions can only be visited once, whereas hotels can be included multiple times. The tour consists of D connected trips, which start and end at one of the available hotels. The maximum trip length for each trip d is defined by parameter T_d. Within this context, the problem aims to find the tour with the maximum total score in

P. Päprer (✉) · B. Zipfel
Department of Business Administration, esp. Industrial Management,
TU Dresden, Dresden 01069, Germany
e-mail: paul.paepprer@tu-dresden.de
URL: https://tu-dresden.de/bu/wirtschaft/bwl/lim

B. Zipfel
e-mail: benedikt.zipfel@tu-dresden.de

O. Grothe et al. (eds.), *Operations Research Proceedings 2022*, Lecture Notes in Operations Research, https://doi.org/10.1007/978-3-031-24907-5_48

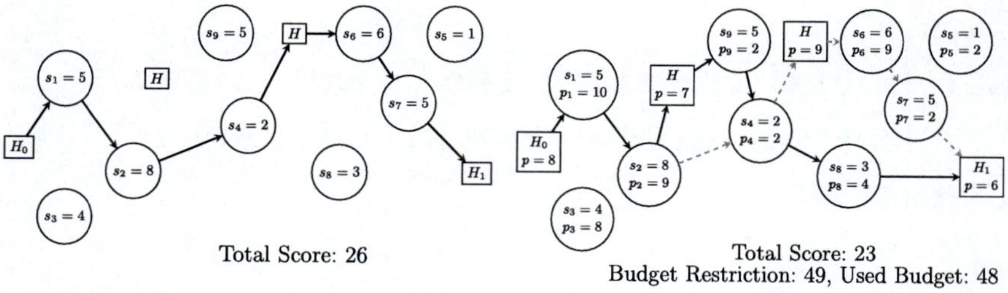

(a) Exemplary solution of the OPHS (b) Exemplary solution of the OPHSBC

Fig. 48.1 Illustration of exemplary tours of the OPHS and the OPHSBC

accordance with the individual trip lengths [1]. Initially motivated by tourists' travel planning, who try to include as many points of interest as possible in their time-limited itinerary, the problem can be applied to other decision problems like routing for maintenance technicians or truck drivers. To give a better understanding of the problem configuration, Figure 48.1a presents an exemplary tour with two trips in a graph with four hotels and nine attractions.

Given this problem configuration for the original OPHS, we consider a new variant, which introduces a cap on the available budget B for the overall tour. Consequently, every hotel and attraction gets assigned a price p_i. We refer to this variant as the orienteering problem with hotel selection and budget constraint (OPHSBC). The extension does not only increase the applicability of the OPHS in the context of travel planning but also its practicality in other applications. For example, maintenance technicians may have limited capacity for operating resources and materials in their vehicles. Figure 48.1b shows a tour resulting from the OPHSBC based on the previous example. The budget cap is set to 49, which makes the solution of the OPHS in Figure 48.1a infeasible. Accordingly, the tour gets adjusted in both trips to meet the required budget restriction.

The OPHSBC has not yet been studied in the literature. In [1], Divsalar et al. consider the basic OPHS for the first time. The authors propose an algorithm to construct initial solutions and a variable neighborhood search. Further, a memetic algorithm for the OPHS is presented in [2]. Sohrabi et al. present a Greedy Randomized Adaptive Search Procedure (GRASP) and an Ant Colony Optimization Procedure in [3] and [4], respectively. To the best of our knowledge, only one extension to the original OPHS exists today, which takes into account the opening hours of attraction by considering time windows [5].

Similar approaches to the considered OPHSBC exist in the area of the Team Orienteering Problem (TOP), where all trips have the same start and end point [6]. Thus, the resulting tour does not consist of consecutive trips. In this research field, capacitated TOP (CTOP) is probably the problem with most similarities [7]. Based on the presented observations, the contribution of this study is as follows:

- We present a new budget constraint for the OPHS, which enforces the trade-off between profit and the consumption of resources.
- We propose a multi-start variable neighborhood search (MSVNS) with adaptive sorting procedures, which is inspired from the relevant literature.
- We analyze the impact of budget variations on the objective value.

A Multi-start Variable Neighborhood Search

A mathematical model for the OPHS is given in [1]. To model the OPHSBC, we adapted this formulation and add Eq. (1), which ensures that the cumulated prices of all attractions and hotels visited does not exceed the budget limit.

$$p_0 + \sum_{d=1}^{D} \sum_{i=0}^{H+N} \sum_{j=0}^{H+N} x_{ijd} \cdot p_j \leq B \tag{1}$$

The OPHSBC is a variant of the original OPHS and therefore builds a generalization of the common OP. Due to the NP-hardness of the OP, OPHSBC is also NP-hard [6]. Based on the complexity of the considered problem configuration, we propose a heuristic solution approach presented to generate good solutions with appropriate computational effort. Our algorithm consists of two phases: a construction phase and an improvement phase.

Algorithm 1: Overall Solution Approach

```
1  function ConstructionPhase(data)
2      rate all possible hotel pairs
3      evaluate all feasible hotel sequences
4      build feasible solutions
5  return Set S of starting solutions
6  function ImprovementPhase(S, data)
7      S* = ∅
8      for s ∈ S do in parallel
9          s* = VariableNeighborhoodSearch(s, data)
10         S* = S* ∪ s*
11 return best solution s* ∈ S*
```

Construction Phase. The construction phase is built based on the algorithm of [1] with problem-specific adjustments. As can be seen in lines one to five of Algorithm 1, it is a three-step process that tries to find good, feasible, and diversified starting solutions. First, we rate all possible hotel pairs to hint at how good a trip between those hotels can presumably be in terms of the score. In contrast to [1], we refer to a virtual trip with an average value for the available time and budget when solving the classical orienteering problem between two hotels. Furthermore, our heuristic orders all attractions by a rating. Using this rating, nodes with the best trade-off between score, distance, and price get inserted first. As a result, all hotel pairs have a potential score. Second, we evaluate all feasible hotel sequences by adding up the potential scores of the previous step to receive an estimated score for each sequence. Afterward, the best hotel sequences are passed to the last step. Here we build feasible solutions by ordering the attractions by score and inserting them in the best possible positions of the tour with regards to the time consumption. Finally, the outcome of the construction phase is set $\mathcal{S}$ of feasible starting solutions. Each solution is a permutation of numbers, each of which represents an attraction or a hotel.

Improvement Phase. After constructing initial solutions, we aim to improve these solutions with a multi-start variable neighborhood search (MSVNS). The procedure of the MSVNS is shown in lines six to eleven of Algorithm 1. By multi-start, we mean that we initiate an independent run of the VNS for each $s \in \mathcal{S}$. The best found solution for each run is stored within set $\mathcal{S}^*$. VARIABLENEIGHBORHOODSEARCH represents the function of our VNS procedure. The heuristic is developed using several different shaking mechanisms and neighborhoods. Regarding the latter, we consider three different neighborhoods, which are all based on the Extract-k-Insert principle. In this neighborhood, we first remove k connected attractions from the solution. Afterward, we try to insert as many previously excluded nodes as possible. Within the VNS, we consider the neighborhoods with $k = 1, 2, 3$. In every iteration, the procedure randomly considers one of the neighborhoods Extract-2-Insert and Extract-3-Insert, calculating the probabilities for the choice adaptively depending on the success rate. After this, neighborhood Extract-1-Insert always tries to improve the solution by applying smaller changes. When inserting the excluded attractions, they are ordered by score. In case several nodes have the same score values, it is necessary to break the ties. Therefore, we use another adaptive procedure, which analyses the current tour's tightness regarding its trip length and budget consumption. If the time condition is the bottleneck, attractions with the same score are ordered by increasing distances to the current tour. If the current cost consumption prevents a better score, we sort the attractions with identical scores by increasing prices. In the shaking step, we consider three procedures, one of which is randomly applied to the current solution in each iteration.

1. Make a random move in one Extract-k-Insert neighborhoods ($k = 1, 2, 3$).
2. Reverse a random partial chain of three attractions and delete excess nodes.
3. Remove all attractions, change one hotel, and refill the tour with attractions.

After each iteration, a solution has been changed using one shaking mechanism as well as two local search neighborhoods. If the objective value of the resulting new solution is better than the objective value of the current solution, it is always accepted to become the new base solution for the next iteration. To enable new solution space areas in further iterations, we also accept solutions that are slightly worse compared to the current objective value. This is handled by multiplying a factor r to the current best objective value. The VNS terminates, if there are $i_{\max}$ iterations without improving the overall best solution.

Computational Experiments

We implement the MIP formulations of the OPHS by [1] and the MIP formulation of the OPHSBC with GUROBI 9.5.1 using its PYTHON API. We also use PYTHON for the implementation of the proposed MSVNS. All tests are run on an Intel® Xeon® Gold 6136 with 3.00 GHz clock speed and 128 GB RAM. As test data, we use SET1, SET2, and SET3 from [1], who provide their data online. To obtain suitable problem instances for the OPHSBC, we choose an integer price $p_i \in [0, 100]$ for each attraction, and a price $p_i \in [100, 150]$ for each hotel. Further, we solve each instance of the original OPHS and calculate the consumed budget using the aforementioned prices. Then we reduce this budget by 20%.

We split our computational experiments into two parts. First, we compare the solutions of the original OPHS and the OPHSBC that result from solving the GUROBI model with a time limit of 7200 s. Table 48.1 presents the aggregated test results for each set of data. The table shows the properties of the data first. Furthermore, we distinguish in column *opt.* whether instances were solved optimally or not. Columns #, Gap, and $\overline{RT}$ describe the number of optimally and not optimally solved instances, the average Gap compared to the upper bound, and average runtime. For every data set the number of non-optimally solved instances for the OPHSBC is higher than for the OPHS. Furthermore, the average computational runtime for the optimally solved instances of the OPHSBC is higher in five out of seven sets. These facts indicate that the additional restriction of the OPHSBC may make the problem harder to solve. However, a general statement about the solution behavior needs more evaluation since introducing a budget constraint can also reduce the solution space and thus make the problem easier to solve.

Second, we evaluate the performance of the proposed MSVNS in comparison with the MIP. Deduced from preliminary testing, we choose parameter values $r = 0.85$ and $i_{\max} = 100$ for the MSVNS. The number of starting points is set to 4. Table 48.2 displays again aggregated results for both solution approaches. To compare the procedures, the table describes for the MIP in columns z, RPD, and $\overline{RT}$ the average values for the objective value, the relative percentage deviation, and the wall clock runtime, respectively. In terms of the MSVNS, we show the same values, additionally distinguishing the best and average results with stars and bars, respectively. It can be seen that the MSVNS yields in better results than the MIP with lower computational effort for all sets. By using a more runtime-efficient programming language than PYTHON, the computational effort of MSVNS may even be reduced by a factor of up to 30 [8].

Table 48.1 Comparison of the original OPHS and the extended OPHSBC

Data set	$\|H\|$	D	*opt.*	OPHS			OPHSBC		
				#	*Gap*	$\overline{RT}$ (in s)	#	*Gap*	$\overline{RT}$ (in s)
SET1 1-2	3	2	True	27	0.00	110.75	24	0.00	197.88
			False	8	0.08	7200.00	11	0.07	7200.00
SET1 2-3	4	3	True	28	0.00	467.09	23	0.00	1010.41
			False	7	0.04	7200.00	12	0.07	7200.00
SET1 3-4	5	4	True	27	0.00	657.06	22	0.00	926.31
			False	8	0.04	7200.00	13	0.09	7200.00
SET2 5-3	7	3	True	27	0.00	781.65	26	0.00	847.17
			False	8	0.06	7200.00	9	0.06	7200.00
SET2 6-4	8	4	True	28	0.00	1282.94	18	0.00	968.12
			False	7	0.02	7200.00	17	0.08	7200.00
SET3 10-4	12	4	True	5	0.00	1377.34	4	0.00	2073.01
			False	17	0.15	7200.00	18	0.10	7200.00
SET3 12-5	14	5	True	3	0.00	2668.76	2	0.00	876.85
			False	19	0.17	7200.00	20	0.12	7200.00

Table 48.2 Aggregated results of the computational study

Dataset	$\|H\|$	D	MIP			MSVNS					
			z	RPD (in %)	$\overline{RT}$ (in s)	z^*	$\overline{z}$	RPD^* (in %)	$\overline{RPD}$ (in %)	RT^* (in s)	$\overline{RT}$ (in s)
SET1 1-2	3	2	670.11	0.00	2398.55	671.63	670.63	0.20	0.10	14.38	18.30
SET1 2-3	4	3	659.63	0.00	3132.56	660.46	659.62	0.12	0.04	18.01	23.66
SET1 3-4	5	4	642.17	0.00	3256.54	652.97	650.14	1.25	0.94	18.68	24.36
SET2 5-3	7	3	666.77	0.00	2480.75	667.69	665.46	0.07	−0.20	17.18	24.02
SET2 6-4	8	4	641.57	0.00	3995.03	651.74	643.69	1.33	0.04	19.58	25.39
SET3 10-4	12	4	924.41	0.00	6267.82	952.91	942.28	2.97	1.81	116.41	166.24
SET3 12-5	14	5	896.00	0.00	6625.17	927.00	915.33	3.66	2.21	120.10	162.24

Conclusion

This study extends the classical OPHS with an additional budget constraint and proposes a new MSVNS to solve adjusted benchmark instances. Our computational test results indicate that the problem may become harder to solve due to the additional budget constraint. Future work should investigate this observation more precisely. Further, the evaluations show that we can achieve better results than the commercial solver GUROBI with less computational effort with our heuristic. Future studies can take this work as a base for additional extensions with practical relevance. Another interesting point would be developing other neighborhoods that simultaneously address all restrictions.

References

1. Divsalar, A., Vansteenwegen, P., & Cattrysse, D. (2013). A variable neighborhood search method for the orienteering problem with hotel selection. *International Journal of Production Economics, 145*(1), 150–160.
2. Divsalar, A., Vansteenwegen, P., Sörensen, K., & Cattrysse, D. (2014). A memetic algorithm for the orienteering problem with hotel selection. *European Journal of Operational Research, 237*(1), 29–49.
3. Sohrabi, S., Ziarati, K., & Keshtkaran, M. (2020). A greedy randomized adaptive search procedure for the orienteering problem with hotel selection. *European Journal of Operational Research, 283*(2), 426–440.
4. Sohrabi, S., Ziarati, K., & Keshtkaran, M. (2021). ACS-OPHS: Ant colony system for the orienteering problem with hotel selection. *EURO Journal on Transportation and Logistics, 10*(100), 036.
5. Divsalar, A., Vansteenwegen, P., Chitsaz, M., Sörensen, K., & Cattrysse, D. (2014). Personalized multi-day trips to touristic regions: A hybrid GA-VND approach. In *European Conference on Evolutionary Computation in Combinatorial Optimization* (pp. 194–205). Springer.
6. Vansteenwegen, P., & Gunawan, A. (2019). *Orienteering problems*. Springer.
7. Ben-Said, A., El-Hajj, R., & Moukrim, A. (2019). A variable space search heuristic for the capacitated team orienteering problem. *Journal of Heuristics, 25*(2), 273–303.
8. Lion, D., Chiu, A., Stumm, M., & Yuan, D. (2022). Investigating managed language runtime performance: Why JavaScript and Python are 8x and 29x slower than C++, yet Java and Go can be faster? In *2022 USENIX Annual Technical Conference (USENIX ATC 22)* (pp. 835–852).

Part XI
Logistics

Chapter 49
A Generalized Approach for Train Marshalling

Elias Dahlhaus

Abstract In a distributer freight train, its cars have to be sorted in such a way that during the trip of the train, only cars on the rear the train are left at any intermediate stop. The sorting requirements of the freight cars of a distributer train will be described in this paper. Then an algorithm is introduced that minimizes the number of tracks that are necessary to get the freight cars sorted in a hump yard, such that the sorting requirements are fulfilled. The algorithm is a dynamic programming approach and generalizes the algorithm of Falsafain and Tamannaei.

Keywords Train marshalling · Dynamic programming · PQ-tree

Introduction

The train marshalling problem in hump yards is a quite important problem to sort the cars of a distributer train that serves several stations to deliver cars. In the mathematical literature, the following problem is considered [2, 4, 7, 11]. We are given an inbound sequence of cars and, for each such car, its final destination. The problem is to transfer the inbound sequence into an outbound sequence, such that cars of the same destination appear consecutively and the number of classification tracks that are needed to transfer the inbound sequence to the outbound sequence in one sorting step are minimized.

The problem is also interesting if the number of available tracks is fixed. In Hansmann [9] and Jacob et al. [10] it has been shown, how to transfer an inbound sequence to an outbound sequence in logarithmically many steps dependent on the number of tracks that would be needed for sorting in one step.

Above train marshalling problem is only a special case of the real problems one has to deal with.

1. Each car has to be left in a final destination track

E. Dahlhaus (✉)
Department of Computer Science, Darmstadt University of Technology, Darmstadt, Germany
e-mail: dahlhaus@algo.informatik.tu-darmstadt.de

O. Grothe et al. (eds.), *Operations Research Proceedings 2022*, Lecture Notes in Operations Research, https://doi.org/10.1007/978-3-031-24907-5_49

2. Some customers would like that the cars left in the destination track appear in a fixed sequence
3. Destination stations are served in a fixed sequence and when a destination is reached, cars determined for that destination have to appear at the rear of the train.

These aspects show that.

1. Groups of cars may be divided into subgroups that should appear consecutively
2. Subgroups of a group may have to appear in a fixed sequence.

In Dahlhaus [3] (and mentioned also in [5]) it has been shown how to model the sorting requirements as described above by directed PQ-trees. They are root directed trees with P-nodes and Q-nodes. Children of P-nodes may be permuted in any how. The sequence of children of Q-nodes is fixed. Originally, PQ-trees were introduced by Booth and Lueker [1] to model different representations of interval graphs.

In this paper, we show how to extend the algorithm of [7] (that considers only disjoint subsets of cars that have to be grouped) to directed PQ-trees to transfer an inbound sequence into an outbound sequence that satisfies the sorting requirements with a minimum number of classification tracks. In "Transfering the Inbound Sequence of Cars into the Required Outbound Sequence" its is described, how to transfer an inbound sequence to an outbound sequence in a hump yard. In "Modelling the Sorting Requirement", it is described, how the sorting requirements are modelled by directed P-Q-trees. In "The Optimization Algorithm", the algorithm to find an optimal outbound sequence is introduced. In "Conclusions", the applicability of the algorithm will be discussed. This includes also the possible integration of known heuristics into the algorithm. An extension or a use of the approach in [2] and [8] to directed PQ-trees might also be of interest.

Transfering the Inbound Sequence of Cars into the Required Outbound Sequence

Here we assume that each car has a fixed inbound position and a fixed outbound position. To transfer the inbound sequence of cars into the outbound sequence of cars, cars to be put into a certain track have to have consecutive outbound positions and increasing inbound positions. Let π be the permutation that maps the outbound position of each car to its inbound position. Then we consider the maximal subintervals of outbound positions where π is monotonically increasing, These subintervals are also called **chains**. Two consecutive chains are separated by a pair $(i, i + 1)$, where $\pi(i) > \pi(i + 1)$. Such a pair is called a **jump**. Note that the number of tracks that are needed to transfer the inbound sequence of cars into its outbound sequence is the number of the chains of the corresponding permutation and that the number of jumps is the number of chains–1.

Modelling the Sorting Requirement

Grouping Freight Cars by Their Final Destination Tracks, Final Destination Station etc

At the lowest level, cars are grouped by their final destination tracks. The customer might require or not require that the cars are ordered in a fixed sequence. Next any harp of destination tracks that is served from one direction is collected to one group. The tracks of a harp of parallel tracks can be served in any how. One level higher, one has to distinguish between harps or tracks to be served in the direction of the distributer train and those harps or tracks to be served against the direction of the distributer train. Since cars are pushed (and not drawn) into their final destination tracks, the group of cars to be pushed in direction of the distributer train has to appear behind the group of cars to be pushed against the direction of the distributer train. In the next level above, there is the group of cars that have a certain destination station. Destination stations have to be ordered in the reversal sequence as they are served by the distributer train if there are no branchings. In the level above, cars are grouped by the branches of the destination stations.

Modelling the Sorting Requirements by Directed PQ-Trees

There might be even more complicated structures. But in any how,

1. Two groups are either disjoint or one group is contained in the other
2. The immediate subgroups of a group are in a fixed order or they can be ordered in any how.

That means that the groups are ordered tree like. The children of any group are its immediate subgroups. The leaves are the single cars. There are two kinds of inner nodes.

1. The P-nodes: children are ordered in any how
2. The Q-nodes: the ordering sequence of children is fixed. Note that the orderings of the children of all nodes induce an ordering of the leaves. Following the notation of [1], P-nodes are expressed by cycles and Q-nodes are expressed by boxes.

We consider the following toy example (Figs. 49.1 and 49.2).

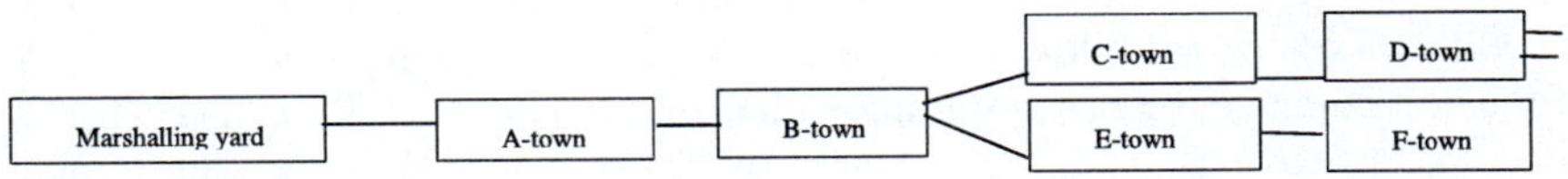

Fig. 49.1 A toy example

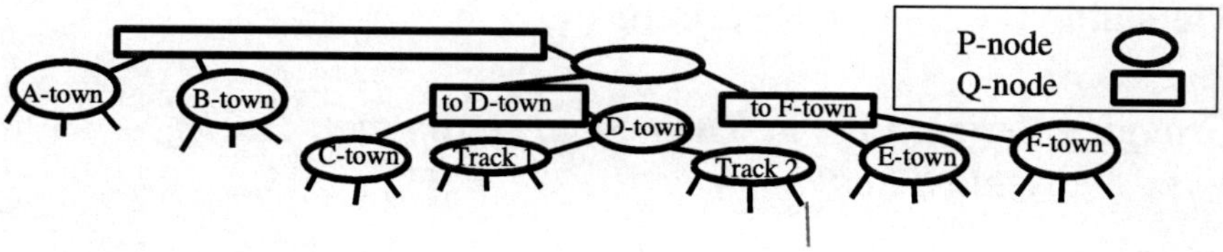

Fig. 49.2 The directed PQ-tree

There is a train from the marshalling yard to B-town and a connecting train to D-town via C-town and a connecting train to F-town through E-town. D-town has two parallel destination tracks. All other stations have one destination track. One gets the following PQ-tree.

A less trivial example can be found in [3].

The Optimization Algorithm

There are several papers considering the case that the directed PQ-tree consists of a P-node with P-nodes as children that have leaves as children. Note that finding the optimal outbound ordering compatible with the given directed PQ-tree is NP-complete [4]. There are several dynamic programming approaches that are exponential in the number of the children of the upper P-node, i.e. exponential in the number of car groups to be put together. One approach is that of [7] as sketched below.

We are given a set {1,…,n} of cars (1, … n, are the inbound positions) and a set **B** of pairwise disjoint subsets of {1,…,n} representing the groups of cars that have to be put together. It is to remark that this corresponds to a PQ-tree that has a P-node as root and each child is again a P-node that correspond to a set B ∈ **B**. The children of such a P-node are the elements of B. For each **B ‘** ⊂ **B** and for each i = 0,1,…,n, we determine the **minimum number** F(**B ‘**,i) **of tracks (chains)**, that is needed to transfer i together with the cars of **B ‘** to a sequence starting with i, such that all elements(cars) belonging to the same B∊ **B ‘** appear consecutively. For formal reasons, a car at position 0 is provided that does not belong to any B in **B** and remains at the beginning also of the outbound sequence. The algorithm of Falsafain and Tamannaei [7] determines F(**B ‘**,i), i.e. the minimum number of tracks to transfer i together with the cars of. **B ‘** to a sequence starting with i, such that all elements(cars) belonging to the same B∊ **B ‘** appear consecutively, as follows.

All **B ‘** ⊂ **B** are sorted with respect to the size in increasing order and F(**B ‘**,i) is determined for all **B ‘** in sorted order and for all inbound positions I as follows. If **B ‘** = Ø then F(**B ‘**,i) = 1. Otherwise, for any B ∈ **B ‘**, the last element i ‘ = i ‘(B,i) of the best outbound sequence starting with i followed by B is determined. That is either the last element in B if i is smaller than all elements of B or the last j ≤ i that is in B. Afterwards the number of tracks is determined that are needed to have first I, then B and then **B ‘**-{B}, i.e. F(**B ‘**-{B}, i ‘(B,i)) if i is smaller than all elements of B, otherwise it is F(**B** '-{B}, i '(B,i)) + 1. The B ∊ **B** ' is selected such that the number of tracks is minimum.

For PQ-trees in general, we proceed as follows.

Nodes are denoted by N, N1, N2. Inbound positions are denoted by i, j, k. Subsets of the children of a certain node are denoted by **B** '.

1. We always can transform a directed PQ-tree into a directed PQ-tree with the property that each Q-node has only two children
2. For each node N and each i, we consider any sorting starting with i and followed by the leaves being descendents of N, such that the sequence of leaves is appropriate to the PQ-subtree with N as its root. We pick a such a sorting, such that in the first range, the number of tracks (jumps) is minimized and in the second range, the inbound position of the last element of the resulting outbound sequence is minimized. These parameters are denoted by Track(N,i) and Last(N,i) respectively. We can also minimize the number of jumps instead to minimize the number of tracks. The minimum number of jumps for N and i is denoted by Jump(N,i). We denote the pair (Jump(N,i), Last(N,i)) by F(N,i)
3. If N is a P-node, we proceed in the same spitit as in the algorithm of Falsafain and Tamannaei [7] and consider any subset **B'** of the children of N and extend F to pairs (**B'**,i).

The algorithm is now as follows.

- Sort the nodes of the PQ-tree in reverse order with respect to the distance of the root or determine a postorder sorting on the nodes of the PQ-tree
- Execute, for each PQ-tree node in sorted order and for each i = 0,.....,n (each outbound position including 0).

    1. For a leaf node N in inbound position j, Jump(N,i) = 0, if i < j and Jump(N,i) = 1 otherwise. In both cases, Last(N,i) = j. F(N,i) is set to be (Jump(N,i), Last(N,i))
    2. For a Q-node N with children N1 and N2, Jump(N,i) = Jump(N1,i) + Jump(N2,Last(N1,i)) and Last(N,i) = Last(N2,Last(N1,i)). F(N,i) is set to be (Jump(N,i), Last(N,i))
    3. For each subset **B'** of the children of the P-node N in sorted order with respect to the size, and each inbound position i, we proceed as follows.
        a. If **B'** = {N'} has only one element then F(**B'**,i) = (Jump(**B'**,i), Last(**B'**,i) = F(N',i).
        b. Otherwise, for each N' ∊ **B'**, execute c.-f.
        c. we provide a new Q-node N"(N') with N' as its first child and a Q-node $N_{\mathbf{B}'-\{N'\}}$ with the elements of **B'**-{N'}as its children as its second child.
        d. We determine as in 2. F(N"(N), i) = (Jump(N"(N), i), Last(N"(N), i)), where Jump(N"(N), i) = Jump(N',i) + Jump(**B'**-{N'}, Last(N',i)) and Last(N"(N), i) = Last(**B'** –{N'},Last(N',i))).
        e. F(**B'**,i) is the F(N"(N), i) = (Jump(N"(N), i), Last(N"(N), i)), such that Jump(N"(N), i) is minimal in the first range and Last(N"(N), i) is minimal

in the second range. chosen, such that Jump(**B'**, i) is minimal in the first range and the value of Last(**B'**,i) is minimal in the second range.

f. If **B'** is the set of all children of N, F(N,i) = F(**B'**,i).

We observe the following.

Lemma 1 *F(N,i) as computed by the algorithm is the optimum.*

Sketch of proof Consider any optimal outbound sequence. When we consider a node N, we assume that all nodes left from N have been transformed into an outbound sequence that is determined by the algorithm. If the number of jumps under N does not decrease then the inbound position of the last car under N does not increase. Otherwise the number of jumps of the next node to be considered might increase by 1. In both cases, the overall number of jumps does not increase. *Q.E.D.*

The algorithm runs in time exponentially dependent on the maximum number of children of a P-node and polynomially dependent on the number of cars (leaves) if the number of children of any P-node is bounded by a fixed number.

Conclusions

The last algorithm becomes efficient if the number of children of any P-node is restricted by a small number. In practice, P-nodes with a large number of children appear only in the case that final destinations are harps of many parallel tracks or if many cars have to be left at the same track. That means that these P-nodes are of level at most two from the bottom. In so far it appears to be possible that heuristics as in [6] can be adapted to the PQ-tree model. This has to be considered in a separate paper.

References

1. Booth, K. S., & Lueker, G. S. (1976).Testing for the consecutive ones property, interval graphs, and graph planarity using PQ-tree algorithms. *Journal of Computer and System Sciences. 13*(3), 335–379
2. Brueggeman, L., Fellows, M., Fleischer, R., Lackner, M., Komusiewicz, C., Koutis, Y., Pfandler, A., & Rosamond, F. (2012). Train marshalling is fixed parameter tractable. In E. Kranakis, D. Krizanc, & F. Luccio (Eds.), *Fun with Algorithms* (pp. 51–56). Berlin, Heidelberg: Springer Berlin Heidelberg
3. Dahlhaus, E. (2009). The structure of sequencing requirements of distributer freight trains, Proceedings of EURNEX-ZEL 2009, Zilina
4. Dahlhaus, E., Horak, P., Miller, M., & Ryan, J. (2000). The train marshalling problem. *Discrete Applied Mathematics, 103*(1–3), 41–54.
5. Dahlhaus, E., Manne, F., Miller, M., & Ryan, J. (2000). Algorithms for combinatorial problems related to train marshalling. In *Proceedings of AWOCA 2000*, Hunter Valley, pp. 7–16
6. Dörpinghaus, J., & Schrader, R. (2018) A graph-theoretic approach to the train marshalling problem. In *Proceedings of the Federated Conference on Computer Science and Information Systems* pp. 227–231

7. Falsafain, H., & Tamannaei, M. (2020). A novel dynamic programming approach to the train marshalling problem. *IEEE Transactions on Intelligent Transportation Systems, 21*, 701–710.
8. Ganbaatar, E. (2014). Implementierung von Algorithmen zur Optimierung von Rangierabläufen in Ablaufberganlagen, Master Thesis (German)
9. Hansmann, R. (2010). *Optimal sorting of rolling stock*. Doctoral Thesis Braunschweig University of Technology
10. Jacob, R., Marton, P., Maue, J., & Nunkesser, M. (2011). Multistage methods for freight train classification. *Networks, 57*, 87–105.
11. Rinaldi, F., & Rizzi, R. (2017). Solving the train marshalling problem by inclusion-exclusion. *Discrete Applied Mathematics, 217*, 685–690.

Chapter 50
A Genetic Algorithm for the Multi-compartment Vehicle Routing Problem with Stochastic Demands and Flexible Compartment Sizes

Shabanaz Chamurally and Julia Rieck

Abstract The multi-compartment vehicle routing problem (MC-VRP) consists of designing a set of routes to perform the collection of different product types from customer locations with minimal costs. The MC-VRP arises in several practical situations, such as selective waste collection or different color of glass collection. Compartment sizes can be either set as fixed or as flexible. Often in practice, the collection quantity from customers is stochastic in nature, that is, the exact value is not available during route planning and is known only once the vehicles are at the customers' locations. Our work introduces the MC-VRP with stochastic customer demands and with flexible compartment sizes. We propose a genetic algorithm (GA) to solve this problem and investigate the benefits of setting the compartment sizes to be flexible instead of fixed with pre-defined sizes. By using flexible compartment sizes, the GA shows an overall average improvement of 7.8%, compared to the state-of-the-art approach for fixed compartment sizes.

Keywords Multi-compartment · Stochastic demands · Flexible compartment sizes · Genetic algorithm

Introduction

In the multi-compartment vehicle routing problem (MC-VRP), supplies of different product types have to be collected from a set of customers and transported to a central depot. The problem consists of designing a set of routes to perform the collection of different product types from the customers with minimal costs. The products in this problem are incompatible with each other (e.g., different-colored of glass waste, temperature sensitive food or building materials) and must be transported separately. Vehicles with multiple compartments allow products to be transported together and thus offer a greater flexibility in routing decisions. Using multi-compartment vehicles

S. Chamurally (✉) · J. Rieck
Operations Research Group, Institute of Business Administration and Information Systems, University of Hildesheim, Universitätsplatz 1, Hildesheim 31141, Germany
e-mail: chamurally@bwl.uni-hildesheim.de

O. Grothe et al. (eds.), *Operations Research Proceedings 2022*, Lecture Notes in Operations Research, https://doi.org/10.1007/978-3-031-24907-5_50

has particular advantages when products of different types have to be picked up from individual customers (see [4], for a more recent review).

Compartment sizes in the MC-VRP can be either fixed or flexible [1]. If flexible compartment sizes are used, the vehicle can be divided into multiple compartments. Thus, the number of compartments also remains flexible. To solve such problems, [2] developed a genetic algorithm (GA) which showed promising results. In practice, the collection quantity from customers is most often stochastic in nature, that is, the exact value is not available during route planning and is known only once the vehicles are at the customers' locations. Therefore, there is a strong motivation to extend the MC-VRP to handle this type of uncertainty, since neglecting this stochastic aspect can lead to substantial increase in cost. Incorporating the uncertainty in the customer demands to MC-VRP leads to the MC-VRP with stochastic demands (MC-VRPSD). One common way to handle the uncertainties in customers' request quantity is to implement a-priori policies, where the vehicles visit customers in a pre-defined order and return to the depot for replenishment or unloading of products, when a *route failure* occurs. Particularly, a route failure happens when the capacity of a vehicle is inadequate to fully collect the product quantities from the next customer. Mendoza et al. [3] explored methods to approximate the expected cost of an a-priori policy for the MC-VRPSD and reported that a *take-all* approximation yields good estimates of the expected cost. The *take-all* approximation assumes that even if a route failure occurs while servicing a customer i, the whole demands for all products of i are picked up before returning to the depot. However, after unloading the compartments at the depot, the vehicle must return back to i before travelling to the next customer.

In what follows, we consider the MC-VRPSD and for practical as well as economical reasons, we also take the flexible compartment sizes (MC-VRPSD-FCS) into account. Previous research has shown the successful application of GA in solving the MC-VRP-FCS and, hence, we solve our problem using a GA as in [2]. Moreover, we analyse the benefits of using flexible compartment sizes over fixed compartment sizes.

Problem Description and Formulation

The MC-VRPSD-FCS can be formally defined as follows: Let an undirected, weighted graph $G = (V, E)$ be given, where $V = \{0, \ldots, n\}$ is the vertex set consisting of the depot (node 0) and n customer locations, and $E = \{(i, j) : i, j \in V, i < j\}$ represents the set of edges which can be traversed between locations. A travel cost $c_{ij} \geq 0$, $(i, j) \in E$, and a distance $d_{ij} \geq 0$ are assigned to each edge. There exists a set P which contains the different product types that must be transported in independent compartments. For product p, customer v has an independent stochastic demand $\xi_{v,p}$ that follows a known probability distribution with mean $\mu_{v,p}$ and standard deviation $\sigma_{v,p}$. The actual demand realizations are non-negative and less than the vehicle capacity Q. However, the exact values of the demands are only known when the vehicle arrives at the customer location. All vehicles are identical in capacity and the fleet size is unlimited. Each vehicle can be further divided into a limited number of compartments $m \leq |P|$ and the size of each compartment can be set arbitrarily between 0 and Q. The vehicles can carry all the different types of products, but in separate compartments, which is decided based on the assignment of the customer demands to vehicles. Since a vehicle might not carry all product types at the same time, a customer can be visited by more than one vehicle. Moreover, the total length of each route should not exceed a maximum distance L. We assume that the variable travel costs are strongly correlated with the travel distance.

The MC-VRPSD-FCS can be formulated as a two-stage stochastic programming with recourse (SPR) model. In the first stage (planning stage), a set R of a-priori routes is designed, where each route r consists of a sequence of customers and starts and ends at the depot. In addition, it must be determined which vehicle serves which customers and which product types are assigned to which vehicle.

In the second stage (execution stage), each route r is executed until a route failure occurs, that is, until the capacity of the vehicle is exceeded. Then, the vehicle is loaded up to its capacity at the current customer v and afterwards a recourse action takes place. Here, the vehicle drives back to the depot, is unloaded there and returns to customer v empty. The solution of the second-stage is defined as the set of routes traveled by the vehicles and consists of the costs for the planned routes and the recursive trips.

The objective of the considered MC-VRPSD-FCS is to assign all supplies to multi-compartment vehicles and to determine a set of routes so that the expected costs of the transportation plan $\mathbb{E}[C(R)]$ in the first stage is minimized. Thus, we obtain:

$$\mathbb{E}[C(R)] = \sum_{r \in R} l_r + \sum_{r \in R} \mathbb{E}[G_r(\overrightarrow{\xi})], \tag{1}$$

where l_r is the planned route length (planned cost) which is the sum of the distance covered by the arcs traversed, $\overrightarrow{\xi}$ is the demand realization, and $G_r(\overrightarrow{\xi})$ is the length of the recourse trips caused by route failures. Since the total travel distance of a route r is a random variable, the maximum distance constraints are thus modelled as:

$$l_r + \mathbb{E}[G_r(\overrightarrow{\xi})] \leq L \quad \forall r \in R \tag{2}$$

The expected cost of recourse $\mathbb{E}[G_r(\overrightarrow{\xi})]$ of route r can be computed from the distance between the respective customer $v_i \in r$ and the depot and then back to v_i multiplied by the probability of failure $Pr(v_i)$ while servicing customer v_i:

$$E[G_r(\overrightarrow{\xi})] = \sum_{i=1}^{n_r} 2 \cdot d_{v_i,0} \cdot Pr(v_i) \tag{3}$$

Similar to the multi-compartment scenario as in [3], the failure probability while servicing customer v_i in the MC-VRPSD-FCS depends on the location where the last failure occurred and the sum of demands collected since then. As the compartment sizes are flexible in this problem, we need to ensure that the total expected demand collected does not exceed the vehicle capacity. Let $x_{r,p}$ be a binary variable that represents whether route r has a vehicle carrying product p. Using the *take-all policy* (TAP) as in [3], $Pr(v_i)$ can be computed in terms of the failure probability before servicing customer v_i and after servicing customer v_i and is given as follows:

$$
\begin{aligned}
Pr(v_i) = \sum_{j=0}^{i-1} \Bigg[\prod_{p \in P} Pr\left(\sum_{k=j+1}^{i} \xi_{v_k,p} \leq Q \right) \cdot x_{r,p} \\
- \prod_{p \in P} Pr\left(\sum_{k=j+1}^{i-1} \xi_{v_k,p} \leq Q \right) \cdot x_{r,p} \Bigg] \cdot Pr(v_j), \forall v_i \in r
\end{aligned} \tag{4}
$$

The TAP assumes that even if a route failure occurs at customer v_i, the entire demand of v_i is fulfilled before the vehicle performs the return trip to the depot. Moreover, once the vehicle leaves the depot again, it proceeds to v_i before continuing its planned route. However, the underlying SPR is hard to compute and, hence, is computationally expensive, but as shown in [3], the objectives remain meaningful. In order to solve the MC-VRPSD-FCS, a GA is used, which will be further described in the next section.

Genetic Algorithm

In our GA, a population n_{pop} of individuals (solutions) is generated and each individual has a cost as described in Eq. (1), which has to be minimized. To achieve a high level of diversity, we do not allow duplicates within the population. Two individuals are marked as duplicates if they have the same fitness value. At every generation t, n_{mat} individuals are selected to be part of the mating pool using a tournament selection process. Then, two individuals from the mating pool are chosen, a crossover operator as well as a mutation operator are applied with probabilities p_c and p_m, respectively. The new offsprings are integrated in the current generation with an *elitism* procedure. The GA terminates once a stopping criteria (e.g., number of generation) is satisfied and the fittest individual with the minimum expected transportation costs is selected as the best solution. A minimum expected cost, here, indicates that the solution not only found routes with minimal costs but also routes with minimal number of failures. The following subsections describe the insights of the GA in more detail.

Solution Encoding and Initial Population

The MC-VRPSD-FCS solutions are represented as a multipermutation genotype as in [5], where each permutation contains an ordered set of customers forming a route. Each gene represents a demand of a specific product type at a specific customer location. Due to the flexible compartment sizes of the fleet in this problem, the chosen representation allows a straightforward application of the different crossover and mutation operators. For each demand, it must be identified in which vehicle it will be transported as well as in which order it will be served. The solutions determined which product will transported by which vehicle as well as the capacity that will be allocated to the respective compartments. To accelerate convergence, the initial population is generated based on a randomized best insertion approach as in [3]. For each customer that has not been assigned to a vehicle yet, it is firstly ascertained whether a vehicle exists that has already opened a compartment for the respective product type or whether a new compartment can be opened. Then, the insertion costs of adding the new request to the vehicle are determined in two phases. We first calculate the best K insertions based on the fixed costs l_r. Second, the insertion costs of the K insertions are computed using Eq. (1) to identify the best insertion position.

Crossover Operator

In the crossover phase, genetic materials are exchanged between two parents p_1 and p_2. A route is randomly selected from p_1 and p_2, respectively. Then, random subsets of customers s_1 and s_2 from each route are chosen and integrated into the other parent. Before including the donated parts in either p_1 or p_2, it must be decided if the inclusion of all genes from either s_1 or s_2 might lead to violations of certain constraints, such as the distance constraints. This indicates whether the whole subsets (s_1 or s_2) or only parts can be added to the other parent. Any customer from the subsets, where no available compartment can be opened in the recipient parent, are discarded. The remaining subsets are added to the cheapest insertion point. An example of the crossover operator is displayed in Fig. 50.1, where genetic materials are chosen from the first route in Solution 1 and from the second route in Solution 2. Each demand from the substring from Solution 1 is included in the best position in the second route of Solution 2 and vice-versa provided that the distant constraints are respected in each route.

Mutation Operator and Population Management

In the proposed GA, three mutation operators are considered: request swap, inversion, and cheapest reinsertion which are applied with probabilities p_{swap}, p_{inv}, p_{re},

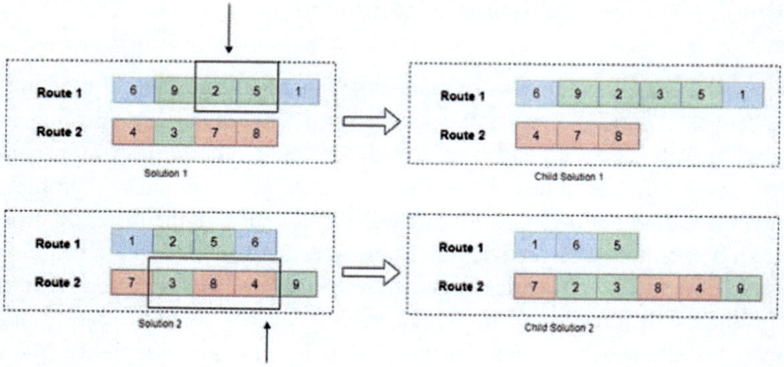

Fig. 50.1 Crossover between two solutions

respectively. In the request swap operator, two genes are randomly selected and, if possible, swapped. In the inversion operator, a sub-string from a route is chosen randomly and the sequence of genes within this sub-string is reversed, while in the cheapest reinsertion operator, a random gene is chosen and reinserted into the solution at the cheapest insertion position. If no crossover occurred between p_1 and p_2, the mutation is performed on p_1 and p_2, otherwise, it is conducted with a probability p_m. To select new individuals for the new generation, an elitism procedure is used. Here, the best x_{old}% from the current generation and the best x_{new}% of the newly created individuals are added and the remainder of the next generation is filled up with randomly selected individuals.

Parameter Tuning

We performed an extensive grid-search parameter tuning approach and found that the GA performs best with the following parameters: $n_{pop} = 80$, $t = 100$, $n_{mat} = 0.6$, $p_c = 0.6$, $p_{mut} = 0.2$, $p_{swap} = 0.3$, $p_{inv} = 0.3$, $p_{re} = 0.4\%$, $x_{old} = 5\%$, and $x_{new} = 5\%$.

Evaluation and Conclusion

In order to show the benefit of using flexible compartment sizes over fixed compartment sizes, we ran the presented GA five times over the 180 instances proposed by [3]. The instances consist of 50, 100, and 200 customers each with three different product types. Moreover, the instances can be grouped into six different groups of 30 problems according to their size and coefficient of variation of the demands (CV). We compared the GA for the MC-VRP-FCS to the results reported in [3] which

Table 50.1 Performance metrics of MC-VRPSD-FCS

Instance	Customers	CV	MA-Fixed	GA-Fixed	GA-FCS		
			Avg. costs	Avg. costs	# Best sol.	Avg. costs	Avg. impr (%)
Set 1	50	0.1	700.5	712.7	30	**602.6**	13.8
Set 2	50	0.3	727.4	742.9	30	**618.4**	14.8
Set 3	100	0.1	1198.5	1262.4	23	**1136.4**	4.7
Set 4	100	0.3	1196.9	1278.4	25	**1127.0**	5.2
Set 5	200	0.1	2007.7	2050.9	23	**1937.1**	3.6
Set 6	200	0.3	2081.2	2147.7	27	**1979.5**	4.6

solved the MC-VRPSD with fixed compartment sizes (MA-Fixed) using a memetic algorithm. Additionally, we also measured the performance of the GA when setting the compartment sizes as fixed.

Table 50.1 shows the benefit of using flexible compartment sizes (GA-FCS) over fixed compartment sizes (MA-Fixed and GA-Fixed). Compared to the average costs of the MA-Fixed approach, we get an improvement of 7.8% on average and an improvement of up to 14.8% in Set 2 in the GA-FCS. The improvement is particularly visible for instances with a low number of customers and CV = 0.3. Furthermore, the table shows that with the GA-FCS, best solutions can be found in most cases (see # Best sol.). The results also demonstrate that the GA we implemented still has room for improvement. But it performs well nevertheless, with an average positive deviation of 3.7% from the MA-Fixed approach and within acceptable computation time.

Conclusion and Outlook

This paper introduces the MC-VRP with stochastic customer demands and with flexible compartment sizes which we solved using a genetic algorithm. Our results show that we can reduce the expected transportation costs if we use flexible compartment sizes over fixed sizes. In some instance settings, using flexible compartment sizes can have improvements of up to 14.8% in the expected transportation costs. Research currently underway will address more robust optimization techniques for the multi-compartment vehicle routing problem with demand uncertainty in practical settings as well as more data-driven approaches to solve such problems.

References

1. Henke, T., Speranza, M. G., & Wäscher, G. (2015). The multi-compartment vehicle routing problem with flexible compartment sizes. *European Journal of Operational Research, 246*(3), 730–743. https://doi.org/10.1016/j.ejor.2015.05.020
2. Koch, H., Tino, H., & Gerhard, W. (2016). A genetic algorithm for the multi-compartment vehicle routing problem with flexible compartment sizes. Working Paper Series.
3. Mendoza, J. E., Castanier, B., Guéret, C., Medaglia, A. L., & Velasco, N. (2011). Constructive heuristics for the multicompartment vehicle routing problem with stochastic demands. *Transportation Science, 45*(3), 346–363. https://doi.org/10.1287/trsc.1100.0353
4. Ostermeier, M., Henke, T., Hübner, A., & Wäscher, G. (2021). Multi-compartment vehicle routing problems: State-of-the-art, modeling framework and future directions. *European Journal of Operational Research, 292*(3), 799–817. https://doi.org/10.1016/j.ejor.2020.11.009
5. Pereira, F. B., Tavares, J., Machado, P., & Costa, E. (2002). GVR: A new genetic representation for the vehicle routing problem. In *Irish Conference on Artificial Intelligence and Cognitive Science* (pp. 95–102). https://doi.org/10.1007/3-540-45750-X_12

Chapter 51
Benefits of Proactive Transshipments for an Automotive Manufacturer Under Emission Constraints

Bastian Vorwerk, Christian Weckenborg, and Thomas S. Spengler

Abstract The high number of transports in industry causes a significant proportion of the emissions that need to be reduced regarding global warming. A part of the transports is caused by lateral transshipments, which occur when parts are distributed among different locations within one echelon for various reasons. To be able to reduce the number of transports, proactive transshipments can be used. In proactive transshipments, parts are transshipped at predetermined points in time before their demand occurs. In the scientific literature, the consideration of emissions, as well as a differentiation of vehicle types, are neglected in planning models to proactive transshipments. Our planning model adopts emission limits and provides a detailed consideration of different vehicle types for the execution of transshipments. We decide about the types and amount of vehicles used between locations in different periods. An illustrative example is presented, comparing the costs and emissions between proactive and reactive transshipments with and without emission limits. We find that emission limits can influence vehicle type selection and the products to be transshipped during proactive transshipments.

Keywords Lateral transshipments · Multi-location inventory · Proactive · Reactive

Introduction

In recent years, the societal and economic focus has continued to shift towards reducing emissions due to their impact on climate change. The transport sector can be identified as one of the major drivers of emissions, as it accounted for around 21.6% of all emissions globally in 2018, of which 29.4% result from freight transport by

B. Vorwerk (✉) · C. Weckenborg · T. S. Spengler
Institute of Automotive Management and Industrial Production, Technische Universität Braunschweig, 38106 Braunschweig, Germany
e-mail: bastian.vorwerk1@volkswagen.de

B. Vorwerk
Volkswagen Group, 38440 Wolfsburg, Germany

O. Grothe et al. (eds.), *Operations Research Proceedings 2022*, Lecture Notes in Operations Research, https://doi.org/10.1007/978-3-031-24907-5_51

road [1, 2]. For this reason, emissions reduction plays a decisive role in logistics. In transport logistics, there are repeated cases of poorly utilized transports, which are not only detrimental to the environment but also economically disadvantageous for the companies. Without including empty transports, the average load capacity utilization of all transports in Germany was 55.9% in 2021 [3]. In traditional systems, items are transshipped from one echelon to the next, while more flexible systems also allow transshipments within an echelon in form of lateral transshipments. These transshipments are intended to ensure that stockout situations can be quickly eliminated and demand can be satisfied by transshipping items from warehouse locations or retailers to other warehouse locations or retailers. These lateral transshipments are supposed to result in a lower inventory level, lower costs, and a higher satisfaction of demand [4]. However, such lateral transshipments can also be driven by changes in demand or limited capacity at the warehouse locations, and cause both emissions and costs for the company [5]. Costs and emissions are largely dependent on the type of vehicle carrying out the transshipment. Consequently, they should be considered in the associated planning models. For this reason, we develop a model for planning proactive and reactive lateral transshipments considering multiple vehicle types and emission limits and investigate the effect of proactive and reactive transshipment as well as emission limits on costs and emissions using an illustrative example. To this end, we will introduce the characteristics of reactive and proactive lateral transshipments and refer to the associated literature in Sect. "Lateral Transshipments". Section "Model" discusses our modeling approach and in Sect. "Illustrative Example" we provide insights through an illustrative example. The paper ends with a conclusion in Sect. "Conclusion".

Lateral Transshipments

In the review article by Paterson et al., lateral transshipments are differentiated into reactive and proactive transshipments. Reactive transshipments react to demand that causes real or potential stockout situations by transshipping items from warehouses with sufficient stock [4]. This type of transshipment can take place at any time when actual or potential stockout situations occur [4]. Proactive transshipments, on the other hand, are restricted to take place at predetermined times, where items can be redistributed within an echelon before their demand occurs, to avoid future stockout situations [4, 6]. Due to the predetermined time, such transshipments can be organized in advance with the lowest possible costs by transshipping all items that will be needed in the future together and not, as in the case of reactive transshipments, carrying out several individual transports after an actual or potential stockout situation occurs, which can, for example, save transportation cost [4, 5].

In both reactive and proactive transshipments, decisions must be made about the time of transshipment, the quantity of items as well as the location from which the items are to be transshipped, while taking into account the capacity of the vehicles to execute the transshipment [4].

There is a lot of research already, investigating different settings of lateral transshipments, with different objectives such as reducing costs or improving service levels, but mostly without taking environmental aspects into account [4–6]. The existing models differ in the number of echelons, items, and locations or the cost structure of transshipments, as in some cases a distinction is made between flexible item-dependent and fixed vehicle-dependent costs [4, 5]. Furthermore, hardly any other model takes up the capacities of vehicles and examines the advantages of a heterogeneous fleet [7]. According to the current literature, however, there is no model that takes both vehicle and location capacities into account while allowing transshipments with a heterogeneous fleet within an echelon in order to examine their environmental impact [7–9]. In the next section, we present our model considering the addressed characteristics.

Model

This section discusses the required sets, indices, parameters, decision variables, constraints, and the objective function of our mathematical model. In total, the model consists of four sets. The set $i, j \in N$ indicates the warehouse locations occurring in the model. Each location i has a specific capacity c_i^{L} for storing items (in m^3). The set $x \in P$ represents the set of all item types. Each item type is assigned a volume v_x applicable per unit. The holding cost rates h_{ix} depend on the location i and the item type x and apply per period. The set $l \in V$ specifies the transport vehicle types. The vehicle types l differ in their volume c_l^{V}, fixed emissions E_{jil}^{fix}, and fixed costs R_{jil}^{fix} between two locations j and i. In addition to the fixed costs for the vehicles, there are handling cost rates R_{jix}^{u} in which costs for picking and storing items are included. The handling cost rate depends on the item type x and the starting and destination locations j and i. The fourth set $t \in T$ represents the progression of time in periods. The incoming goods A_{ixt} and the demand D_{ixt} depend on the location i, item type x, and the period t.

The decision variables $z_{xjit} \in \mathbb{N}_0$ indicate the number of items of type x that are transshipped between locations j and i within a period t. The decision variables $w_{jilt} \in \mathbb{N}_0$ indicate the number of vehicle types l operated between locations j and i in period t. The decision variables $s_{ixt} \in \mathbb{N}_0$ show the stock level for each item type x at the respective locations i within a period t before the transshipments were realized in the period and the decision variables $y_{ixt} \in \mathbb{N}_0$ show the stock level for each item type x at the respective locations i within a period t after the transshipments were realized.

The objective function shown in Formula (51.1) minimizes the total costs resulting from transportation cost $R_t^{\mathrm{T}} \in \mathbb{R}_0^+$ and holding cost $R_t^{\mathrm{H}} \in \mathbb{R}_0^+$ over all periods t. The transportation cost R_t^{T} and the emissions $E_t^{\mathrm{T}} \in \mathbb{R}_0^+$ are based on the types and amount of vehicles used between locations in different periods shown in Formula (51.2) and (51.4). The transportation cost R_t^{T} are additionally influenced by the handling cost rates per transshipped item R_{jix}^{u}. The holding cost R_t^{H} are defined by the number of parts in the warehouse location and the specific holding cost rate h_{ix} for the item type x at location i for each period t shown in Formula (51.3). The consideration of emissions is given by a constraint, in which the maximum emissions output over all periods EL is defined. In the case of reactive transshipments, the timing and items of transshipments are fixed, as only items are allowed to be transshipped in the period in which they are needed at another location, which is why there is no freedom to decide on the transshipment of further items.

$$\min \left(\sum_{t \in T} R_t^{\mathrm{T}} + \sum_{t \in T} R_t^{\mathrm{H}} \right) \tag{51.1}$$

$$R_t^{\mathrm{T}} = \sum_{x \in P} \sum_{u \in U} \sum_{j \in N} \sum_{i \in N | i \neq j} z_{xujit} R_{jix}^{\mathrm{u}} + \sum_{j \in N} \sum_{i \in N | i \neq j} \sum_{l \in V} w_{jilt} R_{jil}^{\mathrm{fix}} \quad \forall t = 1, \ldots, |T| \tag{51.2}$$

$$R_t^{\mathrm{H}} = \sum_{i \in N} \sum_{x \in P} \sum_{u \in U} s_{ixut} h_{ix} \quad \forall t = 1, \ldots, |T| \tag{51.3}$$

$$E_t^{\mathrm{T}} = \sum_{j \in N} \sum_{i \in N | i \neq j} \sum_{l \in V} w_{jilt} E_{jil}^{\mathrm{fix}} \quad \forall t = 1, \ldots, |T| \tag{51.4}$$

Overall, there are three categories of constraints. In the first category, it is ensured that the demand at each location can be satisfied in every period. For this purpose, a constraint set ensures that the stock of an item type x at location i is greater than the demand for the item type x in that period t shown in Formula (51.5). In the second category, a stock balance equation is used to determine the stock level of the respective periods. If the demand cannot be satisfied without transshipment, the model is forced to transship items. The demand D_{ixt}, incoming goods A_{ixt}, and the number of items that are transshipped z_{xjit} are deducted or added to the stock level of each location i in every period t. In the third category, capacity compliance is ensured. The sum of the volumes of all items in a location i must not exceed the capacity of the locations c_i^{L}. For the vehicles, the volume of all items v_x to be transshipped must be less than the sum of the capacity of the vehicles c_l^{V} used between two locations i and j within the respective period t.

$$\begin{aligned} y_{ixut} &= s_{ixut-1} + \sum_{j \in N} z_{xujit} + A_{ixut} - \sum_{j \in N} z_{xuijt} \\ &\forall i \in N | i \neq j; x \in P; u \in U; t = 1, \ldots, |T| \end{aligned} \tag{51.5}$$

In the proactive transshipment model, items that are required at a location in later periods can already be transshipped in earlier periods. To cover for reactive transshipments in the model, however, additional constraints are introduced only in the reactive model ensuring that only those items whose demand cannot be satisfied in a location and period can be transshipped in exactly the quantity required for demand fulfillment. The additional conditions result in the model for reactive transshipments being a quadratically constrained model, as seen in Formula (51.6), (51.7), and (51.8). It can be seen that the Big M method was introduced to limit the transshipment behavior.

$$0 \geq (s_{ixut-1} + A_{ixut} - D_{ixut}) \cdot q_{ixut} \quad \forall i \in N; x \in P; u \in U; t = 1, \ldots, |T| \tag{51.6}$$

$$q_{ixut} \cdot Big\_M \geq \sum_{j \in N | j \neq i} z_{xujit} \forall i \in N; x \in P; u \in U; t = 1, \ldots, |T| \tag{51.7}$$

$$(D_{ixut} - s_{ixut-1} - A_{ixut}) \cdot q_{ixut} \geq \sum_{j \in N \mid j \neq i} z_{xujit} \quad \forall i \in N;\ x \in P;\ u \in U;\ t = 1,\ \ldots,\ |T| \tag{51.8}$$

The resulting model covering proactive transshipments classifies as a mixed-integer linear programming model and the model covering reactive transshipments as a mixed-integer quadratically-constrained model. They are implemented in Python and solved using the Gurobi 9.5.1 Python API on a standard machine with i5-8250U @ 1.60 GHz CPU and 8 GB RAM. For the illustrative example reported in the next section, the reactive model comprises around 6,000 quadratic and 12,270 linear constraints with 15,554 integer and 2,222 binary variables. The proactive model includes 8271 linear constraints with 11,554 integer variables. Both models have 30 continuous variables. The models are solved optimally in less than 1 s.

Illustrative Example

In this section, we present an illustrative example where the data for warehouses and items is derived from an example in the automotive industry. In total, there are 100 different item types with different initial stock levels in the example, which are divided between two warehouse locations. The locations differ in capacity and holding cost rates for the different items, the first location having a lower capacity than the second location. A general overview of the illustrative example is shown in Fig. 51.1.

Three different types of vehicles are available for lateral transshipments between the two locations, which differ in their capacity and emissions. The emissions and costs for transportation are based on real data and increase with the available capacity of the vehicles. Based on the average fuel consumption for the individual vehicle types, the emissions output is calculated with the help of a conversion factor [3, 10]. Both the demand and the incoming goods of items to the respective locations are given deterministically over the periods. The majority of demand and incoming goods occur in the first location. Overall, a planning horizon of 10 periods is considered.

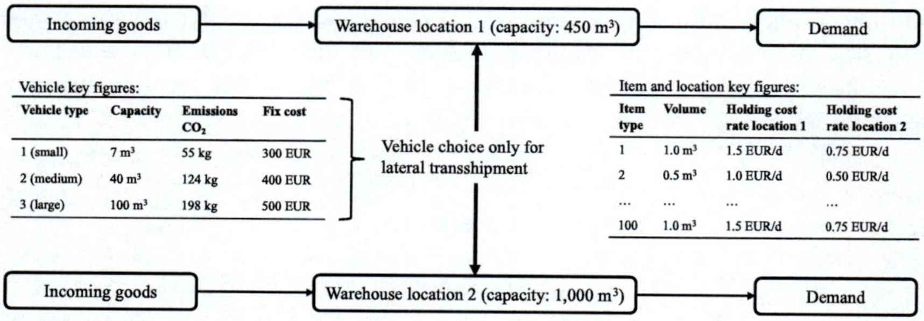

Fig. 51.1 Illustrative example

We use the models from Sect. "Model" to calculate the cost-optimal satisfaction of demand by reactive and proactive transshipments. The solutions with and without an emission limit are compared. The emission limits correspond to the lowest level of emissions for which feasible solutions were found. An overview of the results can be found in Table 51.1.

Table 51.1 Results of the illustrative example

Restriction	Total cost (EUR)	Transshipment cost (EUR)	CO_2 emissions (kg)	Choice of transportation vehicles	Number of transshipped items	Average utilization of vehicles (%)
Reactive *no EL*	14,935	6,230	1,086	11 (small) 4 (medium) 0 (large)	133	48
Reactive $EL < 1,042$	15,577	6,830	1,041	17 (small) 1 (medium) 0 (large)	133	69
Proactive *no EL*	11,504	2,780	450	1 (small) 0 (medium) 2 (large)	148	53
Proactive $EL < 430$	11,586	2,860	429	2 (small) 1 (medium) 1 (large)	136	74

The result in Table 51.1 shows the significant cost savings through proactive transshipment compared to reactive transshipment. The total costs consist of the transport and holding costs as described in Formula (51.1). As the holding costs remain almost the same in all four variants and the transport costs vary greatly, Table 51.1 specifically shows the transport costs as part of the total costs. This shows that the transport costs are the decisive factor for the cost savings. The difference in emissions between the four experiments is particularly interesting. Given the strictest feasible emission limits, the emissions can be reduced by 4.1% and 4.6% compared to the unrestricted cases for reactive and proactive transshipments, respectively. However, the associated costs increase by 4.3% and 0.7%, respectively. The emissions savings can be explained by the choice of vehicle types. For example, to transport a volume of 45 m^3, one large or one small and one medium vehicle can be selected. In the proactive transshipment without emission limits, a large vehicle was chosen because it is cheaper than a small and a medium vehicle in combination. In this example, however, the emissions of the small and medium-sized vehicles are lower in total than those of a large vehicle. Therefore, this variant is selected when the emission limit is introduced. Another effect is that the average utilization increases. Instead of a large vehicle with low utilization, smaller vehicles with higher utilization are used. The division of the items between two vehicles as well as the capacity limitation and the high utilization of Location 1 have an impact on the number of items to be transshipped. In this example, all items that are needed from Location 2 in Location 1 over all periods are already in Location 2 at the beginning of the planning horizon and could therefore already be transshipped in the first period in the case of proactive transshipment. However, the capacity of Location 1 is not sufficient to accommodate all items from Location 2 in the first period. To release capacity, items that are not needed at Location 1 are transshipped together with the items that are needed from Location 1 at Location 2.

Conclusion

In conclusion, the illustrative example shows that proactive transshipment can lead to significant cost savings and emission reductions compared to reactive transshipments. It also shows that an emission limit can decisively change the decision on vehicle types and the decision on when to transship items. In this context, a conflict of objectives between cost savings and emissions savings can be identified. These findings were reached with deterministic knowledge of future demand. In practice, the incoming goods and the demand are usually unknown, therefore a suitable procedure for demand forecasting must be introduced in the future to be able to realistically depict the conditions from practice. In addition, the conflict of objectives between cost savings and emission limits should be examined in more detail in the future. In the illustrative example, it is assumed that no influence can be taken on inventory levels and incoming goods. This assumption could be changed in further research so that the best trade-off between the number of transports and inventory levels can be analyzed.

References

1. Ourworldindata Homepage. https://bit.ly/3bVgLef. Last accessed Mar 02 2023.
2. Statista Homepage. https://bit.ly/3AAelwd. Last accessed Mar 02 2023.

3. KBA Homepage. https://bit.ly/3nFK739. Last accessed Mar 02 2023.
4. Paterson, C., Kiesmüller, G., Teunter, R., & Glazebrook, K. (2011). Inventory models with lateral transshipments: A review. *European Journal of Operational Research, 210*, 125–136.
5. Meissner, J., Senicheva, O (2018) Approximate dynamic programming for lateral transshipment problems in multi-location inventory systems. *European Journal of Operational Research, 265*, 49–64.
6. Glazebrook, K., Paterson, C., Rauscher, S. (2015). Benefits of hybrid lateral transshipments in multi-item inventory systems under periodic replenishment. *Production and Operations Management, 24*, 311–324.
7. Asghari, M., Mirzapour Al-e-hashem, S. M. J. (2021). Green vehicle routing problem: A state-of-the-art review. *International Journal of Production Economics, 231*, 107899.
8. Mirzapour Al-e-hashem, S. M. J., Rekik, Y. (2014). Multi-product multi-period inventory routing problem with a transshipment option: a green approach. *International Journal of Production Economics, 157*, 80–88.
9. Kaviyani-Charati, M., Ameli, M., Heidarzadeh Souraki, F., Jabbarzadeh, A. (2022). Sustainable network design for a non-profit food bank supply chain with a heterogeneous fleet under uncertainty. *Computers & Industrial Engineering, 171*, 108442.
10. Webfleet Homepage. https://bit.ly/3IgsmRt. Last accessed Mar 02 2023.

Chapter 52
Different MIP Formulations for a Dynamic Lot-Sizing Model with Rework of Defectives

Steffen Rudert

Abstract This paper discusses a basic dynamic lot-sizing model for a single item with rework of defectives. Due to the imperfect production process, some fraction of the generated items is not of sufficient quality. After rework, these goods serve the same demand as the initial perfect quality items; both are called serviceables.The internal processes production and rework are conducted independently from another at different or same periods. The basic model is proven to be NP-hard. We present three main unique characteristics that describe the specific model behavior observed in the optimal solutions: production only, multiple rework, and overproduction of serviceables. Different MIP formulations are derived to analyse the effects of these three characteristics on the optimal solutions that exclude each of these characteristics from the basic model. Afterward, computations for given data sets are conducted, using all different MIP formulations. It can be shown that production only occurs most frequently and has the highest effect on the total cost.

Keywords Dynamic lot-sizing · Defectives · Rework · MIP formulations

Introduction

Nowadays, companies face a drastic scarcity of resources while at the same time are confronted with stricter government regulation and increased public awareness regarding environmental issues. One way to reduce their ecological footprint is to consider product returns in their manufacturing systems explicitly. While there are vast activities in the scientific literature to incorporate external returns under the term of remanufacturing, only little attention is paid to internal returns that can arise from an imperfect production process that produces perfect quality items and defectives.

Since the seminal work of [1], many extensions of the single-item dynamic lot-sizing problem have been published [2]. Teunter et al. [3] introduced the economic lot-sizing problem with remanufacturing for the case of joint and separate setups

S. Rudert (✉)
Faculty of Business and Economics, TU Dresden, Dresden 01069, Germany
e-mail: steffen.rudert@mailbox.tu-dresden.de

O. Grothe et al. (eds.), *Operations Research Proceedings 2022*, Lecture Notes in Operations Research, https://doi.org/10.1007/978-3-031-24907-5_52

for new production and remanufacturing. [4] showed that both problems are NP-hard. A multicommodity formulation for the same problem and new additional valid inequalities were presented by Cunha and Melo [5]. Kilic and van den Heuvel [6] used a decomposition approach to develop a polynomial-time heuristic.

For internal returns, [7] examined a multiple-product model with limited capacity. [8] proposed a late acceptance hill-climbing matheuristic for another model that also considers rework and lifetime constraints for defective items. [9] incorporated internal and external returns and applied a Wagner/Whitin-based solution procedure.

This paper introduces the basic lot-sizing model, including the corresponding MIP formulation in Section "The Basic Model and the Specific Characteristics of the Optimal Solutions". Afterward, the three specific characteristics of the optimal solutions are presented and followed by additional MIP conditions to exclude each from the basic model. This is used in Section "Numeric Study" to show each characteristic's frequency and cost effect by using a commercial solver, and Section "Conclusions" completes the paper by providing conclusions.

The Basic Model and the Specific Characteristics of the Optimal Solutions

Our model represents an imperfect production process where some fraction of the generated items do not meet the quality requirements. Perfect quality items store at the serviceables inventory and the others at the defectives inventory. After rework, these goods serve the same demand as the initial perfect quality items. Both processes, production and rework, are carried out internally on different machines. Thus, no capacity restrictions apply. The MIP for the basic model reads as follows:

$$\min \sum_{t=1}^{T} h_s \cdot I_{s,t} + h_d \cdot I_{d,t} + y_t \cdot R_p + z_t \cdot R_r \tag{52.1}$$

subject to

$$I_{s,t} = I_{s,t-1} + (1-\beta) \cdot p_t + r_t - d_t \quad \forall t = 1, \dots, T \tag{52.2}$$

$$I_{d,t} = I_{d,t-1} + \beta \cdot p_t - r_t \quad \forall t = 1, \dots, T \tag{52.3}$$

$$p_t \leq d_{t,T} \cdot y_t \quad \forall t = 1, \dots, T \tag{52.4}$$

$$r_t \leq d_{t,T} \cdot z_t \quad \forall t = 1, \dots, T \tag{52.5}$$

$$I_{s,0} = I_{d,0} = I_{d,T} = I_{s,T} = 0 \tag{52.6}$$

$$p_t, r_t, I_{s,t}, I_{d,t} \geq 0, d_t > 0 \quad \forall t = 1, \dots, T \tag{52.7}$$

$$y_t, z_t \in \{0; 1\} \quad \forall t = 1, \dots, T \tag{52.8}$$

The basic model minimises the total cost that comprises the holding and the setup costs (52.1). Holding a serviceable for one period is associated with costs of h_s and for a defective item h_d. The setup cost for production is R_p, and for rework, R_r. The stock levels for serviceables $I_{s,t}$ and for defectives $I_{d,t}$ at the end of each period t are determined by the inventory balance equations (52.2, 52.3) and are empty at the beginning and the end of the planning horizon (52.6). If there is production at period t ($p_t > 0$), the binary variable is $y_t = 1$ and $y_t = 0$ otherwise. The same is true for the rework amount r_t and the corresponding binary z_t (52.4, 52.5, 52.8). β denotes the proportion of defective units during production, d_t the demand of period t, and $d_{t,T}$ the sum of demands from period t to the end of the planning horizon T. Shortages or backorder are not allowed and all demand must be met by inventory, production, rework, or a combination of them. Consequently, all variables are non-negative (52.7). The decision variables are p_t, r_t, y_t and z_t, while $I_{s,t}$ and $I_{d,t}$ directly result from them. All others are model parameters. Besides, it can be proven that the basic model is NP-hard [10].

The *standard case* of one production and rework lot is to produce in $t = k$ the sum of the demands from $t = k$ to $t = l$: $p_k = d_{k,l}$. As a certain proportion of these items is defective (βp_k), one rework is carried out at $k \leq m \leq l$, and the number of reworked items is $n_m = \beta p_k = \beta d_{k,l}$. When the stocks of serviceables and defectives are empty at period l, a new production lot will start: $I_{s,l} = 0$ and $I_{d,l} = 0$.

In addition to the *standard case*, the following subsections describe the three specific characteristics of the model. They differ from the *standard case* as they optimise the production schedule and reduce the total costs. By presenting additional conditions to the basic MIP formulation, each of them can be excluded from the model. Thereby, it is possible to examine each characteristic's frequency and cost effect.

Production Only (PO)

In general, rework of defectives is not mandatory for each single production lot. It can be cost-efficient to fulfill the demand by only producing new items and storing defectives until the next production starts. This is called *production only* and is especially favourable for low holding costs of defectives and/or high rework setup costs. Rework will be carried out later to guarantee that the stock of defectives will be empty at the end of the planning horizon. If *production only* is used, there are cost savings by skipping the rework setup cost of R_r but also extra cost for holding all defectives from the start to the end of this lot.

By applying the following additional conditions to the MIP formulation of the basic model, the occurrence of *production only* can be excluded.

$$p_t \leq d_{t,T} \cdot g_{t-1} \quad \forall t = 2, \ldots, T \tag{52.9}$$

$$g_t \leq z_t \quad \forall t = 1 \tag{52.10}$$

$$g_t \leq g_{t-1} - y_t + z_t \quad \forall t = 2, \ldots, T \tag{52.11}$$

Production only means that there are two successive production runs without rework in between. To restrict this, production in period t is now only possible according to the status of the additional control variable of the previous period g_{t-1}. For $g_{t-1} = 1$, production is permitted: $p_t \leq d_{t,T} \cdot g_{t-1} = d_{t,T}$. For $g_{t-1} = 0$, production is not permitted: $p_t \leq d_{t,T} \cdot g_{t-1} = 0$, see (52.9).

The status of the control variable g_t considers three inputs: rework z_t, production y_t, and the control status of the previous period g_{t-1}, see (52.11). If there is only rework at one period ($z_t = 1$), the status of the control variable will be $g_t = 1$: $g_t \leq g_{t-1} - y_t + z_t = 0 - 0 + 1 = 1$. This status will be transferred to the next period if there are no changes: $g_t \leq g_{t-1} - y_t + z_t = 1 + 0 + 0 = 1$.

If there is production without rework at one period, the status of the control variable will be $g_t = 0$. Production is only allowed if $g_{t-1} = 1$ and (52.11) shows: $g_t \leq g_{t-1} - y_t + z_t = 1 - 1 + 0 = 0$. Also here, this status will be transferred to the next period if there are no changes: $g_t \leq g_{t-1} - y_t + z_t = 0 + 0 + 0 = 0$. To work correctly, g_t must be initialised by (52.10) in period $t = 1$.

Multiple Rework (MR)

Contrary to the previous characteristic, *multiple rework* uses several rework activities between two successive production operations. This policy is interesting when defectives store at lower costs than serviceables ($h_s > h_d$) and rework setup costs are lower than production setup costs. The cost savings originate from the cost difference of $h_s > h_d$ and extra costs from one or more additional rework setups R_r.

The following additional conditions for the MIP formulation of the basic model exclude the occurrence of *multiple rework.*

$$r_t \leq d_{t,T} \cdot y_t \quad \forall t = 1 \tag{52.12}$$

$$r_t \leq d_{t,T} \cdot (y_t + g_{t-1}) \quad \forall t = 2, \ldots, T \tag{52.13}$$

$$g_t \leq 1 - z_t \quad \forall t = 1, \ldots, T \tag{52.14}$$

$$g_t \leq y_t \quad \forall t = 1 \tag{52.15}$$

$$g_t \leq g_{t-1} + y_t \quad \forall t = 2, \ldots, T \tag{52.16}$$

To guarantee that there are not multiple reworks between two successive production runs, rework at period t is only possible if there is production at the same period ($y_t = 1$) or the status of the control variable of the previous period is $g_{t-1} = 1$, see (52.12, 52.13). The status of g_t is determined by (52.14) and (52.15, 52.16).

To restrict the occurrence of multiple rework, (52.14) changes the status of the control variable to $g_t = 0$ when there is rework at one period: $g_t \leq 1 - z_t = 1 - 1 = 0$. Contrary to this, (52.16) would allow status $g_t = 1$ if there is production at one period: $g_t \leq g_{t-1} + y_t = 0 + 1 = 1$ or $g_t \leq g_{t-1} + y_t = 1 + 1 = 2$. (52.15) does the same and initialises g_t at $t = 1$.

If there is production and rework at one period, (52.14) and (52.16) must hold simultaneously, and consequently, $g_t \leq 0$ applies. It should also be noted explicitly that rework is unnecessary before the next production can start.

Overproduction of Serviceables(OP)

The optimal policy for the basic planning problem of this paper does not consider the so-called zero inventory property. This means there may be perfect quality items left over from the previous period ($I_{s,k-1} > 0$), although production runs at the actual period. These items were taken from the last lot size to help optimise the upcoming lot's production schedule by enabling later rework, for example. In this case, rework would have to be carried out earlier due to the lack of serviceables at some period between $t = k$ and $t = l$. Using the leftover items $I_{s,k-1}$ from the previous lot, rework can be postponed, and savings apply when defectives store at a lower cost than serviceables. Additional costs arise from storing the leftover items.

To prevent the *overproduction of serviceables*, the additional conditions for the MIP formulation of the basic model are used.

$$p_t \leq d_{t,T} \cdot g_t \quad \forall t = 2, \ldots, T \tag{52.17}$$

$$g_t \leq 1 - \frac{I_{s,t-1}}{d_{t,T}} \quad \forall t = 2, \ldots, T \tag{52.18}$$

Production is only possible by the permission of the control variable $g_t = 1$, see (52.17). The status of g_t is determine exclusively by the serviceables inventory level of the previous period $I_{s,t-1}$, see (52.18). Production is permitted ($g_t \leq 1$) if the stock is empty ($I_{s,t-1} = 0$): $g_t \leq 1 - \frac{I_{s,t-1}}{d_{t,T}} = 1 - \frac{0}{d_{t,T}} = 1$. For $I_{s,t-1} > 0$, production is restricted ($g_t < 1$ and as g_t is binary $g_t = 0$): $g_t \leq 1 - \frac{I_{s,t-1}}{d_{t,T}} < 1$.

Numeric Study

In the following, computations have been carried out for all four MIP formulations from Section "The Basic Model and the Specific Characteristics of the Optimal Solutions": the basic model and excluding PO, MR, and OP, respectively. All tests were performed on a Windows Server 2012 R2 with Intel(R) Xeon(R) CPU E5-4627 v2 @ 3.3 GHz processors with 32 cores, of which four were used for the computations, 768GB RAM, and Gurobi 8.1 as the MILP solver.

A whole set of different problem instances was solved to obtain highly meaningful results. The model parameters and the demand patterns are chosen based on [3] and are as follows: $\beta = 1\%;\ 5\%;\ 10\%;\ 20\%;\ 30\%$, $h_s = 1$, $h_d = 0.1;\ 0.5;\ 1.0;\ 1.5$, $R_p = R_r = 50;\ 100;\ 250;\ 500;\ 750;\ 1000;\ 1500;\ 3000$. There are 10 demand series for each of these parameter constellations, which is two for each of the five different demand patterns, according to the demand function given in [3]. Thus, there are a total of $5 \cdot 4 \cdot 8 \cdot 8 = 1280$ test instances. Consequently, a total of $10 \cdot 1280 = 12800$ data sets were used as the input for the computation of each MIP formulation.

First, we examine the frequency of the *standard case* and of the characteristics at all data sets for the basic model. The *standard case* of one production combined with one rework and no carryover of any items from one lot to another is present in 22.0%. PO can be observed in 76.5% of all cases. There are fewer occurrences of MR (1.7%), and OP occurs even less frequently (1.1%).

For increasing values of β, MR and OP occur more frequently while the opposite is true for PO. All three characteristics appear less frequently for rising values of h_d. For $h_d \geq h_s$, MR and OP cannot be observed at all. Rising values of R_p lead to less occurrence of PO and OP while more cases of MR appear. Finally, increasing values of R_r show more of PO but less of MR and OP.

Next, when the specific MIP formulations exclude the characteristics, it is possible to calculate the cost delta when applying each characteristic. The average cost per instance is 5402, and PO shows the highest cost saving per test instance of 1108, while MR shows only 82 and OP 27. For PO, this means that if the *production only* policy is available and is used for the optimal solution of a single instance, the total costs are 1108 less compared to a policy where *production only* is unavailable. The cost delta of MR increases very strongly with an increased defective rate β and would yield substantially higher values if β would exceed 30%, which is the maximum used in the numeric study here.

Conclusions

Internal returns have not yet gained much attention in the literature on dynamic lot-sizing. We presented a basic model that includes the rework of defective items originating from an imperfect production process. Optimal solutions show three specific characteristics in addition to the *standard case*, which is similar to the Wagner-Whitin model. *Production only* occurs most frequently and strongly reduces total costs compared to the *standard case*. *Multiple rework* and the *overproduction of serviceables* are rarely present and reduce costs to a far less extent. The evaluation of this paper provides a guideline for developing solution procedures as all three characteristics can be used to build problem-specific algorithms.

References

1. Wagner, H. M., & Whitin, T. M. (1958). Dynamic version of the economic lot size model. *Management Science, 5*, 89–96.
2. Brahimi, N., Absi, N., Dauzèré-Pérès, S., & Nordli, A. (2017). Single-item dynamic lot-sizing problems: An updated survey. *European Journal of Operational Research, 263*(3), 838–863.
3. Teunter, R. H., Bayindir, Z. P., & van den Heuvel, W. (2006). Dynamic lot sizing with product returns and remanufacturing. *International Journal of Production Research, 44*(20), 4377–4400.
4. Helmrich, M. J. R., Jans, R., van den Heuvel, W., & Wagelmans, A. P. (2014). Economic lot-sizing with remanufacturing: Complexity and efficient formulations. *IIE Transactions, 46*, 67–86.
5. Cunha, J. O., & Melo, R. A. (2016). A computational comparison of formulations for the economic lot-sizing with remanufacturing. *Computers and Industrial Engineering, 92*, 72–81.
6. Kilic, O. A., & van den Heuvel, W. (2019). Economic lot sizing with remanufacturing: Structural properties and polynomial-time heuristics. *IISE Transactions, 51*(12), 1318–1331.
7. Goerler, A., & Voß, S. (2016). Dynamic lot-sizing with rework of defective items and minimum lot-size constraints. *International Journal of Production Research, 54*(8), 1–14.
8. Goerler, A., Lalla-Ruiz, E., & Voß, S. (2020). Late acceptance hill-climbing matheuristic for the general lot sizing and scheduling problem with rich constraints. *Algorithms, 13*, 1–26.
9. van Zyl, A., & Adetunji, O. (2022). A lot sizing model for two items with imperfect manufacturing process, time varying demand and return rates, dependent demand and different quality grades. *Journal of Remanufacturing*. https://doi.org/10.1007/s13243-022-00110-z
10. Rudert, S., & Buscher, U. (2022). On the complexity of the economic lot-sizing problem with rework of defectives. Dresdner Beiträge zur Betriebswirtschaftslehre, Nr. 182/22, Dresden: TU Dresden. https://doi.org/10.25368/2022.322

Chapter 53
Manipulating Waiting-Plus-Detour-Time Mechanisms for Pickup and Delivery Problems

Martin Damyanov Aleksandrov

Abstract We consider routing problems where agents have preferences over pickup and delivery travel options. We look at the class of mechanisms that maximise social welfare. We study computing outcomes with such mechanisms. We also show that agents can manipulate such mechanisms. In response, we study computing pure Nash equilibria induced by such mechanisms. Finally, we analyse the price of anarchy for such mechanisms, which quantifies the welfare loss in an equilibrium.

Keywords Logistics · Game theory · Pure Nash equilibria

Introduction

According to the Federal Statistical Office of Germany,[1] nearly 20% of all Germans (i.e. $\approx$ 16 mils) are disabled and nearly 10% of them (i.e. $\approx$ 1.6 mils) live with progressive dementia. Depending on their medical conditions, these people have different preferences about when and how urgently they need to visit medical facilities. In such settings, it is perhaps not surprising that people rank minimising their waiting plus detour times as one of the most important criteria: see e.g. [10]. Satisfying preferences about waiting plus detour times is also crucial for achieving customer satisfaction in various taxi, ride-sharing, and car-sharing settings because waiting and detouring for shorter times is often more convenient for customers and, additionally, customers pay normally lower costs for such travel options in such settings.

We present a formal model for such problems, where agents have quasi-linear valuation functions over their waiting plus detour times. For example, an agent might be indifferent between travel options where their waiting plus detour time is at most 15 min, in which case their value is zero, but they may strictly prefer each of these options to an option where their waiting plus detour time is strictly

[1] https://www.destatis.de/.

M. D. Aleksandrov (✉)
Freie Universität Berlin, Berlin, Germany
e-mail: martin.aleksandrov@fu-berlin.de

O. Grothe et al. (eds.), *Operations Research Proceedings 2022*, Lecture Notes in Operations Research, https://doi.org/10.1007/978-3-031-24907-5_53

more than 15 min, in which case their value is strictly negative. We look at the class of mechanisms that maximise social welfare, which is the sum of the valuations. For such mechanisms, we study computing (1) individual valuations (Theorem 1), (2) group valuations (Theorem 2), (3) possible manipulations (Theorem 3), (4) pure Nash equilibria (Theorem 4), and (5) the price of anarchy (Theorem 5).

Related Work

Maximising and manipulating the social welfare in our setting relate to the Travelling Repairman Problem (TRP) and the Hamiltonian Path Problem (HPP), respectively. Afrati et al. [1] showed that the TRP is NP-hard. Garey and Johnson [4] showed that the HPP is NP-hard. Ma et al. [7] designed an incentive-aligned mechanism for ride-sharing. Rheingans-Yoo et al. [11] presented a mechanism that encourages drivers to report their location preferences truthfully. By comparison, we focus on the customer side of the problem. Interestingly, in our setting, *any* welfare-maximising mechanism can be manipulated. Waiting- and detour-time customer objectives have received a limited attention: see e.g. [12]. Nucamendi et al. [9] studied minimising the total waiting time. Gschwind and Drexl [5] considered detour time constraints. By comparison, we combine waiting and detour times into a natural objective such as maximising the social welfare.

Formal Preliminaries

We write $N = (L, E)$ for a *network* with $|L| \geq 2$ and $E \subseteq L \times L$. We assume that each $l_i \in L$ is reachable from each $l_j \in L \setminus \{l_i\}$ through some path $(s_1 = l_j, \ldots, s_a = l_i)$ of length $a \in \mathbb{N}_{\geq 2}$, where each $s_k \in L$ and each $(s_k, s_{k+1}) \in E$. We consider set $V = \{v_1, \ldots, v_n\}$ of $n \in \mathbb{N}_{\geq 1}^{<\infty}$ *vehicles*, where v_i begins at $b_i \in L$, ends at $e_i \in L$, and has capacity $q_i \in \mathbb{N}_{>0}$, as well as set $R = \{r_1, \ldots, r_m\}$ of $m \in \mathbb{N}_{\geq 1}^{<\infty}$ *requests*, where $r_j = (p_j, d_j)$ is for transporting agent j from pick-up $p_j \in L$ to drop-off $d_j \in L \setminus \{p_j\}$. For each v_i and $(l, l') \in L \times L$, we write $t(i, l, l') \in \mathbb{R}_{\geq 0}^{\leq \infty}$ for the travel time v_i needs when moving from l to l', where $t(i, l, l') = \infty$ if $(l, l') \notin E$ and else $t(i, l, l') < \infty$. We write T_i for the *measure* $[t(i, l, l')]_{|L| \times |L|}$. We let $\mathcal{I} = (N, V, R, [T_1, \ldots, T_n])$ denote an *instance*. Plan $\mathcal{P} = \{\mathcal{R}_1, \ldots, \mathcal{R}_n\}$ for $\mathcal{I}$ is a set of routes, where route $\mathcal{R}_i = (b_i, s_1(i), \ldots, s_{a_i}(i), e_i)$ is a path of length $(a_i + 2)$, where $a_i \in \mathbb{N}_{\geq 0}$, each $s_l(i) \in L$, $(b_i, s_1(i)) \in E$, each $(s_l(i), s_{l+1}(i)) \in E$, and $(s_{p_i}(i), e_i) \in E$. We consider *feasible plans* for instance $\mathcal{I}$, where all requests are serviced: see e.g. [2].

Pick a feasible plan $\mathcal{P}$ for instance $\mathcal{I}$. The waiting time $w_{ji}(\mathcal{P})$ is the time agent j waits for vehicle v_i to pick them up from their origin p_j: $w_{ji}(\mathcal{P}) = s(i, b_i, s_1(i)) + [\sum_{s_l(i) \in \mathcal{W}_{ji}, s_l(i) \neq p_j} t(i, s_l(i), s_{l+1}(i))]$, where $s(i, b_i, s_1(i))$ is the shortest travel time for vehicle v_i along a path from b_i to $s_1(i)$ and $\mathcal{W}_{ji}$ is the sub-path of $\mathcal{R}_i$ from $s_1(i)$

to p_j. The detour time $d_{ji}(\mathcal{P})$ is the additional travel time agent j spends onboard of vehicle v_i instead of travelling directly from their origin p_j to their destination d_j: $d_{ji}(\mathcal{P}) = [\sum_{s_l(i)\in\mathcal{D}_{ji}, s_l(i)\neq d_j} t(i, s_l(i), s_{l+1}(i))] - s(i, p_j, d_j)$, where $\mathcal{D}_{ji}$ is the sub-path of $\mathcal{R}_i$ from p_j to d_j and $s(i, p_j, d_j)$ is the shortest travel time for vehicle v_i along a path from p_j to d_j. We let $x_{ji}(\mathcal{P}) = 1$ if agent j is serviced by v_i in $\mathcal{P}$ and, otherwise, $x_{ji}(\mathcal{P}) = 0$. We suppose that agent j has some *private threshold value* $\tau_j \in \mathbb{R}_{\geq 0}$ for their waiting plus detour time. However, in our model, agent j can report any *threshold value* $\tau'_j \in \mathbb{R}_{\geq 0}$. Next, for a given reported τ'_j, we let agent j have a *valuation* $\tilde{u}_{ji}(\mathcal{P}, \tau'_j)$ that is a quasi-linear function with respect to (wrt) τ'_j.

More specifically, we let $\tilde{u}_{ji}(\mathcal{P}, \tau'_j) = 0$ if agent j is serviced by v_i and $w_{ji}(\mathcal{P}) + d_{ji}(\mathcal{P}) \leq \tau'_j$; $\tilde{u}_{ji}(\mathcal{P}, \tau'_j) = \tau'_j - (w_{ji}(\mathcal{P}) + d_{ji}(\mathcal{P}))$ if agent j is serviced by v_i and $w_{ji}(\mathcal{P}) + d_{ji}(\mathcal{P}) > \tau_j$; $\tilde{u}_{ji}(\mathcal{P}, \tau'_j) = 0$ if agent j is serviced by some v_k with $k \neq i$. Thus, we let $u_{ji}(\mathcal{P}) = \tilde{u}_{ji}(\mathcal{P}, \tau_j)$ denote their *private valuation* in $\mathcal{P}$. We look at mechanisms that maximise the *(social) welfare* wrt *(reported) profile* $(\tau'_j)_m$, i.e. return $\mathcal{P}_{\max}((\tau'_j)_m) = arg\,max_{\mathcal{P}:\text{ feasible for }\mathcal{I}} \sum_{v_i\in V}\sum_{r_j\in R} x_{ji}(\mathcal{P}) \cdot \tilde{u}_{ji}(\mathcal{P}, \tau'_j)$.

Maximising Outcomes

Let agents report their private threshold values truthfully. We consider the MAXIMISEOUTCOMES problem: given $\mathcal{I}$, $(\tau_j)_m$, $h \in \mathbb{N}_{\geq 1}$, $F_1, \ldots, F_h \subseteq V$, $C_1, \ldots, C_h \subseteq R$, and $k_1, \ldots, k_h \in \mathbb{R}^{\leq 0}_{>-\infty}$, does it exist a feasible plan $\mathcal{P}$ for instance $\mathcal{I}$ with $\sum_{v_i\in F_g}\sum_{r_j\in C_g} x_{ji}(\mathcal{P}) \cdot u_{ji}(\mathcal{P}) \geq k_g$ for each $g \in \{1, \ldots, h\}$?

This decision problem is relevant whenever the fleet of vehicles is split into sub-fleets and each sub-fleet is assigned to service requests whose locations belong to some, possibly different, subset of L, e.g. regions, age groups. We look at two extreme cases of the MAXIMISEOUTCOMES problem.

MAXIMISEVALUATION is MAXIMISEOUTCOMES with $h = 1$, $F_1 = V$, $C_1 = \{r_j\}$, and $k_1 = \kappa$. This decision problem is relevant whenever we care a lot for a given agent j. For example, in the context of dispatching ambulances, we may want to bring a given patient much quicker to a hospital than all other patients.

Theorem 1 *For given agent j and $\kappa \in \mathbb{R}^{\leq 0}_{>-\infty}$, deciding whether the private valuation of agent j is at least κ in some feasible plan (i.e.* MAXIMISEVALUATION*) can take $O(n \cdot (|E| + |L| \cdot \log |L|))$ time.*

Proof Pick instance $\mathcal{I}$ and agent j. Their waiting plus detour time in a feasible plan is minimised when we send the fastest vehicle $v_i \in F_1$ from b_i to p_j via the quickest path for v_i and then from p_j to d_j via the quickest path for v_i. For each $v_k \in F_1$, we can compute these two paths in $O(|E| + |L| \cdot \log |L|)$ time by making two calls to the improved version of the Dijkstra's algorithm—see e.g. [3]—with inputs N, b_k, p_j, T_k, and N, p_j, d_j, T_k, respectively. After that, we can select in $O(n)$ time the fastest vehicle $v_i \in F_1$. Thus, given τ_j, the private valuation of agent j is the greatest possible they can receive in a feasible plan for $\mathcal{I}$. If it is at least κ then the answer is "yes". Otherwise, the answer is "no" and no other procedure can give an answer "yes". □

MAXIMISEWELFARE is MAXIMISEOUTCOMES with $h = 1$, $F_1 = V$, $C_1 = R$, and $k_1 = \kappa$. This is relevant whenever we care a lot for all m agents. For example, in the context of dispatching ambulances, we may want to bring all patients to the hospitals as quickly as possible.

Theorem 2 *For given m agents and $\kappa \in \mathbb{R}^{\leq 0}_{>-\infty}$, deciding whether the welfare is at least κ in some feasible plan (i.e.* MAXIMISEWELFARE*) is* NP-*hard.*

To prove Theorem 2, we reduce from the TRP. By Theorem 2, it follows that the more general problem MAXIMISEOUTCOMES is NP-hard as well.

Possible Manipulations

If a given agent reports truthfully that they can wait for say 8 min then maximising the welfare may force them to wait for 9 min because doing so may assign them to a vehicle that has to pick up someone else before them. However, if the agent underreports their private threshold value by say 8 min then they can force the mechanism to send a vehicle to their location in the next 3 min and, thus, strictly increase their private valuation: see Example 1.

Example 1 Let us consider locations $A(0, 0)$, $B(0, 3)$, $C(4, 0)$, and $D(2, 3)$. Suppose that there is one vehicle v_1 that begins at B, ends at D, has a capacity of two, and has a minute metric defined as: $t(1, A, B) = t(1, B, A) = 3$, $t(1, A, C) = t(1, C, A) = 4$, $t(1, B, C) = t(1, C, B) = 5$, $t(1, A, D) = t(1, D, A) = \sqrt{13}$, and $t(1, C, D) = t(1, D, C) = \sqrt{13}$. Also, suppose that there is one agent 1 that submits $r_1 = (A, C)$ and another agent 2 that submits $r_2 = (C, A)$. Consider the following two feasible plans: $\mathcal{P}_1 = \{(B, C, A, C, D)\}$ and $\mathcal{P}_2 = \{(B, A, C, A, D)\}$. For plan $\mathcal{P}_1$, we have $w_{11}(\mathcal{P}_1) + d_{11}(\mathcal{P}_1) = 9$ and $w_{21}(\mathcal{P}_1) + d_{21}(\mathcal{P}_1) = 5$. For plan $\mathcal{P}_2$, we have $w_{11}(\mathcal{P}_2) + d_{11}(\mathcal{P}_2) = 3$ and $w_{21}(\mathcal{P}_2) + d_{21}(\mathcal{P}_2) = 7$.

Suppose that the private threshold value of agent 1 is $\tau_1 = 8$ and the one of agent 2 is $\tau_2 = 0$. Thus, supposing that agents report these values truthfully, we derive $\tilde{u}_{11}(\mathcal{P}_1, \tau_1) + \tilde{u}_{21}(\mathcal{P}_1, \tau_2) = (8 - 9) + (0 - 5) = -6 > -7 = (0) + (0 - 7) = \tilde{u}_{11}(\mathcal{P}_2, \tau_1) + \tilde{u}_{21}(\mathcal{P}_2, \tau_2)$. Therefore, each maximizing mechanism returns $\mathcal{P}_1$. However, if agent 1 reports $\tau_1' = 0$ then they force any such mechanism to return $\mathcal{P}_2$ because $\tilde{u}_{11}(\mathcal{P}_1, \tau_1') + \tilde{u}_{21}(\mathcal{P}_1, \tau_2) = (0 - 9) + (0 - 5) = -14 < -10 = (0 - 3) + (0 - 7) = \tilde{u}_{11}(\mathcal{P}_2, \tau_1') + \tilde{u}_{21}(\mathcal{P}_2, \tau_2)$ holds. Thus, agent 1 receives a greater private valuation: $u_{11}(\mathcal{P}_2) = 0 > -1 = (8 - 9) = u_{11}(\mathcal{P}_1)$. Finally, we note that the travel time decreases from $(13 + \sqrt{13})$ in $\mathcal{P}_1$ to $(11 + \sqrt{13})$ in $\mathcal{P}_2$. □

In Example 1, the travel time decreases by 2 min due to strategic behaviour. We might therefore be interested in deciding manipulations that induce the lowest travel time. For this reason, we study the general POSSIBLEMANIPULATION problem: given $\mathcal{I}$, $(\tau'_j)_m$, and agent j, is there $\tau''_j \in \mathbb{R}_{\geq 0}$ such that $u_{jk}(\mathcal{P}_{\max}((\tau'_1, \ldots, \tau''_j, \ldots, \tau'_m))) > u_{ji}(\mathcal{P}_{\max}((\tau'_j)_m))$ holds? To prove Theorem 3, we reduce from the HPP.

Theorem 3 *Deciding whether a given agent can report a threshold value, changing the outcome of a given welfare-maximising mechanism, and thus strictly increasing their own private valuation (i.e.* POSSIBLEMANIPULATION*) is* NP*-hard.*

Pure Nash Equilibria

Agents can use joint strategies to manipulate welfare-maximising mechanisms, according to which not just one but several of them deviate from reporting their private threshold values truthfully. We next look at the equilibrium points—see e.g. [8]—of these strategies.

More formally, for any given welfare-maximising mechanism and instance $\mathcal{I}$, we say that the reported profile $(\tau'_j)_m$ is a *pure Nash equilibrium (PNE)* iff, for every agent $j \in \{1, \ldots, m\}$ and $\tau''_j \in \mathbb{R}_{\geq 0}$, $u_{jk}(\mathcal{P}_{\max}((\tau'_1, \ldots, \tau''_j, \ldots, \tau'_m))) \leq u_{ji}(\mathcal{P}_{\max}((\tau'_j)_m))$ holds, where agent j is serviced by v_k and v_i, respectively.

Theorem 4 *Computing pure Nash equilibria is at least* coNP*-hard.*

Proof We start with verifying pure Nash equilibria. By definition, $(\tau'_j)_m$ is a pure Nash equilibrium if and only if, for every agent $j \in \{1, \ldots, m\}$, it is not the case that there exists $\tau''_j \in \mathbb{R}_{\geq 0}$ such that $u_{jk}(\mathcal{P}_{\max}((\tau'_1, \ldots, \tau''_j, \ldots, \tau'_m))) > u_{ji}(\mathcal{P}_{\max}((\tau'_j)_m))$ holds. A decision problem is coNP-hard if and only if its negated version is NP-hard: see e.g. [4]. By using this Turing reduction and the result in Theorem 3, it follows that verifying a pure Nash equilibrium requires m calls to a coNP-hard oracle and, therefore, computing one such profile is at least coNP-hard. □

The Price of Anarchy

The price of anarchy—see e.g. [6]—quantifies the welfare loss due to strategic behaviour. In our setting, for some $\epsilon < 0$ that is very close to 0, the price is the ratio between the smallest maximum welfare value in any pure Nash equilibrium and the maximum welfare value in the truthful profile $[\min_{(\tau'_j)_m:\,\text{PNE}} \sum_{i \in V} \sum_{j \in R} x_{ji}(\mathcal{P}_{\max}((\tau'_j)_m)) \cdot u_{ji}(\mathcal{P}_{\max}((\tau'_j)_m)) + \epsilon]/[\sum_{i \in V} \sum_{j \in R} x_{ji}(\mathcal{P}_{\max}((\tau_j)_m)) \cdot u_{ji}(\mathcal{P}_{\max}((\tau_j)_m)) + \epsilon]$.

If a given agent reports a high threshold value truthfully then they might be forced to wait even longer, and if they report a low threshold value strategically then they might be serviced within their private threshold value in every pure Nash equilibrium, but thus make another agent with a low private threshold value wait for a very long time and cause arbitrarily welfare loss. As a ratio between two negative values, the price can thus go to ∞ in some edge cases.

Theorem 5 *The price of anarchy can be approaching* ∞.

Conclusions

We studied maximising the social welfare in routing problems where customers have preferences for their waiting and detour times in travel options. Our preliminary results have at least three practical implications. Firstly, in practice, agents may give up manipulating any welfare-maximising mechanism because deciding whether such manipulations are beneficial for them may take more time than they can wait for and detour in vehicles. Secondly, in practice, encouraging agents to be truthful eliminates any possibility of welfare losses with such mechanisms. Thirdly, in practice, we may need a lot of time to maximise the welfare exactly because there are many customers and, for this reason, we might wish to use approximation algorithms for this task. Finally, as a response to these observations, our future work includes approximating the welfare in simulated and real-world environments.

Acknowledgements Martin Aleksandrov was supported by the DFG Individual Research Grant on "Fairness and Efficiency in Emerging Vehicle Routing Problems" (497791398).

References

1. Afrati, F., Cosmadakis, S., Papadimitriou, C. H., Papageorgiou, G., & Papakostantinou, N. (1986). The complexity of the travelling repairman problem. *RAIRO—Theoretical Informatics and Applications—Informatique Théorique et Applications, 20*(1), 79–87. http://www.numdam.org/item/ITA_1986_20_1_79_0/
2. Aleksandrov, M. D. (2021). Fleet fairness and fleet efficiency in capacitated pickup and delivery problems. In *Proceedings of the 32nd IEEE Intelligent Vehicles Symposium (IV21)*, Nagoya, Japan, July 11–17, 2021 (pp. 1156–1161). IEEE Xplore. https://doi.org/10.1109/IV48863.2021.9576002
3. Fredman, M., & Tarjan, R. (1984). Fibonacci heaps and their uses in improved network optimization algorithms. In *25th Annual Symposium on Foundations of Computer Science* (pp. 338–346). https://doi.org/10.1109/SFCS.1984.715934
4. Garey, M. R., & Johnson, D. S. (1979). *Computers and intractability: A guide to the theory of NP-completeness*. W. H. Freeman. https://doi.org/10.5555/574848
5. Gschwind, T., & Drexl, M. (2019). Adaptive large neighborhood search with a constant-time feasibility test for the dial-a-ride problem. *Transportation Science, 53*, 480–491. https://doi.org/10.1287/trsc.2018.0837
6. Koutsoupias, E., & Papadimitriou, C. (1999). Worst-case equilibria. In C. Meinel, & S. Tison (Eds.), *Proceedings of 16th Annual Symposium on Theoretical Aspects of Computer Science*. Lecture Notes in Computer Science (Vol. 1563, pp. 404–413). Springer. https://doi.org/10.1007/3-540-49116-3_38

7. Ma, H., Fang, F., & Parkes, D. C. (2020). Spatio-temporal pricing for ridesharing platforms. *SIGecom Exchange, 18*(2), 53–57. https://doi.org/10.1145/3440968.3440975
8. Nash, J. (1950). The bargaining problem. *Econometrica, 18*(2), 155–162. https://doi.org/10.2307/1907266
9. Nucamendi, S., Cardona-Valdes, Y., & Angel-Bello Acosta, F. (2015). Minimizing customers' waiting time in a vehicle routing problem with unit demands. *Journal of Computer and Systems Sciences International, 54*(6), 866–881. https://doi.org/10.1134/S1064230715040024
10. Paquette, J., Bellavance, F., Cordeau, J. F., & Laporte, G. (2012). Measuring quality of service in dial-a-ride operations: The case of a Canadian city. *Transportation, 39*(3), 539–564. https://doi.org/10.1007/s11116-011-9375-4
11. Rheingans-Yoo, D., Kominers, S. D., Ma, H., & Parkes, D. C. (2019). Ridesharing with driver location preferences. In *Proceedings of the Twenty-Eighth International Joint Conference on Artificial Intelligence, IJCAI-19, International Joint Conferences on Artificial Intelligence Organization* (pp. 557–564). https://doi.org/10.24963/ijcai.2019/79
12. Vidal, T., Laporte, G., & Matl, P. (2020). A concise guide to existing and emerging vehicle routing problem variants. *European Journal of Operational Research, 286*(2), 401–416. https://doi.org/10.1016/j.ejor.2019.10.010

Chapter 54
The Grey Zone Two-Echelon Vehicle Routing Problem with Customer-to-Parcel Locations and Low-Pollution Vehicles for Inner-City Logistics

Edgar Ricardo Silva Russi, Nacima Labadie, and Caroline Prodhon

Abstract This study addresses the two-echelon vehicle routing problem with grey zones and Customer-to-Parcel (C2P) stations. This problem arises in the search for new sustainable delivery schemes for e-commerce and last-mile distribution in urban areas and aims at reducing last-mile transportation costs. This study proposes a literature review on the subject, and a mixed integer linear programming (MILP) formulation to model and solve small instances of the described problem.

Keywords 2-Echelon vehicle routing problem · Synchronization · Customer to parcel · City logistics · Last-mile delivery · Sustainable logistics

Introduction

Last-mile logistics has gained greater importance as the demand for *Business-2-Consumer* products in urban areas has increased with the strong growth of e-commerce driven by a wider access to internet services, both for customers and for companies that buy and sell their products through this channel [1]. This generates a huge potential for consolidation and coordination of distribution flows, which play a key role in multi-echelon distribution systems [2] and can help to reduce traffic volume by improving the use of transportation resources [3]. To address last-mile particular challenges, the transition from the use of Internal Combustion Engine Vehicles (ICEV) to Alternative Fuel Vehicles (AFV), the implementation of access restrictions into urban areas and the integration of Customer-to-Parcel stations (e.g. lockers and pick-up points) have become commonly adopted measures. Therefore,

E. R. Silva Russi (✉) · N. Labadie · C. Prodhon
Laboratoire d'Optimisation des Systèmes Industriels (LOSI), Université de Technologie de Troyes, 12 rue Marie Curie, CS 42060, 10004 Troyes Cedex, France
e-mail: edgar.silva_russi@utt.fr

N. Labadie
e-mail: nacima.labadie@utt.fr

C. Prodhon
e-mail: caroline.prodhon@utt.fr

O. Grothe et al. (eds.), *Operations Research Proceedings 2022*, Lecture Notes in Operations Research, https://doi.org/10.1007/978-3-031-24907-5_54

we introduce the Grey Zone Two-Echelon Vehicle Routing Problem with Customer-to-Parcel Locations and Low-pollution Vehicles for Inner-city Logistics. Different from the classical Two-Echelon Vehicle Routing Problem (2E-VRP), clients may be served via either C2P stations or home delivery in a specified time window; furthermore we consider satellites as possible C2P stations and grey-zone customers on city borders. The contributions of this paper are threefold, as described in the following sections: First, a review of the existing literature is given in Section "Literature Review", then Section "Problem Description and Mathematical Formulation", defines the problem and presents a mathematical formulation to model it. Finally, the results and conclusions are given in Sections "Results" and "Conclusion and Future Research".

Literature Review

Regarding innovative notions into 2E-VRP models intended for last-mile logistics, [6] introduced the notion of grey zones in their multi-objective problem denoted 2eVRPSyn. Customers in such zones are accessible by fleets of the two levels, giving more flexibility to their model. They proposed a large neighborhood search embedded in a heuristic rectangle/cuboid splitting to efficiently solve the problem. Moreover, [5] introduced the 2-Echelon Production Routing Problem with Cross-docking Satellites (2E-PRPCS), it features production and inventory decisions and it is solved by a Branch-and-Cut algorithm coupled with a first-solution matheuristic while [4] introduced a new variant called the Two-Echelon multiple-trip Vehicle Routing Problem with Satellite Synchronization (2E-MTVRP-SS) with time windows and service times.

On the other hand, to address e-grocery last-mile challenges, [15] proposed a Decision Support System (DSS) for a model in which the second echelon deliveries are performed through either C2P stations or electric cargo bikes.[16] developed an state of the art Hybrid Immune Algorithm (HIA) to solve a multi-objective Two-Echelon Location-Routing Problem with Mixed Vehicles and Mixed Satellites (2E-LRP-MVMS) in which the second echelon deliveries are performed through either C2P stations or Autonomous Delivery Robots (ADRs). Further, [16] modeled the Two-Echelon Vehicle Routing Problem with Mixed Vehicles (2E-VRP-MV) where Autonomous Delivery Vehicles (ADVs) serve the clients in the second echelon. To solve it, a two-step clustering-based hybrid Genetic Algorithm and Particle Swarm Optimization (C-GA-PSO) algorithm are presented. Moreover, as environmental factors had gained more importance in last-mile delivery schemes, [7] introduced the 2-Echelon Electric Vehicle Routing Problem with Time Windows (2e-EVRP-TW). The model takes into account charging stations, charging times and time windows. To solve it, a heuristic based on Clarke and Wright and a Variable Neighborhood Search (VNS) are presented. For supplementary information on 2E-VRP refer to [8].

Furthermore, it was until 2018 when [9] introduced the Multi-Depot Two-Echelon Vehicle Routing Problem with Delivery Options for Last Mile Distribution (MD-

TEVRP-DO) that the concept of delivery options through C2P stations was incorporated into a 2E-VRP. Moreover, [10] proposed a Simulated Annealing Algorithm to minimize the transportation cost in their conception of 2E-VRP with locker facilities; while [11] introduced the Two-Echelon Vehicle Routing Problem with Coverage Options (2E-VRP-CO) where last mile deliveries are performed via C2P stations and cargo bikes which is then solved thanks to an Adaptive Large Neighborhood Search (ALNS). Similarly, [12] considered the synchronization of cargo bikes and vans in their two-echelon model solved by a Greedy Randomized Adaptive Search heuristic with path re-linking for real-world data of the city of Vienna. Later, [13] introduced the concept of occasional drivers in a Two-Echelon Vehicle Routing Problem with Time Windows, Coverage Options, and Occasional Drivers (a crowd-shipping concept first introduced by [14] that can provide potential advantages by its implementation). For further information on the new developments in last-mile logistics, see [17].

Problem Description and Mathematical Formulation

The problem that arises involves two different fleets of vehicles making deliveries in two different echelons. In our particular case, we consider that the vehicles start and end their routes at their respective depot. ICEVs deliver the goods from the depot of the first echelon, where the entire stock of goods is located, to the first level customers within their time windows and/or to the exchange points, also called satellites. These last are also used as Customer-to-Parcel (C2P) stations where customers can pick up their products directly. AFVs start from the second echelon depot without any goods in cargo so they must immediately meet ICEVs at the satellites where they will pick up the goods to start their deliveries. Then the AFVs serve both direct delivery customers within their respective time windows and second-level C2P stations where customers can retrieve their goods. Due to the costs associated with the use of vehicles and satellites, the waiting time of vehicles at these locations is limited to a specific value and is minimized by the inclusion of the waiting costs in the objective function (54.1) of the MILP model. For this reason, the arrival of the ICEVs and AFVs at the satellites must be approximately at the same time so that the transfer of the goods is as fast as possible and both the ICEV and the AFV can continue their deliveries. Once an AFV has finished its deliveries it can rejoin an ICEV at a satellite to retrieve new goods and continue a new route as long as the maximum route time represented by the delivery operator's working day is not exceeded. Likewise, the same satellite can be used by different AFVs as long as the ICEV's maximum waiting time at the satellite is not exceeded. In the traditional formulation of the 2E-VRP the customers of each echelon are pre-allocated to the first or second echelon, however as [6] shows, this pre-allocation can lead to poor quality results because customers close to the satellites could be served by either an ICEV or an AFV. Thus, [6] introduced the grey zone, an area in which customers are not predefined and can therefore be served by any vehicle. In this model, grey zones will be taken into account only for home delivery customers located on the city borders. On the other hand, C2P stations are increasingly used to eliminate the risk of unattended deliveries and to optimize delivery costs thanks to the possibility of accumulating multiple demands at the same station. Similarly, these stations benefit customers by allowing them to pick up their deliveries at times that are convenient for them at stations close to their homes. Each of these stations has a specific capacity and coverage radius. The mathematical model is inspired on [6] but it considers time windows and waiting time on client nodes, C2P stations for both levels, and location decisions for satellites and stations (Table 54.1).

The objective function (54.1) takes into account the fixed cost of vehicles, the cost per distance, the vehicle operating cost per hour for each level, the cost of opening satellites and *C2P* stations, and the waiting costs at satellites and client nodes. The classic restrictions of the two-echelon models were formulated (i.e. restrictions on flow, exit and return to the depots, use of vehicles, and service to home-delivery customers for each level).

$$\min \sum_{i\in v^{all}} \sum_{j\in v^{all}} \sum_{k\in F} [(T(i,j) + ST(i)) * CT(k) + (Dis(i,j) * CD(k))] * X(i,j,k) + \sum_{i\in V^{all}} \sum_{k\in F} W(i,k) * CT(k) + \sum_{e\in E} \sum_{j\in v^{all}} \sum_{k\in F} X(e,j,k) * FC(k) + \sum_{i\in c2p} Y(i) * CU(i) \tag{54.1}$$

Also, temporal constraints have been considered to guarantee the service to the clients within their time window. Similarly, for node routing (54.2), the time between nodes, the service time, and the waiting time at satellites and at client's nodes (in case of arrival before the lower limit of the time window) have been considered. Finally, (54.3) limits the maximum duration of the routes.

$$T(j,k) \geq (T(i,k) + Time(i,j) + ST(i) + W(i,k)) - M* (1 - X(i,j,k)) \quad \forall i \in v^{1,2}, j \in v^{1,2}, k \in F^{1,2} \tag{54.2}$$

$$T(e^f, k) \leq TMAX \quad \forall e^f \in E^f, k \in F \tag{54.3}$$

Subsequently, constraints (54.4)–(54.6) guarantee that the vehicle load contains the demands of both home-delivery customers and C2P stations. Furthermore, capacity constraints have been formulated for all vehicles and C2P stations.

$$U(j,k) + D(j) \leq U(i,k) + M * (1 - X(i,j,k)) \forall i \in v^{all}, j \in chd, k \in F \tag{54.4}$$

$$U(j,k) + \sum_{c\in cloc^2} S(c,j)*D(c) \leq U(i,k) + M * (1 - \sum_{s} X(i,j,k)) \quad \forall i \in v^{all}, j \in loc, k \in F^2 \tag{54.5}$$

Table 54.1 Notation

Sets	
v^{all}	Set of all nodes
E/E^{f}	Set of the real and the cloned depots of each Echelon
$chd/chd^{0,1,2}$	Set of all home delivery clients/set of home delivery clients in the Grey Zone (0), 1st Echelon (1) and 2nd Echelon (2)
$c2p$	Set of all possible locations for lockers and satellites
sat/sat'	Set of real and cloned satellites each physical satellite (*sat*) is duplicated *n* times (once for each 2nd and Grey Zone customer).
loc	Set of possible location for lockers
$cloc/cloc^{1,2}$	Set of all *c2p* clients/set of *c2p* clients from the 1st (1) and 2nd Echelon (2)
$v^{0,1,2}$	Set of nodes that can be visited by all vehicles (0), only by 1st Echelon vehicles (1) and only by 2nd Echelon vehicles (2)
$F/F^{1;2}$	Fleet of all vehicles/fleet of 1st (1) and 2nd Echelon (2) vehicles
Parameters	
M	Big number/sum of all customer demands
TMAX/WMAX	Maximum route duration/"(TMAX/10)" as a value of the waiting time allowed at satellites
Time/Dis(i,j)	Travel time and distance from node (*i*) to node (*j*)
Q(k)/QLoc(c2p)	Capacity of vehicle (*k*) and capacity of the (*c2p*) station
CT/CD(k)	Cost of using the vehicle (*k*) per unit of time and distance
CF(k)	Fixed costs of using the vehicle (*k*)
CU/Radius(c2p)	Opening cost and Covering radius of (*c2p*) station
ST/D(i)	Service time and demand in node (*i*)
EA/LA(chd)	Earliest and latest arrival time to home delivery clients (*chd*)
Variables	
X(i,j,k)	Binary variable, 1 if the arc (*i,j*) is crossed by the vehicle (*k*)
T(i,k)	Arrival time of vehicle (*k*) at node (*i*)
Y(c2p)	Binary variable, 1 if the (*c2p*) station is opened, 0 otherwise
U(i,k)	Load of vehicle (*k*) at node (*i*)
W(i,k)	Waiting time of vehicle (*k*) at node (*i*)
S(cloc,c2p)	Binary variable, 1 if the (*c2p*) station serves the client (*cloc*)

$$U(j,l) + U(j,k) + \sum_{c \in cloc^1} S(c,j) * D(c) \leq U(i,k) + M * (1 - X(i,j,k)) \quad \forall i \in v^{all}, j \in sat', k \in F^1, l \in F^2 \tag{54.6}$$

Moreover, (54.7) ensure that if a satellite is used by an ICEV, it must also be used by an AFV. Likewise, the waiting times are calculated for home delivery customers and satellites, whereas (54.8) determine the maximum waiting time allowed on satellites. Finally, (54.9) ensure that vehicles do not contain cargo when returning to their respective depot at the end of their routes, and (54.10) guarantee that AFVs do not contain any load when visiting a satellite.

$$\sum_{i \in v^{1,2}} X(i, s, k) = \sum_{j \in v^{1,2}} X(s, j, k) \quad \forall s \in sat', k \in F^{1,2} \tag{54.7}$$

$$W(s, k) \leq \sum_{j \in v^{1,2}} (X(s, j, k)) * WMAX \quad \forall s \in sat', k \in F^{1,2} \tag{54.8}$$

$$U(i, k) \leq M * (1 - X(i, e^f, k)) \quad \forall e^f \in E^f, i \in v^{1,2}, k \in F^{1,2} \tag{54.9}$$

$$U(i, k) \leq M * (1 - X(i, s, k)) \quad \forall i \in v^2, k \in F^2, s \in sat' \tag{54.10}$$

Then, C2P stations constraints were defined (i.e. opening stations, allocation of customers and service at the opened ones). Subsequently (54.11) limit to one station per client and (54.12) guarantee that for opened stations, customers are within the coverage radius.

$$\sum_{j \in c2p} S(i, j) = 1 \quad \forall i \in cloc \tag{54.11}$$

$$S(i, j) * Ds(i, j) \leq R(j) \quad \forall i \in cloc, j \in c2p \tag{54.12}$$

Results

See Table 54.2.

Our model was implemented in GUROBI 10.0 coded in Python 3.8.8 and tested under an Intel(R) Core (TM) i7-VPRO CPU 2.3 GHz 16 GB RAM is efficient on instances of up to 25 nodes. For all the C2P clients, the problem is first defined as an assignment problem and then a routing problem for the opened stations, which are accessible at all times (no time windows) and create more flexibility for routing. Similarly, grey zone customers add further flexibility to the model, as they can be part of any vehicle itinerary. However, home-delivery clients reduce model flexibility as their arrival times must be within their time windows. It is relevant to highlight that the definition of the maximum waiting time allowed plays a determining factor in the correct synchronization of the vehicles at the satellites and, consequently, in the route calculation. As well, it is noted that if the model does not allow vehicles to arrive before the lower limit of the time window (which generates waiting times at the customer nodes) the results tend to increase the use of vehicles, to create longer waiting times at the satellites and, in certain cases, due to the inflexibility of the problem, to end up in a non-feasible zone.

Table 54.2 Computational results

chd^0	chd^1	chd^2	F^1	F^2	sat'	loc	$cloc^{1/2}$	$CV^{1/2}$	v^{all}	WT	Ex.Time	Objective	GAP %
1	1	2	1	1	2	1	1/1	75/50	15	15	0.5061	2 575	0.00
1	1	2	2	2	2	1	1/1	30/10	15	45	0.8160	6 935	0.00
1	1	2	3	3	2	1	1/1	30/10	15	45	0.8755	5 680	0.00
2	2	2	1	2	2	2	2/2	50/35	20	15	275.85	4 465	0.00
2	2	2	1	2	2	2	2/2	40/20	20	15	236.41	5 335	0.00
2	2	2	1	2	2	2	2/2	35/15	20	20	96.328	6 220	0.00
3	1	2	3	3	2	2	2/2	40/20	20	15	1351.7	2 335	0.00
4	0	2	3	3	2	2	2/2	40/20	20	25	301.59	4 950	0.00
2	4	2	3	3	2	2	3/3	75/50	24	0	3982.0	5 230	0.00
4	2	2	3	3	2	2	3/3	75/50	24	–	9000.0	4 200	10.5
4	4	4	3	3	2	2	3/3	75/50	28	–	9000.0	5 900	14.4
4	4	4	3	3	2	2	3/3	50/25	28	–	9000.0	7 059	23.8
4	4	4	3	3	2	2	3/3	45/25	28	–	9000.0	7 135	23.8
4	4	4	3	3	4	2	3/3	75/50	32	–	9000.0	5 375	18.0
4	4	4	3	3	4	2	3/3	45/25	32	–	9000.0	7 135	24.9
4	4	4	3	3	4	2	3/3	50/35	32	–	9000.0	6 800	30.0

Conclusion and Future Research

Our model aims to offer a last mile delivery scheme, where the costs associated with waiting times at customer nodes and satellites are minimized through synchronization at the satellites and waiting costs reflected in the objective. Moreover, our model includes delivery options through the integration of C2P stations in addition to traditional home deliveries under time windows. Our mathematical formulation is efficient for small size instances, however due to the NP-hard nature of the problem, it is necessary to build approximate algorithms for solving large size and real-city instances to compare the performance of our model under conditions closer to those faced by real last mile delivery services.

References

1. Ismail, S. B., Legras, F., & Coppin, G. (2006). Synthèse du problème de routage de véhicules. Research Report. HAL, vol. RR-2011-03-LUSSI (pp. 1–52). Institut TELECOM.
2. Lewczuk, K., Zak, J., Pyza, D., $ Jacyna-Golda, I. (2013). Vehicle routing in an urban area: Environmental and technological determinants. *WIT Transactions on the Built Environment, 130*, 373–384.
3. Gonzalez-Feliu, J., Perboli, G., Tadei, R., & Vigo D. (2008). The two-echelon capacitated vehicle routing problem. HAL.
4. Grangier, P., Gendreau, M., Lehuédé, P., & Rousseau, L. (2016). An adaptive large neighborhood search for the two-echelon multiple-trip vehicle routing problem with satellite synchronization. *In European Journal of Operational Research, 254*, 80–91.
5. Qiu, Y., Zhou, D., Du, Y., Liu, J., Pardalos, P. M., & Qiao, J. (2021). The two-echelon production routing problem with cross-docking satellites. *Transportation Research Part E: Logistics and Transportation Review, 147*, 102210.
6. Anderluh, A., Nolz, P. C., Hemmelmayr, V. C., & Crainic, T. G. (2021). Multi-objective optimization of a two-echelon vehicle routing problem with vehicle synchronization and 'grey zone' customers arising in urban logistics. *European Journal of Operational Research, 289*, 940–958.
7. Anıl, A. M., Kalayci, C. B., Blum, C., & Polat, O. (2022). Variable neighborhood search for the two-echelon electric vehicle routing problem with time windows. *Applied Sciences, 12*.
8. Sluijk, N., Florio, A. M., Kinable, J., Dellaert, N., & Van Woensel, T. (2022). Two-echelon vehicle routing problems: A literature review. *European Journal of Operational Research*.
9. Zhou, L., Baldacci, R., Vigo, D., & Wang, X. (2018). A multi-depot two-echelon vehicle routing problem with delivery options arising in the last mile distribution. *European Journal of Operational Research, 265*, 765–778.
10. Redi, A. A. N. P., Jewpanya, P., Kurniawan, A. C., Persada, S. F., Nadlifatin, R., & Dewi, O. A. C. (2020). A simulated annealing algorithm for solving two-echelon vehicle routing problem with locker facilities. *Algorithms, 13*, 1–14.
11. Enthoven, D. L. J. U., Jargalsaikhan B., Roodbergen K. J., uit het Broek, M. A. J., & Schrotenboer A. H. (2020). The two-echelon vehicle routing problem with covering options: City logistics with cargo bikes and parcel lockers. *Computers & Operations Research, 118*, 104919.
12. Anderluh, A., Hemmelmayr, V. C., & Nolz, P. C. (2017). Synchronizing vans and cargo bikes in a city distribution network. *Central European Journal of Operations Research, 25*, 1005–1016.
13. Yu, V. F., Jodiawan, P., Hou, M.-L., & Gunawan, A. (2021). Design of a two-echelon freight distribution system in last-mile logistics considering covering locations and occasional drivers. Transportation Research Part E: Logistics and Transportation Review, 154, 102461.

14. Macrina, G., Di Puglia Pugliese, L., Guerriero, F., & Laganà, D. (2017). The vehicle routing problem with occasional drivers and time windows. In A. Sforza, & C. Sterle (Eds.), *Optimization and decision science: Methodologies and applications*, Vol. 217. Springer Proceedings in Mathematics & Statistics.
15. Leyerer, M., Sonneberg, M., & Breitner, M. H. (2018). Decision support for urban E-grocery operations completed research. In *24th Americas Conference on Information Systems. AMCIS 2018.*
16. Liu, D., Deng, Z., Mao, X., Yang, Y., & Kaisar, E. I. (2020). Two-echelon vehicle-routing problem: Optimization of autonomous delivery vehicle-assisted E-grocery distribution. *IEEE Access, 8*, 108705–108719.
17. Boysen, N., Fedtke, S., & Schwerdfeger, S. (2021). Last-mile delivery concepts: A survey from an operational research perspective. *Spectrum, 43*, 1–58. OR.

Part XII
Mobility and Traffic

Chapter 55
A General Framework to Evaluate Different Rebalancing Operations Strategies in One-Way Car Sharing Systems

Selin Ataç, Nikola Obrenović, and Michel Bierlaire

Abstract Car sharing systems (CSSs) are one of the environmentally beneficial solutions in urban transportation. However, the operators still struggle to make these systems profitable. One of the main contributors in operational cost is rebalancing operations. Therefore, it is important to identify strategies that are tailored according to the needs of the considered system. To overcome this challenge, this work proposes a simulation-optimization framework that compares different rebalancing operations strategies in one-way station-based car sharing systems in terms of cost and level of service. The simulation module utilizes the Multi-Agent Transport Simulation Toolkit (MATSim) whilst the rebalancing operations are determined in the optimization module. The framework allows us to explore the different uncertainties that can occur in the system, such as fluctuations in trip demand thanks to the MATSim. The results of the framework help the operator to better analyze the system and the best rebalancing strategy under different scenarios.

Keywords Car sharing systems · Optimization · Agent-based simulation

Introduction and Literature Review

Car sharing systems are considered to be one of the sustainable mobility solutions. The higher parking and vehicle utilization, the more can environment benefit from its usage. From the user perspective, it becomes attractive as they share the fixed costs of owning a car, such as insurance, maintenance, and parking, with other system

S. Ataç (✉) · M. Bierlaire
Transport and Mobility Laboratory (TRANSP-OR), École Polytechnique Fédérale de Lausanne, Lausanne 1015, Switzerland
e-mail: selin.atac@epfl.ch

M. Bierlaire
e-mail: michel.bierlaire@epfl.ch

N. Obrenović
BioSense Institute, University of Novi Sad, Dr Zorana Dinđića 1, Novi Sad 21000, Serbia
e-mail: nikola.obrenovic@biosense.rs

O. Grothe et al. (eds.), *Operations Research Proceedings 2022*, Lecture Notes in Operations Research, https://doi.org/10.1007/978-3-031-24907-5_55

users. On the other hand, this requires smart decisions at every decision level, i.e., strategic, tactical, and operational. This paper focuses on the tactical and operational level decisions [1]. We kindly ask the reader to refer to [1] for the car sharing system terminology.

Initial systems are formed as round-trip, and later with the technology, they are replaced with station-based one-way and free-floating systems. However, increasing user flexibility leads to more complex challenges for the operator. These include rebalancing operations, trip demand forecasting, and disaggregate mode choice at the operational level.

Rebalancing operations are applied in systems, where one-way trips are allowed, to reduce the vehicle imbalance in the service area. The rebalancing operations can be static or dynamic. In the former, the operations are conducted at night or when the system is low in operation [2], whereas in the latter, they are done during the system operating hours. In general, the network is expanded to a time-space graph to represent the dynamic structure [3, 4], which increases the computational complexity of the problem. Therefore, the works propose heuristic algorithms to overcome the burdens of the computational complexity.

Obtaining and utilizing disaggregate data for trip demand forecasting is effortful. It requires a detailed survey, analysis, and computational time, whereas using such data is essential to reflect the heterogeneity of the population and see the direct effects on individuals. Traditional four-step trip-based models (FSMs), that include trip generation, trip distribution, mode choice, and traffic assignment, cannot answer complex questions as they are static and sequential. Therefore, the literature proposes transport simulation toolkits that are activity-based multi-agent platforms to be able to analyze each agent. Some examples to such toolkits are the Multi-agent Transport Simulation Toolkit (MATSim), SimMobility, and mobiTopp [5].

In the literature, most works focus on one specific subject rather than having a holistic approach. Furthermore, although activity-based multi-agent transport simulation toolkits can handle the disaggregate data, they lack the representation of the supply side. To fill this gap, we introduce a framework, that consists of three main components: the agent-based transport simulator, rebalancing operations optimization that follows a rebalancing operations strategy, and choice modeling that affects the plans of the agents. We use this framework to identify the best rebalancing strategy in combination with agent-based modeling for a one-way station-based car sharing system with operator-based rebalancing operations solutions, which is not studied in the literature, to the best of our knowledge. This way both supply and demand sides of car sharing systems are considered. We utilize MATSim as a transport simulator because of the possibility to simulate car sharing transport mode [6]. The disaggregate nature of MATSim allows a detailed analysis regarding the most suitable rebalancing operations strategy.

Methodology

The proposed framework is presented in Fig. 55.1. In our case, the transport simulation refers to MATSim but any other transport simulation toolkit can be used. We kindly ask the reader to refer to [7] for further details on MATSim.

MATSim receives the daily plans of the agents (i.e., the start and end times of each activity, the transport mode, and purpose of the trip) as well as the system parameters (i.e., location of stations and facilities, car sharing system membership information, initial vehicle and parking configuration, socio-economic characteristics of the agents, and public transport schedule). Then, MATSim simulates the given day, calculates the utilities of each agent, each agent replans their day according to their utilities in the previous iteration and the given day is simulated once again until the pre-specified number of iterations, I, is reached. We refer to this loop iterations as *inner-loop iterations* and present it in red arrows in Fig. 55.1.

The output of the simulation gives the realized car sharing trips, which helps us to compute the final state of the vehicles and parking spots. This information is passed to the rebalancing operations optimization module and the initial vehicle configuration of the following day is determined according to the rebalancing strategy. The initial vehicle and parking configuration is modified accordingly and the feedback loop is then closed by triggering the next iteration of the *outer-loop iterations*, which are shown in black arrows in Fig. 55.1. The change in initial configuration modifies the choices of the agents in the next outer-loop iteration. Here, an outer-loop iteration corresponds to a one simulated day and run for pre-specified number of times, O, to observe the convergence.

Within MATSim, the generalized cost of car sharing travel from activity $q-1$ to activity q is shown in Eq. (55.1) [7].

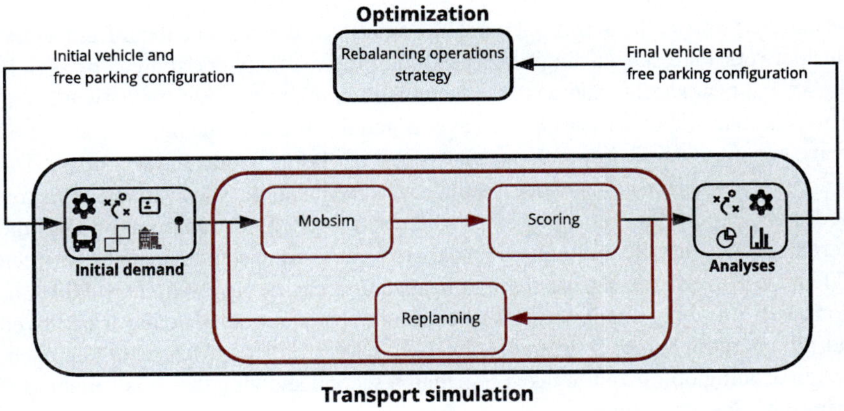

Fig. 55.1 The framework

$$U_{trav,q,cs} = \alpha_{cs} + \beta_{c,cs} \cdot c_t \cdot t_r + \beta_{c,cs} \cdot c_d \cdot d + \beta_{t,walk} \cdot (t_a + t_e) + \beta_{t,cs} \cdot t \tag{55.1}$$

The first term, α_{cs}, refers to the alternative specific constant of the car sharing alternative. The second term relates to the time whilst the third is the distance dependent component of the fee. The access and egress times to and from the stations are considered in the fourth term and the coefficient of the last term represents the marginal utility of an additional unit of time spent on traveling with car sharing, where t is the actual in vehicle travel time.

For the other modes such as walking, private car, public transportation, and bike, the utility of traveling is shown in Eq. (55.2).

$$U_{trav,q,mode} = \alpha_{mode} + \beta_{c,mode} \cdot c_d \cdot d + \beta_{t,mode} \cdot t \tag{55.2}$$

For the current state of research, we test two different rebalancing strategies. In the first strategy, *do nothing*, the final configuration of the cars from the previous iteration is taken as an initial configuration for the following iteration. In the second strategy, *rebalance*, we follow a heuristic approach. The minimum vehicle inventory reached per station during the day is defined as the minimum number of vehicles required for that station. If the total number of available vehicles is more than the total number of minimum required vehicles per station, we sequentially distribute the excess number of vehicles among stations. Finally, this obtained configuration becomes the initial configuration for the following *outer-loop* iteration.

Results

We use the Sioux Falls, South Dakota, USA case study to experiment our framework. This network represents a simplified version of the real network and can be seen in Fig. 55.2. 24 car sharing stations are placed at each intersection of the network. The provided plans file consists of 84,110 agents and the 100% of the population is used for the experiments. There are three event types, i.e., home, work, and secondary. The four facility types are home, work, secondary, and education. The available transport modes are car, public transport, bike, walk, and one-way car share. Although the literature states that the willingness to walk to a car sharing station changes between 400 and 800 m [8], we set the search distance of a car sharing vehicle to 200 m as Sioux Falls is a very small network. We assume that static rebalancing is deployed and the operations happen instantaneously. As we are using a heuristic approach, the rebalancing optimization takes less than a second showing that it is suitable for operational level decisions.

For preliminary experiments, we run 100 inner-loop, and 10 outer-loop iterations. When we look at Figs. 55.4 and 55.3, we see that the transport simulation converges at around 70 iterations. This observation is important as determining the cut-off itera-

Fig. 55.2 Sioux Falls network

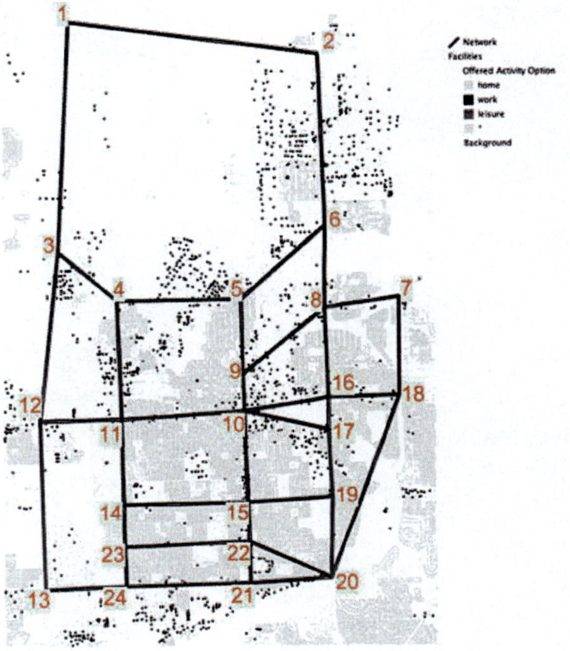

Fig. 55.3 Mode statistics

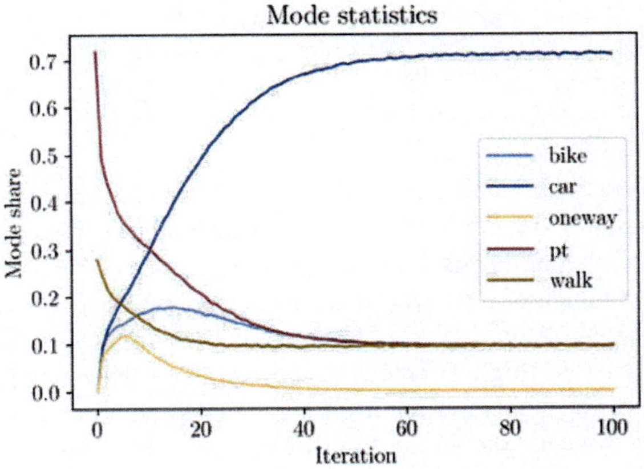

tion number saves considerable amount of computation time. Furthermore, Fig. 55.4 gives us some insights on the mode share. The respective mode shares for the modes car, public transport, bike, walk, and one-way car share are 70.9%, 9.6%, 9.6%, 9.7%, and 0.2%. Regarding the trip purpose, we observe that 56% of the activities are work related whereas 44% are secondary activities such as from home to secondary or vice versa.

Figure 55.5 shows the results for both strategies. The outer-loop iterations are depicted in the x-axis and the y-axis shows the number of rentals at each outer-loop

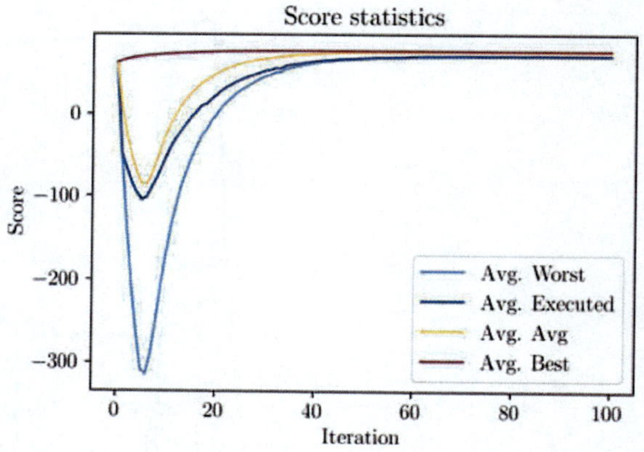

Fig. 55.4 Score statistics

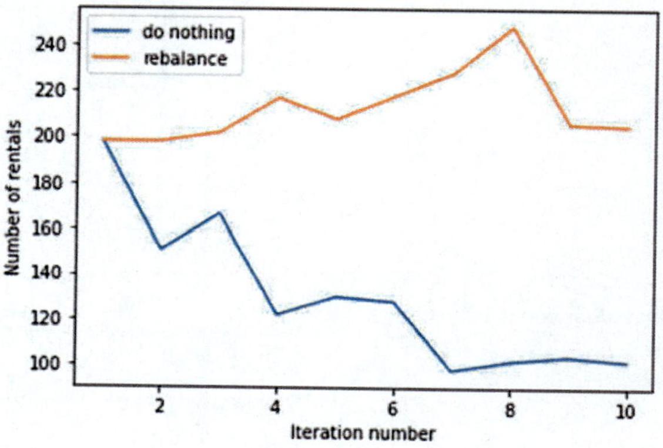

Fig. 55.5 Number of rentals both strategies

iteration for both strategies. We see that *rebalance* strategy is more stable than the *do nothing* strategy. Also, in line with the intuition, the *rebalance* strategy leads to higher number of rentals than the *do nothing* strategy. The fluctuations for each strategy can be explained by the fact that unnecessary rebalancing operations are conducted which leads to few number of parking spots in some specific stations where the drop-off demand is high. For the *do nothing* strategy, the trend of number of rentals is negative, i.e., the number of rentals is less and less with increasing number of iterations. This is expected due to the fact that the pick-up stations have less and less vehicles and drop-off stations do not have enough parking spots to serve the drop-off demand.

After analyzing the fluctuations, we observe that the number of rentals tend to decrease for some number of iterations after reaching a relatively high number of iterations for *do nothing* strategy whereas for *rebalance* strategy, the behavior is opposite, i.e., the number of rentals tend to increase after reaching a relatively low value. This also results in the positive and negative trends of the strategies.

Conclusion and Future Work

In this work, we introduce a holistic framework that aims to compare different rebalancing strategies involving agent-based transport simulation, rebalancing operations optimization, and choice modeling. Later, we present a case study based on Sioux Falls, USA, and preliminary results for two rebalancing strategies. Future work includes investigating the results with higher number of outer-loop iterations to see the convergence and generalize results. We also plan to include more sophisticated rebalancing strategies as well as simplistic approaches such as equal distribution of vehicles. Furthermore, incorporating user-based rebalancing strategies, where operator offers incentives to the users for specific trips, would be interesting as the choices of the users would depend on the pricing. Finally, as transportation involves discrete choice by its nature, we aim at including a choice model that takes pricing into consideration in the framework.

References

1. Ataç, S., Obrenović, N., & Bierlaire, M. (2021). Vehicle sharing systems: A review and a holistic management framework. *EURO Journal on Transportation and Logistics, 10*, 100033. https://doi.org/10.1016/j.ejtl.2021.100033
2. Kypriadis, D., Pantziou, G., Konstantopoulos, C., & Gavalas, D. (2018). Minimum walking static repositioning in free-floating electric car-sharing systems. In *2018 21st International Conference on Intelligent Transportation Systems (ITSC)*. https://doi.org/10.1109/ITSC.2018.8569912
3. Gambella, C., Malaguti, E., Masini, F., & Vigo, D. (2018). Optimizing relocation operations in electric car-sharing. *Omega, 81*. https://doi.org/10.1016/j.omega.2017.11.0
4. Zhao, M., Li, X., Yin, J., Cui, J., Yang, L., & An, S. (2018). An integrated framework for electric vehicle rebalancing and staff relocation in one-way carsharing systems: Model formulation and Lagrangian relaxation-based solution approach. *Transportation Research Part B: Methodological, 117*. https://doi.org/10.1016/j.trb.2018.09.014
5. Vosooghi, R., Puchinger, J., Jankovic, M., & Sirin, G. (2017). A critical analysis of travel demand estimation for new one-way carsharing systems. In *2017 IEEE 20th International Conference on Intelligent Transportation Systems (ITSC)*, Vol. 47. https://doi.org/10.1109/itsc.2017.8317917
6. Balać, M., Ciari, F., & Axhausen, K. W. (2015). Carsharing demand estimation: Zurich, Switzerland, area case study. *Transportation Research Record, 2563*. https://doi.org/10.3141/2536-02
7. Horni, A., & Nagel, K., Axhausen, K. W. (2016). *The multi-agent transport simulation MATSim*. Ubiquity Press. https://doi.org/10.5334/baw
8. Shaheen, S., Cano, L., & Camel, M. (2016). Exploring electric vehicle carsharing as a mobility option for older adults: A case study of a senior adult community in the San Francisco Bay Area. *International Journal of Sustainable Transportation, 10*. https://doi.org/10.1080/15568318.2014.962675

Chapter 56
A Multi-criteria Assessment Framework for Zero-Emission Vehicles from a Customers' Perspective

Paul Fabianek and Reinhard Madlener

Abstract On the basis of economic and user-relevant criteria, this paper proposes an assessment framework for zero-emission vehicles. This framework enables a transparent evaluation of different zero-emission vehicles (ZEVs) from the perspective of (potential) customers, based on the Analytic Hierarchy Process (AHP) approach for a multi-criteria decision analysis. The relevant criteria for the evaluation were derived from literature and from semi-structured interviews. These interviews were held with individuals having driving experience with both battery electric and fuel cell electric vehicles. An AHP survey was also conducted with ZEV drivers and ZEV-interested individuals for the criteria weights to be determined. Seven criteria were found to be particularly relevant for evaluating zero-emission vehicles: *total cost*, *range*, *charging/refueling time*, *charging/refueling infrastructure availability*, *greenhouse gas emissions*, *spaciousness*, and *driving dynamics*. The assessment framework includes value scores representing the degree to which a specific ZEV satisfies a given quality criterion. The proposed framework combines these scores with the criteria weights derived in the AHP. The framework is useful for the design of ZEVs by vehicle manufacturers.

Keywords Analytic Hierarchy Process · Transportation · OR in Sustainability

P. Fabianek (✉) · R. Madlener
Institute for Future Energy Consumer Needs and Behavior (FCN), School of Business and Economics/E.ON Energy Research Center, RWTH Aachen University, Aachen, Germany
e-mail: Paul.Fabianek@rwth-aachen.de

R. Madlener
e-mail: RMadlener@eonerc.rwth-aachen.de

R. Madlener
Department of Industrial Economics and Technology Management, Norwegian University of Science and Technology (NTNU), Trondheim, Norway

O. Grothe et al. (eds.), *Operations Research Proceedings 2022*, Lecture Notes in Operations Research, https://doi.org/10.1007/978-3-031-24907-5_56

Introduction

In recent years, sales of zero-emission vehicles (ZEVs), especially battery electric vehicles (BEVs), have increased rapidly in all major vehicle markets [1]. This global trend is driven by policy goals such as the European Green Deal [2], which aims to reduce emissions of greenhouse gases and local pollutants in the transport sector. These targets demand strong monetary and non-monetary governmental incentives that foster this mobility transformation. For a successful, sustainable, and widely accepted mobility transformation to be created, the individual mobility needs of users must be reliably served. Consequently, users must be at the center when designing ZEVs and their ecosystems. Only when the user needs are fulfilled to a high degree can high user satisfaction be achieved. In this way, it is possible to convince potential users of ZEVs, e.g., BEVs or fuel cell electric vehicles (FCEVs).

Therefore, we examine which passenger vehicle purchase criteria are relevant from the perspective of ZEV drivers and ZEV-interested individuals.[1] These criteria were selected through interviews conducted with German and Swiss drivers (N = 7) of FCEVs and BEVs. We then aimed to determine the weights of these criteria by using the Analytic Hierarchy Process (AHP)—a multi-criteria decision analysis approach. The weights of these criteria were obtained with an AHP online survey (N = 569) distributed through various channels (forums, podcasts, newsletters) on the topic of zero-emission mobility. With the goal of creating an assessment framework for ZEVs, scales were developed for each criterion. These so-called 'value scores' represent the degree to which a specific ZEV satisfies the defined criteria.

Despite the early market phase of ZEVs, customer decisions, especially in the context of BEVs, have already been intensively studied. Mainly using discrete choice experiments [3], the preferences for different BEV attributes or purchase criteria have been investigated. A comprehensive literature review was conducted [4].

Using AHP, the purchase of vehicles has been studied in the literature by the following selection of authors: Based on a survey of dealership experts, [5] analyzed vehicle purchase decisions. Canadian drivers' preferences for vehicle attributes were examined [6]. The interaction of environmental attributes and consumer vehicle purchase decisions was investigated [7] and vehicle purchase decision-making during COVID-19 [8].

Unlike previous studies that solely concentrated on determining the decision-relevant criteria and their respective weights [6, 7] while comparing specific vehicles (alternatives) [5, 8], we chose a different approach. In the study at hand, a flexible assessment framework is created with which any ZEV can be evaluated from the customer's point of view. Such an evaluation framework was not found in the literature on ZEVs—neither for BEV nor for FCEV.

The purpose of the assessment framework is to facilitate the evaluation and comparison of ZEV models across technologies with each other from a customers'

[1] ZEV drivers are individuals who have gained experience with BEVs or FCEVs through car sharing, ownership, company vehicles, or rental cars. ZEV-interested individuals keep themselves up to date via podcasts, forums, and newsletters.

perspective. Moreover, with a differentiated consideration of the weighted purchase criteria and value scores, it is possible to design ZEVs that are tailored to specific customer needs.

This paper first introduces the methodology used to identify and assess the criteria and value scores for the multi-criteria assessment framework (Sect. "Multi-Criteria Assessment Framework"). In Sect. "Introduction", passenger vehicle purchase criteria are derived through the analysis of the interviews conducted. Section "Multi--Criteria Assessment Framework" introduces the value scores used for each criterion, allowing for the comprehensive characterization of particular ZEVs. Section "Results and Discussion" presents the criteria weights obtained from the AHP. Section "Conclusion and Outlook" presents the conclusion of this study, proposing more in-depth research approaches.

Multi-criteria Assessment Framework

The creation of this assessment framework required the methodology presented in [9, 10]. As shown in Fig. 56.1, we first derived the relevant criteria for the multi-criteria assessment from the interviews, which took place from October to November 2021. Following [11], we conducted semi-structured interviews via video conferencing with an average duration of 33 min each. It was essential to us that the interviewees have experience with both BEVs and FCEVs, as these technologies are primarily associated with ZEVs. However, to date, only few FCEVs are in use, thus complicating the search for individuals with practical experience. To select and define the relevant evaluation criteria, we evaluated the interviews and related literature (such as [12]) on vehicle purchase decisions.

Criteria weights for ZEVs were obtained through applying AHP [13]; in a participatory multi-criteria decision analysis with stakeholders. AHP was chosen because it provides an intuitive and easily understandable method for determining criteria weights [9]. This reduces methodological comprehension problems and thus participation hurdles. Since the focus was on the general passenger vehicle purchase criteria, care was taken to ensure that the questions were technology-independent and free of brand framing. We asked ZEV drivers and ZEV-interested individuals x to compare every possible criteria pair in isolation on a scale from 1 = *'criteria A equally important as B'* to 9 = *'A significantly more important than B'* in an online survey. We received such pairwise comparisons from 569 individuals.

On the basis of the maximum eigenvalue, we performed a mathematical test for the consistency of the individuals' choices. The consistency value ($CR \leq 0.2$) defined

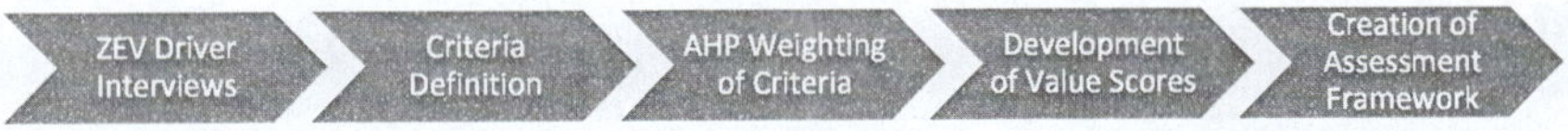

Fig. 56.1 Procedural approach to the creation of an assessment framework, adopted from [9]

in [14] was met by 244 respondents. The results of the comparisons between all n criteria i and j are stored in the evaluation matrix $A_x = (a_{ij})$ for these 244 participants x:

$$A_x = \begin{pmatrix} a_{11} & \cdots & a_{1n} \\ \vdots & \ddots & \vdots \\ a_{n1} & \cdots & a_{nn} \end{pmatrix}, \quad \forall x \in \{1, 2, \ldots, 244\} \tag{56.1}$$

For a_{ij} it holds

$$\begin{aligned} &\forall i = 1, \ldots, n \quad \forall j = 1, \ldots, n : a_{i,j} > 0 \\ &\forall i = j : a_{i,j} = 1 \\ &\forall i = 1, \ldots, n \quad \forall j = 1, \ldots, n : \; a_{i,j} = a_{j,i}^{-1}. \end{aligned}$$

Subsequently, the priority vector is calculated with the eigenvalue method. The aggregation of individual valuations (A_x) was accomplished via the geometric mean of the individual evaluation matrices. The group of ZEV drivers and ZEV-interested individuals can thus be divided into subgroups based on socio-demographic information or information on their driving behavior. Each (sub-) group can be interpreted as a synergistic individual and can be analyzed separately [15]. The eigenvalue method was implemented in a MATLAB environment and was applied to calculate criteria weights from the perspective of the ZEV drivers and ZEV-interested individuals and their subgroups.

For each criterion i, a spectrum of value scores v_{ic} is described with values ranging from 0 (criterion not satisfied) to 10 (criterion perfectly satisfied). With these value scores and the criteria weights w_i, ZEVs c can be evaluated from the point of view of (potential) users (cf. [9]):

$$ZEV \text{ score} = \sum_{i=1}^{n} w_i \cdot v_{ic}. \tag{56.2}$$

Results and Discussion

The following section presents the results of the sub-steps for establishing the assessment framework according to Eq. (56.2).

Interviews

The criteria selected and mentioned in the interviews as relevant for the assessment of ZEV are listed and defined in Table 56.1. These criteria were defined in dialogue

Table 56.1 Assessment criteria for passenger vehicle purchase decisions derived from interviews

Criterion	Definition
Total costs	Total costs of operating a ZEV (including investment costs, fuel/charging costs, repair and maintenance costs, and insurance)
Range	Maximum distance within which ZEVs with an initially full battery/tank can be moved without interruptions for recharging/refueling
Charging/refueling time	Time needed to fully charge/fuel the vehicle on the way
Availability of infrastructure	Number of charging points or fuel dispensers, permanent accessibility, and even utilization, among other things, improve availability
Greenhouse gas emissions	Greenhouse gas emissions generated during the production and usage of ZEVs
Spaciousness	Space in the ZEV interior (passenger area, luggage trunk)
Driving dynamics	Driving dynamics measured by acceleration and top speed

with the interviewees. Participants were asked to name ZEV-specific criteria. For these criteria to be weighted, pairwise comparisons of the criteria were integrated into an online survey.

Criteria that have been considered relevant in the literature on consumer preferences for BEVs [4] are: *purchase price*, *operation cost*, *driving range*, *charging time*, *engine power*, *acceleration time*, *maximum speed*, *CO_2 emissions*, *brand*, *warranty*, and *charging availability*. Our criteria selected are mostly consistent with the criteria found in the literature or summarize these. Criteria such as *design*, *warranty*, or *brand* were not included, as they were not considered as necessarily related to zero-emission drive technology. In the interviews, spaciousness was explicitly mentioned as relevant to cross-technology assessments because hydrogen tanks and batteries have strong differences in space requirements that affect the space inside the vehicle.

Criteria Weights

In this survey, 569 ZEV-interested individuals with a driver's license (including 399 ZEV drivers) participated. About 90% of the participants are male and about 70% are older than 40 years. Over 60% have a college degree, some of which even hold a doctorate. About 25% of the participants had no previous professional contact with hydrogen, (electric) mobility, energy, or technology. About 85% have a net household income of more than €2500. The typical consumer interested in ZEV is therefore rather well-off, educated, predominantly male, and middle-aged or older. Similar observations regarding the composition of samples of drivers and people interested in electrified vehicles have been made in previous studies [12, 16]. Table 56.2 shows

Table 56.2 Relative importance of criteria

Criterion	Weight/relative importance [%]	
	ZEV drivers (*N* = 179)	ZEV-interested (*N* = 65)
Total costs	10.6^{+}	12.5^{+}
Range	14.7^{+}	15.4^{+}
Charging/refueling time	14.9^{+}	13.4^{+}
Infrastructure availability	14.4^{+}	17.1^{++}
Greenhouse gas emissions	23.3^{+++}	17.9^{++}
Spaciousness	12.7^{+}	16.2^{++}
Driving dynamics	9.4	7.5

Notes: Relative importance > 20% (+++), 15–20% (++), 10–15% (+)

the analysis of the pairwise AHP comparisons in the form of the priority vector for ZEV drivers and interested parties.

Overall, ZEV drivers and ZEV-interested individuals show similar tendencies regarding the weighting of the purchase criteria. *Greenhouse gas emissions* is the most important and *driving dynamics* is the least important criterion for both groups. *Total costs* also play a minor role across groups. *Greenhouse gas emissions* are clearly more important to drivers than to interested individuals, while costs are slightly more important to the latter. The great interest in the ecological added value of ZEV appears to have motivated drivers to actively use ZEV despite possible additional costs. Moral licensing could also influence this prioritization. *Spaciousness* is less important to ZEV drivers than to ZEV-interested individuals. While *range, charging or refueling time*, and *availability of charging infrastructure* are of similar importance to ZEV drivers, *the availability of charging infrastructure* stands out among ZEV-interested individuals. In the survey, respondents were explicitly asked about ZEV purchase barriers. One of the biggest barriers is the insufficient availability of ZEV infrastructure. Thus, individuals to whom the security of an available infrastructure is particularly important could still be deterred from using ZEVs. This consideration could explain, among other things, why those interested in ZEVs attach great importance to the availability.

Value Scores

Table 56.3 provides an overview of all criteria showing the corresponding measuring units of their value scores. On the basis of data and values from literature and market research, the scales of the value scores are created. Common minimum and maximum

Table 56.3 Value scores and measuring units

Criterion	Measuring unit	Criterion	Measuring unit
Total costs	€	Greenhouse gas emissions	kg_{CO2_eq}/100 km
Range	km (WLTP)	Infrastructure availability	Number of charging/filling stations
Charging/refueling time	Min	Driving dynamics	0–100 km/h in seconds; max. km/h
Spaciousness	Liter		

levels of the criteria serve as the extrema of the scales (e.g., *range*: < 100 km indicates a value score of 0; *range* ≥ 700 km indicates a value score of 10). Linear interpolation is performed between the extrema to create the scale from 0 to 10. For reasons of space, no concrete value scores are presented here. The exact values can be found in [17].

Conclusion and Outlook

In this study, we identified relevant ZEV assessment criteria. We derived weights for these criteria by applying an AHP that is based on an online survey. By defining their respective value scores, we formed a multi-criteria assessment framework. This framework assesses any ZEV from the perspectives of (potential) customers, without the need for repeated involvement of the surveyed and interviewed individuals. The weighting of the criteria was shown here for ZEV drivers and ZEV-interested individuals. By dividing the sample by socio-demographic data and by data on mobility and behavior, the specific prioritizations of these subgroups can be determined. Therefore, this study contributes to a better understanding of the purchasing behavior and prioritizations of ZEV drivers and ZEV-interested individuals. The framework is useful for the design of ZEVs by vehicle manufacturers, e.g., in the field of battery-electric and hydrogen mobility.

In this short paper, the AHP priority vectors were analyzed for ZEV drivers and ZEV-interested individuals only. A resolution of the total sample by socio-demographic data and mobility behavior data is also possible. The preferred passenger vehicle segments of the survey participants were also queried. In addition, it was determined which ZEV drive technologies the survey participants have already had experience with. Taking these data into account, the analysis of the criteria weights seems particularly interesting and promising. These evaluations will be included in future publications and will provide more detailed insights into the prioritizations of ZEV purchase decisions.

References

1. Global EV Outlook. (2021). https://iea.blob.core.windows.net/assets/ed5f4484-f556-4110-8c5c-4ede8bcba637/GlobalEVOutlook2021.pdf. Last accessed Mar 10 2022
2. A European Green Deal. https://ec.europa.eu/info/strategy/priorities-2019-2024/european-green-deal/. Last accessed Mar 10 2022
3. Mandys, F. (2021). Electric vehicles and consumer choices. *Renewable and Sustainable Energy Reviews, 142*, 1–9.
4. Liao, F., Molin, E., & van Wee, B. (2017). Consumer preferences for electric vehicles: A literature review. *Transport Reviews, 37*(3), 252–275.
5. Byun, D. (2001). The AHP approach for selecting an automobile purchase model. *Information & Management, 38*(5), 289–297.
6. Vrkljan, B. H., & Anaby, D. (2011). What vehicle features are considered important when buying an automobile? An examination of driver preferences by age and gender. *Journal of Safety Research, 42*(1), 61–65.
7. Fujita, K. S., Yang, H., Taylor, M., & Jackman, D. (2022). Green light on buying a car: How consumer decision-making interacts with environmental attributes in the new vehicle purchase process. *Transportation Research Record, 2676*(7), 743–762.
8. Raza, S., Masmoudi, M. (2020).Consumer vehicle purchase decision-making during COVID-19. In: *2020 International Conference on Decision Aid Sciences and Application* (pp 692–696) Sakheer, Bahrain.
9. Fabianek, P., Will, C., Wolff, S., & Madlener, R. (2020). Green and regional? A multi-criteria assessment framework for the provision of green electricity for electric vehicles in Germany. *Transportation Research Part D: Transport and Environment, 87*, 1–22.
10. Fabianek, P., Madlener, R. (2022). A Multi-Criteria Assessment Framework for Evaluating the User Experience at E-Vehicle Charging Stations in Germany. *Institute for Future Energy Consumer Needs and Behavior*. FCN Working Paper No. 6/2022. RWTH Aachen University, May.
11. Wilson, C. (2013). Interview techniques for UX practitioners.1st edn. Elsevier B.V.
12. Hackbarth, A., & Madlener, R. (2016). Willingness-to-pay for alternative fuel vehicle characteristics: A stated choice study for Germany. *Transportation Research Part A: Policy and Practice, 85*, 89–111.
13. Saaty, R. W. (1987). The analytic hierarchy process—what it is and how it is used. *Mathematical Modelling, 9*(3–5), 161–176.
14. Saaty, T. L. (1991). Some mathematical concepts of the analytic hierarchy process. *Behaviormetrika, 18*(29), 1–9.
15. Aull-Hyde, R., Erdogan, S., & Duke, J. M. (2006). An experiment on the consistency of aggregated comparison matrices in AHP. *European Journal of Operational Research, 171*(1), 290–295.
16. Haustein, S., & Jensen, A. F. (2018). Factors of electric vehicle adoption: A comparison of conventional and electric car users based on an extended theory of planned behavior. *International Journal of Sustainable Transportation, 12*(7), 484–496.
17. Fabianek, P., Madlener, R. (2023). Assessing Zero-Emission Vehicles from the Customer's Perspective using a Multi-Criteria Framework. *Institute for Future Energy Consumer Needs and Behavior*. FCN Working Paper. RWTH Aachen University, in prep.

Chapter 57
A Mutation Based Modular Evolutionary Scheme for Integrated Timetabling and Vehicle Scheduling With headways and Connection Quality Criteria

Lucas Mertens, Bastian Amberg, and Natalia Kliewer

Abstract We propose an adaptive modular evolutionary scheme for optimizing the NP-Hard integrated timetabling and vehicle scheduling problem (TTVSP) in public bus transit. Various heterogeneous mutation operators are utilized within the scheme. Depending on their impact on the computed timetable and corresponding bus schedule, each mutation operator will be adaptively applied with a weighted probability in future iterations. To validate the solution quality and runtime of the heuristic, we exactly solve the TTVSP considering vehicle fix and operational costs. The experiments show that the heuristic computes a cost-optimal solution in a reasonable time. We further increased the solution quality relating to service quality by extending the utilized approach considering both short connecting times between different lines and good headways.

Keywords Public bus transport · Timetabling · Vehicle scheduling

Introduction

Public transport planning is a traditionally sequentially executed multi-level process [2]. Each planning step covers a different time horizon and is advancing for shorter periods. Strategic planning covers the longest horizon and can be divided into network design and line planning. Following these steps, tactical planning is concerned with frequency setting and timetabling. Operational planning is executed for short periods and is mainly divided into vehicle scheduling, crew scheduling, and crew rostering. In our study, we focus on an integrated optimization of timetabling and vehicle scheduling, resulting in a bridge between tactical and operational public transport planning. Based on the line network, passenger demands, service times, vehicle capacities, and frequency specifications such as headways between service trips, the *timetabling problem* (TT) aims to compute a seasonal, weekday, and holiday-related timetable. The timetable defines each service trip and their arrival and departure at

L. Mertens (✉) · B. Amberg · N. Kliewer
Department of Information Systems, Freie Universität Berlin, Berlin, Germany
e-mail: lucas.mertens@fu-berlin.de

O. Grothe et al. (eds.), *Operations Research Proceedings 2022*, Lecture Notes in Operations Research, https://doi.org/10.1007/978-3-031-24907-5_57

each stop. Given this timetable, the *vehicle scheduling problem* (VSP) is concerned with generating a cost-minimal vehicle schedule. Service trips are assigned to vehicles and complemented by deadhead trips without passengers to plan pull-outs from depots, pull-ins to depots, and trips between different lines. Traditionally, the TT is solved first, and the computed service trips are used as an input for the VSP. By integrated optimizing both planning steps, the timetable leading to the lowest cost vehicle schedule can be computed. However, solving the *integrated timetabling and vehicle scheduling problem* (TTVSP), the solution space and hence the complexity increases significantly. An extensive overview of optimization approaches for the TTVSP is given by [1]. Only a few publications (e.g. [4, 5]) utilize a true integration considering the entire solution space and are not relying on multiple iterations between both planning steps. Even though integrated approaches for solving the TTVSP have been proposed, to the best of our knowledge, neither of the publications cover the full complexity of both real-world TT and VSP yet. E.g., both [4] and [5] consider one vehicle type, and do not allow any interlining. In this short paper, we include line synchronization, interlining, and deadheading in TTVSP. We sketch the underlying principle and first results of an *adaptive modular extendable evolutionary scheme* (AMEES) for solving the TTVSP, considering these requirements for real-world instances. In the following section the AMEES is described, and an exact MIP serving as a benchmark for solving the TTVSP is formulated. The results and possibilities to extend the new approach are evaluated in Section "Experiments". Section "Outlook" summarizes our work and gives an outlook on future research.

Integrated Timetabling and Vehicle Scheduling

We propose an AMEES to solve the TTVSP for real-world instances in reasonable computational time, which is able to consider a diversity of cost and service quality criteria in addition to nominal costs such as vehicle fix and operational costs. The AMEES evolutionarily improves a population of individuals, with an individual representing a TTVSP solution, i.e., a timetable and corresponding vehicle schedule. An individual's fitness is defined by the costs of serving the timetable and can be further specialized, e.g., with boni for short connections between different lines reached within a specified time interval and for good headways. Headways are considered good if they are clock headways and if their prior and following headway have the same frequency. Since crossover operations on promising individuals in an evolutionary scheme frequently lead to invalid or deteriorated solutions, the main novelty of the AMEES for public transport planning is the inclusion of an extendable variety of numerous mutation operators (MO). In the following, the design of the AMEES solving the TTVSP is described. First, we outline the structure and MOs directed at optimizing the timetable, followed by the specifications for the vehicle schedule.

The timetable is represented as a binary list of possible service trips. Each possible service trip is created for each minute a bus could start a trip. The trip schedule contains n values, where n equals the total number of possible service trips. A one in the list represents a planned trip, otherwise a zero is set. A simple MO could, e.g., flip a random binary value within the trip schedule, resulting in planning or unplanning a trip. A pool of 23 timetabling MOs is rapidly applied to each individual. MOs aiming to improve the timetable are separated into four classes: *Planner*, *Unplanner*, *Allocator*, and *Switcher*. Several Planner and Unplanner aim at planning or unplanning particular trips to improve the timetable. Switcher specifically change the value of two trips and Allocator of multiple trips simultaneously. The class of Planner, e.g., contains five distinct MOs: *PassengerMissingPlanner*, *MaxHeadwayViolatedPlanner*, *ServiceTimeStartPlanner*, and *ServiceTimeEndPlanner*. Whereas the PassengerMissingPlanner plans a new trip to transport all required passengers, the MaxHeadwayViolatedPlanner schedules a trip such as the maximal headways are satisfied. Both the ServiceTimeStartPlanner and ServiceTimeEndPlanner plan trips to cover the service times of a line.

The vehicle schedule is represented as a vector of vehicle rotation objects. Each rotation contains a list of the covered trips, deadhead trips, and its associated depot. A pool of 12 MOs aiming to improve an individuum's vehicle schedule is divided into 5 classes: *Veh-Planner*, *Veh-Unplanner*, *Veh-Trip-Allocator*, *Veh-Merger*, and *Veh-Splitter*. New vehicle rotations are created via Veh-Planner to cover unassigned trips. Veh-Unplanner, on the other hand, unschedule redundant vehicles and allocate the now unassigned trips to existing rotations. Resulting in lower operational costs, Veh-Trip-Allocator reallocate trips between rotations. Veh-Merger reduce the total number of required buses by merging trips of two rotations into a single one. Long idle times or too much deadheading can lead to inconveniently planned vehicle rotations. Veh-Splitter divide these into separate new rotations. E.g., 2 MOs are classified as Veh-Splitter: *VDeadheadReductionSplitter* and *VTimeReductionSplitter*. If a rotation switches between lines frequently, the VDeadheadReductionSplitter separates it into at least two rotations that switch less between lines. Reducing long idle times in a rotation, the VTimeReductionSplitter splits it at these waiting times into separate rotations.

Figure 57.1 displays the solution process within the AMEES. First, an initial population of individuals with no service trip or vehicle is created (1). Several construction operators (CO) are applied, resulting in various individuals (2). COs, e.g., schedule a sequence of multiple trips with a fixed headway and assign these to vehicles. Since individuals are not crossed over, every iteration within the AMEES starts with

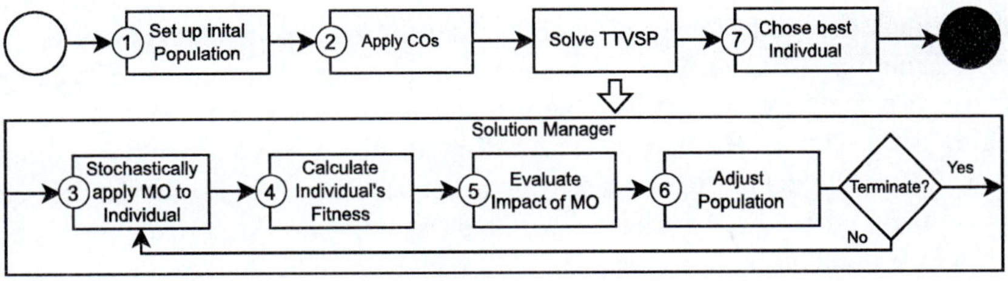

Fig. 57.1 Process of solving the TTVSP with the AMEES

stochastically applying a MO (3). At first, there is no information about an MO's performance. Hence, every MO is applied to every individual. An individual's validity is determined by a quantified violation against the minimal required headway, maximal allowed headway, and passenger demand constraints. The individual's fitness is defined by the nominal costs, boni for good headways and connections reached and a weighted validity. After calculating every individual's fitness (4), the impact of each MO on the fitness is measured and tracked (5). If applying a specific MO results in an improved fitness, the MO will be adaptively applied with a higher probability on individuals with a similar fitness value in future iterations. However, not only the sole performance of a MO is evaluated, but their synergies. Isolated applied, MOs classified as, e.g., *Veh-Splitter*, increase the nominal costs, hence decrease the individual's fitness. However, a collaborative sequencing of heterogeneous MOs might frequently lead to an improved solution. Applying a *Veh-Merger* subsequently to a *Veh-Splitter*, can lead to a better allocation of trips within the vehicle schedule and to a superior solution after all. These promising sequences of combining multiple heterogeneous MOs are elaborated and applied to individuals within the same generation. After applying and measuring the impact of the MOs, the diversity of the population, derived from each individual's fitness value, is evaluated and its composition and size adjusted (6). Identical as well as individuals with low fitness are removed. Keeping a constant population size, the individuals with the highest fitness are duplicated. However, if the fitness of every individual converges, the population size is gradually enlarged to enable an increased diversity within the population. Applying MOs to a broader range of individuals requires more computational time but increases the likeliness of escaping possible local optima. Reaching a termination criteria (e.g. time, a predefined amount of valid individuals, or passing an objective value), the fittest individual is chosen (7). The fitness function can be easily adjusted, and new MOs can be integrated without the need to adjust prior implementations or the solution structure.

An exact solution serves as a benchmark to evaluate the quality of the computed solutions by the AMEES. The corresponding mathematical model is based on a time-space network (TSN) formulation according to [3]. The TSN consists of nodes N representing the time and location of each stopping point and arcs representing every possible trip T, deadhead trips D, and waiting times W. Binary and integer decision variables set the flow for each arc. The objective function minimizes the total costs of all used arcs. The required passenger service is defined within tracking gates. Tracking gates are virtual points between two stops and define the number of passengers that must be transported within an hour, as well as a minimal required and maximal allowed headway between two departures. One tracking gate can set requirements for different time periods. If multiple lines path the same tracking gate, requirements are set for each line simultaneously. Let G represent the set of tracking gates. The set M_g represents each discrete minute and S_g each hour within each $g \in G$. $H^{min}_{g,m}$ defines the trips starting at each minute $m \in M_g$ for the defined minimal required headway associated with the tracking gate. A service time of one hour, with a minimal required headway of 3 minutes would, e.g., result in 58 sets of 3 trips with the following departure times: [6:00;6:01;6:02], [6:01;6:02;6:03], ..., [6:57;6:58;6:59]. A maximum of one service trip of each set is allowed to be planned to guarantee that buses do not depart within a too-small frequency. Similarly, $H^{max}_{g,m}$ represents a set of trips within a $g \in G$ starting at each minute $m \in M_g$ for the defined maximal allowed headway. Furthermore, the set E^{in}_n contains each arc from T, D, and W going into and E^{out}_n leaving a node $n \in N$. If multiple vehicle types are considered, c_t represents the passenger capacity for each possible service trip and is fixed otherwise. $R_{g,s}$ includes each trip associated to a $g \in G$ at each hour $s \in S_g$. The parameter $p_{g,s}$ sets the lower bound of required passengers to be transported for each $g \in G$ at each hour $s \in S_g$. Aiming to minimize the overall costs, the constraints of the model can be formulated as follows:

$$\sum_{t \in H^{min}_{g,m}} x_t \leq 1 \qquad \forall g \in G, \forall m \in M_g \tag{57.1}$$

$$\sum_{t \in H^{max}_{g,m}} x_t \geq 1 \qquad \forall g \in G, \forall m \in M_g \tag{57.2}$$

$$\sum_{t \in R_{g,s}} c_t x_t \geq p_{g,s} \quad \forall g \in G, \forall s \in S \tag{57.3}$$

$$\sum_{i \in E^{in}_n} x_i - \sum_{j \in E^{out}_n} x_j = 0 \qquad \forall n \in N \tag{57.4}$$

$$x_t \in 0, 1 \quad \forall t \in T, \;\; x_i \in \mathbb{N} \quad \forall i \in D \cup W \tag{57.5}$$

For each tracking gate, constraints (57.1) guarantee that no more than one trip within the set of minimum required headways is executed. Similarly, (57.2) ensure that at least one trip departs within the defined maximum allowed headway. Constraints (57.3) guarantee that at least the required amount of passengers are transported. Lastly, the flow-conservation constraints (57.4) guarantee that every vehicle entering a node also leaves it.

Experimental Results

To validate the proposed AMEES, the computed results are compared with the results of the exact approach for different instances. The exact approach is implemented in Python and solved with Gurobi v 9.1.2. The AMEES is implemented in C++. All computational tests are carried out on an Intel(R) Core(TM) i9-10900 CPU with 2.8GHz and 32GB RAM. First, three real-world-inspired instances of increasing complexity (I1, I2, and I3) are exactly and heuristically solved. Following, a fourth instance (I4) is considered to evaluate the applicability of the AMEES for a large problem setting. I1 covers only trips of a single line, with a total of six stops and two tracking gates. I2 considers an additional line with 8 stops, passing through two new tracking gates. Each tracking gate defines requirements considering headways and passenger demands for three different time periods. I3 considers a total of 33 stops and 5 lines starting at 5 a.m. and ending at 2 a.m. the next day. I3 includes a

Table 57.1 Comparison between the exact approach and AMEES

	TTVSP exact				TTVSP AMEES (20 runs)		
Instance	Min #Trips	Min #Buses	Time opt.	Total time	Total time (min)	Avg. time opt. (s)	Min/Max time opt. (s)
I1	125	5	3.4 min	4.5 min	5	0.8	0.7/1.2
I2	260	11	30.2 min	30.2 min	5	11.2	9.8/16.0
I3	356	16	–	24 h	5	23.5	11.1/32.9

Table 57.2 Improved headways and connections considering additional quality criteria

	AMEES		AMEES with quality criteria	
Instance	#Connections Avg/Min/Max	#Good Headways Avg/Min/Max	#Connections Avg/Min/Max	#Good Headways Avg/Min/Max
I1	0	16.4/14/19	0	42.4 /39/45
I2	0	26.6/23/29	0	181.2/168/188
I3	84.4/78/92	129.4/120/144	138.2/121/163	203.7/190/211

total of 8 tracking gates and 4 intersecting lines. Exactly solving the TTVSP leads to a minimal amount of trips and a minimal number of required buses shown in the second and third columns of Table 57.1. Column *Time Opt.* indicates the time until the exact approach finds an optimal solution, while *Total Time* includes the time needed to prove optimality. The AMEES is executed 20 times and terminates after a time limit of 5 minutes. The column *Avg. Time Opt.* indicates the average time the AMEES required to compute the best solution. The objective value of these solutions unanimously equaled the exactly computed optimum. The last column of Table 57.1 displays the minimal and maximal time the computation of the best individual took in all executions of the AMEES. Already for very small instances, the AMESS outperforms the runtime of the exact approach. With growing complexity, the runtime of the exact approach is increasing crucially and still holds an optimality gap of 5% after 24h runtime for I3. For the same instance, the AMEES converged to its best computed solution in an average of under 30 s.

The AMEES can be extended to compute good headways as well as connections reached. So far, an exact model can not consider these additional requirements in reasonable computation time. The left half of Table 57.2 shows the average, minimal and maximal number of connections reached and good headways of the 20 prior evaluated computations. We included the additional quality criteria in the objective function of the AMEES and added 6 MOs, such as the *GoodHeadwayPlanner* that schedules trips with a regular headway. As seen in the right half of Table 57.2, 20 new executions of the enhanced AMEES with only one minute runtime improved both quality criteria significantly.

To further validate the applicability of the AMEES, we consider an additional large instance (I4) that cannot be handled in a reasonable time by the exact approach.

I4 consists of 25 lines, with 165 stops, 50 different routes, and 40 tracking gates. This affects, in particular, the trip schedule representation in the AMMEES. The binary trip schedule representing each possible trip linearly increases with the total service time covered by each line. A total service time of 9480 minutes is covered in I3, and the same amount of binary decision variables (9480) are required to model the corresponding trip schedule. I4 covers the fivefold service time and requires exactly five times the amount of trip schedule decision variables (47400). The AMEES was executed 20 times with a time limit of 30 minutes. Each execution resulted in a valid timetable and vehicle schedule. The best solution of each execution exhibits a similar convergence behavior and required between 4300 and 5000 mutations in an average of 22 minutes until no further improvements are computed. On average 98.3 vehicles are required to cover 2189.7 service trips. The best solution required 97 vehicles to cover 2179 service trips, and the solution with the highest costs required 101 vehicles to cover 2214 service trips.

Conclusion and Outlook

We proposed an AMEES to solve the TTSVP heuristically. Compared to the exact approach, the nominal costs computed by the AMEES converge to the same optimal value. No additional trips are planned or buses required. In addition, significantly less computational time is needed. We extended the AMEES to compute good headways and connections reached. This resulted in a significant improvement of the enhanced quality criteria within a short computation time. For a large real-world-inspired instance, multiple runs of the AMEES converged to solutions of similar quality within reasonable computational time. However, there is still a quality gap between these solutions, and additional MOs have to be elaborated to achieve a lower variance. This work lays the foundation for further investigating the performance of the AMEES for scheduling in public transport. Additionally, we aim to extend the AMEES to integrate a third planning step, i.e., the crew scheduling.

References

1. Carosi, S., Frangioni, A., Galli, L., Girardi, L., & Vallese, G. (2019). A matheuristic for integrated timetabling and vehicle scheduling. *Transportation Research Part B: Methodological, 127*, 99–124.
2. Desaulniers, G., & Hickman, M. D. (2007). Public transit. *Handbooks in Operations Research and Management Science, 14*, 69–127.
3. Kliewer, N., Mellouli, T., & Suhl, L. (2006). A time-space network based exact optimization model for multi-depot bus scheduling. *European Journal of Operational Research, 175*(3), 1616–1627.
4. Ibarra-Rojas, O. J., Giesen, R., & Rios-Solis, Y. A. (2014). An integrated approach for timetabling and vehicle scheduling problems to analyze the trade-off between level of service and operating costs of transit networks. *Transportation Research Part B: Methodological, 70*, 35–46.

5. Weiszer, M., Fedorko, G., & Čujan, Z. (2010). Multiobjective evolutionary algorithm for integrated timetable optimization with vehicle scheduling aspects. *Perner's Contacts, 5*(4), 286–294.

Chapter 58
A New Flow-Based Location and Capacity Model for Profit-Oriented Refueling Station Network Transformation

Tjard Bätge, Christian Weckenborg, and Thomas S. Spengler

Abstract The availability of refueling stations for alternative fuels is of high importance for the penetration of alternative fuel vehicles (AFV) in today's markets and is an important lever for sustainable mobility. Therefore, refueling station network deployment strategies are currently of great interest to researchers. Several quantitative location models have been specifically developed, adapted, and applied to determine optimal refueling station locations and network build-up strategies for AFV refueling from an overall system's perspective. However, the network operator's intrinsic economic objectives and its preference for transforming the existing refueling station network rather than building up new refueling stations are subordinate in most research. That is why a new formulation of a flow-based location model is introduced in this study that not only decides on the locations of new alternative refueling stations but also on the provisioned capacities of multiple fuel types at each newly opened and each already existing refueling station in a multi-period planning horizon. This new approach uses a profit-oriented, cash flow-based objective function to transform the existing refueling station network over time. A case study depicting German highways is presented to validate the new model and the computational results are discussed.

Keywords Alternative fuel vehicle · Facility location · Refueling station · Capacity planning · Multi-period · Network transformation · Refueling infrastructure

Introduction

Recently, the usage of alternative fuel vehicles (AFV) is widely discussed as a potential means to reduce greenhouse gas emissions in the mobility sector. One idea under

T. Bätge (✉) · C. Weckenborg · T. S. Spengler
Institute of Automotive Management and Industrial Production, Technische Universität Braunschweig, Mühlenpfordtstr. 23, 38106 Braunschweig, Germany
e-mail: t.baetge@tu-braunschweig.de

O. Grothe et al. (eds.), *Operations Research Proceedings 2022*, Lecture Notes in Operations Research, https://doi.org/10.1007/978-3-031-24907-5_58

discussion and the focus of this study is the replacement of conventional heavy-duty diesel trucks with hydrogen fuel cell trucks for long-haul transportation.

The usage of AFV, however, also necessitates the existence of a corresponding alternative refueling station network. Most countries' current conventional public-access refueling infrastructure is operated by multiple profit-oriented companies each managing its network of refueling stations. Equipping existing conventional stations with alternative refueling systems as well as the build-up of new alternative refueling stations comes with high investments [1]. Thus, the network operators need to decide how to optimally invest in the transformation of their refueling station network if they want to secure their market share in the emerging refueling markets for AFV. Because of the dynamic nature of the emerging vehicle markets with coexisting fuel technologies and a present conventional vehicle stock, these investment decisions need to consider (i) different fuels at (ii) different capacity levels and (iii) a strategic planning horizon.

Several quantitative location models have been specifically developed, adapted, and applied to determine optimal refueling station locations and network build-up strategies for AFV refueling from an overall system's perspective, e.g. [2–4]. However, (iv) the network operator's intrinsic economic objectives and (v) its preference for transforming the existing refueling station network rather than building up new refueling stations are subordinate in most research. Moreover, most research neglects the coexistence of multiple fuel types. However, these requirements need to be considered to support the investment decisions of a single refueling network operator in a competitive environment. That is why the main contribution of this research is the development of a new dynamic, multi-fuel, flow-based refueling station location and transformation model. The model accounts for the interdependencies between the company's decisions, the competitor's decisions, and the AFV market by assuming a simple oligopoly-like refueling market. The model is used to investigate how different competitor actions influence the decision-making of a major refueling station network operator.

In the following, the literature is reviewed in Sect. "Literature Review" and the new model is introduced in Sect. "Methodology". Subsequently, a case study depicting German highways is presented in Sect. "Case Study", the case study's results are discussed in Sect. "Results", and the research is concluded in Sect. "Conclusions".

Literature Review

To optimally allocate alternative refueling stations, Kuby and Lim [5] developed the Flow Refueling Location Model (FRLM), where demand is modeled as traffic flow passing facilities on origin-destination paths while considering vehicle range. A widely used and more efficient formulation of the FRLM has been found by Capar et al. [6].

There are many extensions and applications of FLRM formulations. Tafakkori et al. [7] consider multiple fuel types (i), while Wang and Lin [8] consider multiple charging station types. Rose et al. considered node capacity restrictions (ii) and applied the model to heavy-duty fuel cell trucks for long-haul transportation in Germany [1, 4]. This model has then been extended by a dynamic perspective (iii) to generate step-by-step infrastructure construction plans [3]. Zhang et al. also developed a multi-period capacitated FRLM to find optimal locations for electric vehicle charging stations. Different approaches to model the dynamics of the electric vehicle market are presented [2]. Multiple fuels, however, are not considered in [1–4].

Bersani et al. [9], as well as Crönert and Minner [10], investigate alternative refueling station placement with intrinsic economic objectives in competitive environments (iv). However, both models are static and study only one fuel type. Multiple articles, e.g. [9], consider existing conventional refueling stations as possible alternative refueling station locations. However, the transformation of technical equipment or local restrictions are not explicitly modeled (v).

To the best of the authors' knowledge, no model within the literature on alternative refueling station location modeling accounts for all defined requirements. That is why a new refueling station location and transformation model that extends the model of [6] by capacity and demand variables for multiple fuels is introduced in the next section.

Methodology

The following section introduces the sets, indices, decision variables, constraints, and the objective of the new dynamic, multi-fuel, flow-based refueling station location and transformation model. The foundation of the mathematical model is a traffic network consisting of nodes $i, j, k, m \in \mathrm{N}$, $i, j, k \in \mathrm{L} \subseteq \mathrm{N}$, and $i, j, k \in \mathrm{CL} \subseteq \mathrm{N}$ as well as arcs as roads. L and CL mark the positions of the investigated company's and all competitors' (potential) fuel stations, respectively. At each operated refueling station, different refueling systems can be built. Each system dispenses a specific fuel type $f, alt \in \mathrm{F}$. Following the notation and logic of [6], the traffic flows through the network on roundtrip-paths $q \in \mathrm{Q}$ along the arcs $a_{j,k} \in \mathrm{A}_q$. As vehicles with different fuels have different ranges, the potential fuel stations on paths q are given as $i, m \in \mathrm{K}_{j,k}^{q,f}$ for each arc $a_{j,k}$ and fuel type f. The periods of the planning horizon are denoted as $t \in \mathrm{T}$.

The model's objective, Eq. (58.1), is to support a refueling station network operator regarding investment decisions for the transformation and operation of its existing network. For the economic focus of the associated decisions, the net present value (NPV) is maximized:

$$\max NPV = \sum_{t\in T}\sum_{i\in L}\sum_{f\in F}(p_f - so_f)\cdot 365\cdot \boldsymbol{quantity}_{f,i,t}\cdot(1+h)^{-t}$$
$$-\sum_{t\in T}\sum_{i\in L}(oper_i^{\text{station}}\cdot \boldsymbol{z}_{i,t}^{\text{station}} + constr_i^{\text{station}}\cdot \boldsymbol{k}_{i,t}^{\text{station}}$$

$$+ \boldsymbol{deconstr}_i^{\text{station}} \cdot \boldsymbol{dk}_{i,t}^{\text{station}}) \cdot (1+h)^{-t}$$
$$- \sum_{t \in T} \sum_{i \in L} \sum_{f \in F} (oper_f^{\text{system}} \cdot \boldsymbol{z}_{f,i,t}^{\text{system}} + constr_f^{\text{system}} \cdot \boldsymbol{k}_{f,i,t}^{\text{system}}$$
$$+ deconstr_f^{\text{system}} \cdot \boldsymbol{dk}_{f,i,t}^{\text{system}}) \cdot (1+h)^{-t} \tag{58.1}$$

On the one hand, the refueling station network operator needs to decide which stations $\boldsymbol{z}_{i,t}^{\text{station}} \in \{0, 1\}$ as well as how many and which refueling systems $\boldsymbol{z}_{f,i,t}^{\text{system}} \in \mathbb{N}_0^+$ to operate on these stations in each period. Therefore, stations and systems can be constructed ($\boldsymbol{k}_{i,t}^{\text{station}} \in \{0, 1\}$, $\boldsymbol{k}_{f,i,t}^{\text{system}} \in \mathbb{N}_0^+$) or deconstructed ($\boldsymbol{dk}_{i,t}^{\text{station}} \in \{0, 1\}$, $\boldsymbol{dk}_{f,i,t}^{\text{system}} \in \mathbb{N}_0^+$). Construction and deconstruction are associated with payments ($constr_i^{\text{station}}$, $deconstr_i^{\text{station}}$, $constr_f^{\text{system}}$, $deconstr_f^{\text{system}}$). For the operation of stations and systems, payments per period occur ($oper_i^{\text{station}}$, $oper_f^{\text{system}}$). All payments are discounted with the interest rate h. A balance constraint ensures that the number of systems of fuel type f operated in period t equals the sum of systems operated in the previous period $t-1$ and systems constructed and deconstructed at the beginning of period t at each station $i \in$ L. The stations are balanced similarly. Another constraint accounts for the existing initial situation. The available space at each refueling station $i \in$ L must not be exceeded by the required space of the operated systems.

On the other hand, the fuel sales contribute to net incoming payments via their specific price p_f and specific payments for provision so_f. Each refueling station's quantity of sales $\boldsymbol{quantity}_{f,i,t} \in \mathbb{R}_0^+$ of fuel type f is restricted by the installed capacity and by the demand $\boldsymbol{demand}_{f,i,t} \in \mathbb{R}_0^+$, which is generated by the passing vehicles. As a Cournot oligopoly-like market is assumed for this strategic setting, the refueling needs that are generated by vehicles traveling on the arcs $a_{j,k} \in \mathrm{A}_q$ on paths $q \in$ Q are split into equal parts $\boldsymbol{x}_{i,(j,k),f,t}^q \in \mathbb{R}_0^+$ (in distance units) over the refueling stations i in operation. The total demand at each refueling station $\boldsymbol{demand}_{f,i,t}$ (in fuel units) can now be calculated as product of refueling needs $\boldsymbol{x}_{i,(j,k),f,t}^q$, number of vehicles, and fuel efficiency over paths $q \in$ Q and arcs $a_{j,k} \in \mathrm{A}_q$. The number of cars of a specific fuel type f depend on its realized market share $\boldsymbol{s}_{q,f,t}^{\text{real}} \in \mathbb{R}_{[0,1]}$. Exogenous knowledge of a potential market share is assumed. However, conventional truck owners are assumed to only switch to an AFV if they perceive the alternative refueling options to be sufficient. Thus, the realization of potential market shares depends on the travelability of the vehicles' paths, i.e., the coverage $\boldsymbol{y}_{f,q,t} \in \{0, 1\}$. Following the formulation of [6], a path q is covered if all its arcs $a_{j,k} \in \mathrm{A}_q$ are covered by either own or competitors' refueling stations. Exogenous knowledge of the competitors' actions is assumed. If a path is not travelable for vehicles of fuel type f, a predefined share of the traffic will instead use a different vehicle type with different fuel alt. If neither vehicle types can travel this path, these vehicles do not generate refueling demand.

Case Study

As the focus of this study is the replacement of conventional heavy-duty diesel trucks with hydrogen fuel cell trucks for long-haul transportation, the case study is depicting one network operator with refueling stations along two major highways in Germany, Autobahn A2 and A7. The 70 nodes and 76 arcs of this cross-like traffic network are analogous to the real-world service areas and road segments. Each of the two highways is represented by one OD-round trip. The hydrogen refueling station and truck parameters are based on [11], the remaining parameters were appraised realistically.

The potential hydrogen fuel cell truck market share is assumed to follow a generalized logistic function reaching its maximum growth in 2035 and converging to 100%. As only two fuels are examined, the diesel share is the reversed function. The planning horizon is 20 years or periods, respectively. Four competition scenarios (I–IV) are investigated. Scenario I, early and instantaneous build-up of hydrogen infrastructure at all competitor stations 5 years from the baseline year 2022. Scenario II, build-up 10 years after baseline, and Scenario III, late build-up 15 years after baseline. Scenario IV, no competitor hydrogen refueling infrastructure build-up.

The model classifies as a mixed-integer quadratically-constrained program. It has been implemented in Python 3.9.12 and solved using Gurobi 9.5 on a virtual 64-bit Windows 10 system with 4 cores Intel Xeon Platinum 8180 @2.50 GHz CPU and 32 GB RAM. The case study models comprise around 23,000 linear, 630 quadratic, and 81,600 general constraints as well as 87,100 variables, 7,500 of which are integer and 3,700 are binary. Three scenario instances (I, II, and III) were solved optimally in about 1 min, 6 min, and 2 h. Instance IV was terminated after 72 h with an optimality gap of 5.83%. The large runtime results from increasing symmetries as competition decreases.

Results

The optimal investment and operation plans have been calculated for each scenario of the case study. The results show that the company's decisions strongly depend on the timing of the competitors' actions. Figure 58.1 illustrates the NPV and the number of diesel and hydrogen systems operated on the company's refueling stations in each period.

The transformation with the highest NPV can be realized in Scenario IV without competition in the hydrogen refueling market. The NPV will be lower the earlier the competitors enter the market. All transformation plans include both hydrogen and diesel refueling systems. Hydrogen systems are preferably built along the A2 first, where traffic volume is higher. In Scenario III and in Scenario II along the A2, it is optimal to install hydrogen systems before the competitors act (market leader strategy). As a result, groups of specific refueling stations are simultaneously

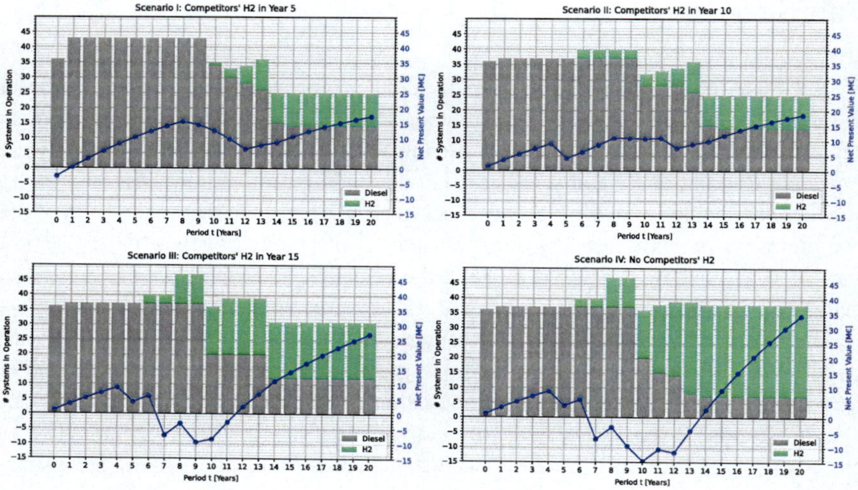

Fig. 58.1 NPV and station configurations for the 4 scenarios

equipped with hydrogen systems. These stations are spatially dispersed along the highway to ensure travelability. Whereas in Scenario I and in Scenario II along the A7, travelability is ensured by the competitors' refueling stations and a market follower strategy is adopted. Depending on their sales quantity, scattered stations are successively equipped with hydrogen systems. In contrast to the market leader strategy, not every operated station is equipped with a hydrogen system in the final period, where the fuel cell vehicle market share is > 95%. This might be perceived as a lack of service by the customers.

Conclusions

The presented model has shown to generate useful investment and operation plans for the transformation of a multi-fuel refueling station network. The results can support operators' decision-making for different competition scenarios. Future research should apply the model to larger case studies and improve computational performance.

Acknowledgements This work was part of the research project THEWA, which was funded by the Ministry of Science and Culture of Lower Saxony (Germany). The authors are responsible for the contents of this publication.

References

1. Rose, P. K., Nugroho, R., Gnann, T., et al. (2020). Optimal development of alternative fuel station networks considering node capacity restrictions. *Transportation Research Part D: Transport and Environment, 78*, 1–16. https://doi.org/10.1016/j.trd.2019.11.018
2. Zhang, A., Kang, J. E., & Kwon, C. (2017). Incorporating demand dynamics in multi-period capacitated fast-charging location planning for electric vehicles. *Transportation Research Part B: Methodological, 103*, 5–29. https://doi.org/10.1016/j.trb.2017.04.016
3. Böhle, A. (2021). Multi-period optimization of the refuelling infrastructure for alternative fuel vehicles. *Junior Management Science, 6*, 790–825. https://doi.org/10.5282/JUMS/V6I4PP673-699
4. Rose, P. K., & Neumann, F. (2020). Hydrogen refueling station networks for heavy-duty vehicles in future power systems. *Transportation Research Part D: Transport and Environment, 83*, 1–17. https://doi.org/10.1016/j.trd.2020.102358
5. Kuby, M., & Lim, S. (2005). The flow-refueling location problem for alternative-fuel vehicles. *Socio-Economic Planning Sciences, 39*, 125–145. https://doi.org/10.1016/j.seps.2004.03.001
6. Capar, I., Kuby, M., Leon, V. J. Tsai, Y.-J. (2013). An arc cover–path-cover formulation and strategic analysis of alternative-fuel station locations. *European Journal of Operational Research, 227*, 142–151. https://doi.org/10.1016/j.ejor.2012.11.033
7. Tafakkori, K., Bozorgi-Amiri, A., & Yousefi-Babadi, A. (2020). Sustainable generalized refueling station location problem under uncertainty. *Sustainable Cities and Society, 63*, 1–14. https://doi.org/10.1016/j.scs.2020.102497
8. Wang, Y.-W., & Lin, C.-C. (2013). Locating multiple types of recharging stations for battery-powered electric vehicle transport. *Transportation Research Part E: Logistics and Transportation Review, 58*, 76–87. https://doi.org/10.1016/j.tre.2013.07.003
9. Bersani, C., Minciardi, R., Sacile, R., Trasforini, E. (2009). Network planning of fuelling service stations in a near-term competitive scenario of the hydrogen economy. *Socio-Economic Planning Sciences, 43*, 55–71. https://doi.org/10.1016/j.seps.2008.02.001
10. Crönert, T., & Minner, S. (2021). Location selection for hydrogen fuel stations under emerging provider competition. *Transportation Research Part C: Emerging Technologies, 133*, 1–17. https://doi.org/10.1016/j.trc.2021.103426
11. H2 Mobility. (2022). Overview—Hydrogen refuelling for heavy duty vehicles. https://h2-mobility.de/wp-content/uploads/2022/03/H2M_Overview_HDV_Refuelling_2022.pdf. Last accessed 2022/07/07

Chapter 59
Bidirectional Green Waves for Major Road Axes by Adjusting Separate Left-Turn Phases

Christian Liebchen

Abstract Planning so-called green waves along major road axes is a well-established target for traffic engineers. This is mainly for two reasons: a smooth traffic flow quality, and less air pollution. For one-way road axes (e.g., the Avenues in Manhattan), this is a trivial downstream task. For bidirectional arterials, there is a well-known necessary condition for establishing a green wave in both directions: The driving times between two subsequent crossings must be integer multiples of half of the cycle time of the signal programs at the nodes. In this presentation, we propose an integer linear optimization model to establish fixed-time green waves in both directions that are as long and as wide as possible, even in the situation where the above-mentioned driving time condition is not fulfilled. In particular, we are considering an arterial along whose nodes separate left-turn signal groups are realized. In our computational results, we show that scheduling left-turn phases before or after the straight phases can reduce waiting times along the arterial. Moreover, we show that there is always a solution with green waves in both directions that are as long and as wide as possible, where absolute priority is put on just one direction. Only when considering prioritized parts of a green band (e.g. some first few seconds), then an ideal green wave into one direction can provide suboptimal quality compared to optimizing both directions together. Finally, we validate the nominal solution quality according to the objective function values with the results of corresponding runs of the well-established traffic flow simulation tool SUMO.

Keywords Traffic light coordination · Synchronization · Phase sequencing · Integer programming · Simulation

C. Liebchen (✉)
Technische Hochschule Wildau, Engineering and Natural Sciences, Hochschulring 1, 15745 Wildau, Germany
e-mail: liebchen@th-wildau.de

O. Grothe et al. (eds.), *Operations Research Proceedings 2022*, Lecture Notes in Operations Research, https://doi.org/10.1007/978-3-031-24907-5_59

Introduction

Despite the availability of traffic-responsive control of street crossings, fixed-time controlled signal programs can still be a matter of choice, in particular, during times that show high traffic volumes in various directions and dense networks with short distances between junctions. When considering one isolated junction (or node), the following decisions are to be taken: definition of the phases (signal groups showing green at the same time), sequence of the phases, and lengths of the phases. Counting intermediate times as “phases”, too, the sum of the lengths of all phases yields the *cycle length*, after which the same signal program repeats in a fixed-time control. When coordinating adjacent junctions, a common cycle length might be set first.

We are considering a road arterial along which both directions deserve attention. If the junctions have only two phases, so-called *green waves* can only be established into both directions, if the driving time between two consecutive junctions equals half of the common cycle length, or some integer multiple of it. This condition is only rarely met in practice. Hence, even with the best possible coordination (or synchronization) of the relative offsets (or shifts) of the signal programs of the junctions, it is not possible to plan green waves into both directions for the entire *platoons* of cars.

This is why we are going beyond the pure coordination of predefined signal programs. Rather, we are optimizing the sequences of the phases, and even select among a predefined small set of alternative phases at each junction. According to [2], our goal is to minimize the number of stops of the cars that travel along the entire arterial.

Mathematical optimization had been applied to the pure coordination of fixed signal programs of traffic lights even for intermeshed street networks [7], involving sophisticated objective functions to measure the waiting times of arriving cars appropriately ([8], in particular Sect. 1.3.5) and making use of time-expanded networks [3].

Zhang et al. [9] include decisions of phase sequencing along two arterials similar to the one that we are working with. But they also take the freedom to adjust the progression speeds in order to compute (at least narrow) green bands along the *entire* arterial into both directions. Köhler and Strehler [4] also select appropriate sequences for the phases. An instance, for which an optimum solution is found within a few seconds involves seven junctions, but six junctions are just connected by a unidirectional traffic flow. For a larger and intermeshed network of 32 signalized intersections, optimality gaps of 20 percent remain even after hours of computation.

With this background, our contribution lies in the combination of:

- sticking to compact mathematical models (see also [6]) and optimization methods,
- not only computing optimal relative shifts of fixed signal programs, but including phase sequencing and partial selection among some pre-defined alternative phases,
- at the price of limiting the use case to arterials (with separate left-turn signal groups) instead of scheduling general road networks,
- finally evaluating the planned results by a microscopic traffic flow simulation.

Our computational results show that allowing for selecting among different strategies for the left-turn phases can be highly beneficial. This is supported by an evaluation using the microscopic traffic flow simulation software tool SUMO [1].

General Setting and Optimization Model

We are considering an arterial along which the left-turn traffic is secured by separate signal groups. According to [2], our goal is to schedule the green phases such that the cars that are driving along the entire arterial are facing as little stops as possible. This is advantageous for both, the quality of the traffic flow, and environmental impacts.

Instead of only coordinating fixed signal programs being pre-defined at each junction, our optimization model selects among the phases that are displayed in Fig. 59.1.

As a result of traffic flow analysis, assume the straight traffic requires a green duration of 50 s per cycle (i.e. 90 s), and left-turn traffic 15 s. These durations can be composed of three sequences I–III of the phases (a) to (d): I (a) 15 s, (b) 35 s, (c) 15 s; II: (c) 15 s, (b) 35 s, (a) 15 s; III: (b) 50 s, (d) 15 s (alternatively, the other way round).

Due to space limitations, we are only able to sketch the optimization model "as is", without discussing any of its simplifying assumptions. We are considering an ordered set J of junctions along the arterial. The two directions of the arterial are $D = \{\text{WE}, \text{EW}\}$, where driving from node j to $j + 1$ is from west to east and takes d_{ij} seconds. The cycle time, which applies to each junction, is denoted by T (e.g. 90 s).

As in [8], the model that we are using is based on the Periodic Event Scheduling Problem (PESP, [7]), see also [5]. There are two types of vertices, namely the beginning time of a green phase (GB) at junction j in direction d, whose time point within the cycle time T is the value of the variable $\pi_{j,\,d,\,\text{GB}} \in [0,T)$. In Fig. 59.2, these vertices are displayed in grey. For sake of convenience, we also introduce variables $\pi_{j,\,d,\,\text{AB}} \in [0,T)$, which model the beginning of the platoon arriving (AB) at junction j (white vertices).

There are three types of arcs along which we are measuring time durations:

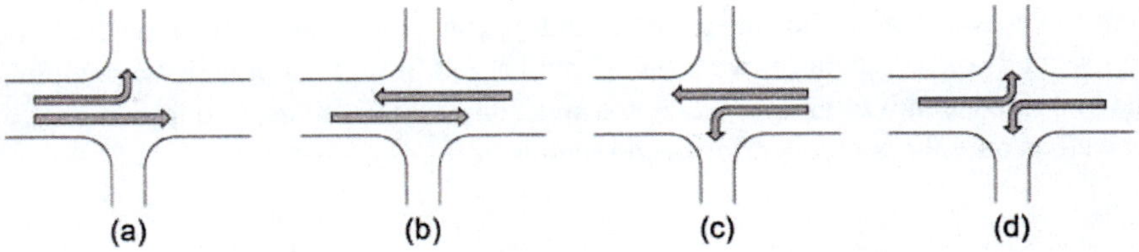

Fig. 59.1 Reference [6]. There are four relevant basic phases for the through-traffic along the arterial together with separate left-turn signal groups. The phases for the crossing traffic are not displayed, but are assumed to be connected (see also, Fig. 59.2 in [9])

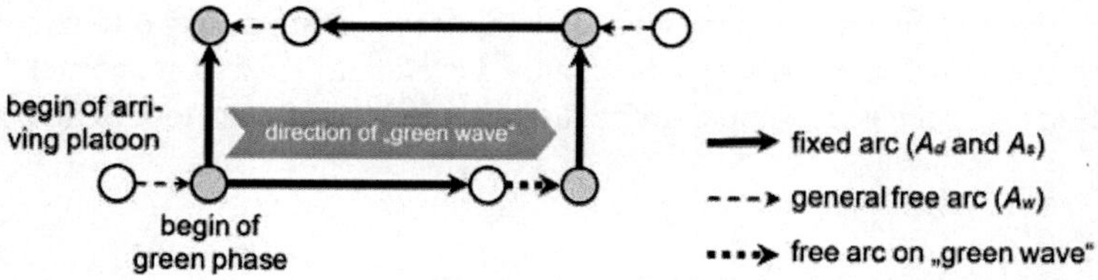

Fig. 59.2 Reference [6]. Highlighting the arcs with possible slack along an arterial, where the four leftmost vertices belong to junction j, while the four rightmost vertices belong to junction $j + 1$

- driving activities A_{d} (from green begin to arrival begin at next junction),
- offset (shift) activities A_{s} (between two green begin vertices at the same junction),
- activities A_{w} at the junctions, where cars could face waiting times in case of a stop (from the beginning of a platoon arriving at a junction to beginning of green there).

The time duration $x_a \in [\ell_a, u_a]$ of an arc $a = (v_1, v_2) \in A$ is related to the points of time of the two involved events as follows (see [5, 7]), where $p_a \in \mathbf{Z}$ is a technical variable:

$$x_a = \pi_{v2} - \pi_{v1} + \pi_a \cdot T \tag{59.1}$$

For arcs $a \in A_d$, the lower and upper bounds are $[d_a, d_a]$, and $[0, T - 1]$ for $a \in A_w$. In order to let the values of x_a, $a \in A_s$ select only among values that correspond to feasible phase sequences, i.e., $x_a \in \{-s_j, 0, s_j\}$ (with $s_j = 15$ s in our example), we introduce a ternary variable $c_j \in \{-1, 0, +1\}$—hereby leaving the pure PESP model—and require:

$$x_a = c_j \cdot s_j + T \cdot r_j,\ r_j \in \{0, 1\} \tag{59.2}$$

Finally, for the objective function, let us just point to [6] to indicate that maximizing the portion of a platoon that arrives at some junction and can continue without having to stop can be modeled by a piecewise linear convex function that has to be maximized. This involves additional binary variables to identify the appropriate linear segment in function of the value of the tension variable x_a, $a \in A_w$. Notice that in addition to considering the entire green band, we put some extra priority on the first few seconds of it: Whenever a car had to stop at red, it will continue its trip within the first few seconds of green. Hence, we may consider the full green duration $gd_{j,\,\mathrm{no}}$ as well as only the first few prioritized seconds of it $gd_{j,\,\mathrm{yes}}$.

Computational Results of Optimization and Simulation

We are considering a 3 km long bidirectional arterial with two lanes for driving into both directions each. Along this arterial, there are located eleven junctions. This situation is similar to several major road axes in the city of Berlin, Germany, e.g. the street Prenzlauer Allee. As on several other streets in Berlin, there is a tramway line in the middle between the two roadways for the two directions. In order to provide safe left-turns from the arterial, there are separate green phases, when the tramway has to stop.

We are assuming a cycle time of T = 90 s, as it is often used in Germany. For the straight traffic along the arterial, we are working with a green phase duration of 47 s, plus a three seconds yellow phase. For the left-turns, we schedule 12 s of green, again plus a three seconds yellow phase. In order to demonstrate the full potential of the proposed integer linear optimization model, we avoid the ideal driving times of half the cycle time (i.e., 45 s) and integer multiples of it. In Fig. 59.3, we depict the driving times (in seconds) between two subsequent junctions, where the total of 200 s corresponds to an average flow progression speed of 15 m/s (54 km/h). The interested reader is invited to check that also only few partial sums of these driving times are close to integer multiples of half the cycle time.

Optimization. In our series of optimization runs, we had varied the following parameters, but report only the conceptually simplest as well as the best quality solutions:

- length of the prioritized green duration $gd_{j,\text{yes}}$,
- degree of freedom, whether both directions must have green at the same time (i.e., $c_j = 0$), or if phase sequences can be selected at each junction (i.e., $c_j \in \{-1, 0, +1\}$),
- requirement whether an ideal green wave into one direction (w.l.o.g. WE) has to be established (i.e., $x_a = 0$, for $a \in A_w$ in direction $d =$ WE),
- different global weights for the (un-) prioritized portion of the objective function.

The optimization ran on an Intel Core i5-5200U at 2.20 GHz with 8 GB RAM, the model was built in AIMMS 4.86.6.2 and solved with CPLEX 20.1. The solution times were below one minute (except ID 18 99 min). If we make it ten times more attractive to put priority only on the first 10 or 25 s of the green band (ID 02 and ID 03), the results turn out to be much worse than when putting weights the other way round.

Simulation. To evaluate the results of our mathematical optimization model (Table 59.1), we use the microscopic traffic flow simulation tool SUMO [1]. The traffic flow for the simulation model is 18% for each direction (16 cars per 90 s). The cars are

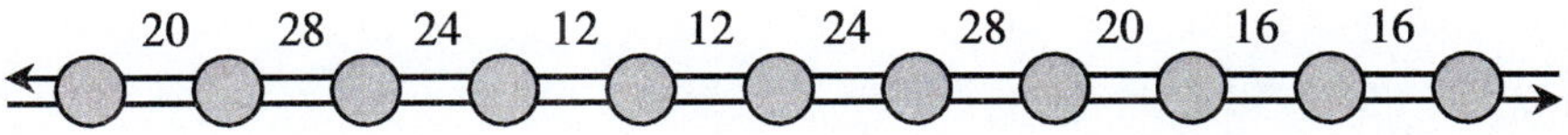

Fig. 59.3 Driving times (in seconds) between two subsequent junctions along the arterial

of two different types: “normal cars” plus some “sporty cars” having speed factors of 0.95 or 1.1, accelerations of 1.5 m/s^2 or 2.5 m/s^2, and decelerations of 3.0 m/s^2 or 4.5 m/s^2, resp., and common speed deviations 0.05. “Solution quality” are the durations of the green bands that do not have to stop, according to the results within the optimization model.

In the simulation, we are reporting the stop and time loss values including the first junction, where the traffic flow arrives uniformly at random. Hence, one may expect already $40/90 = 0.44$ stops there (red duration divided by cycle time). Together with random speed deviations, this explains values of more than 0.5 stops per car even in the direction of an ideal green wave (ID’s 02, 15, 16)—and yield an expected number of only about 1.2 stops per car along the ten subsequent junctions in ID 18. This high quality for the traffic flow can also be observed in the time–space diagram in Fig. 59.4.

Conclusions

From a traffic engineering point of view, the results of our simulation runs clearly support the high benefit of allowing for selecting among different strategies for the left-turn phases at the junctions along a bidirectional road arterial. Potentially, this could also be applied to other speed profiles, such as bicycle pedelecs at, say, 25 km/h. Tuning the performance of solving the integer linear programs, has not been the focus here, but is rather left to possible future merely mathematical contributions.

Table 59.1 Results of the simulation runs (average of five runs with different random seeds)

ID	$gd_{i,yes}$ [s]	Sequence requirement	Green wave WE?	Solution quality [s] for q = yes	Solution quality [s] for q = no	Stops WE	Stops EW	Time loss [s]. WE EW	
00		$c_j = 0$, $\pi_{GB} = 0$	No	(Reference)		4.40	4.39	173	167
03	10	$c_j = 0$	no	200 (100%)	600 (60%)	4.70	3.67	185	147
15	10	$c_j \in \{-1,0,+1\}$	**yes**	172 (86%)	846 (85%)	**0.55**	**3.29**	30	120
02	25	$c_j = 0$	yes	360 (72%)	660 (66%)	0.55	7.55	30	281
08	25	$\mathbf{c_j = 0}$	no	404 (81%)	660 (66%)	**3.36**	**3.41**	**133**	**127**
16	25	$c_j \in \{-1, 0, +1\}$	yes	442 (88%)	846 (85%)	0.57	3.26	31	119
18	25	$\mathbf{c_j \in \{-1, 0, +1\}}$	no	442 (88%)	846 (85%)	**1.49**	**1.64**	**60**	**73**

For $q =$ yes and $gd = 10$ s, 200 s yield 100% (10 junctions, 2 directions), for $gd = 25$ s, 500 s

Fig. 59.4 Illustration of the green bands in both directions for a well-coordinated plan (ID 18) in a time–space diagram (time passing top-down). The first 25 s of the green bands are drawn solid. There is no junction j at which both directions are having green at the same time (i.e. $c_j = 0$)

References

1. Deutsches Zentrum für Luft- und Raumfahrt (DLR) (2022) SUMO—Simulation of Urban Mobility (Version 1.4.0). https://www.eclipse.org/sumo/. Last accessed 2022/07/18
2. Forschungsgesellschaft für Straßen- und Verkehrswesen (FGSV). (2015). Handbuch für die Bemessung von Straßenverkehrsanlagen (HBS) – Teil S: Stadtstraßen. ISBN 978-3-8446-103-3. In German. FGSV-Verlag GmbH.
3. Köhler, E., & Strehler, M. (2015). Traffic signal optimization using cyclically expanded networks. *Networks, 65*(3), pp. 244–261.
4. Köhler, E., & Strehler, M. (2019). Traffic signal optimization: Combining static and dynamic models. *Transportation Science, 53*(1), 21–41.
5. Liebchen, Ch. (2006). Periodic timetable optimization in public transport. dissertation.de – Verlag im Internet.
6. Liebchen, Ch. (2022). Maximizing bidirectional green waves for major road axes. Preprint, Technische Hochschule Wildau. https://doi.org/10.15771/2827. Last accessed 2022/07/19
7. Serafini, P., & Ukovich, W. (1989). A mathematical model for periodic scheduling problems. *SIAM Journal on Discrete Mathematics, 2*(4), 550–581.
8. Wünsch, G. (2008). *Coordination of traffic signals in networks*. Ph.D. thesis, Technical University Berlin, Germany.
9. Zhang, C., Yuanchang, X., Gartner, N. H., Stamatiadis, Ch., & Arsava, T. (2015). AM-band: An asymmetrical multi-band model for arterial traffic signal coordination. *Transportation Research Part C, 58*, 515–531.

Chapter 60
Modeling Uncertainty in the Timetable-Based Railway Network Design Problem

Tim Sander, Nadine Friesen, Karl Nachtigall, and Nils Nießen

Abstract Many European countries plan their railway infrastructure according to strategic timetables, using them as input for further strategic and tactical planning steps, including network design. While both tactical timetabling and network design are well covered by the research, there exists, to the best of the authors' knowledge, no model which focuses on network design based on strategic timetables. In this short paper, we present an overview of our research. The traditional network design problem is extended to incorporate railway-specific features such as headway-based capacity estimations and demand derived from a strategic timetable. This includes trains represented by integral flows with start, destination, and time bounds as well as timetabling constraints for line frequencies and transfers. Since strategic timetabling is done many years in advance, the strategic timetable and as such the demand for the network design problem are subject to uncertainty. To account for this, we aim to calculate networks which are robust towards changing input timetables. We describe two approaches to model this: optimizing the network for a timetable family (a set of discrete scenarios) and varying demands within scenarios, which is modelled by a set of optional trains. The paper describes basic modelling decisions, details the approach to incorporate the uncertainty and shows the main features of the optimization model as well as a case study and further research.

Keywords Railway network design · Strategic railway timetabling · Robust optimization

---

T. Sander (✉) · K. Nachtigall
Traffic Flow Sciences, TU Dresden, 01069 Dresden, Germany
e-mail: tim.sander@tu-dresden.de
URL: https://tu-dresden.de/bu/verkehr/ila/vkstrl

N. Friesen · N. Nießen
Institute of Transport Science, RWTH Aachen, 52074 Aachen, Germany

O. Grothe et al. (eds.), *Operations Research Proceedings 2022*, Lecture Notes in Operations Research, https://doi.org/10.1007/978-3-031-24907-5_60

Introduction

In recent years, many western European railway operators have altered the traditional planning process by including strategic timetabling. These strategic timetables, which are constructed ten to twenty years in advance, are used as input for further strategic and tactical planning steps, including network design. In this paper, we present a railway network design problem where the demand is derived from a strategic timetable. Even though the first strategic timetable has been constructed in Switzerland in the 1980s, the topic has been picked up by academic research only recently, e.g. in [4] or [7]. More research has been conducted focusing on tactical timetables (refer to Chaps. 5 and 6 of [1, 2]).

Railway network design has also been studied in a few publications, including timetable-independent approaches like [6] or microscopic approaches without an implementation like [5]. The integration of these two steps however leaves room for further research. We try to reduce the gap with this paper. Since the strategic timetable is constructed many years in advance, it is subject to uncertainty. We address this uncertainty by using a robust optimization approach based on timetable families. Robust optimization in railway analysis is common, but it is usually applied to the tactical and operational planning stages like tactical timetabling or crew and vehicle scheduling [3].

The remainder of this short paper is structured as follows: in Section "Methodology", we describe the modelling of the network and the input timetable, including a focus on the robust modelling in Section "Considering Robustness". The key properties of the optimization model including the constraints dealing with the robust optimization are presented in Section "Optimization Model". In Section "Case Study", we provide details about the implementation and the case study before concluding the paper in Section "Conclusion and Outlook".

Methodology

Modeling Approach

In our model, we start with a given strategic timetable and use it to calculate a cost-optimal network on which the timetable can be operated. To gain some flexibility for optimization, the timetable is relaxed in several ways:

- the trains can be routed freely from their origin to their destination nodes, with an option to define via-nodes
- only time bounds for the departure at the origin node and the arrival at the destination node are considered.

Using these relaxations, we derive an operational concept which includes a list of trains with start and destination nodes and time bounds. It also includes a list of timing-related connections between pairs of trains which model either frequencies or transfers.

The network is defined as a multi-graph G = (N, A) featuring nodes $i \in N$ and arcs $(i, j, tr) \in A$. Nodes represent stations or junctions and provide the opportunity to change from one arc to another. We also consider node links (i, a, b), which connect two arcs (a, i, tr) and (i, b, tr) with each other, allowing trains to travel from node a to node b through node i. We do not consider limited node capacities at the moment.

The arcs (i, j, tr) are identified by the nodes which they connect (i and j) and the track number tr. One arc represents one track, so multi-track sections are modelled by several parallel arcs. This allows us to model both single-track and multi-track sections. The inclusion of an arc into the solution is controlled by the decision variable $y_{i,j,tr}$. Similar decision variables $l_{i,a,b}$ exist for node links. Both are associated with building costs and are part of the objective function. To estimate the capacity used on a section, we use train-type and train-sequence-dependent minimal headway times (MHT), which have to be respected between two trains that use the same track. In case the capacity is insufficient, the model features three options to extend capacity on arcs:

- including more parallel arcs
- reducing travel times
- reducing headway times.

The last two are modelled using reduction variables r^{time} and r^{MHT}, which are also included in the objective function and come with associated costs c^{time} and c^{MHT} for each unit of time reduction.

Considering Robustness

Because of the long planning horizon and the resulting difficulties to estimate demand and political decisions influencing the timetable, strategic timetables are subject to uncertainties. Since the network designed by our model is based on a strategic timetable, it needs to account for possible changes to this input timetable. We model the uncertain timetables by defining timetable families, which comprise several discrete timetable scenarios with individual lists of trains and connections. Besides, the demand within one scenario is variable.

By defining timetable families, we can model slightly different timetable concepts which may vary in the trains or the timing relationships. We can then calculate a network which enables a specified percentage of the timetable family. By observing this scenario coverage share, we can calculate both full robust networks, which cover all given scenarios and light robust networks, which cover only a certain share of the scenarios.

The variable demand within one scenario is modelled with optional trains. They can be considered in different ways:

- only activate optional trains if they don't require additional infrastructure
- add a penalty to the objective function if a train is not activated and create a trade-off between the capacity to run additional trains and the costs for additional infrastructure
- randomly select a random number of trains from the optional set before the optimization, which then become mandatory.

The optional trains can also be used to evaluate the remaining capacity on the optimized network and to quantify the robustness towards changes in the input timetable.

Optimization Model

We model the timetable-based railway network design problem as a mixed-integer linear program. Our objective function (60.1) minimises infrastructure costs, which contains building costs $f_{i,j,tr}$ and $f_{i,a,b}$ for arcs and node links as well as reduction costs $c_{i,j}^{time}$ and $c_{i,j}^{MHT}$ for reductions of travel and headway times. It also includes penalty terms which are activated if a scenario or an optional train is not included in the final solution. This is modelled using the decision variables o_s^{szo} for scenarios and o_k^{train} for trains.

$$min \sum_{(i,j,tr)\in A} f_{i,j,tr} y_{i,j,tr} + \sum_{(i,a,b)\in L} f_{i,a,b} l_{i,a,b} + \sum_{(i,j)\in A_l} c_{i,j}^{time} r_{i,j}^{time}$$
$$+ \sum_{(i,j)\in A_l} c_{i,j}^{mht} r_{i,j}^{mht} + \sum_{k\in K} pen_k(1 - o_k^{train}) + \sum_{s\in S} pen_s(1 - o_s^{szo}) \qquad (60.1)$$

Because we integrate network design and strategic timetabling as well as observe railway-specific capacity measures by using minimal headway times, the constraint set is extensive and an explanation of the full model would exceed the scope of this short paper. Therefore, we will describe only the constraints dealing with the robustness consideration and leave a full description for a future paper.

$$\sum_{s\in S} \frac{1}{n_s} o_s^{szo} \geq s_s \qquad (60.2)$$

$$o_s^{szo} = 1 \quad \forall s \in S_{mand} \qquad (60.3)$$

$$\sum_{(k,p)\in P_k} p_{k,p} = o_k^{train} \cdot o_s^{szo} \quad \forall k \in K \qquad (60.4)$$

$$o_k^{train} = o_s^{szo} \quad \forall k \in K_{mand,s}, \quad \forall s \in S \qquad (60.5)$$

$$\sum_{k\in K} o_k^{train} \geq n_{k,mand,s} + n_{k,opt,demand,s} \quad \forall s \in S \tag{60.6}$$

The most important aspect of our robust optimization approach, the scenario coverage share s_s, is controlled by (60.2), while (60.3) allows to define certain scenarios as mandatory. To route the trains through the network, we use pre-generated paths for each train. These paths have to include all defined via-nodes for a train and must fulfil the travel time requirements given by the operational concept. Constraint (60.4) makes sure that exactly one path is chosen for each train if both the scenario and the train itself are active. It is important to note that this constraint is linearised in the code. If the correspondent scenario is active, all mandatory trains of this scenario have to be active as well, which is ensured by (60.5). As mentioned in Section "Considering Robustness", we have an option to request a certain amount of optional trains to be included. This is done by constraint (60.6).

Further constraints cover the standard network design aspects by ensuring that trains travel only on arcs and node links included in the solution networks. The integrated timetabling is done by constraints ensuring that given time bounds, travel times, required frequencies and transfer times and minimal headway times are correctly observed. The frequency and transfer constraints are only active if the correspondent scenario is activated. The correct headway time to be observed depends on the train types and the sequence of two trains following each other.

The results include the cost-optimal network with all the arcs and node links necessary to operate the requested timetable scenarios. They also include the routing of each train and a feasible macroscopic timetable for each scenario. Note, that the timetable is not yet optimized. However, it is possible to optimize the timetable in a later optimization step, e.g. by minimizing the total travel time while fixing the infrastructure.

Case Study

To demonstrate the model, prove its functionality and identify performance issues, the optimization model has been tested on a small case study. Several test cases and scenarios have been derived from drafts for the Deutschlandtakt, the strategic timetable concept for Germany. The optimization model has been implemented in Python 3.8 and solved by Gurobi v9.5.1. using a laptop featuring an Intel Core i7-8565U CPU @1.80 GHZ and 16 GB of RAM. The computational results for a test case with ten scenarios are shown in Table 60.1. The scenarios vary in the number, type and route of trains (between 22 and 48) and the amount of frequency and transfer constraints (between 0 and 56). The resulting network has been calculated for varying coverage percentages.

As expected, the infrastructure costs increase with the scenario coverage. It can also be observed, that the demanded coverage share has a significant impact on the model's performance and that computation times are rather long, given that the

Table 60.1 Computational results

Scenario coverage (%)	Objective	Gap (%)	Active scenarios	Runtime (s)
20	407,600	0.00	3/10	502.05
40	411,100	0.00	6/10	930.09
60	411,100	0.00	6/10	1330.84
80	431,900	1.64	8/10	3600.19
100	458,900	0.00	10/10	1890.89

shown example covers only about 12 nodes and 20 arcs. Even though the model is associated with strategic planning, the current runtimes do not allow the calculation of realistically sized instances within reasonable time, especially when multiple scenarios are considered. Therefore, a heuristic approach for calculating a network covering all scenarios has been developed. First, cost-optimal networks are calculated for each timetable scenario independently. Using an adapted version of the optimization model, the feasibility of each timetable scenario on each unique infrastructure resulting from the first step is evaluated. In case no network topology covers all scenarios, the infrastructure covering the largest number of timetable scenarios is then iteratively extended to cover the unfulfilled timetable scenarios. This approach has proven to be faster than the complete optimization model, but still leaves room for further improvements.

Conclusion and Outlook

In this short paper, we gave an overview of our research concerning the integration of strategic railway timetabling and railway network design. We motivated the problem at hand and identified a gap in the literature. We propose a mixed-integer linear program which calculates a cost-optimal railway network based on demand given as an operational concept derived from a strategic timetable, featuring a list of trains and important timetabling aspects such as frequencies and transfers. The model incorporates railway-specific details like track-based routing and a realistic capacity estimation using minimal headway times on both single- and double-track lines. During the optimization, a feasible timetable respecting the key properties given in the operational concept is designed. Further research will focus on performance improvements by studying both heuristic approaches and decomposition techniques. Besides that, an evaluation of the remaining capacity of the networks is planned as well as a sensitivity analysis of the network towards changes in the input parameters.

Acknowledgements Funded by the Deutsche Forschungsgemeinschaft (DFG, German Research Foundation)—Project ID 432300662 and GRK 2236.

References

1. Borndörfer, R., Klug, T., Lamorgese, L., Mannino, C., Reuther, M., & Schlechte, T. (Eds.). (2018). *Handbook of optimization in the railway industry* (Vol. 268). Springer Verlag.
2. Caimi, G., Kroon, L., & Liebchen, C. (2017). Models for railway timetabling optimization: Applicability and applications in practice. *Journal of Rail Transport Planning & Management, 6*(4), 285–312.
3. Lusby, R. M., Larsen, J., & Bull, S. (2018). A survey on robustness in railway planning. *European Journal of Operational Research, 266*(1), 1–15.
4. Polinder, G.-J., Schmidt, M., & Huisman, D. (2021). Timetabling for strategic passenger railway planning. *Transportation Research Part B: Methodological, 146*, 111–135.
5. Schöbel, A., Raidl, G. R., Grujicic, I., Besau, G., & Schuster, G. (2013). An optimization model for integrated timetable based design of railway infrastructure. In *Proceedings of the 5th International Seminar on Railway Operations Modelling and Analysis—RailCopenhagen* (pp. 765–774).
6. Spönemann, J. C. (2013). *Network design for railway infrastructure by means of linear programming* [Doctoral dissertation]. Aachen Technischen Hochschule, Dissertation.
7. Weigand, W., & Heppe, A. (2019). Spurplangestaltung und betriebliche Infrastrukturplanung. In *Handbuch Eisenbahninfrastruktur* (pp. 467–523).

Part XIII
OR in Developing Countries

Chapter 61
Integration of the Multiple Criteria Decision Making Method KEMIRA into a GIS for the Problem of Choosing Suitable Areas for a Given Use

Abdoulaye Ouedraogo and Stéphane Aimé Metchebon Takougang

Abstract Land use management problems generally requires the use of geographic information systems (GIS). To do this, it is first necessary to take into account several often conflicting criteria. However, given the limited capabilities of GIS, the representation of a decision map resulting from the overlaying of more than four conflicting criteria maps is problematic. To overcome this difficulty and enhance the analytical capabilities of GIS, the literature presents a variety of works on the joint use of GIS and multiple criteria decision making (MCDM) methods. In these works, the limitations that emerge are firstly the difficulty of fully integrating a MCDM method into the GIS, secondly the large number of parameters of the MCDM method to be integrated into the GIS, thirdly the aggregation of heterogeneous criteria. In this work we propose an answer to these three concerns through the integration of the KEMIRA method in the GIS open source QGIS and illustrate our approach on a real case study in Burkina Faso.

Keywords MCDM · KEMIRA method · GIS · QGIS · Land use management

Introduction

In this paper, we are interested in the multiple criteria choosing problem statements in land use management which aim at choosing best areas or alternatives in a set $X = \{x^1, x^2, \dots, x^P\}$ taking into account Q criteria and the preferences of Decision Maker (DM). An aggregation model must be constructed to allow the choosing process to yield to a result in accordance with the DM's preferences. Only multiple criteria choosing models based on a utility function will be considered. When using an additive utility function for multicriteria choosing, the analyst must determine weight and threshold parameters. These parameters are used to construct a preference model

A. Ouedraogo · S. A. Metchebon Takougang
LANIBIO, Universite Joseph Ki-Zerbo, 03 BP 7021 Ouagadougou 03, Burkina Faso

S. A. Metchebon Takougang (✉)
Université Aube Nouvelle, 06 BP 9283 Ouagadougou 06, Burkina Faso
e-mail: metchebon@gmail.com

O. Grothe et al. (eds.), *Operations Research Proceedings 2022*, Lecture Notes in Operations Research, https://doi.org/10.1007/978-3-031-24907-5_61

of the decision maker (DM). Generally it is not realistic to assume that the DM would easily provide the values of these parameters. In the case of choosing problems, and precisely concerning the utility functions based models, few authors have proposed methodologies to infer preference parameters [1, 2].

By heterogeneous criteria we mean those for which compensation between strengths and weaknesses of the criteria is not allowed, in contrast to homogeneous criteria where it is allowed. One of the drawbacks of utility functions is most often the total compensation among heterogeneous criteria when it is not allowed. It is the case for instance when dealing with the sustainable management of resources where a strong productivity (economic criterion) can't compensate a lost of biodiversity (environmental criterion). That is why considering the criteria by homogeneous groups of criteria during the process of aggregating the performances of the alternatives would further contribute to legitimize the results of an approach based on the use of utility functions. Such a context is presented by Metchebon Takougang et al. [3], when dealing with the land use problem of assessing the sustainability of resource management practices where criteria derived from economic, social, environmental and governance issues need to be taken into account. Another context showing the need to consider groups of homogeneous criteria that we can mention is that related to the choice of the best crop varieties suitable for a given region where criteria derived from quality, productivity and processing appear. Recently, an iterative approach based on the gradient descend, using increasing functions and thresholds, has been presented by Krylovas et al. [4, 5]. This approach, called KEmeny Median Indicator Ranks Accordance (KEMIRA) method, takes into account the homogeneity of groups of criteria in their aggregation process. This method allows to choose best alternatives.

In spatial decision problem several works promoting the use of Geographical Information System (GIS) and/or remote sensing exist [6, 7]. Particularly concerning the integration of Multiple Criteria Decision Analysis (MCDA) in GIS including their strengths and limitations, a number of works have been listed by Sobrie et al. [8] and Malczewski and Rinner [9]. Some of the limitations that emerge from these works relate to the difficulty of fully integrating an MCDM method into GIS and the large number of parameters, inherent to the MCDM method, that must be taken into consideration. Another limitation to add is the aggregation of heterogeneous criteria which are in most cases aggregated as if they were homogeneous. Our present work aim to give an answer to this three mentioned limitations by integrating the KEMIRA method into the GIS open source QGIS.

The next part of the paper is structured as follows. Firstly we present the KEMIRA method and its operating process. After we briefly expose the strategy adopted to integrate KEMIRA method into the GIS. Finally we illustrate our approach through a real case study of choosing areas of adequate response to the risk of degradation in Burkina Faso.

Mathematical Formulation of KEMIRA Model

We assume that the set of alternative is $X = \{x^1, x^2, \ldots, x^P\}$. For $k \in \{1, \ldots, P\}$, alternative x^k is a real-valued vector of dimension Q. Each component of this vector represents the performance of alternative x^k on a specific criterion. The Q criteria are partitioned into S groups indexed by $\{1, \ldots, S\}$ and the elements of each group are indexed in turn by $\{1, \ldots, n_i\}$ where n_i is the size of group i. We have of course $\sum_{i=1}^{S} n_i = Q$ because the S groups form a partition of the Q criteria. Each criterion is thus indexed by a pair of natural numbers, the first one for the group, and the second one for the criterion within the group. The performance of alternative x^k on criterion (i, j) is thus denoted by $X^k_{i,j}$. To each criterion (i, j), we associate a weight $w_{i,j}$. Our objective is to compute the weight $w_{i,j}$ of each criterion (i, j) and select the best alternatives by formulating and solving an optimization problem.

Criteria Priority and Increasing Functions

We assume that the DM is able to provide a ranking of criteria in each group. Therefore, without loss of generality, suppose that in each group the criteria are ranked in the decreasing order so that the relations (61.1) and the corresponding restrictions on the criteria weights (61.2) hold:

$$\begin{array}{c} (1,1) \succsim (1,2) \succsim \cdots \succsim (1,n_1), \\ (2,1) \succsim (2,2) \succsim \cdots \succsim (2,n_2), \\ \cdots \\ (S,1) \succsim (S,2) \succsim \cdots \succsim (S,n_S). \end{array} \tag{61.1}$$

$$\begin{array}{c} w_{1,1} \geq w_{1,2} \geq \cdots \geq w_{1,n_1}, \\ w_{2,1} \geq w_{2,2} \geq \cdots \geq w_{2,n_2}, \\ \cdots \\ w_{S,1} \geq w_{S,2} \geq \cdots \geq w_{S,n_S}. \end{array} \tag{61.2}$$

So, having the performance $X^k_{i,j}$, $k \in \{1, 2, \ldots, P\}$, $j \in \{1, 2, \ldots, n_i\}$, of alternatives w.r.t. the Q criteria, we normalize them. Here we choose an affine type of normalization and the normalized values are obtained by the relation

$$x^k_{i,j} = \frac{X^k_{i,j} - \min_j X^k_{i,j}}{\max_j X^k_{i,j} - \min_j X^k_{i,j}}. \tag{61.3}$$

In each group we assume that the weights are normalized and we verify:

$$\sum_{j=1}^{n_i} w_{i,j} = 1, \quad \forall i \in \{1, 2, \ldots, S\}. \tag{61.4}$$

We also assume that all criteria are to be maximized. The bigger values of variables $x_{i,j}^k$ represent better satisfaction w.r.t. the considered criterion. Then, for an alternative x^k and for each group i, we compute the weighted average of the performance, denoted by W_i:

$$W_i(x^k) = \sum_{j=1}^{n_i} w_{i,j} \times x_{i,j}^k, \tag{61.5}$$

where weights $(w_{i,j})$ satisfy requirements (61.2) and (61.4).

Process of Selecting the Best Alternatives

For each increasing function W_i we introduce one threshold α_i, $0 < \alpha_i < 1$. Thus, an alternative x^k is selected as better if it satisfies requirements (61.6):

$$W_i(x^k) > \alpha_i \quad \forall i \in \{1, 2, \ldots, S\}. \tag{61.6}$$

In practice the thresholds α_i are the performance levels, set by the DM, that any alternative must meet. So the DM must express his preference on these thresholds in terms of percentage of the best performance respectively in each group of criteria. Formally, for each group i the DM is asked to set a number p strictly between 0 and 100 such that:

$$\alpha_i = p\% \times \max_{k=1}^{P} W_i(x^k), \quad 0 < \alpha_i < 1. \tag{61.7}$$

Objective function and mathematical program Denote B the set of best alternatives and $|B|$ the cardinal of B, i.e., the number of elements in B. $fopt$ denotes the function to be optimized and is defined by $fopt = |B|$. For a given set of weights satisfying relation (61.2) and associated performance levels as stated in inequations (61.6) expressing the preference information of the DM, corresponds a choice of best alternatives whose set is B. The objective function to be maximised then allows the selection process to be stopped when the number of best alternatives, $|B|$, is as large as possible. This statement is formalised by the following mathematical program:

$$\begin{array}{c} \max_{w_{i,j}} fopt = |B| \\ w_{i,j} \text{ satisfying relation (61.2) and (61.6).} \end{array} \tag{61.8}$$

Algorithm for selection problem In this work we have implemented the algorithm proposed by Krylovas et al. [5] to solve the mathematical program (61.8). We stop this algorithm of eliciting the corresponding weights with best alternatives associated when the cardinal of the set of best alternatives representing the objective function $fopt$ is maximal.

Implementation of the KEMIRA Method in the GIS

Figure 61.1 shows an adaptation of the scheme propose by Sobrie et al. [8] that we used to implement KEMIRA Method in the GIS. The main difference between our integration scheme and that of Sobrie et al. is the absence of a criteria weights inference module. Indeed, in our case, the KEMIRA method in the process of selecting the best alternatives also allows to determine the criteria weights.

1st step: Structuring of the problem: identification of criteria which are individually spatialized in the form of corresponding criteria maps built through a GIS. The region under study are partitioned into areas or spatial units.

2nd step: Construction of the multiple criteria map: The position of each spatial unit on a criterion map allowing to deduce the performance of said spatial w.r.t. the corresponding criterion and vice versa. The overlaying of all the criteria maps results to the multiple criteria map.

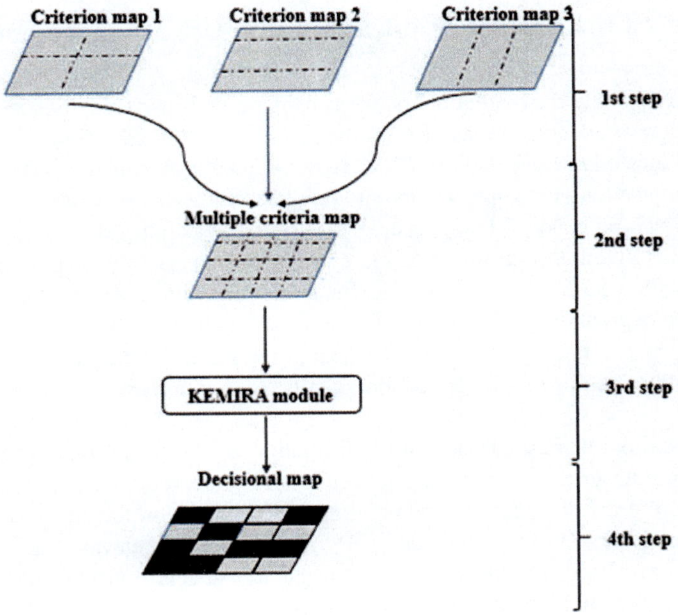

Fig. 61.1 Construction of a decisional map in a GIS

3rd step: Definition of KEMIRA model: the functioning of the KEMIRA method was presented in Section "Mathematical Formulation of KEMIRA Model". The DM expresses his preferences by giving a value to the performance thresholds α_k on each identified group of homogeneous criteria S_k. The DM also expresses his preference by ranking the criteria in each group from best to the worst.

4th step: Generation of the decision map: the KEMIRA model being defined, the decision map is generated with spacial units of two colors. The best spatial unit have all the same color whereas the remaining spatial units have the other color.

Quantum GIS (QGIS)

In this work we have used QGIS 3.14.0 [10] which is an GIS open source allowing to implement new tools in the form of plugging usable as any other basic tool incorporated in the GIS. So KEMIRA method has been implemented in python language directly inside the QGIS 3.14.0. Kernel in such a way to been used as a new decision tool when need.

Application to a Choice Problem in Land Use Management

In order to illustrate the methodology, we consider a real case study concerning the identification of adequate areas of response to the risk of degradation in Loulouka watershed located in Burkina Faso (West Africa). Such areas which is not degraded could be appropriate for a specific activity (e.g. agriculture). This problem has been addressed by Metchebon Takougang et al. [3] using a weak integration of the MCDA method ELECTRE TRI in the GIS. Here we implement a full integration of the MCDA method KEMIRA into the GIS for the identification of areas of adequate response to degradation risk. In the following we highlight the main results of the structuring phase of the problem. For the reader interested by the structuring of the problem, more details on this application can be seen in Metchebon Takougang et al. [3].

The Loulouka watershed has been partitioning in 227 spatial units or areas of 25 ha each. Four groups of criteria has been identified according to the dimensions of limitation of the degradation (soil erosion, biodiversity, soil fertility and agricultural productivity). On this basis ten (10) criteria and their relevant measurement indicators have been developed to evaluate the 227 spacial units or alternatives. The set of ten criteria has been subdivided in four (4) groups, $S_4 = \{(4, 1), (4, 2)\}$, $S_3 = \{(3, 1), (3, 2)\}$, $S_2 = \{(2, 1), (2, 2)\}$, $S_1 = \{(1, 1), (1, 2), (1, 3), (1, 4)\}$:

- S_1 : limiting soil erosion; S_2 : limiting biodiversity loss
- S_3 : soil fertility is maintained; S_4 : Good agricultural productivity is promoted.

The ranking of criteria in each group S_k as well as the performance levels α_k have to be provided by the DM. Note that while it is generally easy for the DM to rank the criteria within each group S_k, the determination of their corresponding thresholds or performance levels α_k is not obvious. For the execution of the KEMIRA algorithm, the parameters were defined as follows:

$$\begin{aligned} &\text{the maximum number of iterations, } \max_{iter} = 800; \\ &\alpha_i = 65\% \times \max_{k=1}^{227} W_i(x^k); \quad \forall k \in \{1, 2, 3, 4\}. \end{aligned} \tag{61.9}$$

$\max_{iter} = 800$ denotes the number of iterations of KEMIRA algorithm executed from which the size of the set of best selected spatial units, $fopt$, is maximal for our case study. Here we obtained $fopt = 56$. The results of Metchebon Takougang et al. [3] showed that, after using a weak integration between ELECTRERI Tri Method and ArcView3.2a GIS, the number of spatial units assigned respectively to the 4th categories of response to the risk of degradation was: 172 for the two worst categories C_1 and C_2 (spatial units belonging to this two categories are considered as degraded), 55 for the two best categories C_3 and C_4 (spatial units belonging to these two categories are considered to be non-degraded). With the parameters of KEMIRA method as state in (61.9), the 56 best selected spatial units sensibly correspond to the 55 non-degraded spatial units obtained by Metchebon Takougang et al. [3].

Conclusion

The new KEMIRA-GIS model that we have proposed provides an answer to the three limitations raised in the introduction. Note that the main difficulty in applying the new KEMIRA-GIS model concerns the determination of the performance threshold α_k for which an indirect method must be developed. In addition, tests on other spatially-referenced selection problems (e.g. selection of green areas with high economic, environmental and cultural potential for sustainability; assessment of the vulnerability of water wells) need to be conducted to confirm its validity.

References

1. Devaud, J., Groussaud, G., & Jacquet-Lagrèze, E. (1980). *UTADIS: une methode de construction de fonctions d'utilite additives rendant compte de jugements globaux* (European working group on MCDA).
2. Siskos, Y., Grigoroudis, E., & Matsatsinis, N. F. (2016). UTA methods. In S. Greco, M. Ehrgott, & J. Figueira (Eds.), *Multiple criteria decision analysis*. International Series in Operations

Research and Management Science (Vol. 233, pp. 315–362). Springer. https://doi.org/10.1007/978-1-4939-3094-4_9

3. Metchebon Takougang, S. A., Pirlot, M., Some, B., & Yonkeu, S. (2015). Assessing the response to land degradation risk: The case of the Loulouka Catchment Basin in Burkina Faso. In R. Bisdorff, L. Dias, P. Meyer, V. Mousseau, & M. Pirlot (Eds.), *Evaluation and decision models with multicriteria. Case studies*. International Handbooks on Information Systems (pp. 341–400). Springer-Verlag. https://doi.org/10.1007/978-3-662-46816-6_12
4. Krylovas, A., Zavadskas, E. K., Kosareva, N., & Dadelo, S. (2014). New KEMIRA method for determining criteria priority and weights in solving MCDM problem. *International Journal of Information Technology and Decision Making, 13*(6), 1119–1134. https://doi.org/10.1142/S0219622014500825
5. Krylovas, A., Dadelo, S., Kosareva, N., & Zavadskas, E. K. (2017). Entropy-KEMIRA approach for MCDM problem solution in human resources selection task. *International Journal of Information Technology and Decision Making, 16*(5), 1183–1210. https://doi.org/10.1142/S0219622017500274
6. Kuter, S., Weber, G. W., & Akyürek, Z. (2016). A progressive approach for processing satellite data by operational research. *Operational Research, 17*(2), 371–393. https://doi.org/10.1007/s12351-016-0229-x
7. Joerin, F., Desthieux, G., Beuze, S. B., & Nembrini, A. (2009). Participatory diagnosis in urban planning: Proposal for a learning process based on geographical information. *Journal of Environmental Management, 90*(6), 2002–2011. https://doi.org/10.1016/j.jenvman.2007.08.024
8. Sobrie, O., Pirlot, M., & Joerin, F. (2013). Integration de la méthode d'aide á la décision ELECTRE TRI dans un système d'information géographique open source. *Revue Internationale de Géomatique, 23*(1), 13–38. https://doi.org/10.1016/S0377-2217(03)00120-6
9. Malczewski, J., & Rinner, C. (2015). *Multicriteria decision analysis in geographic information sciences*. Springer Berlin Heidelberg.
10. Quantum GIS Development Team. *Quantum GIS geographic information system*. http://qgis.osgeo.org

Part XIV
OR in Engineering

Chapter 62
A 2D Convex Shapes Bin Packing Problem in the Production of Laminated Safety Glass

Steffen Goebbels, Thomas Lühring, and Jochen Rethmann

Abstract The discussed two-dimensional nesting problem is motivated by the production of differently shaped tiles of laminated safety glass that can be represented by primitive, convex polygons. Within as few rectangular bins as possible, representing the space of a furnace, tiles must be placed without overlapping. While the primary problem is to minimize the number of occupied bins, distances between adjacent tiles or a tile and an adjacent furnace boundary must be neither too small nor too large to ensure the stability of the furnace filling during a lamination process. To fulfill this condition, a minimum number of additional rectangular support plates must be added. These plates are considered equivalent to tiles when measuring distances. This is a new aspect that, to our knowledge, has not been covered in the literature so far. We represent the problem as a mixed integer linear program based on no-fit polygons and compare results with those of a greedy-type heuristic.

Keywords Nesting problem · 2D-irregular shapes bin packing problem

Introduction

A variety of heuristics and optimization procedures including evolutionary algorithms and simulated annealing strategies have been developed to tackle several problems of nesting polygonal shapes within rectangular spaces, cf. [2]. This paper discusses a variant with additional constraints motivated by the automation of laminated glass tile production by the company HEGLA-HANIC GmbH. The tiles are homogeneous stacks of glass layers and intermediate foils. The composition of the

S. Goebbels (✉) · T. Lühring · J. Rethmann
Faculty of Electrical Engineering and Computer Science, iPattern Institute, Niederrhein University of Applied Sciences, 47805 Krefeld, Germany
e-mail: steffen.goebbels@hsnr.de

T. Lühring
e-mail: thomas.luehring@stud.hn.de

J. Rethmann
e-mail: jochen.rethmann@hsnr.de

O. Grothe et al. (eds.), *Operations Research Proceedings 2022*, Lecture Notes in Operations Research, https://doi.org/10.1007/978-3-031-24907-5_62

layers can be optimized from a material point of view (cf. [9]), but that is not intended here. Rather, the tiles are given as simple, convex 2D polygons, which thus do not have to be generated by guillotine cuts. As a primary optimization goal, the tiles must be arranged on a minimum number of rectangular furnace bins without overlaps. During the laminating process, a plate is placed from above the tiles with great pressure. The tiles must not be too close to each other, but also not too far apart, so that the pressure does not cause any damage. The lower distance bound can be easily achieved by enlarging the polygons through scaling. To fulfill the upper distance condition, rectangular support plates can be added. The secondary optimization goal is to minimize their number. This problem is strongly NP hard since the classical bin packing problem is reducible to it. The survey [10] summarizes modeling techniques for 2D nesting problems. Here, we state the problem as a mixed integer linear program (MILP) based on no-fit polygons. Then we discuss a simple greedy approach.

Mixed Integer Linear Program

Geometric Basics Let simple, convex polygons P_i, $i \in [n] := \{1, \ldots, n\}$, representing tiles be given such that they can be traversed counter-clockwise by following the edges between m_i vertices $\mathbf{v}_{i,1}, \ldots, \mathbf{v}_{i,m_i}$ and back to $\mathbf{v}_{i,m_i+1} := \mathbf{v}_{i,1}$ where $\mathbf{v}_{i,k} = (\mathbf{v}_{i,k}.x, \mathbf{v}_{i,k}.y) \in \mathbb{R}^2$. To guarantee a minimum distance between tiles in the final layout, the original tiles have already been enlarged. We also add N (also enlarged) rectangular support plates P_i, $i \in \{n+1, \ldots, n+N\}$ of not necessarily different size and a rectangle P_{n+N+1} that will be used to limit maximum distances. The model also allows simple, convex polygons instead of rectangles. For each pair $(i, j) \in [n+N] \times [n+N+1]$ with $i < j$ we compute a no-fit polygon (NFP, see [1, 3]) $F_{i,j}$ with vertices $\mathbf{f}_{i,j,1}, \ldots, \mathbf{f}_{i,j,m_{i,j}}, \mathbf{f}_{i,j,m_{i,j}+1} := \mathbf{f}_{i,j,1}$ that are also arranged counter-clockwise. Here, this polygon describes the curve of reference point $\mathbf{v}_{j,1}$ when P_j traverses around the edges of the fixed polygon P_i. Note that with P_i and P_j, also the NFP is simple and convex. This follows directly from the standard algorithm to obtain the shape of an NFP for two convex polygons by orienting P_i counter-clockwise, P_j clockwise, translating all directed edges of both polygons to a single point and then concatenating the edges counter-clockwise giving a polygon $\tilde{F}_{i,j}$ with vertices $\tilde{\mathbf{f}}_{i,j,k}$, see [4]. To obtain the NFP $F_{i,j}$ one only has to translate this shape $\tilde{F}_{i,j}$ according to the reference point and the position of P_i with vector $(\Delta x_{i,j}, \Delta y_{i,j})$,

$$\Delta x_{i,j} := \min_{k\in[m_i]} \mathbf{v}_{i,k}.x - \max_{k\in[m_j]} (\mathbf{v}_{j,k}.x - \mathbf{v}_{j,1}.x) - \min_{k\in[m_{i,j}]} \tilde{\mathbf{f}}_{i,j,k}.x,$$

and $\Delta y_{i,j}$ defined accordingly with x replaced by y. Whereas we restrict ourselves to convex polygons, many algorithms were developed to also compute NFPs for non-convex polygons, see [5, 12] and the literature cited there.

To shift a polygon to a certain position, we use an offset $\mathbf{s}_i = (\mathbf{s}_i.x, \mathbf{s}_i.y)$. Shifted polygons $\mathbf{s}_i + P_i$ and $\mathbf{s}_j + P_j$ do not overlap if and only if $\mathbf{s}_j + \mathbf{v}_{j,1}$ lies outside $\mathbf{s}_i + F_{i,j}$. The NFPs have to be computed in advance. As a result, no intersections need to be calculated later.

MILP Let $B \leq n$ be the maximum number of furnace rectangles (bins) to be considered. To choose B sufficiently small, one can use the number of occupied bins of any feasible solution computed with a heuristic, cf. Section "Greedy Approach". Binary variables $x_{i,k} \in \{0, 1\}$ indicate whether a polygon P_i is placed within the furnace rectangle with index $k \in [B]$ (then $x_{i,k} = 1$) or not ($x_{i,k} = 0$). Then the primary goal of the nesting problem is to minimize the number of occupied bins such that all tiles can be placed without overlaps. The secondary goal is to use a minimum number of support plates to fulfill the maximum distance restriction: With binary variables b_k indicating the use of bin k, the goal is then

$$\text{minimize} \sum_{k\in[B]} b_k + \frac{1}{2N} \sum_{k\in[B]} \sum_{i=n+1}^{n+N} x_{i,k}, \quad \text{s.t.} \; \forall_{k\in[B]} \sum_{i\in[n]} x_{i,k} \leq n \cdot b_k$$

and several further restrictions described in what follows.

The objective function is lower bounded by the area of all tiles divided by the furnace area. All coordinates plus offsets, i.e., coordinates of points $\mathbf{v}_{i,k} + \mathbf{s}_i$, have to be within the range of the furnace rectangle coordinates. Each polygon has to be placed within at most one bin (cf. 62.1): $\forall_{i\in[N+n]} \sum_{k\in[B]} x_{i,k} \leq 1$.

There must be no overlaps between polygons $\mathbf{s}_i + P_i$ and $\mathbf{s}_j + P_j$ placed within the same bin (i.e., $x_{i,k} = x_{j,k} = 1$), i.e., by considering convexity of the NFP $F_{i,j}$, the reference point $\mathbf{s}_j + \mathbf{v}_{j,1}$ must lie in at least one half-plane bounded by a straight line through an edge of the NFP $\mathbf{s}_i + F_{i,j}$ and in which the NFP is not located. For such a half-plane, $y_{i,j,k} \in \{0, 1\}$ is set to one. By applying the inner product "·" and by considering the Hesse normal form of lines (the absolute value of the inner product between a point on a line and a normal of the line is the distance to the origin, here the outer normal of the occupied half plane is chosen to compare signed distances), one gets conditions (cf. [7, 10])

$$\begin{aligned}
&\forall_{i,j\in[n+N],i<j} \forall_{k\in[m_{i,j}]} \forall_{l\in[B]} \\
&\quad \left(\mathbf{f}_{i,j,k+1}.y - \mathbf{f}_{i,j,k}.y, -\mathbf{f}_{i,j,k+1}.x + \mathbf{f}_{i,j,k}.x\right) \\
&\quad \cdot \left[\left(\mathbf{s}_i.x + \mathbf{f}_{i,j,k}.x, \mathbf{s}_i.y + \mathbf{f}_{i,j,k}.y\right) - \left(\mathbf{s}_j.x + \mathbf{v}_{j,1}.x, \mathbf{s}_j.y + \mathbf{v}_{j,1}.y\right)\right] \\
&\quad \leq M(2 - x_{i,l} - x_{j,l}) + M(1 - y_{i,j,k}), \\
&\forall_{i,j\in[n+N],i<j} \sum_{k\in[m_{i,j}]} y_{i,j,k} \geq 1.
\end{aligned}$$

The constant $M > 0$ has to be chosen sufficiently large. For non-convex polygons, checking with convex regions outside the NFP can be done instead of checking with half planes, see [6].

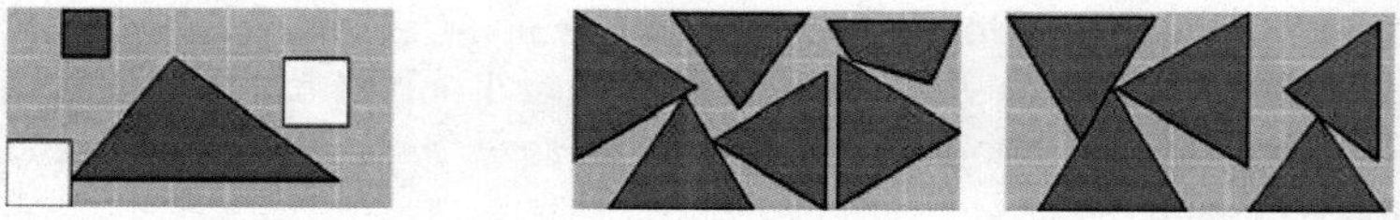

Fig. 62.1 The distance condition (62.2, 62.3) requires that each square of the background grid has to be at least partially covered. Left: a feasible one-bin layout for two tiles (grey) with two support plates (white). Right: an optimal solution using two bins without support plates (instance 3 in Table 62.1)

So far, we have not described how to enable rotations. In the application under consideration, only rotations by multiples of 90° are to be discussed (orthogonal rotation). Rotations by a finite number of angles can be easily represented by adding rotated tile polygons (of different shape) to the list of polygons P_i and by assuring that exactly one rotated instance of a polygon has to be placed in exactly one bin, i.e., for each index set $I \subset [n]$, representing all rotated instances of a tile, we require

$$\sum_{i\in I}\sum_{k\in[B]} x_{i,k} = 1. \tag{62.1}$$

We model a maximum distance condition by placing a grid with g points over all furnace rectangles, i.e., bins, see Fig. 62.1. Let $\mathbf{g}_i \in \mathbb{R}^2$, $i \in [g]$, be offset vectors that shift predefined rectangle P_{n+N+1} to have a center point at a corresponding grid point. The condition is that, for each grid point indexed by $i \in [g]$, in each bin at least one intersection between a placed tile or support plate polygon $\mathbf{s}_j + P_j$ and this shifted rectangle $\mathbf{g}_i + P_{n+N+1}$ has to occur. Such an intersection is indicated by setting a binary variable $z_{i,j,l} \in \{0, 1\}$, $i \in [g]$, $j \in [n+N]$, $l \in [B]$, to one. It occurs if and only if the reference point $\mathbf{g}_i + \mathbf{v}_{n+N+1,1}$ lies inside each half plane that is bounded by a line through an edge of the NFP $\mathbf{s}_j + F_{j,n+N+1}$ and that is occupied by the NFP.

$$\begin{aligned}
&\forall_{i\in[g]}\forall_{j\in[n+N]}\forall_{k\in[m_{j,n+N+1}]}\forall_{l\in[B]}\\
&\quad\left(\mathbf{f}_{j,n+N+1,k+1}.y - \mathbf{f}_{j,n+N+1,k}.y,\ -\mathbf{f}_{j,n+N+1,k+1}.x + \mathbf{f}_{j,n+N+1,k}.x\right)\\
&\quad.[(\mathbf{s}_j.x + \mathbf{f}_{j,n+N+1,k}.x,\ \mathbf{s}_j.y + \mathbf{f}_{j,n+N+1,k}.y)\\
&\qquad - (\mathbf{g}_i.x + \mathbf{v}_{n+N+1,1}.x,\ \mathbf{g}_i.y + \mathbf{v}_{n+N+1,1}.y)]\\
&\geq -M(1 - x_{j,l}) - M(1 - z_{i,j,l}),
\end{aligned} \tag{62.2}$$

$$\forall_{i\in[g]}\forall_{l\in[B]} \sum_{j\in[n+N]} z_{i,j,l} > \sum_{j\in[n+N]} (1 - x_{j,l}). \tag{62.3}$$

If $x_{j,l} = 0$, one can choose $z_{i,j,l} = 1$, i.e., $\forall_{i\in[g]}\forall_{j\in[n+N]}\forall_{l\in[B]}\, z_{i,j,l} \geq 1 - x_{j,l}$.

Certain solver heuristics appear to work better if shifted rectangles $\mathbf{g}_i + P_{n+N+1}$ slightly overlap such that placement in overlap regions is preferred.

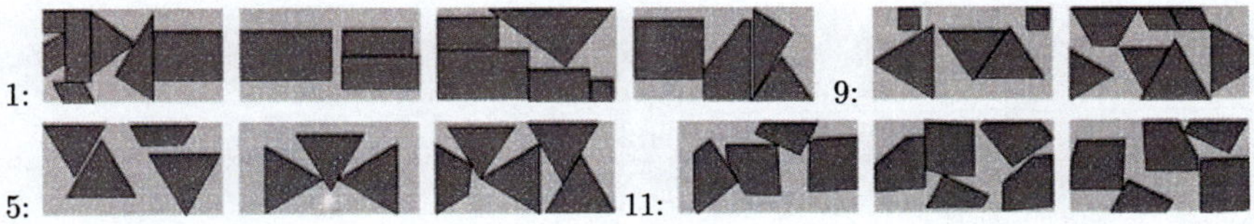

Fig. 62.2 Feasible solutions of four problem instances computed by CPLEX on 12 threads within a limit of one hour elapsed time (instances 1, 5, 9, and 11 in Table 62.1)

Greedy Approach

In up to 10,000 (nearly) random orders (permutations), we iteratively position the tiles in a bottom-left strategy. Then, among all results with the smallest number of bins, we select a result that intersects with the largest number of rectangles $\mathbf{g}_i + P_{n+N+1}$ so that a small number of support plates is needed. Motivated by the instability of the problem, this stochastic experiment replaces a local search to find a good order. To further reduce the required number of support plates, tiles small enough to fit into the distance rectangle P_{n+N+1} are always placed at the end of each permutation so that they can be inserted into empty distance rectangles with priority. We basically use steps 1–5 of the genetic algorithm in [8] in the implementation of the bottom-left strategy. However, we do not only attach to the last placed tile polygon but to all polygons. We shift each attached polygon as far as possible to the left and to the bottom by using a binary search for feasible positions that also allows to fill gaps. After placing the tiles, support plates are added to fulfill (62.2, 62.3). As long as each support plate fits into the rectangle P_{n+N+1} of the distance condition, and if enough support plates are provided, this is always possible. Since we use distance rectangles that slightly overlap, we greedily search for a vertex of these rectangles that is covered by a maximum number of so far empty distance rectangles. Then we place a support plate there (if it fits). Finally, we remove some of the support plates by re-arranging tiles: For each rectangle $\mathbf{g}_i + P_{n+N+1}$ in which a support plate is placed, we try to shift a tile from the left or from the bottom to the border of this rectangle such that (62.2, 62.3) holds without the support plate.

Results

Results for exemplary problem instances[1] provided by HEGLA-HANIC GmbH are listed in Table 62.1. Small instances can be solved with our MILP to optimality, e.g., see Fig. 62.1. However, for most instances up to 20 tiles, CPLEX 12.8 was able to find feasible (but not necessarily optimal) solutions within 60 min, cf. Fig. 62.2. The greedy approach found feasible solutions for all instances within less than two

[1] Data available at https://www.hs-niederrhein.de/fileadmin/dateien/FB03/Personen/goebbels/Publikationen/dataset.zip.

Table 62.1 Results of the MILP and the greedy strategy (10,000 permutations): a best feasible solution of the MILP is considered if the time limit of 60 min is exceeded

		MILP (CPLEX, 12 threads)			Greedy strategy (one thread)		
Instance	Tiles	Bins	Support	Time (s)	Bins	Support	Time (s)
1	19	4	0	Exceeded	4	0	13
2	47	–	–	Exceeded	9	8 ($\leq$ 5)	86
3	11	2	0	18	2	1 (0)	5
4	12	–	–	Exceeded	4	4 ($\leq$ 3)	15
5	12	3	0	Exceeded	3	3 (0)	7
6	12	–	–	Exceeded	8	7 ($\leq$ 5)	20
7	28	–	–	Exceeded	9	3 ($\leq$ 1)	18
8	36	–	–	Exceeded	7	5 ($\leq$ 4)	51
9	12	2	0	Exceeded	2	0	6
10	14	–	–	Exceeded	4	1 (0)	8
11	14	3	0	Exceeded	3	1 (0)	12
12	46	–	–	Exceeded	16	17 ($\leq$ 9)	73

minutes when working with 10,000 permutations, but in most cases 1000 permutations led to similar results in a fraction of time. For instances that could be solved with the MILP, the greedy heuristic was able to obtain the same number of bins as the MILP and reduce the number of support plates to one on average, while the feasible solutions found by MILPs within the time limit had zero support plates on average. For some instances, we could further reduce the support plates manually, see upper bounds for the optimum in brackets.

Conclusions

Although the greedy approach often fails to find a minimum number of support plates, it is apparently sufficient for practical use. Future work may test other strategies. For example, the assignment to bins could be separated from the placement of tiles and support plates within the bins in a branch-and-bound approach. The prerequisites of the framework in [11] are fulfilled. Grouping tiles into classes could help doing the tile assignment.

References

1. Adamowicz, M., & Albano, A. (1976). Nesting two-dimensional shapes in rectangular modules. *Computer-Aided Design, 8*, 27–33.
2. Albano, G. (2010). Optimal allocation of two-dimensional irregular shapes using heuristic search methods. *IEEE Transactions on Systems, Man, and Cybernetics, 10*, 242–248.
3. Art, R. (1966). *An approach to the two dimensional irregular cutting stock problem* (Technical Report IBM Cambridge Scientific Centre 36-Y08).
4. Cunninghame-Green, R. (1989). Geometry, shoemaking and the milk tray problem. *New Scientist, 1677*, 50–53.
5. Dean, H. T., Tu, Y., & Raffensperger, J. F. (2006). An improved method for calculating the no-fit polygon. *Computers and Operations Research, 33*, 1526–1539.
6. Fischetti, M., & Luzzi, I. (2009). Mixed-integer programming models for nesting problems. *Journal of Heuristics, 15*(3), 201–226.
7. Gomes, A., & Oliveira, J. (2006). Solving irregular strip packing problems by hybridising simulated annealing and linear programming. *European Journal of Operational Research, 171*, 811–829.
8. Junior, B. A., Pinheiro, P. R., & Saraiva, R. D. (2013). A hybrid methodology for tackling the irregular strip packing problem. *IFAC Proceedings Volumes, 46*(7), 396–401.
9. Kaveh, A., Dadras, A., & Geran Malek, N. (2019). Robust design optimization of laminated plates under uncertain bounded buckling loads. *Structural and Multidisciplinary Optimization, 59*(3), 877–891.
10. Leao, A. A., Toledo, F. M., Oliveira, J. F., Carravilla, M. A., & Alvarez-Valdés, R. (2020). Irregular packing problems: A review of mathematical models. *European Journal of Operational Research, 282*(3), 803–822.
11. Rebennack, S., Kallrath, J., & Pardalos, P. M. (2009). Column enumeration based decomposition techniques for a class of non-convex MINLP problems. *Journal of Global Optimization, 43*(2–3), 277–297.
12. Rocha, P. (2019). Robust NFP generation for nesting problems. arXiv preprint arXiv:1903.11139v1

Chapter 63
Learning Strategies for Outsourcing Problems With asymmetric Information and Uncertain Execution

Alexander Herbst

Abstract In this contribution, we consider an outsourcing problem based on a specific principal–agent relationship with hidden characteristics. Under the assumption that the principal knows the probability distribution on the agent's discrete type space, a standard solution technique for the resulting contracting problem is stochastic optimization on the set of incentive compatible menus of contracts from which the agent can choose a single contract according to the take-it-or-leave-it principle, respectively. Admittedly, this approach neglects any sort of uncertainties in the post-contract phase which is not realistic in many practical environments like production and logistics. To address this issue, we present a novel and holistic problem formulation that links the contracting phase to an uncertain execution phase in a logistical context containing the possibility of renegotiating contracts as a reaction to environmental changes. Since the resulting problem has the character of a multi-round game, we apply well-known concepts from the trendy AI-area of Deep Reinforcement Learning to exploit clever contracting strategies for the principal. Finally, we evaluate our approach inside a computational study.

Keywords Agent systems · Artificial intelligence · Transportation

Introduction and Problem Description

We consider the general problem from [1] where we want to process $Q \in \mathbb{N}$ units of some good and have the opportunity to outsource a partial quantity $q \in [0; Q]$ to an external service provider while the remaining quantity $Q - q$ has to be handled by own resources.

Within this setting we adapt the common assumption that both principal and agent are faced with linear cost terms $\theta_p x$ and $\theta_a x$ for processing an arbitrary quantity x, respectively, where the principal has total knowledge about his own cost factor θ_p but

A. Herbst (✉)
Technology Management, University of Siegen, Unteres Schloss 3, 57072 Siegen, Germany
e-mail: alexander.herbst@uni-siegen.de
URL: https://www.wiwi.uni-siegen.de/technologiemanagement/

O. Grothe et al. (eds.), *Operations Research Proceedings 2022*, Lecture Notes in Operations Research, https://doi.org/10.1007/978-3-031-24907-5_63

can only restrict the agent's cost factor θ_a to a set $\Theta_a = \{\underline{\theta}_a; \overline{\theta}_a\}$ [2]. As an extension to [2] and similar to [1] we consider a time aspect as well, i.e. we assume speed parameters $\delta_p, \delta_a > 0$ such that x is processed in $\delta_p x$ time units by the principal and in $\delta_a x$ time units by the agent. In contrast to [1] we assume that δ_p and δ_a are always completely observable for the principal and that Θ_a is discrete as above (instead of $\Theta_a = [\underline{\theta}_a; \overline{\theta}_a]$).

We claim that a cooperation only happens on the basis of a contract $\mathcal{C} := (q, p)$ with some quantity $q \in [0; Q]$ and payment of $p \in \mathbb{R}^+$ monetary units. For θ_a fixed we want to define the agent's utility $\mathcal{U}^a_{\theta_a}(\mathcal{C})$ for accepting contract $\mathcal{C}$ as a quasi-linear function of quantity q and payment p [1, 2].

Definition 1 (*Utility of the agent*) For a contract $\mathcal{C} := (q, p)$ we define

$$\mathcal{U}^a_{\theta_a}(\mathcal{C}) := p - \theta_a q.$$

In order to deal with uncertainty about the true type θ_a it is an essential approach to design contract alternatives for each possible scenarios and equip them with specific incentives that rationally affect the agent in our interest [1–3].

Definition 2 (*Incentive compatibility*) A menu of contracts $\{\mathcal{C}(\theta_a) : \theta_a \in \Theta_a\}$ is called incentive compatible if the following expression holds:

$$\mathcal{U}^a_{\theta_a}(\mathcal{C}(\theta_a)) \geq 0 \quad \wedge \quad \theta_a \in \arg\max_{\theta' \in \Theta_a} \mathcal{U}^a_{\theta_a}(\mathcal{C}(\theta')) \qquad \forall \theta_a \in \Theta_a.$$

The overwhelming majority of principal-agent approaches put the principal's preferences in focus [1–4], often meaning that the optimal incentive compatible menu of contracts measured by some utility of the principal is of interest. For our setting such a menu of contracts contains two contracts $\{\mathcal{C}(\underline{\theta}_a); \mathcal{C}(\overline{\theta}_a)\}$ from which $\mathcal{C}(\underline{\theta}_a)$ is chosen if $\theta_a = \underline{\theta}_a$ and $\mathcal{C}(\overline{\theta}_a)$ is chosen if $\theta_a = \overline{\theta}_a$ [2]. The payment $p(\underline{\theta}_a)$ or $p(\overline{\theta}_a)$ is transferred immediately and both principal and agent start operating according to their speed factors δ_p and δ_a [1].

However, many similar approaches end at this point as they consider deterministic circumstances after some contract is negotiated making further investigations irrelevant [1–4]. The closeness to reality of those simplified models can be doubted. As a main contribution of this work we consider the possibility of specific parameter changes in an uncertain execution phase and the option of renegotiation as a reaction to these.

Deterministic Execution Phase

For now, let us assume that the four problem parameters θ_p, θ_a, δ_p and δ_a remain constant after the initial negotiation phase. According to [2] the utility function of the principal should be of the general form

$$\mathcal{U}(q, p) := S(q) - p \tag{63.1}$$

where $S(\cdot)$ is a concave function that values the intrinsic preferences of the principal for an agreed quantity q and p monetary units have to be paid to the agent. Such a function lives up to the principle of a decreasing marginal utility and ensures the existence of a global maximum.

A specific representation of $S(\cdot)$ in (63.1) which charges up the principal's preference for an early and uniform processing of the total quantity Q against the internal costs for precessing $Q - q$ units by using own resources is introduced in [1]:

$$S(q) := R - c \cdot \left(\frac{\delta_p + \delta_a}{2Q} q^2 - \delta_p q + \frac{\delta_p}{2} Q \right) - \theta_p (Q - q). \tag{63.2}$$

Within this setting the optimal incentive compatible menu of contracts (from the viewpoint of the principal) can be stated explicitly.

Theorem 1 *Let* $\Theta_a := \{\underline{\theta}; \overline{\theta}\}$ *with* $\mathfrak{v}$ *being the probability that* $\underline{\theta}$ *is the true value of* θ_a*. For* $\mathcal{U}$ *as above and* $\Delta\theta := \overline{\theta} - \underline{\theta}$*, the incentive compatible menu of contracts* $\{C(\underline{\theta}); C(\overline{\theta})\} := \{(q(\underline{\theta}), p(\underline{\theta})); (q(\overline{\theta}), p(\overline{\theta}))\}$ *maximizing the expected utility* $\mathfrak{v} \cdot U(C(\underline{\theta})) + (1 - \mathfrak{v}) \cdot U(C(\overline{\theta}))$ *of the principal is then given by*

$$q(\underline{\theta}) = \frac{Q\left(c\delta_p + \theta_p - \underline{\theta}\right)}{c(\delta_p + \delta_a)}, \qquad q(\overline{\theta}) = \frac{Q\left(c\delta_p + \theta_p - \overline{\theta} - \frac{\mathfrak{v}}{(1-\mathfrak{v})} \Delta\theta\right)}{c(\delta_p + \delta_a)},$$

$$p(\underline{\theta}) = \underline{\theta} \cdot q(\underline{\theta}) + \Delta\theta \cdot q(\overline{\theta}), \qquad p(\overline{\theta}) = \overline{\theta} \cdot q(\overline{\theta}).$$

Proof The theorem follows directly from the general optimization problem for two possible cost parameters $\{\underline{\theta}; \overline{\theta}\}$ in [2] by concretely setting $S(q)$ as in (63.2). □

Uncertain Execution Phase and Renegotiation

In what follows we assume that the parameters δ_p, δ_a, θ_p and θ_a can possibly change over time due to specific uncertainties in the execution phase. In this context, imaginable real world events could be changing traffic scenarios like completely free highways on the one hand or jams which can possibly occur in waves of different lengths [5] on the other hand.

We want to assume that parameter changes can only occur at discrete points in time $t = 0, 1, 2, 3, \ldots$ and that for each time stamp t, the principal has complete knowledge about his own parameter constellation $(\delta_p^{(t)}, \theta_p^{(t)})$ as well as the agent's speed parameter $\delta_a^{(t)}$ (for example by observing the current traffic situation) for the current time interval $[t; t + 1]$. As an example from literature, [6] utilizes variable-length

Markov chains to extract traffic scenarios from real world data. For our model, however, we will use a simpler approach incorporating a Markov chain with stationary transitions and small state space [7].

More precisely, let us consider three possible speed scenarios (normal traffic, low traffic and high traffic) for both the principal and the agent by defining the parameter sets $\Delta_p = \{\underline{\delta}_p; \hat{\delta}_p; \overline{\delta}_p\}$, $\underline{\delta}_p < \hat{\delta}_p < \overline{\delta}_p$, and $\Delta_a = \{\underline{\delta}_a; \hat{\delta}_a; \overline{\delta}_a\}$, $\underline{\delta}_a < \hat{\delta}_a < \overline{\delta}_a$. We initialize $\delta_p^{(0)} = \hat{\delta}_p$ and $\delta_a^{(0)} = \hat{\delta}_a$ by default and assume that the probabilities for a parameter transition into one direction are given by $\mathfrak{p}_p$ and $\mathfrak{p}_a$, with $0 < \mathfrak{p}_p, \mathfrak{p}_a < 1/2$. This means for example that if $\delta_p^{(t)} = \hat{\delta}_p$ for some t, the probability for a transition to $\delta_p^{(t+1)} = \underline{\delta}_p$ or $\delta_p^{(t+1)} = \overline{\delta}_p$ is given by $\mathfrak{p}_p$, respectively, which conversely implies $\delta_p^{(t+1)} = \hat{\delta}_p$ with probability $1 - 2 \cdot \mathfrak{p}_p$. On the other hand, if $\delta_p^{(t)}$ already equals one of both extreme values, i.e. $\underline{\delta}_p$ or $\overline{\delta}_p$, only a transition to $\hat{\delta}_p$ is allowed to happen with probability $\mathfrak{p}_p$, whereas $1 - \mathfrak{p}_p$ delivers the probability for the parameter to remain constant. The exactly same behavior is assumed for $\delta_a^{(t)}$ with probability $\mathfrak{p}_a$.

Regarding the costs, we define a set of three possible scenarios $\Theta_p = \{\underline{\theta}_p; \hat{\theta}_p; \overline{\theta}_p\}$ for the principal which are directly linked to their speed parameter, i.e. only the pairs $(\underline{\delta}_p, \underline{\theta}_p)$, $(\hat{\delta}_p, \hat{\theta}_p)$, $(\overline{\delta}_p, \overline{\theta}_p)$ can occur. For the agent, on the other hand, we continue to be guided by Section "Introduction and Problem Description", where the information asymmetry between principal and agent lies in the principal's fuzzy knowledge about the true value of $\theta_a \in \Theta_a := \{\underline{\theta}_a; \overline{\theta}_a\}$. In what follows, we assume that the probability $\mathfrak{v}$ for $\theta_a^{(t)} = \underline{\theta}_a$ depends on the current speed parameter $\delta_a^{(t)}$, i.e. we have to consider three probability values $\underline{\mathfrak{v}}$, $\hat{\mathfrak{v}}$ and $\overline{\mathfrak{v}}$, one for each element of Δ_a.

Regarding uncertainties for $t = 1, 2, 3, \ldots$ we want to provide the principal with the option of renegotiating a currently active contract. In this context we define q_t as the remaining quantity the agent has to supply at time $t > 0$, i.e. $q_t = \max\{0; q_{t-1} - 1/\delta_a^{(t-1)}\}$ holds if we consider the same contract for $t - 1$ and t.

Claim For $t > 0$, q_t and $\theta_a^{(t)}$ given, the agent accepts a contract update (q_t^{new}, p_t) from the principal if and only if it leads to a non-decreasing utility, i.e.

$$\mathcal{U}^a_{\theta_a^{(t)}}(q_t^{new}, p_t) \geq \mathcal{U}^a_{\theta_a^{(t)}}(q_t, 0) \quad \Longleftrightarrow \quad p_t - \theta_a^{(t)} q_t^{new} \geq -\theta_a^{(t)} q_t. \tag{63.3}$$

Learning Renegotiation Strategies

We can interpret the problem resulting from the model synthesis in Section "Uncertain Execution Phase and Renegotiation" as a single player multi-stage game with incomplete information [8] (due to the random parameter switches) from the viewpoint of the principal. The contract offers are the game actions and the game terminates when the whole quantity Q is processed. One could object at this point that the interpretation as a two-person game would be more obvious, since we are actually dealing with two different participants—the principal and the agent—with their own decision-making possibilities and opposing objectives. However, we assume that at any time $t = 0, 1, 2, \ldots$ and for each realization of $\theta_a^{(t)} \in \{\underline{\theta}_a; \overline{\theta}_a\}$ the agent chooses the action that optimizes their own easily calculable utility. For this reason, the actions of the agent appear to the principal as the realization of random scenarios from the environment which can be specified exactly in advance.

The interpretation as a one-person game also allows us to apply standard methods from the AI-area of Deep Reinforcement Learning as Actor-Critic, DDPG or Deep Q-Learning [9]. We decided to set up a Deep Q-Learning environment in Python (version 3.7.5) by using the packages `gym` to define the actual game environment and `keras-rl` to build and train the neural network. In what follows, the concrete definitions of state and action space (which both have to be discrete for Deep Q-Learning) as well as the step-wise rewards are explained.

State space: Consists of tuples $(Q_t, q_t, \delta_p^{(t)}, \theta_a^{(t)})$ where $Q_t \in \{0; \ldots; Q\}$ is the current total quantity, $q_t \in \{0; \ldots; Q_t\}$ is the current outsourced quantity, $\delta_p^{(t)} \in \Delta_p$ and $\delta_a^{(t)} \in \Delta_a$. The size of the state space is $|\delta_p| \cdot |\delta_a| \cdot (Q+1)(Q+2)/2$.

Action space: Consists of tuples (a, s), where $a \in \{-K; \ldots; 0; \ldots; K\}$ aims at an update of the current outsourced quantity q_t, i.e. $q_t^{new} = q_t + k \cdot a$ for some $k \in \mathbb{N}$, and $s \in \{0; 1\}$ defines the "scope" of acceptance by the agent, meaning that we choose $p_t = \underline{\theta}_a(q_t^{new} - q_t)$ if $s = 0$ and $p_t = \overline{\theta}_a(q_t^{new} - q_t)$ if $s = 1$. Comparing (63.3) one can check that for $s = 0$, q_t^{new} is accepted if $\theta_a^{(t)} = \underline{\theta}_a$ or $q_t^{new} < q_t$ (p_t is then a redemption from the agent) and for $s = 1$, q_t^{new} is accepted if $\theta_a^{(t)} = \overline{\theta}_a$ or $q_t^{new} > q_t$. A special case is the start of the game where we take the initial contract menu $\{\mathcal{C}(\underline{\theta}_a); \mathcal{C}(\overline{\theta}_a)\}$ from Theorem 1 and possibly update its quantities by some a, a' as above. After the game starts an action (a, s) can indeed only be applied to the currently active contract. The size of the action space is $4 \cdot K + 1$.

Rewards: The principal's reward R_t for the period $[t; t+1]$ consists of a performance term in the spirit of $\phi \circ \tau$ in Section "Introduction and Problem Description" (with $R = 0$), the costs for using own resources in $[t; t+1]$ and an eventual transfer p_t from/to the agent. Let

$$R_t = -\frac{c}{Q}\left(\int\limits_{S_t} \alpha_{\delta_p^{(t)}, \delta_a^{(t)}}(Q_t, q_t; x)\, dx + Q_{t+1}\right) - \theta_p^{(t)} \cdot \min\left\{Q_t - q_t; 1/\delta_p^{(t)}\right\} - p_t$$

with $S_t = \{x : 0 \le \alpha_{\delta_p^{(t)}, \delta_a^{(t)}}(Q_t, q_t; x) \le 1\}$ being the processed units in $[t; t+1]$.

Computational Study

In this last section we present a concrete neural network implementation which realizes the reinforcement learning approach for smart renegotiation strategies. As parameters for our problem we chose $Q = 100$, $c = 5$, $\Delta_p = \Delta_a = \{1/3; 1/2; 1\}$, $\mathfrak{p}_p = \mathfrak{p}_a = 1/5$, $\Theta_p = \{1; 3/2; 2\}$ and $\Theta_a = \{1; 2\}$ with $\underline{\mathfrak{v}} = 3/4$, $\hat{\mathfrak{v}} = 1/2$ and $\overline{\mathfrak{v}} =$

$1/4$. We take the initially best contracts $\mathcal{C}(1)$ and $\mathcal{C}(2)$ as computed in Theorem 1 as starting point, resulting in initial quantities $q(1) = 60$ and $q(2) = 20$ due to $\delta_p^{(0)} = \delta_a^{(0)} = 1/2$.

Using the respective formula from Section "Learning Renegotiation Strategies", we got a state space of size $3 \times 3 \times 101 \times 102/2 = 46{,}359$. The entire space is of course difficult to explore due to its extent, but we can assume that the neural network can provide good approximations for a variety of states. For the actions we achieved good results by setting $k = 2$ and $K = 3$ (compare Section "Learning Renegotiation Strategies"), resulting in an action space size of $4 \times 3 + 1 = 13$. The neural network itself was basically built in the following linear manner by using the package `keras` [10]: one `Embedding` layer mapping each state to a vector of size 20, two `Dense` layers of size 32 with `relu` activation and one linear `Dense` output layer predicting the Q-values of the actions.

We trained our model for 10^6 episodes with a learning rate of 10^{-3} and an ϵ-greedy policy [9] (ϵ was set to 0.1 by default). Overall, the training took about 310 min and yielded an average reward of − 1464 on 100 test instances. We compared this value to a benchmark reward resulting from the original setting where only an initial menu of contracts but no renegotiation was allowed. By taking the same 100 sequences of random parameters $\delta_p^{(t)}$, $\delta_a^{(t)}$, $\theta_p^{(t)}$ and $\theta_a^{(t)}$ we obtained an average reward of − 1695, meaning that the renegotiation strategies from our model led to a significant increase in the principal's average utility.

Conclusion

The general usefulness of renegotiating contracts during an uncertain execution phase of principal-agent models is obvious whereas optimal strategies for the principal are hard to compute due to a possibly large "game tree". In this contribution, however, we were able to generate beneficial renegotiation strategies for a specific outsourcing problem by training a neural network. The general approach interpreting the multi-round principal-agent setting as a single player game from the principal's viewpoint and exploiting deep reinforcement learning techniques on it can naturally be transferred to similar problems as well.

References

1. Herbst, A. (2022). Pooling of contracts for outsourcing problems with two-dimensional asymmetric information. In *Selected Papers of the International Conference of the Swiss, German and Austrian Operations Research Societies* (pp. 311–317). Springer. https://doi.org/10.1007/978-3-031-08623-6
2. Laffont, J. J., & Martimort, D. (2002). *The theory of incentives: The principal-agent model*. Princeton University Press. https://doi.org/10.2307/j.ctv7h0rwr

3. Kerkkamp, R. B. O., van den Heuvel, W., & Wagelmans, A. P. M. (2019). Robust pooling for contracting models with asymmetric information. *European Journal of Operational Research, 273*(3), 1036–1051. ISSN: 0377-2217. https://doi.org/10.1016/j.ejor.2018.08.041
4. Fayezi, S., O'Loughlin, A., & Zutshi, A. (2012). Agency theory and supply chain management: A structured literature review. *Supply Chain Management, 17*(5), 556–570. https://doi.org/10.1108/13598541211258618
5. Hegyi, A., Netten, B. D., Wang, M., Schakel, W., Schreiter, T., Yuan, Y., van Arem, B., & Alkim, T. (2013). A cooperative system based variable speed limit control algorithm against jam waves—An extension of the SPECIALIST algorithm. In *16th International IEEE Conference on Intelligent Transportation Systems (ITSC 2013)* (pp. 973–978). https://doi.org/10.1109/ITSC.2013.6728358
6. Antoniou, C., Koutsopoulos, H. N., & Yannis, G. (2007). Traffic state prediction using Markov chain models. In *2007 European Control Conference (ECC)* (pp. 2428–2435). https://doi.org/10.23919/ECC.2007.7069053
7. Freedman, D. (1983). *Markov chains*. Springer. https://doi.org/10.1007/978-1-4612-5500-0
8. Fudenberg, D., & Tirole, J. (1991). *Game theory* (1st ed., Vol. 1). MIT Press Books, The MIT Press.
9. Fenjiro, Y., & Benbrahim, H. (2018). Deep reinforcement learning overview of the state of the art. *Journal of Automation, Mobile Robotics and Intelligent Systems, 12*(3), 20–39. https://doi.org/10.14313/JAMRIS_3-2018/15
10. Chollet, F. (2018). *Deep learning with Python*. Manning Publications.

Chapter 64
Temperature-Based Trajectory Planning for Surfaces in Wire-Arc Additive Manufacturing

Johannes Schmidt and Armin Fügenschuh

Abstract In Wire-Arc Additive Manufacturing, the desired workpiece is built layer-wise by a weld source moving freely over a substrate plate, either welding or transiting. The main issue with this manufacturing technique is the temperature distribution within the workpiece since the large thermal gradients caused by the welding process lead to thermal stress. We consider the trajectory planning problem of finding an optimal welding trajectory for a given two-dimensional layer. It is formulated as a mixed-integer linear problem (MILP) searching the welding path with the most homogeneous temperature distribution during manufacturing. The heat conduction, the weld source, and the heat radiation are incorporated into the model, together with two coupled time discretizations to accelerate the solution process. For two example surfaces, the computed optimal trajectory is compared to commonly used strategies like raster, zigzag or spiral paths.

Keywords Wire-arc additive manufacturing · Mixed-integer linear programming · Heat equation · Finite element method

Introduction

In the process of Wire-Arc Additive Manufacturing (WAAM), the desired workpiece is split up into slices and built up layer-by-layer. The welding head can move freely over the clamped substrate plate to deposit droplets of metal wire molten by an electrical arc or a laser. Also, transfer moves without welding, called transits, are possible. One of the main factors for the quality of the resulting workpiece are the high thermal gradients caused by the weld source leading to strain or even cracks. Especially for filled surface structures consisting of many weld beads, this must

J. Schmidt (✉) · A. Fügenschuh
Brandenburg University of Technology Cottbus-Senftenberg, Platz der Deutschen Einheit 1, 03046 Cottbus, Germany
e-mail: johannes.schmidt@b-tu.de

A. Fügenschuh
e-mail: fuegenschuh@b-tu.de

O. Grothe et al. (eds.), *Operations Research Proceedings 2022*, Lecture Notes in Operations Research, https://doi.org/10.1007/978-3-031-24907-5_64

be taken into account. Thus, careful planning of the welding trajectory is essential for high-quality workpieces. Today, the trajectories for surfaces are planned using standard patterns like rastering, zig-zag, contour, or spiral paths, which are chosen by different criteria without optimization [4]. An extensive review of WAAM can be found in [6].

We formulate the problem of finding a feasible weld source trajectory that maximizes the quality of the finished workpiece as a mixed-integer linear problem (MILP), incorporating a detailed calculation for the temperature distribution of the constructed part into the optimization. To model the heat transmittance, heat conduction, radiation, and the heat input by the weld source are taken into account using a two-dimensional heat equation with Robin boundary, which is discretized by a finite element method to fit the chosen framework. A similar approach for space discretization with a heuristical solution method is presented in [2]. Furthermore, we use different time discretizations regarding the space and the time dimension to achieve a flexible level of detail for the computed temperature. For a set of example instances, the presented approach is compared to the standard patterns rastering, zig-zag, and spiral paths, applying the standard mixed-integer solver CPLEX. Furthermore, the effect of the different time discretizations is examined on this instances.

Mathematical Model

In this work, we consider a single two-dimensional layer with a bulky structure of a sliced workpiece, i.e., the layer is a surface where multiple weld beads are placed next to each other. The velocities $v^w \in \mathbb{R}_+$ and $v^m \in \mathbb{R}_+$ of the weld source while welding or transiting are respectively known. Then, for a given discrete time step length $\Delta t \in \mathbb{R}_+$, the layer is covered by a set $\mathcal{V}$ of non-overlapping quadratic pixels of side length $v^w \Delta t$ and each pixel $i \in \mathcal{V}$ is identified with a node located in its center $(x_i, y_i) \in \mathbb{R}^2$. Let denote $n = |\mathcal{V}|$ in the following. Defining an edge $(i, j) \in \mathcal{W}$ for every pair of squares $i, j \in \mathcal{V}$, $i \neq j$, with a common side, the considered layer is described by an undirected grid graph $G = (\mathcal{V}, \mathcal{W})$.

To calculate the temperature distribution within the layer, a time-expanded version of the graph G is necessary. Due to the choice of the pixel's side length, the weld source requires one time step to reach the next node while welding. Furthermore, for an accordingly chosen weld bead width, the area covered by a pixel is filled with material if the weld source reaches the node in its center. Thus, the complete layer is processed if every node $i \in \mathcal{V}$ is visited exactly once. For transit moves, the number of required time steps is given by $\tau^m_{i,j} = \left\lceil \frac{d^e_{i,j}}{v^m \Delta t} \right\rceil$, where $d^e_{i,j}$ is the Euclidean distance between the nodes $i, j \in \mathcal{V}$. Let $\mathcal{U} = \{(i, j) \in \mathcal{V} \times \mathcal{V} \mid i \neq j\}$ denote the set of all possible transit moves and $\omega \in \mathbb{N} \cup \{0\}$ their preprocessed minimal necessary number.

By applying a second discretization scheme for the time discretization of the temperature calculation, we achieve a more flexible approach where the level of detail of the temperature distribution can be adjusted to the requirements of the application avoiding unnecessarily complex optimization models. Therefore, a number $n_c \in \mathbb{N}$ is chosen and the time step length $\Delta t_c = n_c \Delta t$ is set for this second, more coarse, time discretization taking only multiples of n_c into account.

Since the number of time steps to process the complete layer $T^w \in \mathbb{N}$ varies with the chosen transit moves, which are not known a priori, it affects the number of variables and constraints in the resulting optimization model. Thus, a good upper bound of T^w is crucial to avoid unnecessarily complex models. In the following, we consider the rather coarse approximation $T^w = n - 1 + \omega \max_{(i,j)\in\mathcal{U}} \tau^m_{i,j}$. Regarding the second time discretization, the last time step has to be a multiple of n_c to allow a correct computation. Thus, the maximum number of time steps is given by $T^{max} = \min\{kn_c \mid k \in \mathbb{N}, kn_c \geq T^w\}$. This yields the discrete time horizons $\mathcal{T} = \{1, \dots, T^w\}$ for the space discretization regarding the trajectory and $\mathcal{T}^c = \{n_c, 2n_c, \dots, T^{max}\}$ for the time discretization of the temperature calculation. In the following, the abbreviations $\mathcal{T}_0^c = \mathcal{T}^c \cup \{0\}$, $\mathcal{T}_0 = \mathcal{T} \cup \{0\}$, $\mathcal{T}^- = \mathcal{T} \setminus \{T^w\}$, $\mathcal{T}_0^- = \mathcal{T}_0 \setminus \{T^w\}$, and $\mathcal{T}^{end} = \{n - 1 + \omega, \dots, T^w\}$ are used. Note, that the last abbreviation $\mathcal{T}^{end}$ describes the set of all time steps, in which the manufacturing process can end. Incorporating the processing time, we define sets $\mathcal{W}^* = \{(i, t_i, j, t_j) \in \mathcal{V} \times \mathcal{T}_0^- \times \mathcal{V} \times \mathcal{T} \mid (i, j) \in \mathcal{W}, t_j = t_i + 1\}$, $\mathcal{U}^* = \{(i, t_i, j, t_j) \in \mathcal{V} \times \mathcal{T}^- \times \mathcal{V} \times \mathcal{T}^- \mid (i, j) \in \mathcal{U}, t_j = t_i + \tau^m_{i,j}\}$ and introduce two sorts of binary variables $w_{i,t_i,j,t_j} \in \{0, 1\}$ for $(i, t_i, j, t_j) \in \mathcal{W}^*$ and $u_{i,t_i,j,t_j} \in \{0, 1\}$ for $(i, t_i, j, t_j) \in \mathcal{U}^*$, indicating if the weld source moves from node $i \in \mathcal{V}$ to node $j \in \mathcal{V}$ while welding or transiting, respectively. Furthermore, auxiliary binary variables $u_i^+ \in \{0, 1\}$, indicating if the trajectory starts at node $i \in \mathcal{V}$, $u_{i,t}^- \in \{0, 1\}$, indicating if the trajectory ends at node $i \in \mathcal{V}$ at time step $t \in \mathcal{T}^{end}$, and

$$w_{i,t} = \begin{cases} u_i^+, & i \in \mathcal{V}, t = 0 \\ \sum\limits_{h,t_h:(h,t_h,i,t)\in\mathcal{W}^*} w_{h,t_h,i,t} + \sum\limits_{h,t_h:(h,t_h,i,t)\in\mathcal{U}^*} u_{h,t_h,i,t}, & i \in \mathcal{V}, t \in \mathcal{T} \end{cases}, \tag{64.1}$$

indicating if the weld source is above node $i \in \mathcal{V}$ at time step $t \in \mathcal{T}_0$, are defined. The problem of finding a feasible trajectory is then given by

$$\sum_{j,t_j:(i,0,j,t_j)\in\mathcal{W}^*} w_{i,0,j,t_j} = u_i^+, \quad \forall i \in \mathcal{V} \tag{64.2a}$$

$$\sum_{t\in\mathcal{T}_0} w_{i,t} = 1, \quad \forall i \in \mathcal{V} \tag{64.2b}$$

$$\sum_{(i,t_i,j,t_j)\in\mathcal{U}^*} u_{i,t_i,j,t_j} = \omega, \tag{64.2c}$$

$$\sum_{j,t_j:(i,t,j,t_j)\in\mathcal{W}^*} w_{i,t,j,t_j} + \sum_{j,t_j:(i,t,j,t_j)\in\mathcal{U}^*} u_{i,t,j,t_j} = w_{i,t} \quad \forall i \in \mathcal{V}, t \in \mathcal{T} \setminus \mathcal{T}^{end}, \tag{64.2d}$$

$$\sum_{j,t_j:(i,t,j,t_j)\in\mathcal{W}^*} w_{i,t,j,t_j} + \sum_{j,t_j:(i,t,j,t_j)\in\mathcal{U}^*} u_{i,t,j,t_j} = w_{i,t} + u^-_{i,t}, \quad \forall i \in \mathcal{V}, t \in \mathcal{T}^{end}, \tag{64.2e}$$

$$\sum_{i\in\mathcal{V}} u^+_i = 1, \tag{64.2f}$$

$$\sum_{i\in\mathcal{V}} \sum_{t\in\mathcal{T}^{end}} u^-_{i,t} = 1. \tag{64.2g}$$

The trajectory has to start somewhere (64.2a) and every node must be visited (64.2b). Only the minimal number of necessary transit moves is allowed (64.2c). Constraints (64.2d) and (64.2e) guarantee the continuity of the trajectory and the start and end node have to be unique (64.2f) and (64.2g).

Regarding the calculation of the temperature distribution, a two-dimensional heat equation of the form

$$\frac{\partial\theta}{\partial t}(x,y,t) = \alpha\left(\frac{\partial^2\theta}{(\partial x)^2}(x,y,t) + \frac{\partial^2\theta}{(\partial y)^2}(x,y,t)\right) + q(x,y,t)$$
$$\forall (x,y) \in \Omega, t \in [0,T], \tag{64.3a}$$

$$\frac{\partial\theta}{\partial n}(x,y,t) = \kappa^e\left(\varphi^{add} - \theta(x,y,t)\right) \quad \forall (x,y) \in \partial\Omega, \forall t \in [0,T], \tag{64.3b}$$

$$\theta(x,y,0) = \theta^{init}(x,y) \quad \forall (x,y) \in \Omega, \tag{64.3c}$$

is used with thermal diffusivity $\alpha \in \mathbb{R}_+$, weld source term $q : \Omega \times [0,T] \to \mathbb{R}_+$, and artificial cooling parameters $\kappa^e, \varphi^{add} \in \mathbb{R}_+$. The heat conduction is described in (64.3a), the Robin boundary condition (64.3b) represents a linear approximation of the non-linear radiation term, and the initial temperature distribution is given in (64.3c). For the computation of the temperature, we introduce variables $\theta_{i,t} \in \mathbb{R}_+$ for the temperature of node $i \in \mathcal{V}$ at time step $t \in \mathcal{T}^c_0$ and $q_{i,t} \in \mathbb{R}_+$ for the temperature gain by the weld source of node $i \in \mathcal{V}$ at time step $t \in \mathcal{T}^c$. Following the approach in [5], the partial differential equation system (64.3) is discretized using a finite element method with the node set $\mathcal{V}$ as discretization points, linear triangle elements, the shape of the considered layer as its boundary, the time step length Δt_c, and an implicit time approach, yielding the linear equation system

$$(M + \Delta t_c K)\vec{\theta}_{t+1} = \Delta t_c(\vec{q}_{t+1}\vec{f}^H + \kappa^e\varphi^{add}\vec{f}^R) + M\vec{\theta}_t \qquad \forall t \in \mathcal{T}^c \setminus \{T^{max}\}, \tag{64.4}$$

with mass matrix $M \in \mathbb{R}^{n\times n}$, stiffness matrix $K \in \mathbb{R}^{n\times n}$, temperature vector $\vec{\theta}_t = \left(\theta_{1,t} \cdots \theta_{n,t}\right)^\top$, weld source vector $\vec{q}_t = \left(q_{1,t} \cdots q_{n,t}\right)^\top$, and load vectors

$f^H, f^N \in \mathbb{R}^n$. Regarding the weld source, we follow the piece-wise constant approximation approach in [1] with $K^w \in \mathbb{N}$ intervals $\mathcal{P}_1, \ldots, \mathcal{P}_{K^w}$, related factors $1 = \kappa_1^w \geq \kappa_1^w \geq \ldots \geq \kappa_{K^w}^w \geq 0$, and temperature gain $\varphi^w \in \mathbb{R}_+$ and adapt it to the used framework of a coarse time discretization for the temperature calculation. Thus, the variables $q_{i,t}$ are given by

$$q_{i,t} = \sum_{p=0}^{n_c-1} (\kappa^c)^p \sum_{k=1}^{K^w} \sum_{\substack{j \in \mathcal{V} \setminus \{i\} \\ d^e_{i,j} \in \mathcal{P}_k}} \kappa_k^w \varphi^w w_{j,t-p} \qquad \forall i \in \mathcal{V}, \forall t \in \mathcal{T}^c, \tag{64.5}$$

where $\kappa^c \in \mathbb{R}_+$ is a cooling factor applied to the previous time steps. At the initial time step, the start node of the welding trajectory affects the initial temperature. Since the parameter φ^w must be estimated for every value of n_c, a new parameter $\varphi_0^w \in \mathbb{R}_+$ is introduced to achieve the same initial temperature distribution independently of n_c. This leads to the additional constraints

$$\theta_{i,0} = \theta_i^{init} + \sum_{k=1}^{K^w} \sum_{\substack{j \in \mathcal{V} \setminus \{i\} \\ d^e_{i,j} \in \mathcal{P}_k}} \kappa_k^w \varphi_0^w w_{j,0} \qquad \forall i \in \mathcal{V}. \tag{64.6}$$

In this work, we aim to achieve a most homogeneous temperature distribution within the workpiece to maximize its quality. Therefore, the computed temperature's absolute deviation from a given target temperature $\vartheta^{target} \in \mathbb{R}_+$ should be minimal. To linearize the necessary absolute value function, new variables $\vartheta_{i,t}^+, \vartheta_{i,t}^- \in \mathbb{R}_+$ for the positive and negative deviation of the temperature of node $i \in \mathcal{V}$ at time step $t \in \mathcal{T}_0^c$ are introduced. This gives the linear objective function

$$\min \quad \sum_{t \in \mathcal{T}_0^c} \sum_{i \in \mathcal{V}} (\vartheta_{i,t}^+ + \vartheta_{i,t}^-), \tag{64.7}$$

with the additional constraints

$$\vartheta_{i,t}^+ - \vartheta_{i,t}^- = \theta_{i,t} - \vartheta^{target} \qquad \forall i \in \mathcal{V}, \forall t \in \mathcal{T}_0^c. \tag{64.8}$$

Computational Results

The model consisting of the constraints (64.1)–(64.6) and (64.8) with the objective function (64.7) was implemented in Python 3.8 and all instances were solved using IBM ILOG CPLEX 20.1.0.0 [3] on a Mac Pro with an Intel Xeon W running 32 threads parallel at 3.2 GHz clock speed and 768 GB RAM. In the CPLEX settings, the branching priority based on increasing cost per coefficient count was enabled. For the approximation of the weld source, the intervals and parameters obtained in [1] are

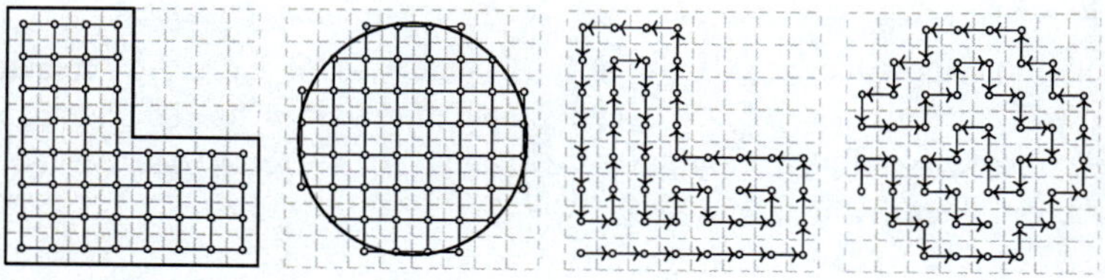

Fig. 64.1 Discretization of the instances and the respective obtained optimal trajectories. Both model variants found the same solution for the respective surface

Table 64.1 Objective values and computation times of both model variants for all instances and the comparison to the standard patterns

		Raster	Zig-zag	Spiral	Optimization	
					$n_c = 1$	$n_c = 2$
Angle	Value	934,012	934,442	714,966	711,597	392,581
	Time (s)				205.88	66.08
Circle	Value	1,115,886	1,116,044	841,977	833,399	461,795
	Time (s)				560.19	337.36

used. The parameters φ^w, φ^{add}, κ^e, and κ^c are chosen as the solution of a least-squares problem for every value of the parameter n_c based on a simulated instance. Their values for $n_c = 1$ are $\varphi^w = 1514$ K, $\varphi^{add} = 709.73$ K, $\kappa^e = 0.1831$, and $\kappa^c = 1$, while for $n_c = 2$ they are given by $\varphi^w = 915$ K, $\varphi^{add} = 711.03$ K, $\kappa^e = 0.2246$, and $\kappa^c = 0.8576$. Furthermore, the thermal diffusivity is set to $\alpha = 3.774$ mm/s, the initial temperature of all nodes $i \in \mathcal{V}$ is set to $\theta_i^{init} = 500$ K, the initial temperature gain is $\varphi_0^w = 1514$ K, and as time step length we use $\Delta t = 0.5$ s.

In the experiments, we consider a workspace of 8×8 pixels and choose a subset of them to describe an angle and a circle as example instances. Both instances are solved for the values $n_c \in \{1, 2\}$ with the respective parameters and compared to the results of the standard patterns raster, zig-zag, and spiral path considering $n_c = 1$. The discretization and the obtained optimal trajectories are displayed in Fig. 64.1, while detailed information about the computational performance and the comparison can be found in Table 64.1.

Comparing the objective values of the solutions, the raster and the zig-zag pattern yield nearly the same results due to their similar structure, while the spiral path produces much better solutions. The computed optimal solution is slightly better than the spiral path and the resulting trajectories have only slight adjustments compared to a spiral. For every instance, the optimal solution for both values of n_c is the same. Regarding the computation times, the model with $n_c = 2$ outperforms the standard approach using only a single time discretization.

Conclusions and Future Work

In this work, we derived an MILP formulation for the trajectory planning problem for WAAM including a detailed calculation of the temperature distribution. Applying two coupled time discretizations yields a less complex model with faster computation times. The computational results on example instances show the applicability of our approach to the standard mixed-integer solver CPLEX and its benefits compared to the use of a single time discretization. In our future work, we will extend this approach to layers with arbitrary geometry, also including thin-walled structures, and derive better bounds for T^w. Furthermore, it remains to formulate criteria for applying a second time discretization to avoid changing the optimal solution by larger discretization errors.

References

1. Bähr, M., Buhl, J., Radow, G., Schmidt, J., Bambach, M., Breuß, M., & Fügenschuh, A. (2021). Stable honeycomb structures and temperature based trajectory optimization for wire-arc additive manufacturing. *OPTE, 22*(2), 913–974.
2. Ferreira, R. P., & Scotti, A. (2021). The concept of a novel path planning strategy for wire + arc additive manufacturing of bulky parts: Pixel. *Metals, 11*(3), 498.
3. International Business Machines Corporation. (2021). *IBM ILOG CPLEX optimizer reference manual*. https://www.ibm.com/docs/en/icos/20.1.0
4. Jiang, J., & Ma, Y. (2020). Path planning strategies to optimize accuracy, quality, build time and material use in additive manufacturing. *Micromachines, 11*(7), 633.
5. Taler, J., & Ocłoń, P. (2014). Finite element method in steady-state and transient heat conduction. In *Encyclopedia of thermal stresses* (Vol. 4, pp. 1604–1633).
6. Treutler, K., & Wesling, V. (2021). The current state of research of wire arc additive manufacturing (WAAM): A review. *Applied Sciences, 11*(18), 8619.

Part XV
Pricing and Revenue Management

Chapter 65
A Conceptual Framework for Studying Self-learning Agents in Recommerce Markets

Rainer Schlosser and Alexander Kastius

Abstract In many markets, customers as well as retailers look for increased sustainability. Recommerce markets—which offer the opportunity to trade in and resell used products—are constantly growing and help to use resources more efficiently. To additionally manage the trade in and resell prices for used product versions is challenging for retailers as substitution and cannibalization effects have to be taken into account. An unknown customer behaviour as well as competition with other merchants regarding both sales and buying back resources further increases the problem's complexity. Data-driven pricing agents offer the potential to find well-performing strategies and satisfy the need for automated decision support, particularly in online markets. As the training of such agents is typically data hungry and too costly to be performed in practice, synthetic test environments are needed to try out and evaluate self-learning pricing agents in different market scenarios. In this paper, we propose a conceptual approach for such a recommerce market simulation framework and its basic components. Further, we discuss requirements and opportunities to study self-learning strategies in synthetic markets.

Keywords Recommerce · Dynamic pricing · Reinforcement learning · Sustainability

Introduction

Sustainability and an efficient use of resources are of general growing interest. Recommerce markets, in which used products are sold, are constantly growing. Such markets follow the concept of a circular economy and help to save resources by giving them a longer lifetime.

R. Schlosser (✉) · A. Kastius
Hasso Plattner Institute, University of Potsdam, Potsdam, Germany
e-mail: Rainer.Schlosser@hpi.de

A. Kastius
e-mail: Alexander.Kastius@hpi.de

O. Grothe et al. (eds.), *Operations Research Proceedings 2022*, Lecture Notes in Operations Research, https://doi.org/10.1007/978-3-031-24907-5_65

Recommerce firms buy in returned articles (such as smart phones, clothes, etc.) from customers or other firms and resell them again to consumers as used products. As the costs for grading, storing, repairing, or refurbishing products are comparably low and the demand for used products is currently increasing, recommerce is a beneficial business model. However, recommerce firms also face challenges, which can be described as follows:

1. To successfully manage trade ins and sales an optimized pricing is essential.
2. Further, when also new product versions are sold, substitution effects between new and used products have to be taken into account.
3. Many recommerce markets are characterized by duopoly or oligopoly competition.
4. The impact and the interaction of own as well as competitors' prices on demand is not easy to anticipate.
5. Usually fully manual pricing decisions are infeasible and automation is required, but effective rule-based pricing strategies are hard to derive.
6. Self-learning data-driven algorithms typically require too much data to be trained in practice.

To tackle these challenges, simulated market environments are key to develop, test, and evaluate the strategic interplay of automated pricing strategies applied in recommerce markets. In addition, the potential performance of self-learning strategies can be compared to rule-based baseline strategies.

In this paper, we propose a conceptual framework for such a recommerce market simulation, including an adjustable customer behavior model as well as the capability to apply rule-based and self-learning agents based on reinforcement learning (RL). Monitoring tools shall allow to analyze each agent's policy and their effects on the overall market and the associated resource flows. With the help of such simulations, we seek to study the competitiveness of self-adapting pricing tools and their long-term impact on market competitors and customers.

This paper is organized as follows. In Section "Related Work", we discuss related work. In Section "Conceptual Framework", we describe our conceptual framework to simulate recommerce markets and to test self-learning algorithms. Concluding remarks are given in the final Section "Conclusions".

Related Work

We shortly discuss related work along the following different streams related to the topic of the paper: circular economy (Section "Circular Economy"), dynamic pricing (Section "Dynamic Pricing"), RL techniques (Section "RL Algorithms"), and market simulations (Section "Market Simulations").

Circular Economy

An entrypoint to study the concepts of circular economy and sustainable markets are given in, e.g., Stahel [17] and Bocken et al. [1]. The overall idea is to save resources, reduce the the use of resources and to avoid waste, which is also closely related to closed loop supply chains, see, e.g., [6, 12]. Further, related aspects are environmental cost [4] or recycling investments [13].

Dynamic Pricing

Selling products on online marketplaces is a classical revenue management application, see, e.g., [20]. A comprehensive overview about the literature in dynamic pricing research is provided by the surveys [3, 10, 18]. The recent work [5] particularly discusses dynamic pricing models under competition on online marketplaces.

RL Algorithms

Simulation-based algorithms allow to heuristically solve large Markov decision problems with incomplete information. An overview of the field of RL is given, e.g., in [19]. Established RL algorithms are, e.g., Deep Q-learning Networks [11], PPO [15], or Soft Actor Critic [7]. Applications of such methods to dynamic pricing problems can, e.g., be found in [8]. Unfortunately, RL algorithms typically require a lot of training data making it less attractive to use them in practice. To overcome this issue, approaches like transfer learning or multitask RL seek to use observable data more efficiently.

Market Simulations

The combined problem of (i) updating prices, (ii) learning demand behaviours, and (iii) identifying strategy equilibria in competitive markets is challenging as information is incomplete and merchants may constantly adapt their strategies. For analyzing and evaluating the complex interplay and long-term behavior of mutual self-adaptive pricing strategies market simulations for classical e-commerce application have been used [9], see [2, 14, 16, 21]. However, the additional buy back option of recommerce markets and associated circular resource flows have not been studied in the mentioned frameworks.

Conceptual Framework

In this section, we describe main components of our recommerce simulation framework (Section "Market Model"), discuss the integration of RL methods within a Markov decision process (MDP) (Section "MDP Framework and Application of RL Agents"), and propose the design of different computational studies to study different market setups (Section "Design of Computational Studies").

Market Model

To mimic real-life recommerce markets, in our market simulation framework, we consider the following main components: (i) supplier(s), (ii) firm(s) including a private data/event store, (iii) a (shared) marketplace, (iv) consumer, (v) resources in use, and (vi) waste (cf. garbage), see Fig. 65.1. The components are connected as follows. Firms set prices for new and used products on their (or a central) marketplace. Arriving consumers decide (whether and) which product to buy from which firm. Bought products are considered as resource in use at the consumer side. They may be sent to garbage after a while or sold back to one of the firms which offer a corresponding buy back price. Firms can also order new resources from their (or a central) supplier at a certain cost (cf. virgin cost). To be able to easily integrate various RL libraries, we follow a standard stationary discrete time setup with infinite horizon.

Note, this basic model sketched above also allows to describe the following special cases: (i) monopoly settings, (ii) scenarios with just one product type, and (iii) classical e-commerce scenarios with no buy back option (cf. linear economy), as well as combinations of those cases (i)–(iii).

Moreover, the basic model can also be extended to capture more complex settings. For instance, for each firm, we may additionally consider a technology state serving as a sustainability image (cf. greenness, signaling, etc.), which increases demand. Further, this state could be stimulated via corresponding investment efforts and otherwise depreciates over time.

Active decisions are made by the firms and the consumers. The pricing decisions of a firm can be organized by a certain rule-based strategy as well as via an RL agent exploiting a firm's gathered partially observable market data. The consumer behavior can be arbitrarily defined, e.g., by generating random numbers of interested customers with heterogeneous preferences and a diversified willingness-to-pay. Besides consumer of myopic type also certain shares of strategic or loyal customers can be defined and considered.

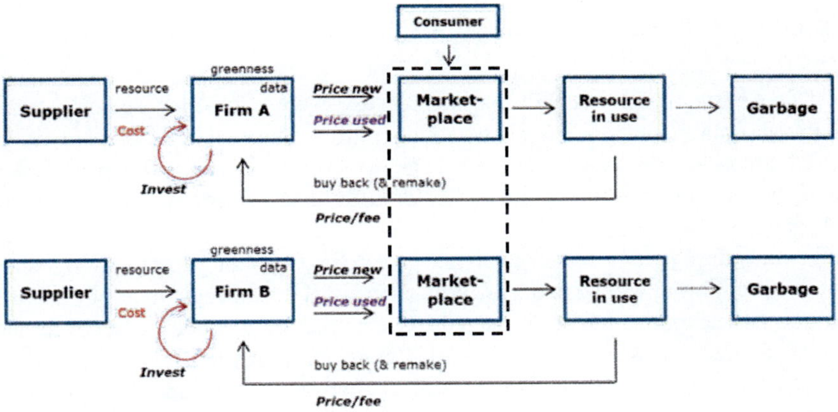

Fig. 65.1 Illustration of the main components of a recommerce market simulation with two competing merchants, cf. Firm A and Firm B

MDP Framework and Application of RL Agents

To train dynamic pricing agents for the proposed market framework by using standard RL algorithms, the formulation of a so-called environment is required, which includes states, actions, reward signals, and state transitions dynamics.

From a firm's, i.e., the agent's, perspective the *state* is characterized by the own inventory level, the current prices of the competitors, and—if considered observable—the amount of resources in use. Further aspects such as, e.g., the firms' current greenness, are also part of the state.

Further, a firm's *action* is a combination of prices for new and used products, the buy back price, and, e.g., a potential investment decision. The *reward signal* of a firm is the aggregated reward associated to realized sales, purchases, and holding costs (within one period); it is characterized by the underlying customer behaviour including the defined arrival stream of interested consumers.

State transitions are organized in an MDP fashion and governed by the evolution of the own inventory level and in particular the (subsequent) price adjustments of all competing firms. This requires that certain, e.g., rule-based, policies are defined for the competing firms. The agent's objective is to find a state-dependent strategy that maximizes expected discounted long-term rewards.

Finally, within the described environment, different standard RL algorithms such as DQN, PPO, or SAC can be applied by using common RL libraries.

Design of Computational Studies

In a first basic model, we will consider an RL agent competing against certain (unknown) rule-based strategies in a duopoly setting. The goal is to compare different RL algorithms by evaluating the RL agent's and the competitor's performance, the associated resource flows, and the required amounts of data to sufficiently train the agent.

To study the agents' realized prices, sales, purchases, holding costs, profits, etc., a suitable monitoring will be required. With such a tool the strategic interplay and the mutual price adaptions can be visualized and analyzed. Further, the impact of certain model parameters (e.g., virgin cost, holding costs, consumer behavior, discounting, etc.) on the long-term performance of all market participants and the associated resource flows (cf. steady state) can be investigated.

In further experiments, we will also study how agents adapt to changing market environments, e.g., (i) if virgin costs are increasing over time or (ii) if the customer behaviour changes, e.g., towards a more sustainable one. Moreover, we will examine the case, in which two self-adapting RL agents compete against each other. Note, in this scenario the Markov property is no longer satisfied and stability issues may arise. In another experiment, we will study how an RL agent's performance is affected by different information structures, i.e., if the number of resources in use or even the competitor's inventory level would be an observable part of the state space. Finally, we also aim to analyze an RL agent's performance in oligopoly settings with different numbers of competitors.

Conclusions

In this paper, we have proposed a market simulation framework for recommerce markets under competition. The framework has modular components, which allow to study different pricing strategies in different market scenarios. The simulation is designed such that self-learning RL algorithms can be easily integrated and compared. In future research, we will implement the proposed concepts and evaluate the proposed numerical experiments. Results might be interesting from both a theoretical and a practical perspective.

References

1. Bocken, N. M., De Pauw, I., Bakker, C., & Van Der Grinten, B. (2016). Product design and business model strategies for a circular economy. *Journal of Industrial and Production Engineering, 33*(5), 308–320.
2. Boissier, M., Schlosser, R., Podlesny, N., Serth, S., Bornstein, M., Latt, J., Lindemann, J., Selke, J., & Uflacker, M. (2017). Data-driven repricing strategies in competitive markets: An interactive simulation platform. In *RecSys 2017* (pp. 355–357).

3. Chen, M., & Chen, Z. L. (2015). Recent developments in dynamic pricing research: Multiple products, competition, and limited demand information. *Production and Operations Management, 24*, 704–731.
4. Commoner, B. (1972). The environmental cost of economic growth. *Population, Resources and the Environment, 3*, 343–363.
5. Gerpott, T., & Berends, J. (2022). Competitive pricing on online markets: A literature review. *Journal of Revenue Pricing Management.*
6. Gönsch, J. (2014). Buying used products for remanufacturing: Negotiating or posted pricing. *Journal of Business Economics,84*, 715–747.
7. Haarnoja, T., Zhou, A., Abbeel, P., & Levine, S. (2018). Soft actor-critic: Off-policy maximum entropy deep reinforcement learning with a stochastic actor. In *ICML 2018*, 10–15, 2018 (Vol. 80, pp. 1856–1865). PMLR.
8. Kastius, A., & Schlosser, R. (2022). Dynamic pricing under competition using reinforcement learning. *Journal of Revenue and Pricing Management, 21*, 50–63.
9. Kephart, J. O., Hanson, J. E., & Greenwald, A. (2000). Dynamic pricing by software agents. *Computer Networks, 32*(6), 731–752.
10. Klein, R., Koch, S., Steinhardt, C., & Strauss, A. (2020). A review of revenue management: Recent generalizations and advances in industry applications. *European Journal of Operational Research, 284*, 397–412.
11. Mnih, V., Kavukcuoglu, K., Silver, D., Rusu, A. A., Veness, J., Bellemare, M. G., Graves, A., Riedmiller, M., Fidjeland, A. K., Ostrovski, G., & Petersen, S. (2015). Human-level control through deep reinforcement learning. *Nature, 518*(7540), 529–533.
12. Savaskan, R. C., Bhattacharya, S., & Van Wassenhove, L. N. (2004). Closed-loop supply chain models with product remanufacturing. *Management Science, 50*(2), 239–252.
13. Schlosser, R., Chenavaz, R., & Dimitrov, S. (2021). Circular economy: Joint dynamic pricing and recycling investments. *International Journal of Production Economics, 236*(108117), 1–13.
14. Schlosser, R., & Richly, K. (2019). Dynamic pricing under competition with data-driven price anticipations and endogenous reference price effects. *Journal of Revenue and Pricing Management, 18*, 451–464.
15. Schulman, J., Wolski, F., Dhariwal, P., Radford, A., & Klimov, O. (2017). Proximal policy optimization algorithms. arXiv preprint arXiv:1707.06347
16. Serth, S., Podlesny, N., Bornstein, M., Lindemann, J., Latt, J., Selke, J., Schlosser, R., Boissier, M., & Uflacker, M. (2017). An interactive platform to simulate dynamic pricing competition on online marketplaces. In *EDOC 2017* (pp. 61–66).
17. Stahel, W. R. (2016). The circular economy. *Nature, 531*(7595), 435–438.
18. Strauss, A. K., Klein, R., & Steinhardt, C. (2018). A review of choice-based revenue management: Theory and methods. *European Journal of Operational Research, 271*, 375–387.
19. Sutton, R. S., & Barto, A. G. (1998). Reinforcement learning—An introduction. In *Adaptive computation and machine learning*. MIT Press.
20. Talluri, K. T., & Van Ryzin, G. J. (2006). *The theory and practice of revenue management.* Springer.
21. van de Geer, R., den Boer, A. V., Bayliss, C., Currie, C. S., Ellina, A., Esders, M., Haensel, A., Lei, X., Maclean, K. D., Martinez-Sykora, A., & Riseth, A. N. (2019). Dynamic pricing and learning with competition: Insights from the dynamic pricing challenge at the 2017 INFORMS RM & pricing conference. *Journal of Revenue Pricing Management, 18*, 185–203.

Chapter 66
Multi-agent Dynamic Pricing Using Reinforcement Learning and Asymmetric Information

Alexander Kastius, Nils Kiele, and Rainer Schlosser

Abstract Self-learning agents can be used in numerous ways for dynamic pricing nowadays. It has been shown, that reinforcement learning can serve as a toolkit to efficiently develop pricing strategies in dynamic environments. In many real-world situations, it can be expected that multiple market participants rely on such self-learning agents to implement pricing decisions. From the view of one agent, this violates the fundamental Markov property, which leads to instability in the learning process. Past publications proposed to rely on asymmetric information to achieve equilibria and usually focused on tabular solutions or solvers. We use multi-agent learning and asymmetric information with function approximation tools for high-dimensional state spaces by exchanging policy information between multiple actors. We discuss possible problems and their solutions and propose a simulation environment for further evaluation of the developed system.

Keywords Dynamic pricing · Reinforcement learning · Multi-agent systems · Asymmetric information

Introduction

Past research has shown, that dynamic pricing serves as a suitable use case for reinforcement learning (RL) [2]. In practice, we often consider market setups, in which many traders exist and each trader is distinguished by more than one feature. Price-relevant features are for example other traders' prices, product features, and estimated inventory sizes. The order of magnitude grows with the number of sellers as well. In such high-dimensional environments, deep RL covers many of the

A. Kastius (✉) · N. Kiele · R. Schlosser
Hasso Plattner Institute, University of Potsdam, Potsdam, Germany
e-mail: Alexander.Kastius@hpi.de

N. Kiele
e-mail: Nils.Kiele@hpi.de

R. Schlosser
e-mail: Rainer.Schlosser@hpi.de

O. Grothe et al. (eds.), *Operations Research Proceedings 2022*, Lecture Notes in Operations Research, https://doi.org/10.1007/978-3-031-24907-5_66

tools necessary to deduct reasonable pricing policies automatically. Given enough information, RL algorithms can deduct which features are relevant and which can be ignored. Dedicated feature selection mechanisms like L1-regularization can increase this effect [1].

Modern RL algorithms can work on continuous state spaces, which can consist of high-dimensional real valued features, and continuous action spaces, which allow an intuitive representation of the allowed pricing span. Many of those algorithms belong to the domain of policy gradient algorithms, which directly improve the agents policy according to an estimation of the expected discounted future reward. The policy gradient theorem and its variations serve as a foundation of those algorithms [9].

Further, past experiments have shown, that instabilities arise when multiple agents in the system learn their policies at the same time [2]. In practice, those instabilities are a problem as they reduce the effectiveness of pricing algorithms and increase the cost of exploration for all algorithms. The cost of exploration is the loss that is inferred due to following non-optimal policies and deviating from possibly optimal policies because exploration needs to be performed to search for better policies. As the general assumption in RL is that the optimal policy cannot be identified, exploration has to be performed at all times.

In commercial environments, the cost of exploration is the element that hinders such algorithms from application. In the past, several experiments have been performed to display the feasibility of the concept of leader-follower games to implement multi-agent pricing setups, see Section "Related Work". In many of those setups, one agent under assessment learns its policies given information about the potential behaviour of the other agents. We seek to reduce the overall cost of exploration for all market participants by reintroducing this concept. Measuring the cost of exploration then provides a possible mechanism to implement such systems in practice, as a fee could be implemented which hands over some of the cost improvements of the leader to the followers, to encourage information exchange in an asymmetric environment. The novelty of the presented system comes from two major aspects: First of all, it incorporates a sophisticated multi-agent deep RL algorithm, whose structure is suited to handle complex environments and possibly improve overall learning performance. As to our best knowledge, this algorithm has not been used yet to tackle a dynamic pricing scenario as the one outlined in this paper. With the cost-of-exploration, as it is described later on, we also provide a tool to measure, if the expected performance increase does actually occur on a global, system-wide, level for all agents.

In Section "Related Work", we will first introduce some examples of such asymmetric setups and leader-follower implementations using RL on various environments. In Section "Reinforcement Learning and Leader-Follower Games", we display the internal mechanisms of one of the algorithms listed in Section "Related Work". After that, in Section "Implications of Asymmetric Information in Pricing", we propose a mechanism to measure the cost of exploration in an ideal environment. The final section concludes.

Related Work

As mentioned in [8], hierarchies naturally arise in multi-agent problems when some agents have the ability to commit to their action before the others or when there is asymmetric information. We focus on these kind of leader-follower hierarchies. On the other hand, the term 'Hierarchical Reinforcement Learning' [6, 7] is used in the literature to refer to a hierarchy of goals that is created by breaking a complex task down into easier subtasks. In the early 2000s, Könönen published papers about learning a Stackelberg equilibrium in asymmetric multi-agent settings, using RL methods. Reference [4] uses a Q-learning approach to solve a hierarchical problem in which a supplier and a broker are learning simultaneously to maximize their own profits. In [3], a gradient-based method is used to determine a Stackelberg equilibrium. However, this approach relies on computing sums which makes it unsuitable for continuous state and action spaces.

The aforementioned works require complete information of the game. Also, evaluation only took place in Single-Leader-Single-Follower (SLSF) settings. Reference [10] however, developed an algorithm where agents learn only using their received rewards (revenue). Also, the algorithm can be applied to a Single-Leader-Multiple-Follower (SLMF) setting. This is achieved by modeling each leaders action as a different subgame the followers have to learn. The leader meanwhile learns at a slower rate, to encourage convergence of all agents to a Stackelberg equilibrium.

So far, we have only looked at discrete action spaces. Reference [8] recently proposed an algorithm called SMARL to solve problems with multiple leaders and followers (MLMF) and continuous state and action spaces. SMARL is an actor-critic method that performs well in multi-agent settings by using a shared policy for all agents during training. In SMARL, mixed cooperative and competitive objectives are possible. In this context, a brief comparison of the different papers is shown in Table 66.1.

Table 66.1 Overview of related work (LF refers to leader-follower)

Paper	Action space	LF type	Agents' information	Problem domain
[3]	Discrete	SLSF	Full information	Grid-world
[4]	Discrete	Mostly SLSF	Full information	Grid-world, pricing
[5]	Discrete	SLSF	Full information	Pricing
[10]	Discrete	SLMF	Reward-based	Economics
[8]	Continuous	MLMF	Reward-based	Highway-driving

Reinforcement Learning and Leader-Follower Games

In single-agent RL, an agent interacts in discrete time-steps with an environment by choosing actions [9]. The environment provides the agent with rewards, conditioned on the state it is in. The optimization goal usually consists of maximizing the expected discounted reward by finding a policy which maximizes this value. For a single agent, the value of a state s (at any time t) is defined as follows, given a policy π and step wise rewards r_{k+t} for step $k+t$:

$$V_\pi(s) = \mathbb{E}\left[\sum_{k\geq 0} \gamma^k r_{k+t} | s_t = s, a_t \sim \pi\right].$$

For a pricing problem, we usually consider the action space to consist of the range of prices any agent is allowed to choose from. Those might be different for each agent, which influences the resulting policies. The state space might consist of several features. This includes current competitor prices and additional features describing, e.g., both quality and availability of offered products. In past experiments, we discovered that discretization of the action space in a pricing problem has disadvantages [2]. Thus, we suggest the solution algorithm proposed in [8] might serve as a foundation of a solution system for pricing systems. In this paper, as previously outlined, a leader-follower game is considered.

The general optimization goal consists of optimizing the value of the starting state s_0 by improving the policy. In multi-agent RL, multiple agents are learning cooperative or competitive behaviour together in the same environment. This requires particular solution methods because using single-agent RL in a multi-agent setting leads to non-stationarity from any agent's perspective.

Leader-follower games model a decision-making problem where the players have asymmetric roles: One agent (the leader) commits to an action first and the other participants (followers) then make their decision on the basis of the leader's action, competitively playing a Nash game. In such asymmetric games, we are interested in finding a Stackelberg equilibrium in which the leader's optimal action maximizes its own reward, knowing the followers will play a Nash game among themselves.

We now summarize the solution method SMARL from [8]. SMARL is a model-free, off-policy actor-critic method. Model free means that no full model of the environment is needed, off-policy means that the data can be generated by any policy, not just the one that is optimized, and actor-critic means that every agent in the system learns both a policy and a value estimator. There are N agents, each of which has a relationship to any other agent: leader, follower, or equal in hierarchy. Each agent with id i has a deterministic policy $\boldsymbol{\mu}_i$, conditioned on parameters θ_i, as well as a centralized Q-function $Q^{\boldsymbol{\mu}}_{S,i}$, with $\boldsymbol{\mu} = \{\boldsymbol{\mu}_1, \ldots, \boldsymbol{\mu}_N\}$, which is only used during training. $Q^{\boldsymbol{\mu}}_{S,i}$ takes as input all agent's observations $\boldsymbol{x} = (o_1, \ldots, o_N)$ as well as the actions $a_1, \ldots, a_N$ of all agents. μ_i is the actor and $Q^{\boldsymbol{\mu}}_{S,i}$ the critic for an agent. Because both are continuous, they are approximated with artificial neural networks.

In this case, an artificial neural network is a parametric non-linear function of arbitrary structure, usually containing several layers of computation, whose parameters can be adjusted to change its output given a specified input and which is differentiable. Random noise is added to $\boldsymbol{\mu}_i$ to enable exploration.

The agent's policies $\boldsymbol{\mu}$ are updated using the Deterministic Policy Gradient Theorem. The gradient of $\boldsymbol{\mu}_i$ is impacted by the actions of its followers $\mathcal{F}(i)$. Let D be a replay buffer containing tuples of the current global state $\boldsymbol{x}$, the next global state $\boldsymbol{x}'$, all agent actions $a_1, \ldots, a_N$, and rewards $r_1, \ldots, r_N$. With $a_i = \boldsymbol{\mu}_i(o_i)$ and $a_F = \boldsymbol{\mu}_f(o_f, a_i), \forall f \in \mathcal{F}(i)$, the gradient $\nabla_{\theta_i} J(\theta_i)$ is equal to

$$\mathbb{E}_{\boldsymbol{x},a\sim D}\left[\nabla_{\theta_i}\boldsymbol{\mu}_i(o_i)\nabla_{a_i}Q^{\boldsymbol{\mu}}_{S,i}(\boldsymbol{x}, a_1, \ldots, a_N) + \nabla_{\theta_i}\boldsymbol{\mu}_i(o_i)\nabla_{a_i}Q^{\boldsymbol{\mu}}_{S,i}(\boldsymbol{x}, a_F)\right].$$

On the other hand, each critic's value function $Q^{\boldsymbol{\mu}}_{S,i}$ is updated using off-policy temporal difference (TD) learning. Let $\boldsymbol{\mu}' = \{\boldsymbol{\mu}_{\theta'_1}, \ldots, \boldsymbol{\mu}_{\theta'_N}\}$ be the set of target policies with parameters θ' and let $Q^{\boldsymbol{\mu}'}_S = \{Q^{\boldsymbol{\mu}'}_{S,1}, \ldots, Q^{\boldsymbol{\mu}'}_{S,N}\}$ be the set of target critics with parameters ω'. The target networks keep delayed copies of the original parameters to stabilize training. With $a'_i = \boldsymbol{\mu}'_i(o_i)$ and $y = r_i + \gamma Q^{\boldsymbol{\mu}'}_{S,i}(\boldsymbol{x}', a'_1, \ldots, a'_N)$, the TD error to be minimized becomes:

$$\mathbb{E}_{\boldsymbol{x},a,r,\boldsymbol{x}'\sim D}\left[\left(Q^{\boldsymbol{\mu}}_{S,i}(\boldsymbol{x}, a_1, \ldots, a_N) - y\right)^2\right].$$

Implications of Asymmetric Information in Pricing

Given the information from the last section, we can consider a multi-agent RL problem in which two agents are optimized, one being the leader, one being the follower. With mutual price updates, we usually do not know which pricing decision will be performed by the respective competitor; in a leader-follower setup, this changes with at least one agent knowing the competitors response in advance. We intend to determine, if this information exchange can improve the overall economic outcome for both involved agents.

We propose to use the decrease in cost-of-exploration for a two-player duopoly. To measure this, several factors have to be taken into account. In a leader-follower game, one policy is known to the other agent. In a duopoly as it was evaluated by e.g., [2], this is not case. In both setups, the demand function is considered unknown to the agents. A solution algorithm can compute the optimal response given the customers choice when the competitor policy and the demand model is known. The cost-of-exploration is considered the difference between the revenue of the optimal response and the actual policy. Given an optimal policy π^* and an evaluation policy π, we consider the current relative cost of exploration as

$$C(\pi) = \mathbb{E}\left[\frac{V_{\pi}(s_0)}{V_{\pi^*}(s_0)}\right].$$

This value has to be considered in two additional dimensions: The value in RL is usually considered long-term, not over a single state. Furthermore, the expected value has to be taken into account, as the actual generated revenue depends on the stochastic demand model. For each step during training, it is possible to compute the optimal policy given the demand model and the current competitor's policy. The expected value of this policy given a fixed starting state can be computed. It can then be compared to the expected value of the actual current policy. This computation will be performed for both an asymmetric setup, where one agent can learn given the competitors price choice, and a setup in which this information is not given in advance.

The goal of future assessments is to evaluate the difference between both, serving our hypothesis that an improvement can be made by decreasing this price. A suitable market model then shares the saved cost-of-exploration between all market participants. Due to increased learning stability, it might be possible that even the transparent market participants will have a decreased cost-of-exploration. In this case, a balanced sharing model of the pricing mechanism has to be taken into account. Furthermore, long-term effects might occur, where different market participants observe varied levels of cost effectiveness at different points during the learning process. This effect becomes stronger, if the market consists of an oligopoly and if the different market participants join the market at different points in time. To evaluate this situation, the market model used by us will need further extensions to cover changes in the market setup during the training process. A toolkit that can perform the proposed evaluation has been developed by us in the past. Future work will incorporate full evaluation of all mechanisms proposed at this point in time.

Conclusions

We have shown that several alternative implementations of RL in leader-follower games do exist and have provided an insight in their mechanisms. Further, we have defined a measurement toolkit, which can develop a notion of what we consider the most crucial key performance indicator of an multi-agent RL setup, excluding the absolute revenue itself. Next, we seek to provide a full measurement of all aspects discussed in Section "Implications of Asymmetric Information in Pricing", including a measurement of absolute performance given different possible implementations of the corresponding algorithms. Such measurements shall provide a user with a notion of how a market can be served with regulatory framework, which in turn, may level out the leader's advantage over other market participants. One possible suggestion for such a mechanism introduces licensing fees for the information based on the reduced loss of exploration for a single participant. If those fees are smaller than the increase in cost of exploration for all market participants, an exchange might be possible without putting a disadvantage on those offering their policies to the public, while every participant can make use of the increased system stability.

References

1. Goodfellow, I. J., Bengio, Y., & Courville, A. C. (2016). Deep learning. In *Adaptive computation and machine learning*. MIT Press.
2. Kastius, A., & Schlosser, R. (2022). Dynamic pricing under competition using reinforcement learning. *Journal of Revenue and Pricing Management, 21*, 50–63.
3. Könönen, V. (2003). Gradient based method for symmetric and asymmetric multiagent reinforcement learning. In *IDEAL 2003, revised papers*. Lecture Notes in Computer Science (Vol. 2690, pp. 68–75). Springer.
4. Könönen, V. (2004). Asymmetric multiagent reinforcement learning. *Web Intelligence and Agent Systems, 2*(2), 105–121.
5. Könönen, V. (2006). Dynamic pricing based on asymmetric multiagent reinforcement learning. *International Journal of Intelligent Systems, 21*(1), 73–98.
6. Kulkarni, T. D., Narasimhan, K., Saeedi, A., & Tenenbaum, J. (2016). Hierarchical deep reinforcement learning: Integrating temporal abstraction and intrinsic motivation. In *NIPS 2016* (pp. 3675–3683).
7. Le, T. P., Vien, N. A., & Chung, T. (2018). A deep hierarchical reinforcement learning algorithm in partially observable Markov decision processes. *IEEE Access, 6*, 49089–49102.
8. Pereira, S. (2020). *Stackelberg multi-agent reinforcement learning for hierarchical environments* [Master's thesis]. http://hdl.handle.net/10012/15851
9. Sutton, R. S., & Barto, A. G. (1998). Reinforcement learning—An introduction. In *Adaptive computation and machine learning*. MIT Press.
10. Tharakunnel, K., & Bhattacharyya, S. (2009). Single-leader-multiple-follower games with boundedly rational agents. *Journal of Economic Dynamics and Control, 33*, 1593–1603.

Part XVI
Project Management and Scheduling

Chapter 67
A Heuristic Bicriteria Scheduling Approach for a Flooring Production Planning Problem

D. Leib, E. Finhold, and T. Heller

Abstract We consider a scheduling problem motivated by a particular process step in the production of rubber flooring. In this step, a set of given jobs of different colors has to pass through a heated pressure roller, one job at a time. Since different jobs require different process temperatures, and heating up the machine is expensive, we want to minimize the total temperature change. On the other hand, we aim at minimizing the alternations between bright and dark colors to minimize cleaning effort. We provide an efficient heuristic for finding all job sequences that are Pareto optimal with respect to the two objectives.

Keywords Bicriteria optimization · Heuristics · Scheduling

Introduction

The production of rubber flooring consists of several process steps. In the calendering process, the material is passed through several heated rollers. As the jobs have to be processed at different temperatures, and heating up, respectively, cooling down the machine is both time consuming and expensive, one naturally aims at minimizing temperature differences between consecutive orders. Further, jobs may come in different colors. In order to minimize cleaning effort and the risk of color contamination between jobs, job orders with smooth color transitions are particularly desirable.

Since the required heating time/costs, respectively, the "costs" of the color transition between two jobs depends on their temperature, respectively, color difference, this can be understood as sequence dependent setup times. The literature on scheduling with sequence dependent setup times is rich in results on different scheduling problems (e.g. [1, 6, 7]). In particular, formulating scheduling problems with sequence dependent setup times as a *traveling salesperson problem (TSP)* has a long history (cf. [2, 4]). A setting similar to our multi-objective problem was studied in [5], where a bicriteria TSP with sequence priorities are introduced. In contrast to these,

---

D. Leib (✉) · E. Finhold · T. Heller
Fraunhofer Institute for Industrial Mathematics ITWM, 67663 Kaiserslautern, Germany
e-mail: dominik.leib@itwm.fraunhofer.de

O. Grothe et al. (eds.), *Operations Research Proceedings 2022*, Lecture Notes in Operations Research, https://doi.org/10.1007/978-3-031-24907-5_67

we will focus on paths instead of tours. For a general introduction to multicriteria optimization, see [3].

In this paper, we present a heuristic approach for computing job sequences that are Pareto optimal w.r.t. temperature and color change costs, i.e., sequences that cannot be improved in one objective without worsening the other, when the number of colors is equal to two, that is, we only differentiate between bright and dark colors.

Definitions and Notation

We first start with some definitions. A *job* j is a pair $(t, c) \in \mathbb{R}_{\geq 0} \times \{0, 1\}$, where t is the *process temperature* and c the *color*, sometimes we address the process temperature and color of a job j as $t(j)$ and $c(j)$, respectively. The set of jobs is denoted by J and is assumed to be finite in size. For technical reasons we assume that the process temperatures of all jobs in J are pairwise different.

A *partially ordered path* in J is a finite sequence $P = (P_1, \ldots, P_k)$ of disjoint subsets $P_i \subset J$ with $c(j) = c(j')$ for $j, j' \in P_i$ and $i = 1, \ldots, k$. It is *Hamiltonian* if $\bigcup_{i=1}^{k} P_i = J$. If $P_i = \{j\}$ is just a singleton we may write $(P_1, \ldots, P_{i-1}, j, P_{i+1}, \ldots, P_k)$ instead of $(P_1, \ldots, P_{i-1}, \{j\}, P_{i+1}, \ldots, P_k)$. A (ordered) *path* p is a partially ordered path $(P_1, \ldots, P_k)$ where all P_i are singletons.

We call a partially ordered path *alternating* if the colors of the jobs in P_i and P_{i+1} differ for $i = 1, \ldots, k-1$. We can associate to each partially ordered path P an alternating partially ordered path $A(P)$ by uniting consecutive sequences of sets P_i of the same color of maximal length in P. A path p is an *ordering* of a partially ordered path P if $A(p) = A(P)$.

We consider two cost functions: The *total temperature change* and *color change count* of a path $p = (j_1, \ldots, j_k)$ are given by $T(p) := \sum_{i=1}^{k-1} |t_{i+1} - t_i|$ and $C(p) := \sum_{i=1}^{k-1} |c_{i+1} - c_i| = |\{i < k | c_i \neq c_{i+1}\}|$ respectively, where $j_i = (t_i, c_i)$. Correspondingly we define $T(P) := \min\{T(p) | p \text{ ordering of } P\}$ and $C(P) := \min\{C(p) | p$ ordering of $P\}$ for partially ordered paths. Note that $C(P)$ is easy to compute as it holds $C(P) = C(p)$ for any ordering p of P. Regarding T we will show later that $T(P)$ can be computed efficiently on partially ordered paths in some specific cases.

Example 1 Let $J = \{(10, 0), (20, 1), (50, 0)\}$ and $P = (\{(10, 0), (50, 0)\}, (30, 1))$. There are two orderings for P, namely $p_1 = ((10, 0), (50, 0), (30, 1))$ and $p_2 = ((50, 0), (10, 0), (30, 1))$. It is $C(P) = c(p_1) = c(p_2) = 1$ and since $T(p_1) = 60 > 50 = T(p_2)$ we have $T(P) = T(p_2) = 50$.

For Hamiltonian paths p, p' we say that p' *dominates* p if $T(p') \leq T(p)$ and $C(p') < C(p)$ or $T(p') < T(p)$ and $C(p') \leq C(p)$. A *Pareto optimal path* is a Hamiltonian (ordered) path which is not dominated by any other Hamiltonian path. Our goal is to find the set of Pareto optimal paths. Since the number of color changes in any path of length n is bounded by $n - 1$ this also applies to the number of Pareto optimal paths.

Swapping Algorithm

The idea of the algorithm is to first compute the Pareto extremes, which are the Pareto optimal paths with minimal and maximal color change count, respectively. Then we iteratively compute elements with intermediate color change count. W.l.o.g. we assume that any (partially ordered) path contains jobs of both color 0 and color 1. Otherwise, we get the unique Pareto optimal path by sorting the jobs by temperature (which can be done in polynomial time).

Computing the Pareto Extremes

We begin with the two extreme Pareto optimal solutions; one with minimal total temperature changes and one with minimal color change count. The former is achieved by simply ordering the jobs by their process temperature:

Lemma 1 *Let p be a path in J, then $T(p) \geq t_{\max} - t_{\min}$ where $t_{\max}$ and $t_{\min}$ are the maximal and minimal temperatures of the jobs occuring in p, respectively.*

The Pareto optimal extreme with exactly one change in color can be found using Lemma 2 below. For $P_i \subset J$ we write $\overrightarrow{P_i}$ for the path resulting from ordering the elements in P_i increasingly with respect to the process temperature and $\overleftarrow{P_i}$ for the corresponding path with decreasing temperature. Then we have:

Lemma 2 *Let $P = (P_1, \ldots, P_k)$ be an alternating partially ordered path, then*

$$T(P) \in \left\{ T(\tilde{P}_1, \ldots, \tilde{P}_k) | \tilde{P}_i \in \{\overrightarrow{P_i}, \overleftarrow{P_i}\} \right\}.$$

Let $P_c \subset J$ be the set of all jobs of color c for $c \in \{0, 1\}$. To get a Pareto optimal element with one color change we need to find an ordering p of (P_0, P_1) with $T(p) = T(P)$. But Lemma 2 shows that p can be chosen as an element in $\{(\overrightarrow{P_0}, \overrightarrow{P_1}), (\overrightarrow{P_0}, \overleftarrow{P_1}), (\overleftarrow{P_0}, \overrightarrow{P_1}), (\overleftarrow{P_0}, \overleftarrow{P_1})\}$ with minimal total temperature change.

Going Up and Going Down

The following lemma shows that if a partially ordered path P has a specific structure, then an ordering p of P with $T(p) = T(P)$ can be computed directly.

Lemma 3 *Let $P = (P_1, \ldots, P_k)$ be an alternating partially ordered path for $k \geq 3$ and assume the sequences $P_1, P_3, P_5 \ldots$ and $P_2, P_4, P_6 \ldots$ of packages of the same color are strictly increasing w.r.t. the process temperature, respectively. Set $p :=$ $(\hat{P}_1, \overrightarrow{P_2}, \ldots, \overrightarrow{P_{k-1}}, \hat{P}_k)$ where*

$$\hat{P}_1 = \begin{cases} \overrightarrow{P}_1, & \frac{|t_{\max}(P_1) - t_{\min}(P_2)|}{|t_{\min}(P_1) - t_{\min}(P_2)|} \leq 1 \\ \overleftarrow{P}_1 & else. \end{cases} \quad and\ \hat{P}_k = \begin{cases} \overrightarrow{P}_k, & \frac{|t_{\max}(P_{k-1}) - t_{\min}(P_k)|}{|t_{\max}(P_{k-1}) - t_{\max}(P_k)|} \leq 1 \\ \overleftarrow{P}_k & else. \end{cases},$$

then $T(p) = T(P)$.

Let $C_{\max}$ be the color change count of the Pareto optimal path which is sorted by process temperature as in Section "Computing the Pareto Extremes" and let $p^{C_{\max}}$ be the corresponding path. Since there are no Hamiltonian paths with a smaller total temperature change due to Lemma 1, there is no Pareto optimal path with a higher color change count. Thus all Pareto optimal elements have a color change count between $C_{\max}$ and 1. We initialize a result dictionary that will keep track of one representative for each of those color change counts containing both Pareto extremes initially, where we let p^1 be the Pareto optimal path with a color change count of 1. Paths of intermediate color change count are computed in two ways:

Starting from $p^{C_{\max}}$ we compute new representatives in two ways by *going down* in which we compute new Hamiltonian paths with decreased color change count. Let p be a current path of the result list and $A(p) = (P_1, \ldots, P_k)$.

(D1) Swap the two subsets at both borders, and return the resulting partially ordered paths $\{(P_2, P_1 \cup P_3, \ldots P_k), (P_1, \ldots, P_{k-2} \cup P_k, P_{k-1})\}$.

(D2) Let j, j' be a pair of two consecutive jobs in p of different color with $|t(j') - t(j)|$ minimal and let P_i, P_{i+1} be the sets in P containing j, j', respectively. Return $\{(P_1, \ldots, P_{i-1} \cup P_{i+1}, P_i \cup P_{i+2}, \ldots, P_k)\}$.

Starting from p^1 we may compute new representatives in two ways by *going up* in which we compute new Hamiltonian paths with increased color change count.

(U1) Let $\overrightarrow{P_2} = (j_1^2, \ldots, j_r^2)$ and $\overrightarrow{P_{k-1}} = (j_1^{k-1}, \ldots, j_s^{k-1})$ where $r, s \in \{1, \ldots, n\}$. We construct Hamiltonian partially ordered paths by inserting P_1 into $\overrightarrow{P_2}$ or P_k into $\overrightarrow{P_{k-1}}$ at every inner position of $\overrightarrow{P_2}$ and $\overrightarrow{P_{k-1}}$, respectively, leading to

$$\begin{aligned} &\{(\{j_1^2\}, P_1, \{j_2^2, \ldots, j_r^2\}, P_3, \ldots, P_k), \\ &(\{j_1^2, j_2^2\}, P_1, \{j_3^2, \ldots, j_r^2\}, P_3, \ldots, P_k), \ \ldots \\ &(\{j_1^2, \ldots, j_{r-1}^2\}, P_1, \{j_r^2\}, P_3, \ldots, P_k)\} \cup \\ &\{(P_1, \ldots, P_{k-2}, \{j_1^{k-1}, \ldots, j_{s-1}^{k-1}\}, P_k, \{j_s^{k-1}\}), \\ &(P_1, \ldots, P_{k-2}, \{j_1^{k-1}, \ldots, j_{s-2}^{k-1}\}, P_k, \{j_{s-1}^{k-1}, j_s^{k-1}\}), \ \ldots \\ &(P_1, \ldots, P_{k-2}, \{j_1^{k-1}\}, P_k, \{j_2^{k-1}, \ldots, j_s^{k-1}\})\}. \end{aligned}$$

(U2) Let j, j' be a pair of two consecutive jobs in p of different color with $|t(j') - t(j)|$ minimal and let P_i, P_{i+1} be the sets in P containing j, j', respectively. Return $\{(P_1, \ldots, (P_i \setminus j) \cup \{j'\}, (P_{i+1} \setminus j') \cup \{j\}, \ldots, P_k)\}$ if the color change count increased and nothing, otherwise.

The Hamiltonian partially ordered paths resulting from (D1), (D2), (U1) and (U2) are all alternating and fulfill the requirements of Lemma 3. Thus we can use Lemma 3 to compute an ordering from each of the results with the best possible temperature change values. The result dictionary is then updated with the new paths found where existing entries are replaced if their total temperature change is higher than the ones already computed. We exit the routine if we did not improve the total temperature change for any of the color change counts in the result dictionary after a full cycle of going down and going up. Lastly the Pareto optimal elements in the result dictionary can be computed in linear time as the number of Pareto optimal paths is bounded by $n - 1$.

Algorithm 1 Swapping algorithm

```
p^1, p^{C_max} = get_pareto_extremes(J)
result_dict = {1: p^1, C_max : p^{C_max}}
while true do
  current = p^{C_max}
  for i = C_max, ..., 3 do
    path_list_down = going_down(current)
    update(result_dict, path_list_down)
    current = result_dict[i − 1]
  end for
  current = p^1
  for i = 1, ..., C_max − 2 do
    path_list_up = going_up(current)
    update(result_dict, path_list_up)
    current = result_dict[i + 1]
  end for
  if has_not_improved(result_list) then
    break
  end if
end while
return result_dict
```

Numerical Results

For a numerical analysis of Algorithm 1 we fix the number of jobs to $n = 20$ and apply Algorithm 1 to 1000 randomly generated instances of size n where we generate the jobs uniformly distributed in $[40, 400] \times \{0, 1\}$. We then compare the total temperature changes of the result to the minimal temperature change value possible under all paths with the same color change count, which where computed by enumeration. Let m be a color change count that is contained in the algorithms result dictionary R and let $\hat{P}_m$ be the set of all paths p resulting from Lemma 3 with $C(p) = m$. Set $N_m = |\hat{P}_m|$, $T_m^{\min} := \min\{T(p)|p \in \hat{P}_m\}$, $T_m^{\max} := \max\{T(p)|p \in \hat{P}_m\}$ and $d_m(R) := |\{p \in \hat{P}_m | T(p) < T(R(m))\}|$ where $R(m)$ is the path with color change count m in R. The *accuracy* and *relative temperature change difference* of R are

$$acc(R) := 1 - \frac{1}{|R|}\sum_{m\in R}\frac{d_m(R)}{N_m} \text{ and } rtd(R) := \frac{1}{|R|}\sum_{m\in R}\frac{|T(R(m)) - T_m^{\min}|}{|T_m^{\max} - T_m^{\min}|},$$

respectively. The following table shows the median, 99% quantile and worst values of accuracy and relative temperature change difference of the 1000 runs.

	Mean	99% quantile	Worst
acc	0.9991	0.9921	0.9781
rtd	0.0043	0.0288	0.0514

In 99% of the runs we had an accuracy over 99% and a relative temperature difference of less than 3% to the optimal and the worst run had an accuracy over 97% and less than a 6% difference in relative temperature difference to the optimal values.

Outlook and Conclusion

The main aspect of the presented algorithm was to find qualitative solutions in a reasonable computation time, especially in realtime applications. Future research is directed to the cases of more than two color classes as well as a polynomial time algorithm for finding the Pareto optimal solution for a given color change count. In future work the polynomial time algorithms will be used to give a comparison against the algorithm presented in this paper.

References

1. Allahverdi, A., Gupta, J. N., & Aldowaisan, T. (1999). A review of scheduling research involving setup considerations. *Omega, 27*(2), 219–239.
2. Bagchi, T. P., Gupta, J. N., & Sriskandarajah, C. (2006). A review of TSP based approaches for flowshop scheduling. *European Journal of Operational Research, 169*(3), 816–854.
3. Ehrgott, M. (2005). *Multicriteria optimization* (Vol. 491). Springer Science & Business Media.
4. Potts, C. N., & Strusevich, V. A. (2009). Fifty years of scheduling: A survey of milestones. *Journal of the Operational Research Society, 60*(1), S41–S68.
5. Schmitz, H., & Niemann, S. (2009). A bicriteria traveling salesman problem with sequence priorities. In *Metaheuristics in the service industry* (pp. 1–14). Springer.
6. Yang, W.-H. (1999). Survey of scheduling research involving setup times. *International Journal of Systems Science, 30*(2), 143–155.
7. Zhu, X., & Wilhelm, W. E. (2006). Scheduling and lot sizing with sequence dependent setup: A literature review. *IIE Transactions, 38*(11), 987–1007.

Chapter 68
Propagation and Branching Strategies for Job Shop Scheduling Minimizing the Weighted Energy Consumption

Andreas Bley and Andreas Linß

Abstract We consider a job shop scheduling problem with time windows, flexible energy prices, and machines whose energy consumption depends on their operational state (offline, ramp-up, setup, processing, standby or ramp-down). The goal is to find a valid schedule that minimizes the overall energy cost. To solve this problem to optimality, we developed a branch-and-bound algorithm based on a time-indexed integer linear programming (ILP) formulation, which uses binary variables that describe blocks spanning multiple inactive periods on the machines. In this paper, we discuss the propagation and branching schemes used in that algorithm. The strategies, which are specifically tailored for energy related machine scheduling problems, primarily aim to determine and sharpen the activity profiles of the machines (and thus reduce the number of the inactive block variables) and address the workload profile of the tasks with lower priority. Computational experiments validate the efficiency of those techniques.

Keywords Integer programming · Machine scheduling · Presolving · Branch and bound

Introduction

In recent years, energy awareness and increasing energy prices gained a lot of attention in production planning. Various approaches to incorporate energy consumption or costs into models and solution techniques for machine scheduling problems have been proposed, see for example [6, 10]. Those models explicitly consider different machine states, such as *processing*, *standby* or *off*, the transitions among these states, and the respective energy demands and durations [2, 12]. Problems with time-dependent energy costs are typically modeled using time-indexed formula-

A. Bley · A. Linß (✉)
Institut für Mathematik, Universität Kassel, Heinrich-Plett-Straße 40, 34132 Kassel, Germany
e-mail: andreas.linss@uni-kassel.de

A. Bley
e-mail: andreas.bley@uni-kassel.de

O. Grothe et al. (eds.), *Operations Research Proceedings 2022*, Lecture Notes in Operations Research, https://doi.org/10.1007/978-3-031-24907-5_68

tions, which leads to huge ILP models. Even more, in contrast to classical objectives minimizing completion times, minimizing the energy consumption leads to highly fractional solutions of the linear programming (LP) relaxations. Thus, tailored model reduction and branching techniques are needed to solve those models efficiently.

In this paper, we present such techniques for a variant of the job-shop scheduling problem with flexible energy prices and time windows discussed in [4, 11]. An overview of formulations for the classical job-shop scheduling problem is given in [8]. A survey on alternative modelling and solution approaches for job-shop scheduling can be found in [14].

Problem Description and Formulation

In our problem, the planning horizon $[T] := \{0, \ldots, [T]-1\}$ consists of $[T]$ uniform periods. For each $t \in [T]$, we are given an energy price $C_t \in \mathbb{R}_{\geq 0}$, which is valid during period t. Furthermore, we are given a set of (non-uniform) machines $M := \{1, \ldots, n_M\}$ and a set of jobs $J := \{1, \ldots, n_J\}$. A job $j \in J$ consists of a list of tasks (j, k), $j \in J$, $k \in O_j := [n_j]$, $n_j \in \mathbb{N}$, which must be processed in the order defined by O_j. Each task (j, k) must be setup and processed on a predefined machine $m_{j,k} \in M$. For each task (j, k), we are given its setup time $d^{se}_{j,k} \in \mathbb{N}$ and its processing time $d^{pr}_{j,k} \in \mathbb{N}$. In addition, we are given a release date $a_j \in [T]$ and a due date $f_j \in [T]$ for each job $j \in J$, which apply to the first and the last task of the job, respectively. We let $O := \{(j, k) : j \in J,\ k \in O_j\}$. $O^M_{|_m} = \{(j, k) \in O | m_{j,k} = m\}$ denotes the set of tasks (j, k) on machine $m \in M$.

In each period $t \in [T]$, each machine $m \in M$ must be in one of the operating states *off*, *processing*, *setup*, *standby*, *ramp-up* or *ramp-down*, summarized as $\mathcal{S} = \{off, pr, se, st, ru, rd\}$. A machine is called active if its operating state is *setup*, *processing*, or *standby*, otherwise it is called inactive, with the canonical switches between the states and implications between tasks and machine states. The duration of the *ramp-up* phase, changing from *off* to any state $s \in \{se, pr, st, rd\}$, is $d^{ru}_m \in \mathbb{N}$. The duration of the *ramp-down* phase is $d^{rd}_m \in \mathbb{N}$. For each machine $m \in M$ and state $s \in \mathcal{S}$, D^s_m is the energy demand of machine m in state s.

A feasible solution consists of the start time for each task's processing and a machine state for each machine and each period. Each task is processed non-preemptively and each task's setup immediately precedes (also non-preemptively) its processing. The processing of a task can start only after the processing of its predecessor has been completed, but its setup can already start while the predecessor is processing (on another machine). The start of the first and the completion of the last task of each job must obey this job's release and due dates, respectively. Only one task can be processed or set up on a machine simultaneously. A machine processing or setting up for a task must be in state *processing* or *setup*, respectively. Otherwise, the machine can be active in *standby* or become inactive *ramping down*, being *off*, or *ramping up*, respecting the ramping durations and canonical state relations. At the beginning and the end of the planning horizon, each machine must be *off*. Our goal is to find a solution whose energy cost is minimized.

From the task precedences, the ramping, setup and processing times, and the jobs' release and due dates, we obtain for each task $(j, k) \in O^M_{|m}$ on machine m the earliest period $a_{j,k} := \max\{a_j, d^{ru}_m + d^{se}_{j,0}\} + \sum_{q=0}^{k-1} d^{pr}_{j,q}$ and the latest period $f_{j,k} := \min\{f_j, T - d^{rd}_m\} - \sum_{l=k}^{|O_j|-1} d^{pr}_{j,l}$ when its processing may start.

We use a time-indexed formulation with binary variables to explicitly indicate so-called breaks, i.e., inactive blocks of consecutive *ramp-down–off–ramp-up* periods on the machines, to avoid inequalities describing the ramping mechanism. To use this type of variables also to model the initial and final ramping, we extend the time window for each machine to $T^m_+ := \{-d^{rd}_m, \ldots, T + d^{ru}_m - 1\}$ and enforce that the machine is *off* in periods 0 and T. The energy price is set to $C_t = 0$ for the artificial periods $t \in T^m_+ \setminus T$. For each machine m, B_m denotes the set of all feasible breaks (t_0, t_1), $t_0, t_1 \in [T^m_+]$ and $t_1 - t_0 \geq d^{ru}_m + d^{rd}_m$, where the machine is starting its *ramp-down* at t_0, is *off* from $t_0 + d^{rd}_m$ until $t_1 - d^{ru}_m - 1$, and in *ramp-up* from $t_1 - d^{ru}_m$ to $t_1 - 1$. In total, we have three types of variables:

- $x_{j,k,t} \in \{0, 1\}$ indicating iff task $(j, k) \in O$ starts processing in period $t \in [T]$,
- $z^{st}_{m,t} \in \{0, 1\}$ indicating iff machine $m \in M$ is in state *standby* in $t \in [T]$,
- $z^{m,br}_{t_0,t_1} \in \{0, 1\}$ indicating iff $m \in M$ is in a break from t_0 to t_1, $(t_0, t_1) \in B_m$.

The associated objective coefficients $\hat{c}_{j,k,t}$ and $\hat{c}^{br}_{m,t_0,t_1}$ express the total energy cost induced by the setup and processing of task (j, k) if started in period t and by a break from t_0 until t_1 on machine m, respectively.

We obtain the following integer programming formulation of the problem:

$$\min \sum_{m \in M} \left(\sum_{t \in [T]} \left(C_t D^{st}_m z^{st}_{m,t} + \sum_{(j,k) \in O^M_{|m}} \hat{c}_{j,k,t} x_{j,k,t} \right) + \sum_{(t_0,t_1) \in B_m} \hat{c}^{br}_{m,t_0,t_1} z^{m,br}_{t_0,t_1} \right) \tag{68.1}$$

$$\sum_{t \in \{a_{j,k}, \ldots, f_{j,k}\}} x_{j,k,t} = 1 \quad (j, k) \in O \tag{68.2}$$

$$\sum_{q=0}^{t-d^{pr}_{j,k}} x_{j,k,q} - \sum_{q=0}^{t} x_{j,k+1,q} \geq 0 \quad j \in J,\ k < |O_j|,\ t \in [T] \tag{68.3}$$

$$\sum_{(j,k) \in O^M_{|m}} \sum_{q=\max(t-d^{pr}_{j,k}+1,0)}^{\min(t+d^{se}_{j,k}, T-1)} x_{j,k,q} + z^{st}_{m,t} + \sum_{(t_0,t_1) \in B_m : t \in \{t_0, \ldots, t_1\}} z^{m,br}_{t_0,t_1} = 1 \quad m \in M,\ t \in [T] \tag{68.4}$$

$$\sum_{(-d^{rd}_m, t_1) \in B_m} z^{m,br}_{-d^{rd}_m, t_1} = 1 \quad m \in M \tag{68.5}$$

$$\sum_{(t_0, T+d_m^{ru}-1) \in B_m} z_{t_0, T+d_m^{ru}-1}^{m,br} = 1 \quad m \in M \tag{68.6}$$

$$x_{j,k,t} \in \{0, 1\} \quad (j, k) \in O,\ t \in [T] \tag{68.7}$$

$$z_{t_0,t_1}^{m,br} \in \{0, 1\} \quad m \in M,\ (t_0, t_1) \in B_m \tag{68.8}$$

$$z_{m,t}^{st} \in \{0, 1\} \quad m \in M,\ t \in [T] \tag{68.9}$$

The objective (68.1) describes the total energy cost. Equalities (68.2) ensure that each task is started once. Constraints (68.3) describe the precedence relations between consecutive tasks of each job. Equalities (68.4) enforce that each machine is either processing or setting up a task or in a break or standby in each period. Constraints (68.5) and (68.6) ensure each machine is offline in periods 0 and T.

Propagation and Presolving

In this section, we discuss several preprocessing and propagation techniques to reduce the size of the proposed integer program (68.1)–(68.9) at the root node and within the branch and bound tree. In practice, such reductions are of utmost importance to efficiently solve large models. Many techniques such as detecting dominating columns, bound tightening, and conflict analysis are implemented in general purpose ILP solvers [1]. However, these techniques heavily exploit problem-specific structures, whose (re-)detection from the model is computationally expensive and available only for some very general types of substructures. In our application, where the connection among precedence constraints and time windows plays a key role, problem-specific techniques are necessary.

Precedence constraints and time windows. Throughout this section, we denote by $a_{j,k}$ and $f_{j,k}$ the earliest and the latest period when task $(j, k) \in O$ may start its processing, respectively, at the current branch and bound node. This time window may have holes, when task (j, k) is not allowed to start processing. Pretending that there are no holes in $[a_{j,k}, f_{j,k}]$, we first apply the constraint propagation rules from [5] to detect locally valid precedence constraints and tighten $a_{j,k}$ and $f_{j,k}$.

Conflicts of breaks and tasks. Next, we try to infer implications stemming from overlaps of time windows of tasks and a single break-variable. Clearly, the break $z_{t_0,t_1}^{m,br}$ on machine $m \in M$ with $(t_0, t_1) \in B_m$ cannot participate in any integer feasible solution, if both $t_0 < a_{j,k} + d_{j,k}^{pr}$ and $t_1 \geq f_{j,k} - d_{j,k}^{se}$ hold for an arbitrary task $(j, k) \in O_{|_m}^M$. This condition indicates that the break would conflict with each possible start of processing of task (j, k).

Irrelevant breaks. A break-variable $z_{t_0,t_1}^{m,br}$, $m \in M$ and $(t_0, t_1) \in B_m$, will not be used in any optimal integral solution, if there is a combination of other break- and standby-variables on machine m that exactly cover the periods from t_0 until t_1 with a smaller objective. We detect those cases by enumerating all possible covers with exactly one break. If all energy prices are non-negative, non-artificial breaks before the first and after the last task on a machine are unnecessary. Hence, break-variable $z^{m,br}_{t_0,t_1}$ with $t_0 > 0$ and $t_1 \leq T$ can be eliminated if $t_0 \leq \min_{(j,k)\in O^M_{|m}} (a_{j,k} + d^{pr}_{j,k} - 1)$ or if $t_1 \geq \max_{(j,k)\in O^M_{|m}} (f_{j,k} - d^{se}_{j,k})$.

Clique information. Modern ILP solvers automatically generate clique inequalities using a graph describing pairwise conflicts of binary variables, c.f. [1]. Not all such conflicts are detected automatically. In our code, we explicitly add conflicts for pairs of break-variables if their combined lengths exceed the duration of the overall time window T minus the sum of all task setup and processing actions, the initial ramp-up, and the final ramp-down on the machine. More explicitly, we add all conflicts between break variables $z^{m,br}_{t_0,t_1}$ and $z^{m,br}_{t_2,t_3}$ with $t_3 - t_2 + t_1 - t_0 + 2 > T + d^{ru}_m + d^{rd}_m - \sum_{(j,k)\in O^M_{|m}} (d^{pr}_{j,k} + d^{se}_{j,k})$. In addition, the clique constraints $x_{j,k,t} + \sum_{q=t-d^{se}_{j,k}-d^{pr}_{i,l}+1}^{t+d^{pr}_{j,k}-1+d^{se}_{i,l}} x_{i,l,q} \leq 1$ for all pairs $(j,k), (i,l) \in O^M_{|m}$ and (meaningful) periods $t \in [T]$ are added to the conflict graph.

General presolving for task variables. Eventually, we apply the presolving and constraint propagation rules available in the used ILP solver, see [1], for example.

Branching Scheme

Our branching scheme consists of two rules. Our preferred branching aims to interrupt longer intervals of consecutive fractional inactivity and forces the machine to either ramp-up or ramp-down completely. Both branches sharpen the machine profile and typically increase the dual bound. To enforce integrality of the activity of m in period t', we create two child nodes, one forcing the fractional (in)activity $f_{m,t'} := \sum_{(t_0,t_1)\in B_m : t_0 \leq t' \leq t_1} z^{m,br}_{t_0,t_1}$ to 0 and the other forcing it to 1. The machine $m \in M$ and the interval $[q_0, q_1] \in Q_m$ are chosen to maximize $(q_1 - q_0) \cdot \sum_{t=q_0}^{q_1} f_{m,t}$, where Q_m denotes the set of all consecutive fractionally inactive intervals $[q_0, q_1] \subset [T^m_+]$ on m. The chosen period $t' = (\sum_{t=q_0}^{q_1} t \cdot f_{m,t}) / (\sum_{t=q_0}^{q_1} f_{m,t})$ interrupts the fractional activity of m in $[q_0, q_1]$ and forbids incomplete ramping in favorable periods.

If the first rule does not find an auspicious branching, we employ the branching of [13] to adjust the time windows of single tasks by branching on the assignment constraints (68.2). In contrast to [13], the task $(\hat{k}, \hat{j}) \in O$ to branch on is chosen to maximize $(r(\hat{j},\hat{k}) - l(\hat{j},\hat{k})) \cdot \left(\sum_{t=l(\hat{j},\hat{k})}^{r(\hat{j},\hat{k})} \sum_{(j,k)\in O^M_{|m}} x_{j,k,t}\right)^{-1}$, with $l(j,k) = \arg\min_{t\in[T]} \{x_{j,k,t} > 0\}$ and $r(j,k) = \arg\max_{t\in[T]} \{x_{j,k,t} > 0\}$. This modification prefers tasks that currently have small overlap with others, and so branching on those tasks has a strong effect on the machine activity.

Results

In this section, we compare results obtained by applying the introduced techniques in our branch-and-cut code in SCIP [3] (with GUROBI 9.5.1 [7]), using them to presolve the model solved by GUROBI, and using GUROBI as a standalone solver with default setting and aggressive presolving, but without our presolving methods.

The instances presented here, with real energy prices from March 2021 of Germany/Luxembourg, are derived from the instance la01 [9] by dividing all processing times by 10 and ceiling to get processing and setup times and a manageable time window ($T = 120$). Ramping times are chosen as 2/3, 1, or 1.5 times the mean processing on each machine to derive instances with small (s), medium (m) and large (l) ramping durations, respectively. The energy demands are chosen as $(D_m^{off}, D_m^{ru}, D_m^{se}, D_m^{pr}, D_m^{st}, D_m^{rd}) = (0, 10, 5, 8, 3, 6)$. The resulting full models for s, m, and l have 31,198, 30,998, and 30,698 variables, respectively. Table 68.1 shows the number of variables after presolve (v), the number of branch and bound nodes (NN), the primal-dual gap, and the dual bound after 3600 s.

The results show that our techniques detect more reductions even than the aggressive presolving of GUROBI. Only with our reductions the problem could be solved to optimality within the time limit. Furthermore, our tailored branching substantially reduced the number of branch-and-bound nodes that have been explored. With our branching and presolving SCIP was able so solve the problems as fast GUROBI with only our presolving and its default branching strategies, despite running single-threaded versus GUROBI using up to 8 parallel threads.

Conclusion

We developed preprocessing and branching techniques that enable us to solve time-indexed ILP formulations of job shop scheduling problems involving time windows, machine states and energy costs to optimality. Our reduction and propagation techniques outperform those implemented in standard ILP solvers and can also be applied easily within the branch-and-bound tree. Our experiments show that, in combination with tailored branching strategies, these techniques effectively reduce the number of variables, drive the dual bound and substantially reduce the size of the branch-and-bound trees, especially for instances with relatively large ramping-durations. From our point of view, there is still some need to improve the bounds on the task variables to obtain stronger dual bounds and derive better primal solutions faster.

Table 68.1 Comparison of the effect of our propagation and branching rules

Instance	la01 s				la01 m				la01 l			
	v	NN	Gap	Dual	v	NN	Gap	Dual	v	NN	Gap	Dual
scip + pre + bra	12 k	1.2 k	0.00	124,291	9 k	591	0.00	132,448	5 k	6 k	0.00	146,503
gurobi + pre	14 k	108 k	0.00	124,291	9 k	9 k	0.00	132,448	6 k	113 k	0.00	146,503
gurobi	19 k	157 k	0.16	124,098	23 k	42 k	0.05	132,387	12 k	164 k	0.26	146,162

References

1. Achterberg, T., Bixby, R. E., Gu, Z., Rothberg, E., & Weninger, D. (2020). Presolve reductions in mixed integer programming. *INFORMS Journal on Computing, 32*(2), 473–506.
2. Benedikt, O., Sucha, P., Modos, I., Vlk, M., & Hanzálek, Z. (2018). Energy-aware production scheduling with power-saving modes. In *Integration of constraint programming, artificial intelligence, and operations research* (pp. 72–81). Springer.
3. Bestuzheva, K., Besancon, M., Chen, W. K., Chmiela, A., Donkiewicz, T., van Doornmalen, J., Eifler, L., Gaul, O., Gamrath, G., Gleixner, A., & Gottwald, L. (2021). *The SCIP optimization suite 8.0* (ZIB-Report 21-41). Zuse Institute Berlin.
4. Bley, A., & Linß, A. (2020). Job shop scheduling with flexible energy prices and time windows. In *Operations Research Proceedings 2019* (pp. 207–213). Springer.
5. Brucker, P. (2002). Scheduling and constraint propagation. *Discrete Applied Mathematics, 123*(1), 227–256.
6. Gao, K., Huang, Y., Sadollah, A., & Wang, L. (2020). A review of energy-efficient scheduling in intelligent production systems. *Complex & Intelligent Systems, 6*(2), 237–249.
7. Gurobi Optimization, LLC. (2022). *Gurobi optimizer reference manual.* https://www.gurobi.com
8. Jain, A., & Meeran, S. (1999). Deterministic job-shop scheduling: Past, present and future. *European Journal of Operational Research, 113*(2), 390–434.
9. Lawrence, S. (1984). *Resource constrained project scheduling: An experimental investigation of heuristic scheduling techniques (supplement)* (Technical Report). Graduate School of Industrial Administration, Carnegie-Mellon University.
10. Nolde, K., & Morari, M. (2010). Electrical load tracking scheduling of a steel plant. *Computers & Chemical Engineering, 34*(11), 1899–1903.
11. Selmair, M., Claus, T., Herrmann, F., Bley, A., & Trost, M. (2016). Job shop scheduling with flexible energy prices. In *Proceedings of the 30th ECMS.*
12. Shrouf, F., Ordieres-Meré, J., García-Sánchez, A., & Ortega-Mier, M. (2014). Optimizing the production scheduling of a single machine to minimize total energy consumption costs. *Journal of Cleaner Production, 67*, 197–207.
13. van den Akker, J. M., Van Hoesel, C. P. M., & Savelsbergh, M. W. (1999). A polyhedral approach to single-machine scheduling problems. *Mathematical Programming, 85*(3), 541–572.
14. Zhang, J., Ding, G., Zou, Y., Qin, S., & Fu, J. (2019). Review of job shop scheduling research and its new perspectives under industry 4.0. *Journal of Intelligent Manufacturing, 30*(4), 1809–1830.

Chapter 69
Scheduling Unrelated Parallel Machines with Attribute-Dependent Setup Times: A Case Study

Sven Jäger, Neele Leithäuser, Sebastian Velten, and Christian Weiß

Abstract We study a practical production planning problem that was encountered by an industrial partner. Mathematically, the problem can be described as an unrelated parallel machine scheduling problem with deadlines and sequence-dependent setup times. The setup times depend on certain job attributes (e.g. material, color, size) of the successive products. For all products, dated demands are given that must be met with the means of one or multiple production orders that can be assigned to a set of eligible machines. We describe and evaluate different approaches to solve the scheduling problem. Our final algorithm is a combination of an iterated greedy construction heuristic together with a constraint program to locally improve the solution found by the heuristic. For most of our test instances, our algorithm achieves more than 20% reduction of setup times compared to the original solutions used by our industrial partner.

Keywords Scheduling · Setup times · Unrelated machines · Constraint programming · Heuristics · Iterated greedy

Introduction

In many manufacturing processes, machines must be set up individually for the products that are being produced. The effort caused by a product change often depends on the dissimilarity of the products. In this paper we report on a real world case study

S. Jäger · N. Leithäuser (✉) · S. Velten · C. Weiß
Fraunhofer ITWM, Fraunhofer-Platz 1, 67663 Kaiserslautern, Germany
e-mail: neele.leithaeuser@itwm.fraunhofer.de

S. Jäger
e-mail: sven.jaeger@rptu.de

S. Velten
e-mail: sebastian.velten@itwm.fraunhofer.de

C. Weiß
e-mail: ch.weiss88@gmail.com

O. Grothe et al. (eds.), *Operations Research Proceedings 2022*, Lecture Notes in Operations Research, https://doi.org/10.1007/978-3-031-24907-5_69

that we have conducted with an industrial partner from the metalworking industry. In our real world example, every product is specified by seven attributes. The effort needed to change machine settings depends only on the types of attributes that differ between two successive products produced on the same machine. Crucially, the setup time does not depend on the respective values of the attributes or on the machines they are being manufactured on.

Usually, setups take several hours, but in extreme cases can also need up to two weeks. In relation to this, typical jobs in our setting run for several weeks (but may be defined also for much smaller time scales if necessary). Since setup time is unproductive machine time, it is obviously desired to minimize the setup times. Moreover, different amounts of the various products are demanded at different deadlines. In this study, in order for a solution to be feasible, all demands must be fulfilled at or before their deadlines.

The remainder of this paper is structured as follows: to finish this section we briefly discuss how to model our scheduling problem mathematically, as well as provide a brief review of the related literature. In the Section "Using Mathematical Modelling and Third Party Solvers" we discuss solution approaches involving third party solvers. In the Section "Iterated Greedy Heuristic" we describe a dedicated iterated greedy heuristic. Finally, in the Section "Computational Results" we provide computational results and a short outlook on future research directions.

Problem formulation. Our problem can be modelled as a scheduling problem with unrelated parallel machines and sequence-dependent setup times. For each demand, there is one job with a deadline equal to the deadline of the associated demand. Importantly, jobs may be split across machines in a fashion where the same job may be processed by more than one machine at the same time. This is different from standard parallel machine scheduling problems.

Each machine is available for processing during specific time intervals, and can process only one job at a time. For this case study, we assume that the available intervals are known in advance and that the downtime can be used for setup operations.

Each job is specified by seven attributes and its deadline. For each attribute changing from a job to its successor on the same machine, a certain setup time is required. The setups for multiple changing attributes can be carried out in parallel, so that it can be assumed that for one pair of successive jobs only the maximum setup time implied by any attribute change is needed.

The goal is to minimize the total sum of setup times, while fulfilling all deadlines. Without regard for deadlines this problem is much easier than in the general case with arbitrary sequence-dependent setup times (which is equivalent to solving a travelling salesperson problem). The difficulty thus stems from the combination with the deadlines.

Related work. Scheduling with sequence-dependent setup times has been studied intensively both from a theoretical point of view, see e.g. [6], and from a practical point of view, see the extensive survey by Allahverdi [1], focusing on heuristic approaches. The problem considered in this paper occurs as a subproblem of the problem considered by Fu et al. [2]. However, their practical instances are significantly smaller, so that they are able to use a standard mixed integer linear programming solver as a routine for this subproblem. Silva et al. [9] use techniques similar to ours to solve a scheduling problem where machines have to be equipped with job-specific tools that are also a limited resource.

Using Mathematical Modelling and Third Party Solvers

In this section we describe different approaches we used to try and solve our practical problem with mathematical modelling techniques. In particular, we used mixed integer linear programming (MILP) and constraint programming (CP). We discuss different approaches and stages of our work and point out which turned out to be successful and which did not pass the practice check.

Black Box Solving

Our first approach was to attempt to solve the problem by modelling it as a MILP or a CP and then solve it with a black box solver. To solve the MILP, we used Gurobi [3], while for the CP we used IBM ILOG CP Optimizer [4]. In what follows, due to space limitations, we focus on the constraint programming approach, as it is more important to the steps we describe later on.

The CP is modelled in such a way, that for each job, there are as many interval variables as the job has eligible machines, one for each machine. The length of each interval is variable, but the total amount of product produced for all intervals of one job must meet the demanded amount for the job. Also, each interval belonging to a job must finish before that job's deadline.

Intervals on the same machine may not overlap. This is modelled using a noOverlap-constraint (see [5]). For IBM ILOG CP Optimizer, the noOverlap-constraint is extended in such a way, that sequence dependent setup times can be modelled: two jobs on the same machine, in addition to not overlapping, have to have some fixed time between them that depends on the jobs. We use this IBM ILOG CP Optimizer specific feature to model setup times. During unavailabilities, jobs simply stay on the machine without any amount of product being produced. This behaviour is modelled via the Intensity functions provided by IBM ILOG (see [5]). Note, that if all jobs are known in advance, it is not necessary for jobs to be preempted and later restarted on the same machine. We use this fact in order to simplify our model.

As one might have expected, using these solvers as simple black boxes is not successful for industrial size instances. For most instances, neither the MILP nor the CP solver was able to find a feasible solution within a reasonable time frame. For those instances where it was possible to find a feasible solution, this was usually worse than the solution originally produced by our industrial partner. Additional experiments with smaller instances, as well as a review of the literature (see, e.g., [7]), suggested that the CP approach was more promising for finding feasible solutions quickly, so for our following steps, we dropped the MILP approach from consideration.

Black Box Solving with Starting Solutions

In a next step, we provided the CP with a starting solution, using at first the original solution from our industrial partner. With this help, the CP managed to find additional solutions, however, none of them were much better than the starting solution, in particular for larger instances. For example, within about three hours, for our large four months instance (see Table 69.1), the CP solver only managed to improve the starting solution's setup time by seven minutes, compared to a total setup time of the starting solution of roughly 360 days.

Using the CP Model with Starting Solutions and Localized Search

Finally, we decided to abandon the idea of using the CP to try and optimize the whole instance in one go. Instead, we split up the instance in sets of at most five machines (this value provides the best trade-off between running time and solution quality), and then used the CP to optimize on these subsets only. In particular, our algorithm worked as follows: first, find the machine with the largest amount of setup time. Reorder jobs on this machine to minimize the total setup time, while still fulfilling all deadlines. Then, choose another machine which shares many processable job types with the first machine. Reorder and switch jobs on both machines to minimize setup times, again while still fulfilling all deadlines. Continue to add machines in this manner until at most five machines are optimized at the same time. Then, continue again with a single machine, namely the one with the second largest setup time. Continue in this manner, until each machine was chosen as first machine once.

For each optimization step, we allowed the CP solver to run for four seconds per involved machine, leading to a total of 1 min per each selected first machine. For our large instances, this leads to roughly three hours of computation time, similar to what we allowed the black box CP solver to use.

Using this approach, again with the starting solution as originally used by our industrial partner, we were able to improve, in each instance, the starting solution by roughly 20%. Details can be found in the Section "Computational Results" and Table 69.2.

Iterated Greedy Heuristic

The choice of the starting solution has a large impact on the quality of the schedule resulting from the localized search procedure described in the last section. Since the localized search is able to mend some local deficiencies, it is important that the initial solution has a good global structure. This was one motivation for us to look at algorithms to produce alternative initial solutions for the localized search. In addition, when applying our heuristic to future planning tasks, it is desirable that the planners need not manually provide an initial solution.

In view of their good experimental performance, reported by Ruíz and Stützle [8, 10], we decided to develop an iterated greedy algorithm for this task. The main building block of our algorithm is a list scheduling procedure that obtains a sequence of jobs as input and outputs a schedule possibly violating some deadlines. The iterated greedy algorithm searches for an input order for which the resulting list schedule does not delay jobs and has low total setup time. In every iteration the current input order is modified, based on the currently delayed jobs. Note that the assignment of jobs to machines is not specified in advance but results from the list scheduling procedure. A similar approach was pursued by Silva et al. [9].

In the list scheduling procedure, for each job, in the order of the input sequence, the following two steps are executed: first, the algorithm tries to completely fulfill the demand in time. To this end, as much as possible is scheduled before the deadline on each eligible machine. If the demand cannot be completely satisfied, in a second step, the remaining quantity is produced in one contiguous block on the machine with minimum necessary setup time. An important degree of freedom is the order in which the machines are considered in the first step. For this we have to balance between the speed of the machine, the amount that can be scheduled before the deadline, and the time required to setup the machine for the product. This has to be tuned for the concrete setup and production times occurring in the application.

Now we turn to the procedure for adjusting the job order from one step to the next. In the first step of the algorithm, jobs are ordered in such a way that if all jobs were scheduled on the same machine, setup times would be minimized, i.e., by similarity in job attributes. For the second step, all jobs which are late in the first step are moved to the front, generating two separate groups of jobs. Then each group is individually ordered in the same way as before, to minimize setup times. Then, in each further step, again all jobs which are late in the previous step are moved from their current job group and placed into the group one before that, generating a new group for jobs which already were in the first group. Then, all groups are again ordered as in step one. This is repeated until no job is late or a fixed number of steps has been performed. Note, that for our test instances, based on the actual machine and job data of our industrial partner, this algorithm always found a solution where all deadlines were met.

Computational Results

We tested our algorithm for six real-world problem instances with different numbers of jobs and machines and with different planning horizons. The characteristics of these instances are compiled in Table 69.1. For each instance we compare the schedules obtained by our iterated greedy heuristic (Section "Iterated Greedy Heuristic") to the plans originally implemented by our industrial partner. These original plans are achieved by running first a proprietary scheduling algorithm as part of a market-competitive ERP (enterprise resource planning) software. This scheduling algorithm ignores setup times and instead assumes that all processes are slower by a fixed percentage to allow for setups. Its main goal is fulfilling deadlines. Then, in a second step, the human production schedulers fix the schedules given by the ERP software by hand, mostly focusing on reducing setup times locally.

Additionally, we also subsequently apply the CP-based localized search (see the Section "Using the CP Model with Starting Solutions and Localized Search") to both the original and the iterated greedy schedule. The results are shown in Table 69.2. Naturally, by applying an improvement search to the schedules, they can only get better. Recall that the improvement search from the Section "Using the CP Model with Starting Solutions and Localized Search" requires a starting solution. In particular, for an actual operational roll-out at our industrial partner's site an original solution from our industrial partner is no longer available as starting solution, but it needs to be computed instead. Thus, our full proposed algorithm is to first compute a starting solution via the iterated greedy heuristic and subsequently apply the CP-based localized search to that solution as an improvement heuristic.

Observe that for all instances the combination of iterated greedy and localized search we propose clearly outperforms the original schedules with regard to both total setup time and number of setups. Also note that the relative improvement increases with larger time horizons. For smaller time horizons the improvement is not as significant, which is mostly due to the iterated greedy heuristic providing a worse starting solution than the original solution. In those cases, applying the localized search directly to the original solution would yield even better results, although, as described above, this is not feasible for a future operational algorithm.

Table 69.1 Characteristics of the studied problem instances

Instance	#Jobs	#Machines	Time horizon (months)
S2	98	8	2
S4	252	8	4
S6	346	8	6
L2	3336	284	2
L4	8072	284	4
L6	13,289	284	6

Table 69.2 Results for the studied problem instances

Inst.	Total setup time [machine days]				# Setups			
	`orig`	`itGr`	`orig + locS`	`itGr + locS`	`orig`	`itGr`	`orig + locS`	`itGr + locS`
S2	15.4	17.1	11.9	12.3	45	50	35	36
S4	66.4	81.8	58.3	59.3	114	119	91	93
S6	85.5	41.9	66.7	34.3	170	122	116	100
L2	170.8	200.9	134.8	161.9	562	626	437	499
L4	358.6	301.3	281.8	263.0	1180	926	918	814
L6	564.5	427.0	449.6	373.9	1788	1209	1386	1058

Here, `orig` denotes the original schedule, `itGr` is the schedule obtained by the iterated greedy heuristic described in the Section "Iterated Greedy Heuristic". To both solutions, we also subsequently apply the CP-dependent localized search from the Section "Using the CP Model with Starting Solutions and Localized Search"; the values for those solutions can be found in columns `orig + locS` and `itGr + locS`, respectively

The decreased quality of the iterated greedy heuristic is due to artifacts at the end of the scheduling horizon, which become less meaningful for larger instances. Note that some of these end-of-horizon artifacts are due to the manner in which the instances are truncated: the original schedule is truncated in such a way that it fits exactly the new instance, so it does not lose quality due to instance truncation. Our iterated greedy algorithm, on the other hand, has to plan jobs from scratch and thus has less flexibility towards the end of the schedule than the industrial planners had when creating the original schedule (due to tighter due dates at the end of the instance).

Conclusion

In this paper, we have described a case study of a real life industrial production planning problem and described different stages of developing solution algorithms. Then we have compared different algorithmic solutions to the originally used schedules from our industrial partner. It can be seen that a combination of different algorithmic paradigms (e.g., in this paper, a construction heuristics combined with a CP-based improvement heuristic) is useful when solving practical scheduling problems.

In the future, it will be important to extend our algorithms to an online setting, where jobs and machine availabilities become known over time, and the schedule has to be computed incrementally. To achieve this, it will be especially important to handle end-of-horizon effects as described in the Section "Computational Results".

References

1. Allahverdi, A. (2015). The third comprehensive survey on scheduling problems with setup times/costs. *European Journal of Operational Research, 246*(2), 345–378. https://doi.org/10.1016/j.ejor.2015.04.004
2. Fu, L. L., Aloulou, M. A., & Triki, C. (2017). Integrated production scheduling and vehicle routing problem with job splitting and delivery time windows. *International Journal of Production Research, 55*(20), 5942–5957. https://doi.org/10.1080/00207543.2017.1308572
3. Gurobi Optimization, LLC. (2022). *Gurobi optimizer reference manual.* https://www.gurobi.com
4. IBM Deutschland GmbH. (2022). *IBM ILOG CPLEX optimization studio.* https://www.ibm.com/de-de/products/ilog-cplex-optimization-studio
5. IBM Deutschland GmbH. (2022). *IBM ILOG CPLEX optimization studio reference manual.* https://www.ibm.com/docs/en/icos/20.1.0?topic=interfaces-c-interface-referencemanual
6. Jansen, K., Maack, M., & Mäcker, A. (2019). Scheduling on (un-)related machines with setup times. In *2019 IEEE International Parallel and Distributed Processing Symposium (IPDPS)* (pp. 145–154). https://doi.org/10.1109/IPDPS.2019.00025
7. Pham, T. S., Rousseau, L. M., & De Causmaecker, P. (2022). A two-phase approach for the radiotherapy scheduling problem. *Health Care Management Science, 25*, 191–207. https://doi.org/10.1007/s10729-021-09579-9
8. Ruiz, R., & Stützle, T. (2008). An iterated greedy heuristic for the sequence dependent setup times flowshop problem with makespan and weighted tardiness objectives. *European Journal of Operational Research, 187*(3), 1143–1159. https://doi.org/10.1016/j.ejor.2006.07.029
9. Silva, C., Klement, N., & Gibaru, O. (2018). A generic decision support tool for lot-sizing and scheduling problems with setup and due dates. In E. Viles, M. Ormazábal, & A. Lleó (Eds.), *Closing the gap between practice and research in industrial engineering* (pp. 131–138). Springer International Publishing. https://doi.org/10.1007/978-3-319-58409-6_15
10. Stützle, T., & Ruiz, R. (2018). Iterated greedy. In R. Martí, P. M. Pardalos, & M. G. C. Resende (Eds.), *Handbook of heuristics* (pp. 547–577). Springer International Publishing. https://doi.org/10.1007/978-3-319-07124-4_10

Chapter 70
Storage and Retrieval in Fully Automated Grid-Based Storage Systems

Nicolas Fauvé and Simone Neumann

Abstract In the fast-growing online market, e-grocery providers in particular advertise fast delivery times. These can only be achieved through flexible and compact warehouse solutions close to the customer. Storage providers have recognized this need and therefore offer new types of storage solutions. However, this development has so far received little attention from the scientific community. Therefore, this paper introduces a new type of storage and addresses the modeling of storage and retrieval in terms of cost factors and their composition.

Keywords Warehousing · Compact storage · Grid-based

Introduction

As described by Azadeh et al. [2], new compact storage systems have entered the market and have been successful. Although there are quite a number of publications on the optimization of the storage process for different storage systems or on the coordination of mobile robots, there are so far only a few scientific publications dealing with the kind of systems we describe in this paper, one of which is by Zou et al. [7]. Those systems are interesting for the rising e-commerce (see, e.g., Boysen et al. [3]) due to their high density. The issue of poor accessibility resulting from the density is tried to be addressed by system designs such as AutoStore [1] or Ocado [5].

N. Fauvé
Management Science and Operations Research, Helmut Schmidt University, 22043 Hamburg, Germany
e-mail: nicolas.fauve@hsu-hh.de

S. Neumann (✉)
Sustainable Logistics and Mobility, Universität Hamburg, 20148 Hamburg, Germany
e-mail: simone.neumann@uni-hamburg.de

O. Grothe et al. (eds.), *Operations Research Proceedings 2022*, Lecture Notes in Operations Research, https://doi.org/10.1007/978-3-031-24907-5_70

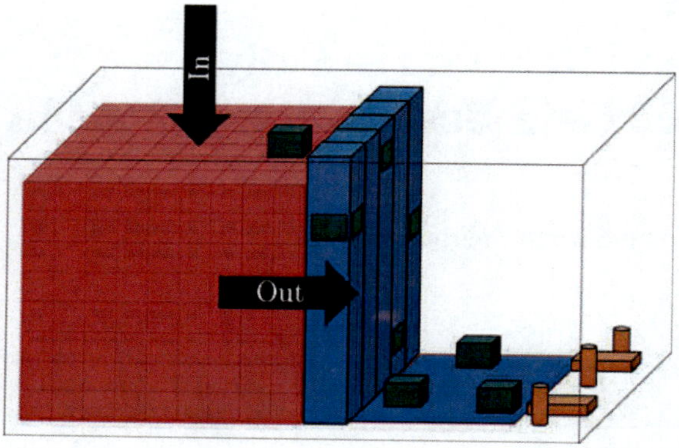

Fig. 70.1 Mode of operation

These designs have the weakness that access is achieved through time-consuming reshuffle processes. To eliminate the disadvantage of having to retrieve unneeded bins during the storage and retrieval process, the following system (see Fig. 70.1) is proposed.

Storage units (we will call them bins) are stacked on top of each other and packed as a cuboid (red). Automated guided vehicles (AGVs) (green) can store bins by being lifted onto the cuboid, driving to the assigned stack and dropping the bin. In addition, AGVs can retrieve a bin by being lifted to the assigned tier and driving into the cuboid after a path has been created. The lifts and the floor (blue) connect the cuboid and pick stations (orange) in which the bins are assigned to orders.

Scope of Paper We focus on one part of the proposed storage system. Our subsystem of interest only includes the handling of the bins with the AGVs in the area of the storage cuboid and the lift. This neglects the possibilities of sequencing bins in the area in front of the cuboid and at the pick stations. Furthermore, we focus on a two-dimensional slice of the storage cuboid (without loss of generality, since the interaction of different slices of the storage cuboid is negligible).

As the proposed storage system is new and the literature on similar systems is scarce, in this paper, we want to lay the foundation for future research by describing in detail how the system works and which relevant costs arise. This is done by determining the cost structure for storing and retrieving a bin at a particular position within a two-dimensional slice of the storage cuboid.

The rest of the paper is structured as follows: In the next Section, we will describe the storage system. Then, we will analyze the cost structure of the storage and retrieval processes in detail and at the end of the paper, we will give some managerial implications and raise some questions for future research.

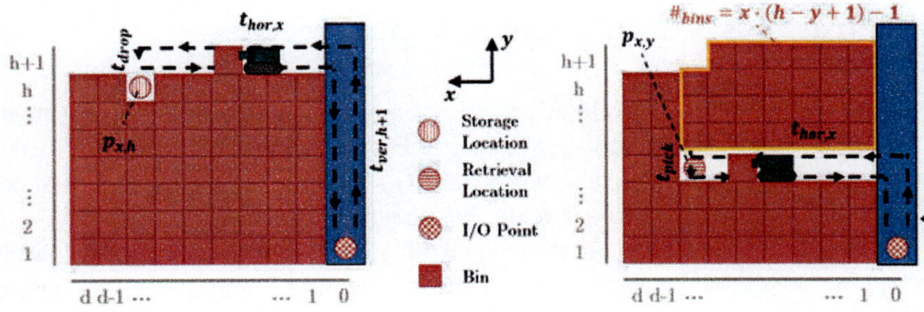

(a) Storage process

(b) Retrieval process

Fig. 70.2 Side view of our system

The Compact Storage System

System Description The two-dimensional slice (see Fig. 70.2) consists of a set $D = \{1, \ldots, d\}$ of stacks aligned in a single row. Each stack holds h bins, resulting in a set of $d \cdot h = |B|$ different bins. The position of a bin is indicated by $p_{x,y}$ with $x \in \{0, \ldots, d\}$ referring to its horizontal position and $y = \{1, \ldots, h + 1\}$ referring to its vertical position. Since the bins do not have a fixed assigned position in the cuboid, it can change constantly over time. AGVs enter and exit the system via the input and output (I/O) point located in position $p_{0,1}$. In $x = 0$ a lift is moving the AGV with the bin to its designated tier y. Within the storage cuboid, the AGV moves the bin sideways. A bin sinks by one tier after a bin of the same stack has been retrieved from below it. In order for an AGV to move sideways, a horizontal path must be created by lifting all blocking bins from the same tier or higher. Empty positions in a stack resulting from the retrieval of a bin are replenished from above. Figure 70.2 shows a storage and a retrieval process.

Assumptions In the following, we consider a simple storage strategy being implemented by refilling the storage cuboid from the top with bins as soon as a storage position is available. Furthermore, the bins move at a constant speed.

We assume that each bin has the same dimension because, in practice, a uniform footprint simplifies bin handling and automation. Moreover, we act on the assumption that the weight of all bins is the same. The energy needed to drop or pick a bin and to lift and lower the AGV can be neglected. Counterweights in the lift compensate for the latter. Moreover, the time needed to create a path can be neglected, since it is overshadowed by the lifting time of the AGV.

Cost Structures of Storage and Retrieval Processes

When calculating the cost of storage and retrieval processes, we can use an objective function taking into account time and/or energy. For each storage or retrieval process, the costs are calculated that occur when the AGV starts at the I/O point, drops or picks the bin, and then drives back to the I/O point.

By doing the retrieval process through creating horizontal aisles, time costs and energy costs are decoupled. For example, a bin in position $p_{2,y}$ with $y < h$ is closer to the I/O point (as it is located at a lower tier) and therefore can be retrieved faster than a bin in position $p_{2,h}$. However, the former requires more mass to be lifted than the latter and therefore its retrieval has a higher energy consumption. A bin in position $p_{2,y}$ requires $2 \cdot (h - y + 1) - 1$ bins to be lifted, while one bin needs to be lifted to retrieve the bin in position $p_{2,h}$. Therefore, we have two different calculations for the time costs and energy costs.

The total time costs $t_{retrieve}$ and t_{store} consist of the following components: the time $t_{hor,x}$ to move the AGV in the x-direction from the stack x to the lift (or vice versa); the time $t_{ver,y}$ to move the AGV in the y-direction from tier y to the I/O point (or vice versa); the time t_{drop} to drop the bin and the time t_{pick} to pick the bin.

The total energy costs $e_{retrieve}$ and e_{store} consist of the following components: the energy $e_{path,x,y}$ needed to create a path for the AGV to retrieve a bin from position $p_{x,y}$ and the energy $e_{hor,x}$ to move the AGV in the x-direction from the stack x to the lift (or vice versa).

Storage Process Since the storage cuboid is assumed to be fully loaded, it can only be replenished once a bin has been removed. This leaves no degree of freedom where to store the bin. As all operations occur on the top tier $h + 1$, costs for a storage process differ only depending on the x-coordinate of the previous retrieval process, as shown in Eqs. (70.1) and (70.2)

$$t_{store} = 2 \cdot (t_{hor,x} + t_{ver,h+1}) + t_{drop} \tag{70.1}$$

$$e_{store} = 2 \cdot e_{hor,x} \tag{70.2}$$

The closer the x-coordinate is to the lift, the lower the energy and time costs are. Since no path is created, overall energy costs are low compared to the energy costs of the retrieval process.

Retrieval Process In the retrieval process, the general structure of energy and time costs differ. While time costs show a linear pattern (see Fig. 70.3), energy costs show a radial one (see Fig. 70.4).

The time costs for the retrieval process are calculated as follows:

$$t_{retrieve} = 2 \cdot (t_{hor,x} + t_{ver,y}) + t_{pick} \tag{70.3}$$

Compared to the storage process, the x-coordinate as well as the y-coordinate are variable. This results in the longest retrieval time being in the top-left corner of the storage cuboid.

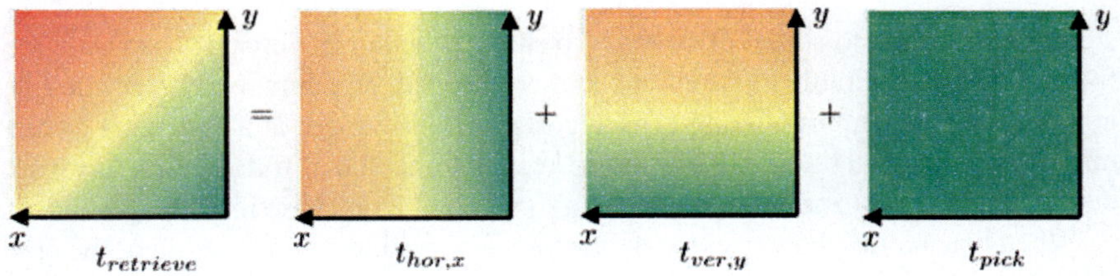

Fig. 70.3 Composition of the time costs of the retrieval process (low time costs in green, high time costs in red)

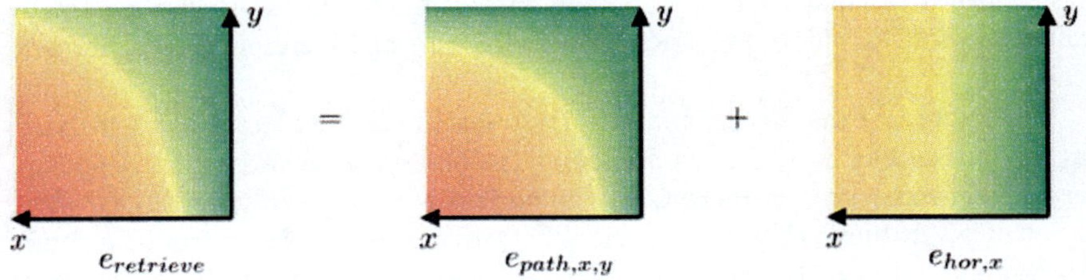

Fig. 70.4 Composition of the energy costs of the retrieval process (low energy costs in green, high energy costs in red)

Concerning the total energy costs, the path creation costs $e_{path,x,y}$ are decisive.

$$e_{store} = 2 \cdot (e_{path,x,y} + e_{hor,x}) \tag{70.4}$$

To retrieve the top right bin ($p_{1,h}$), no path needs to be created, as access to the bin is not blocked by other bins. When retrieving a bin, bins to the right as well as bins above block access, leading to a radial increase of energy costs, with the lowest cost being in the top right corner and the highest cost being in the bottom left corner ($p_{d,1}$) of the storage cuboid. As already mentioned, the energy costs for picking the bin and lifting the AGV can be neglected. The $e_{hor,x}$ component of $e_{retrieve}$ leads to a break in symmetry.

Bounds for the retrieval process Over time, the bin either stays on the top tier or sinks, never worsening the retrieval time. Therefore, we can determine the following bounds on the retrieval times: (I) Upper bound: retrieval time from the top tier and (II) Lower bound: retrieval time from the lowest tier.
The opposite is true for the energy costs. The lower the bin sinks, the higher the energy costs for the retrieval will be, as more bins need to be lifted. (III) Upper bound: energy costs for retrieval from the lowest tier and (IV) Lower bound: energy costs for retrieval from the top tier.

Conclusion and Future Research

In this paper, initial analyses of the time and energy cost structure of a new automated grid-based compact storage system were made, which are necessary for future studies concerning this kind of system. To summarize our findings, we can state the following: it is desirable to store bins that are needed very frequently near the lift, at a high tier (when minimizing energy costs for retrieval) or at a lower tier (when minimizing retrieval time), but one has to keep in mind that as time passes, bins sink down, which leads to an increase in energy costs and a decrease in time.

When considering how to expand the storage cuboid, the composition of the cost structure plays an important role. If the number of tiers h equals the number of stacks d, either adding a tier or a stack can be cost-efficient. Considering the time costs, the speed of horizontal movement or vertical movement is decisive. If the AGV is lifted faster than it can move sideways, adding a tier is time-efficient. Regarding energy costs, the cost for sideways movements are decisive and therefore, adding a tier is energy-efficient.

For future research, in the context of automated compact storage systems such as those described here, various decision questions, either answered independently or simultaneously, are of interest. How to retrieve a set of given bins? How to store a set of given bins? How to rearrange the bins in the cuboid to facilitate future storage and retrieval processes? Which items are to be stored in which bin? At which position in the storage cuboid should a bin be stored (both initially and also during housekeeping)? To answer these, a detailed analysis of storage and retrieval strategies is necessary, existing studies on other storage systems, such as [4, 6], which address scattered storage, could be adapted, and a mathematical model should be set up.

Acknowledgements This research paper is funded by dtec.bw—Digitalization and Technology Research Center of the Bundeswehr which we gratefully acknowledge.

References

1. AutoStore. *Automation made easy*. Available at https://www.autostoresystem.com/system. Accessed: June 28, 2022.
2. Azadeh, K., De Koster, R., & Roy, D. (2019). Robotized and automated warehouse systems: Review and recent developments. *Transportation Science, 53*(4), 917–945. https://doi.org/10.1287/trsc.2018.0873
3. Boysen, N., De Koster, R., & Weidinger, F. (2019). Warehousing in the e-commerce era: A survey. *European Journal of Operational Research, 277*(2), 396–411. https://doi.org/10.1016/j.ejor.2018.08.023
4. Kress, D., Boysen, N., & Pesch, E. (2017). Which items should be stored together? A basic partition problem to assign storage space in group-based storage systems. *IISE Transactions, 49*(1), 13–30. https://doi.org/10.1080/0740817X.2016.1213469
5. Ocado Group. *Building mobile robots using automated solutions*. Available at https://www.ocadogroup.com/technology/blog/life-bot-building-mobile-robot-using-automated-solutions. Accessed: June 28, 2022.
6. Weidinger, F., & Boysen, N. (2018). Scattered storage: How to distribute stock keeping units all around a mixed-shelves warehouse. *Transportation Science, 52*(6), 1412–1427. https://doi.org/10.1287/trsc.2017.0779
7. Zou, B., de Koster, M. B. M., & Xu, X. (2016). Evaluating dedicated and shared storage policies in robot-based compact storage and retrieval systems. In *ERIM report series reference*.

Part XVII
Simulation

Chapter 71
Comparison of Adoption Rates of Hydrogen, Hydrogen-Electric and SAF in the Future Air Transport System with a System Dynamics Model

Chetan Talwar, Imke Joormann, and Thomas S. Spengler

Abstract Aviation has been criticized for its negative climate impact in the past few years due to the emission of harmful greenhouse gasses (GHGs) such as CO_2, NO_x and water vapor that can cause the formation of climate harming ozone and contrails. In recent years, many different technologies have surfaced that have the potential to reduce emissions and replace the existing conventional jet fuel technology. On one hand, H_2 powered aviation just recently regained high attention from the industry, e.g., Airbus launched the ZEROe program where they pledged to develop the world's first zero-emission commercial aircraft by 2035. On the other hand, sustainable aviation fuels (SAF) or biofuels have been identified as an alternative option. Given the different promising future technologies, it is difficult to predict their role in transition pathways that will lead the air transport system towards a more sustainable future. We develop a global scale system dynamics air transport system simulation model and incorporate components like the new potential technologies, production side emissions of new fuels, i.e., SAF and hydrogen, air travel demand, airline industry and aircraft manufacturers. We also include the long, medium and short haul segments of flights. Using this model, we analyze the adoption trends of new technologies and fuels by assessing the amount of fleet operated with them and its effect on emission reduction within each flight segment.

Keywords System dynamics and theory · Dynamical systems · Transportation

C. Talwar (✉) · I. Joormann · T. S. Spengler
Cluster of Excellence—Sustainable Energy Efficient Aviation (SE2A), Technische Universität Braunschweig, Braunschweig, Germany
e-mail: c.talwar@tu-braunschweig.de

Institute of Automotive Management and Industrial Production, Technische Universität Braunschweig, Mühlenpfordtstraße 23, 38106 Braunschweig, Germany

O. Grothe et al. (eds.), *Operations Research Proceedings 2022*, Lecture Notes in Operations Research, https://doi.org/10.1007/978-3-031-24907-5_71

Introduction

Aviation is one of the most important industries on the planet in terms of efficient transportation. Due to the harmful greenhouse gas emissions, IATA passed a resolution for aviation to reach net-zero carbon emissions by 2050. It is widely accepted that this goal cannot be reached by using the current jet fuel technology. On the other hand, new technologies like hydrogen, hydrogen-electric and sustainable aviation fuels (SAFs) have showcased their potential, but it is unclear how the transition to these new technologies would happen if assessed strictly from a usage point of view. Furthermore, it is important to analyze what would be the behavior of the system in different adoption scenarios of these new technologies to find out their emissions reduction potential. To determine the adoption rates, the interaction between different stakeholders of the air transport system needs to be considered to understand the advantages, limitations, and potential delays of the new technologies. For example, when considering the adoption of SAF by airlines, the availability of synthetic and biofuels is a major cause of concern, as there are scalability issues [1]. In this study, a system dynamics air transport system (ATS) model is used to analyze and compare the behavior of the system under different adoption rate scenarios. The dynamics describing the ATS are influenced by feedback loops, time dependencies, stakeholder interactions, decision processes, and non-linearity. Due to its highly interactive and feedback approach, system dynamics is well suited to assess the air transport system. To consider these interactions in a dynamic simulation model, it is imperative that the mutual causality between the variables of interest is clearly defined. In this paper, the feedback structure between the variables in the ATS is described with the help of causal loop diagrams that affect the adoption of novel energy supply systems. In the system dynamics methodology, the causal loop diagram (CLD) is one tool which aids in representing the feedback structure of how different variables of interest in a system are causally interrelated [2]. The different variables from the subsystem interact with each other and affect the adoption rates of new technologies. Representing the system in this way facilitates a better understanding of the initial impression of possibilities with respect to the feedback structure within the system.

Literature Review

The system dynamics (SD) technique has been used by researchers in the commercial jet aircraft industry to demonstrate its effectiveness in forming reliable forecasts, identifying key variables for poor returns and cyclical behavior in the industry [3–5]. These models were later extended to study the potential of reducing CO_2 emissions with the help of alternative fuels in aviation [6]. For this, new feedback mechanisms like alternative fuel production, drop in quota and alternative fuel adoption mechanisms were introduced in the model. Even though the previous studies were robust and insightful, they did not include new technologies such as hydrogen and their role in different flight segments and emission reduction. Furthermore, the role of the new technologies and fuels has not been assessed in the context of recent emission goals set for the aviation industry in a system dynamics model.

Modeling Approach

The system dynamics ATS model developed previously consisted of the interlinked subsystems air travel demand, airline operations, aircraft manufacturers and alternative fuel producers [4, 6]. In this paper, the existing ATS model is used and extended further to analyze the feedback structures that affect the adoption rates of the new technologies.

The previously developed system dynamics ATS model is extended by including hydrogen and SAFs as new technologies. The model includes manufacturing capacity changes, usage adoption mechanism and production capacity changes of the new technologies. The feedback structure of the extended model is represented in Figs. 71.1, 71.2 and 71.3. In the figures, the variables given in roman are the pre-existing variables from the previous model feedback structures. The underlined and italicized variables are the new extensions made to the feedback structures. The bolded variables are the exogenous inputs that affect the feedback mechanism from the outside of the model boundary. The "+" symbol in the diagram indicates the same directional movement between the variables, while the "−" symbol indicates the relationship between variables as moving in opposite directions. The symbol "||" on the connecting lines indicates delays in the transmission of information. Finally, the variables, used more than once in the three diagrams to interlink the subsystems, are marked in green.

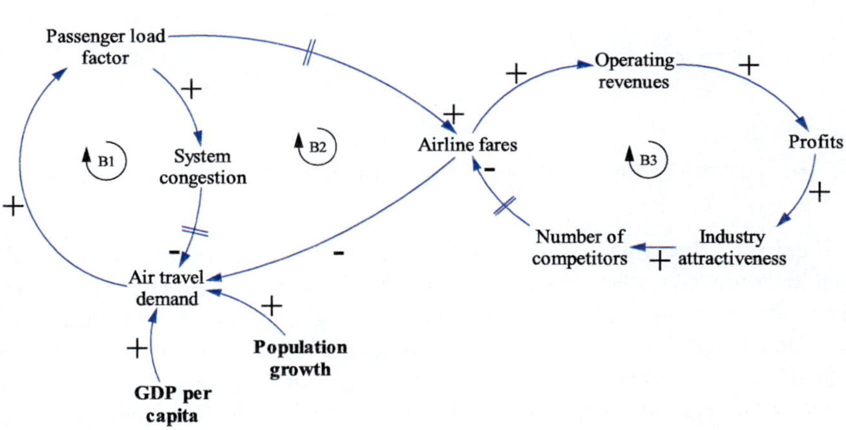

Fig. 71.1 Feedback structure between demand, passenger load factor and fares

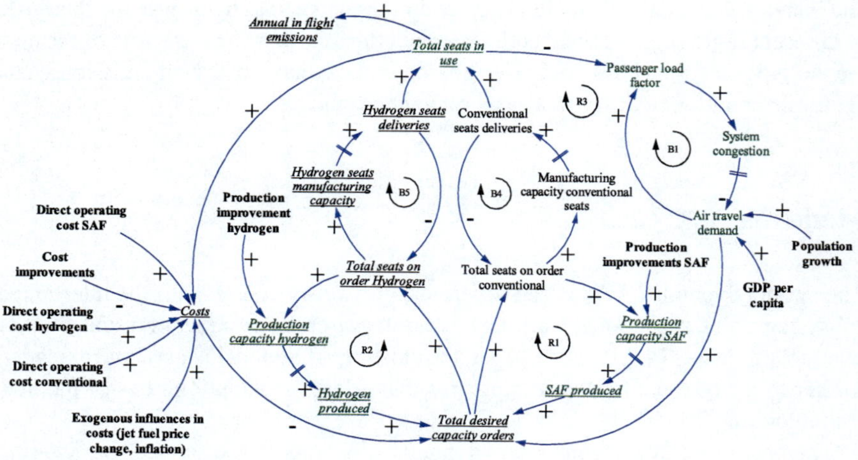

Fig. 71.2 Feedback structure of the airline fleet operation subsystem

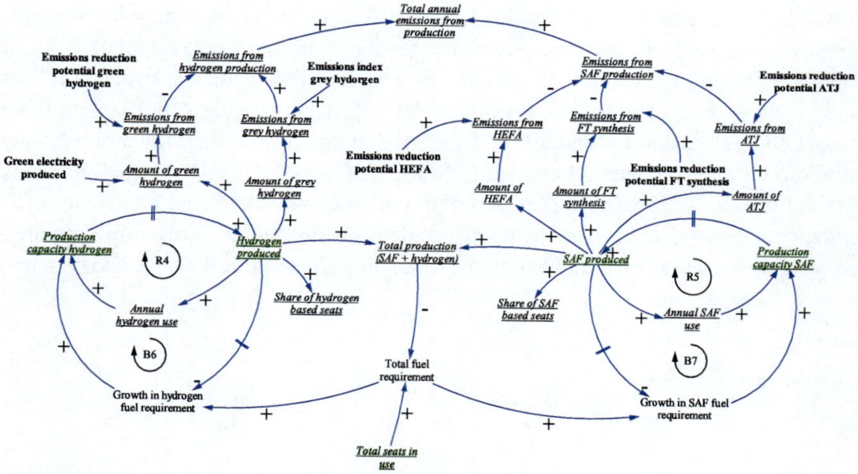

Fig. 71.3 Feedback structure of the fuel production subsystem

Figure 71.1 shows the feedback structure that leads to changes in air travel demand: First, the endogenous factors such as the effect of the passenger load factor, airline fares and profits of the industry and second, the changes in demand due to the exogenous factors like increase in population growth and GDP per capita. The balancing loop B1 shows the balancing effect of an increase in passenger load factor that increases the congestion perceived by the customers, which in turn negatively affects the air travel demand. The balancing loop B2 represents the effect of passenger load factor and ticket prices on the demand. An increase in passenger load factor positively affects airline fares. An increase in airline fares negatively affects the demand as certain customers reduce travel due to high prices. The balancing loop B3 represents the change in the attractiveness of the industry as perceived by the airlines. An increase in airline fares increases the prospected profit due to an increase in operating revenues. The increasing profits prompts other companies to enter the market, which further results in a reduction of fares to gain a higher market share.

Figure 71.2 shows the feedback structure of the airline fleet operation mechanism. The effect of air travel demand and passenger load factor from Fig. 71.1 is also taken into account. The boundaries of the previous model are extended by including feedback mechanisms related to the introduction of SAFs and hydrogen in the system.

An increase in air travel demand increases the total desired capacity orders. This is due to the action taken by airlines to increase their aircraft orders to match the demand. From this point, the extensions made to the boundary are displayed in the form of a decision to choose hydrogen based aircraft or conventional aircraft. The decision to choose one over the other depends on the direct operating costs of each technology and the fuel availability. The cost including fuel, operation, ownership of new aircraft, jet fuel prices, inflation and improvements are treated as exogenous input. The reinforcing loop R2 represents increase in the capacity of hydrogen production as the airlines choose to order more hydrogen based aircraft. With an increase in orders by the airlines, the fuel producers are incentivized to produce more as the expectation of hydrogen fuel based aircraft usage in the future increases. In order to not have unrealistic exponential hydrogen fuel production growth, the reinforcing effect of loop R2 is balanced by the balancing loop B5. In loop B5, the increase in total seats on order hydrogen increases the hydrogen seats manufacturing capacity and the amount of hydrogen seats deliveries. The increase in deliveries negatively affects the total seats on order hydrogen, which represents the adjustments done by the airlines to not over-order the fleet. Similar to loops R2 and B5, for SAF, the increase in reinforcing loop R1 is balanced by the adjustments done by the airlines in balancing loop B4. Lastly, the reinforcing loop R3 represents the steady increase in the fleet size ordered by the airlines as the demand increases depicting the growth of the industry over time.

Figure 71.3 depicts the feedback structure of the fuel production subsystem of the model. In the figure, HEFA, FT synthesis and ATJ are the acronyms used for hydroprocessed fatty acid esters and fatty acids, fischer tropsch synthesis and alcohol to jet SAF production method, respectively. The reinforcing loops R4 and R5 represent the effect of increase over time in usage of hydrogen and SAF on its production capacity changes. An increase in the amount of hydrogen and SAF produced positively affects the annual use of hydrogen and SAF. The increase in use of the new fuels provides a positive incentive for the fuel manufacturers to increase the production capacities, which after certain delays increases the amount of hydrogen and SAF produced in the system. The reinforcing loops R4 and R5 are balanced by the balancing loop B6 and B7, which depict the reduction in growth of hydrogen and SAF fuel requirement if the two technologies are adopted over time. Since the emissions from hydrogen and SAF production depend on the path through which they are produced, the bolded variables represent the emission reduction potential of different paths of production as exogenous input.

Conclusion and Future Work

In this paper, the feedback structures of the air transport system (ATS) were described with the help of causal loop diagrams (CLD). The causal loop diagrams show that if there is an increase in the airline orders of SAF and hydrogen based aircraft then it leads to an increase in SAF and hydrogen fuel production capacity. An increase in airline orders and fuel availability would over time lead to an increase in aircraft manufacturing capacity thereby generating different adoption rates of the new technologies in the system. Comparison of different adoption rates shed light on the future emissions from the system. The future challenges in developing the model lie in gathering exogenous input data, modeling the airline's decision making mechanisms, generating different adoption scenarios and developing a stock and flow model that behaves in accordance with the real system.

Acknowledgements The authors would like to acknowledge the funding by the Deutsche Forschungsgemeinschaft (DFG, German Research Foundation) under Germany's Excellence Strategy—EXC 2163/1-Sustainable and Energy Efficient Aviation—Project-ID 390881007.

References

1. Staples, M. D., Malina, R., Suresh, P., Hileman, J. I., & Barrett, S. R. H. (2018). Aviation CO2 emissions reductions from the use of alternative jet fuels. *Energy Policy, 114*, 342–354. https://doi.org/10.1016/J.ENPOL.2017.12.007
2. Sterman, J. D. (2000). *Business dynamics: System thinking and modelling for a complex world*. McGraw Hill
3. Lyneis, J. M. (2000). System dynamics for market forecasting and structural analysis. *System Dynamics Review, 16*(1), 3–25. https://doi.org/10.1002/(SICI)1099-1727(200021)16:1%3c3::AID-SDR183%3e3.0.CO;2-5
4. Pierson, K., & Sterman, J. D. (2013). Cyclical dynamics of airline industry earnings. *System Dynamics Review, 29*(3), 129–156. https://doi.org/10.1002/sdr.1501
5. Liehr, M., Größler, A., Klein, M., & Milling, P. M. (2001). Cycles in the sky: Understanding and managing business cycles in the airline market. *System Dynamics Review, 17*(4), 311–332. https://doi.org/10.1002/sdr.226
6. Kieckhäfer, K., Quante, G., Müller, C., Spengler, T., Lossau, M., & Jonas, W. (2018). Simulation-based analysis of the potential of alternative fuels towards reducing CO2 emissions from aviation. *Energies, 11*(1), 186. https://doi.org/10.3390/en11010186

Chapter 72
Do Artificial Agents Reproduce Human Strategies in the Advisers' Game?

Maximilian Moll, Jurgis Karpus, and Bahador Bahrami

Abstract Game theory has been recently used to study optimal advice-giving strategies in settings where multiple advisers compete for a single client's attention. In the advisers' game, a client chooses between two well informed advisers to place bets under uncertainty. Experiments have shown that human advisers can learn to play strategically instead of honestly to exploit client behavior. Here, we analyze under which conditions agents trained with Q-learning can adopt similar strategies. To this end, the agent is trained against different heuristics and itself.

Keywords Reinforcement learning · Game theory · Decision making

Introduction: The Advisers' Game

Should we trust the results of search engines and other algorithmic advisers when multiple recommenders compete for our attention? As we will see, the answer depends on when was the last time that we used a particular recommender instead of its competitors. When two advisers compete for a single client's attention, honest advising is good when you already are the client's favored adviser. When you are not the favoured adviser, game theory shows that you should issue advice strategically, not honestly [1]. To show that, we use the advisers' game, which is played by three agents—two advisers and a client. In the beginning, the client chooses an adviser to place a bet on the outcome of a Bernoulli experiment. Then, the two advisers learn the underlying probability and both independently communicate a probability for this outcome to the client, which may or may not be the objective probability that they saw. The client must follow the chosen adviser's recommendation, but observes

M. Moll (✉)
Universität der Bundeswehr München, Werner-Heisenberg-Weg 39, 85577 Neubiberg, Germany
e-mail: maximilian.moll@unibw.de

J. Karpus · B. Bahrami
Ludwig-Maximilians-Universität München, Geschwister-Scholl-Platz 1, 80539 München, Germany

O. Grothe et al. (eds.), *Operations Research Proceedings 2022*, Lecture Notes in Operations Research, https://doi.org/10.1007/978-3-031-24907-5_72

the non-chosen adviser's recommendation, too. Lastly, the lottery is resolved and the result is observed by the client, who can then decide whether to stick with their current adviser or switch advisers for the following round. This is reiterated for multiple rounds.

Preliminary game-theoretic analysis of the problem (the problem turns out to be too complex for a complete analytic solution) and previous experiments with this game have shown that when an adviser is not selected by a human client, it often pays to issue advice strategically, sometimes suggesting to bet on the colour that is less likely to win [2]. We will refer to this strategy as *strategic* from now on. In this paper, we investigate what happens if the adviser is trained by Reinforcement Learning (RL). To avoid the complications of training on thousands of human clients, an established formula for updating the client's trust in an adviser is used instead [1]. The idea of using RL to find optimal strategies in (rational solutions of) games is not new [3–5]. The case here is of particular interest, since the exact game-theoretic solution of the problem is not known [2]. While, in principle, it can be established by backward induction, the length of the game makes this process very complex and thus only approximations of the exact solution are used.

Methods: RL Setup for the Advising Game

In RL a typical problem formulation assumes that the *agent* has no information about the inner workings of the task it is to solve, which is usually encapsulated in the *environment*. Instead it needs to learn to do so by data-based interaction with the latter in a succession of time-steps $t = 0, \dots, T$. In doing so, the environment reports its current *state* $s_t \in \mathcal{S} \subset \mathbb{R}^n$ to the agent, which in turn selects and submits its next *action* $a_t \in \mathcal{A} \subset \mathbb{R}^m$. In response the environment updates itself and returns a new state s_{t+1} and *reward* r_t. The agents goal is to find a *policy* $\pi : \mathcal{S} \to \mathcal{A}$ which maximizes the return $\sum_{t=0}^{T} \gamma^t r_t$ for a given $\gamma \in (0, 1]$. The dynamics of the problem are captured by the state transition $(s_t, a_t) \mapsto (s_{t+1}, r_t)$, which can be probabilistic. The definition of the policy is usually extended to include mappings from the state space to distributions over the action space, from which the action is then being sampled. For more details the interested reader is referred to [6].

In the basic version of the advisers' game, the state consists of an indicator for the selected adviser and the objective probability of winning from placing a bet on the black colour that the advisers observe. Here, the decision was taken to use multiples of $\frac{1}{11}$ to avoid a central 50% and allow for some flexibility while maintaining a reasonably sized state space. To focus training on situations, in which the strategic adviser shows interesting behavior, the possible probabilities were limited to be between $\frac{3}{11}$ and $\frac{8}{11}$. The action consists of the probability that is communicated to the client. Here, we decided to exclude the communication of certainty, leading to the 10 actions $\frac{1}{11}, \dots, \frac{10}{11}$.

After both advisers reported their chosen action, the result is sampled from a Bernoulli distribution according to the probability given in the state and used to update the trust of the client in both advisers. Here, we use the formula proposed in [1]. Initially, the trust in both advisers is equal, i.e. $w = (0.5, 0.5)$. If p_s and p_o are the reported probabilities of the selected respectively other adviser and w_s and w_o are the corresponding trust values, then if the result is 1

$$w_s' = \frac{w_s \cdot p_s^2}{w_s \cdot p_s^2 + w_o \cdot p_o^2}, \quad w_o' = 1 - w_s'$$

and otherwise

$$w_s' = \frac{w_s \cdot (1 - p_s)^2}{w_s \cdot (1 - p_s)^2 + w_o \cdot (1 - p_o)^2}, \quad w_o' = 1 - w_s'$$

The adviser for placing a bet in the following round of the game is then chosen based on the larger trust. If the trust values are equal, the previously chosen adviser is kept. A reward of 1 is given to the agent selected for the next round, and 0 otherwise. In a single iteration of the game this is repeated for 20 rounds, and thus the goal of the agent is to be selected by the client as often as possible during this time frame.

It can be seen that the client internally updates the trust based on its old value and thus implicitly considers the whole history during each iteration. Thus the next state and reward depend on more than the previous state and action. This violates the Markov Property which underpins RL and initial experiments confirmed that this causes issues. However, including a full history in each state would lead to a large state space, which would cause issues for later stages of this investigation, when human clients will be used. As a consequence, the state space was revised to include just a one step history, which indicated which agent was chosen in the previous round or, in the case of the first round, that there is no previous history. This leads to a total of 36 states, the combination of 3 possible previous states, 2 current states, and 6 observable probabilities.

Given the manageable sizes of state and action spaces, tabular Q-learning was implemented for the RL agent. The goal of Q-learning is to find an estimate to the action-value function $Q_t(s, a) = \mathbb{E}_\pi \left[\sum_{\tau=t}^{T-1} r_\tau | s_t = s, a_t = a \right]$, where the expected value is taken with respect to the policy the agent is following. Thus, this function indicates how good a given action in a given state is. It can be shown [8] that we can find the action-value function for the optimal policy not only by iteration, but also that we can use its current version for a bootstrapped value of the return similar to the Bellman principle:

$$Q_t(s, a) \leftarrow Q_t(s, a) + \alpha \left(r + \gamma \max_a Q_{t+1}(s_{t+1}, a) - Q_t(s, a) \right),$$

where α is the learning rate. Typically, Q_t is initialized as $Q_t(s, a) = 0, \forall s, a, t$. In our experiments we set $\gamma = 1$. It should be noted that many implementations ignore the time index and use the same function for all time steps. While not correct in the setting here, it can be done practically, but the values of the function will lose their absolute meaning.

For exploration, two approaches are being compared: The first one, an epsilon-greedy policy, is more commonly used in tandem with Q-learning. It selects the action maximizing $Q_t(s, \cdot)$ with probability ϵ and chooses a random action with probability $1 - \epsilon$. In our experiments, a constant $\epsilon = 0.5$ was chosen. It should be noted that more elaborate cooling schemes are possible and can be beneficial. The second approach applies the softmax function

$$x_i \mapsto \frac{e^{x_i}}{\sum_{j=1}^{m} e^{x_j}}, \quad i = 1, \ldots, m$$

to $Q(s, \cdot)$ and samples the action from the resulting discrete probability distribution. As the differences in value increase, these distributions place more and more mass on the best action making exploration less likely. Thus, a natural cooling scheme occurs.

In addition to the RL agent, the honest and the aforementioned strategic approach are implemented as straightforward heuristics as well.

For the experimental comparisons, we investigated the learning rates $\alpha \in \{0.1, 0.01\}$, number of iterations $N \in \{10{,}000, 50{,}000, 100{,}000\}$ and the two different forms of policy-epsilon-greedy and softmax. For all of these, the simplification of assuming a single Q-function independent of time was taken. For the above mentioned reasons, however, long training is not possible in the softmax case, as the values would become overly large. Thus, for the softmax exploration, the setting with 20 Q-functions is explored as well. Each agent is trained against either the honest, the strategic heuristic or against itself.

For the evaluation the trained agent plays 1000 iterations each against the honest and the strategic adviser respectively and the average return is recorded. In the case of the epsilon-greedy policy, exploration is suppressed and the deterministic policy selecting the best action is used. For the softmax policy the action continues to be sampled to allow for stochastic policies.

Results and Discussion: Emergence of a Better Strategy

The results of the experiments for 100,000 iterations and a learning rate of 0.1 are shown in Table 72.1. For the softmax agents with a single action-value function higher iteration numbers were often not possible due to the aforementioned overflow issue and 5000 iterations had to be used instead.

As can be seen from the results, the agent can only learn to beat the honest heuristic by training against it. This shows the power of being honest, when we are limited to less certain cases. The problem of the strategic heuristic is that often the risk does not pay off. This means the agent can learn a weaker strategy that is good enough to

Table 72.1 Excerpt of results—the training column shows whether the agent was trained against the honest, strategic heuristic or by self-play

Policy	Training	Learning rate	Iterations	Test honest	Test strategic
ϵ-greedy	Honest	0.1	100,000	15.456	12.345
ϵ-greedy	Strategic	0.1	100,000	3.848	14.96
ϵ-greedy	Self	0.1	100,000	6.15	11.673
Softmax single	Honest	0.1	10,000	13.946	14.094
Softmax single	Strategic	0.1	5000	4.477	14.877
Softmax single	Self	0.1	5000	6.373	14.044
Softmax multi	Honest	0.1	100,000	12.236	13.696
Softmax multi	Strategic	0.1	100,000	3.974	14.337
Softmax multi	Self	0.1	100,000	5.7	12.715

The test columns state the performance during the evaluation against the respective heuristic after training. Values larger than 10 mean that the agent outperforms the heuristic

beat the strategic heuristic, but struggles against honest play. The self-play setting, however, requires more analysis, since training the agent on beating earlier versions of itself, as suggested in for example [7], could improve this.

Next, we can look at the strategy that the agent arrived at. Figure 72.1 shows the action-value functions for the softmax agent, as they show the results most clearly. Similar behavior can, however, be observed with all successful agents. First, we analyse behavior when the agent is selected. We can see that it reports the observed probability truthfully, if it has been selected twice (or more) in a row (a). This makes sense, since once repeated trust is established, it is difficult to lose sufficient trust for a change of adviser, so any risk taking is unnecessary.

If the agent is selected in the current but not in the previous round (d), we see that the behavior depends on the observations. If the probabilities are more certain, i.e., closer to 0 or 1, the agent reports them truthfully. This again has the idea to avoid extreme risk of losing trust. For probabilities close to 0.5, however, we can see that the agent is not quite truthful: in one case, the agent reports a more extreme probability, in the other, it reports a probability biased towards the opposite outcome. Both can be seen as the attempt to increase the trust it has gained with fairly low risk.

In the cases in which the agent is not selected, we can observe very strategic behavior. The most extreme case of this happens, when it was not selected for at least two rounds in a row (c). Here we can see, that the agent always advises with near certainty to bet on the colour that is less likely to win, independently of how small the chances are for that to happen. This behavior is reasonable since, were the unlikely event to occur, it would increase the change of trust in the RL agent's favor. At the same time, since there is a level of saturation in the trust formula—increasing numbers of sequential successes will change the trust increasingly little—there is not much to lose.

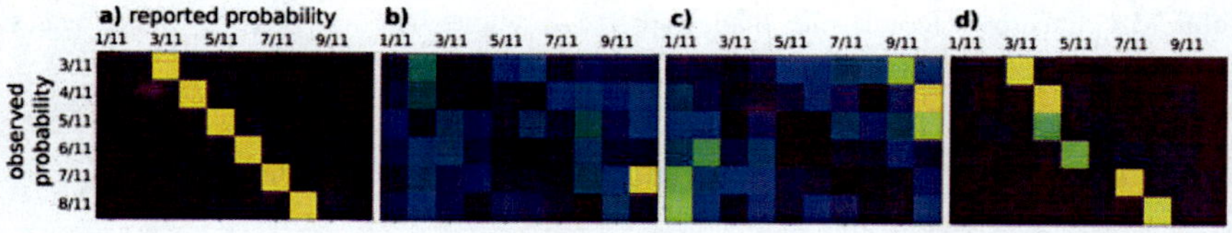

Fig. 72.1 Action-value plots for the four interesting scenarios: **a** selected twice (or more) in a row, **b** selected in the previous but not the current round, **c** not selected for at least two rounds in a row, **d** selected in the current but not the previous round. Brighter colors indicate higher values

In the final case, when the agent was selected in the previous but not the current round (b), we can see again a mixture of behavior. For less decisive probabilities it re-iterates the same strategy as explained in the previous case. However, since the adviser has just changed, the trust values are both still fairly close to 0.5 and thus, taking extreme risks is not justified. That said, the agent does not report extreme probabilities truthfully, but instead exaggerates them. This measure again increases the trust in it over the other agent, if the correct outcome is chosen.

Conclusions

In summary, our agent improves on the strategic heuristic that has been used in the analysis of the advisers' game before. In future work, we will test whether increasing the agent's running memory of past rounds would lead to even better strategies. This memory increase can be contrasted with an approach based on partially-observable Markov decision processes. Another approach will be to replace the simple tables with a neural network including a Long-Short-Term-Memory unit, enabling the agent to maintain a full recollection of the iteration. Finally, we plan to investigate the RL agent's performance in game play with human clients, which replaces the simplistic client formula with a much more complex (and realistic) process.

Acknowledgements Jurgis Karpus was supported by LMUexcellent, funded by the Federal Ministry of Education and Research (BMBF) and the Free State of Bavaria under the Excellence Strategy of the Federal Government and the Länder. B. B. was supported by the Humboldt Foundation and the European Research Council (ERC) under the European Union's Horizon 2020 research and innovation program (819040-acronym: rid-O).

References

1. Hertz, U., Palminteri, S., Brunetti, S., Olesen, C., Frith, C. D., & Bahrami, B. (2017). Neural computations underpinning the strategic management of influence in advice giving. *Nature Communications*, *8*, 2191. https://doi.org/10.1038/s41467-017-02314-5
2. Kruvers, R. H. J. M., Hertz, U., Karpus, J., Balode, M., Jayles, B., Binmore, K., & Bahrami, B. (2021). Strategic disinformation outperforms honesty in competition for social influence. *Nature Communications*, *24*(12), 103505 (2021). https://doi.org/10.1016/j.isci.2021.103505

3. Lanctot, M., Lockhart, E., Lespiau, J. B., Zambaldi, V., Upadhyay, S., Pérolat, J., … , & Ryan-Davis, J. (2019). *OpenSpiel: A framework for reinforcement learning in games*. arXiv preprint arXiv:1908.09453
4. Mukhopadhyay, S.,Tilak, O., & Chakrabarti, S. (2018). Reinforcement learning algorithms for uncertain, dynamic, zero-sum games. In *17th IEEE International Conference on Machine Learning and Applications (ICMLA)* (pp. 48–54). https://doi.org/10.1109/ICMLA.2018.00015
5. Ni, Z., & Paul, S. (2019). A multistage game in smart grid security: A reinforcement learning solution. *IEEE Transactions on Neural Networks and Learning Systems, 30*(9), 2684–2695. https://doi.org/10.1109/TNNLS.2018.2885530
6. Sutton, R. S., & Barto, A. G. (2018). *Reinforcement learning: An introduction*. Cambridge: MIT Press.
7. Vinyals, O., Babuschkin, I., Czarnecki, W. M., Mathieu, M., Dudzik, A., Chung, J., ..., & Silver, D. (2019). Grandmaster level in StarCraft II using multi-agent reinforcement learning. *Nature, 575*(7782), 350–354. https://doi.org/10.1038/s41586-019-1724-z
8. Watkins, C. J. C. H. (1989). *Learning from delayed rewards* (Ph.D. thesis). University of Cambridge, England.

Chapter 73
GTRF: Generalized Trade Reduction Framework for Double-Auction Mechanisms

Jacob Ehrlich, Maximilian Moll, and Stefan Pickl

Abstract In a groundbreaking paper McAfee introduced the Trade Reduction (TR) Mechanism that circumvents the famous Myerson and Satterthwaite impossibility result by sacrificing a small amount of efficiency. Here the author creates order statistics based on the submitted bids and reduces at most the least efficient trade. Based on this principle an alternate mechanism was proposed by Segal-Halevi et al. which extends this to the strongly budget balanced setting. This paper proposes a generalization of these two TR mechanisms to fit into a larger framework that can be implemented based on the market in which the auction is to be applied. Additionally, by taking advantage of the relationship of bid order-statistics a novel mechanism; titled BORS, is revealed to complete the GTRF. Using a simulation based evaluation, performance is characterized across various settings in order to achieve optimized results.

Keywords Auctions/competitive bidding · Simulation

Introduction

Mechanism design exists as a tool to clear complex trading markets. This paper considers two-sided markets consisting of buyers interested in, and sellers in possession of a single identical unit an indivisible g. The goal is to design a mechanism that solves two simultaneous problems with a set of coupled rules. These problems are the Allocation Problem: Who gets to trade in the market, and Payment problem: The fair price to be rendered for a good/service, g. Buyers/sellers have private valuations/reservation of g that act as a maximum/minimum they would be willing to pay/receive. Potential for trade exists when a both sides of the market can obtain a positive gain from a trade, i.e. buyers value g more than the sellers. Mechanisms can posses certain desired economic properties: Individual Rationality (IR), Incen-

J. Ehrlich (✉) · M. Moll · S. Pickl
Universität der Bundeswehr München, 85579 Neubiberg, Germany
e-mail: jacob.ehrlich.1@us.af.mil

O. Grothe et al. (eds.), *Operations Research Proceedings 2022*, Lecture Notes in Operations Research, https://doi.org/10.1007/978-3-031-24907-5_73

tive Compatibility (IC), and Budget Balance (BB). A measure of effectiveness for a mechanism is the markets economic efficiency as determined by the gains from trade (GFT) which sums up all gains from trade across the market. A mechanism that achieves the optimal GFT, thereby leaving no more value in the market to be gained is referred to as ideally efficient. Comparing the achieved GFT from a given mechanism to the ideally efficient mechanism for a market generates a ratio of efficiency. Efficiency can be interpreted as the percentage of ideal GFT a mechanism achieves, where the compliment is the percentage of gains left in the market.

For the bilateral trade setting [7] proved it impossible for a mechanism to be ideally efficient while also maintaining the desired properties of IR, IC, and BB. This impossibility result is circumvented by [6] in what was later deemed the trade reduction (TR) mechanism. McAfee recognized that despite the sacrifice of ideal efficiency, one can still optimize to achieve a high efficiency while satisfying the properties of IR, IC, and BB. This was later extended to the more rigorous strong Budget Balance (SBB) by [10]. This paper aims to introduce an additional mechanism, titled BORS, that is robust to market saturation to the existing menu of TR mechanisms. Additionally a Framework is proposed that generalizes and combines the entire menu allowing a mechanism designer to customize the TR mechanism according the market in which the mechanism is to be implemented.

Taxonomy and Notation

The set of all potential traders are split into two disjoint sets **N** and **M** denoting buyers and sellers respectively. Additionally, define the disjoint sets of winning traders as **T** and loosing traders as $\mathbf{T}'$ as determined by the allocation rule A such that $\mathbf{T} \cup \mathbf{T}' = \mathbf{N} \cup \mathbf{M}$. Buyers strategically submit bids that are organized according to their competitiveness into order statistics $\{b_1, \ldots, b_n\} = \mathbf{b}$ representing their private valuations. Similarly seller bids are organized according to their competitiveness into order statistics $\{s_1, \ldots, s_m\} = \mathbf{s}$ and represent private reservations prices. Consider non-discriminatory pricing rules such that all winning buyers and seller $\in T$ pay p_b and receive p_s. The unique coupling of the allocation $A(\mathbf{b}, \mathbf{s}) = \mathbf{T}$ and a payment rule $P(\mathbf{b}, \mathbf{s}) = \mathbf{p} = (p_b, p_s)$ is defined as a mechanism $X(\mathbf{b}, \mathbf{s}) = (\mathbf{T}, \mathbf{p})$.

Related Work and Extensions

The foundation of this paper stems from the seminal trade reduction (TR) mechanism by [6]. The principle concept behind the McAfee mechanism is to reduce the efficient number of traders by at most the least efficient/favorable trade (LFT). This mechanism achieves IR, IC, BB and allows for asymptotic efficiency to be achieved as at most one trade, the LFT, is reduced. As the number of traders increase, the value lost from the LFT becomes inconsequential to the measure of market efficiency.

Table 73.1 Existing trade reduction mechanisms

Mechanism	Prior-free	Budget	Efficiency
Trade reduction [6]	Yes	Surplus	$1 - \frac{1}{k}$
Random sampling [4]	Yes	Balanced	$1 - O\sqrt{\frac{\ln(k)}{k}}$
VCG-TR [3]	No	Balanced * in expectation	$1 - \frac{1}{k}$ * in expectation
SBBA [10]	Yes	Balanced	$1 - \frac{1}{k}$
DA-NLC [1]	Yes	Balanced	$1 - \frac{1}{k}$

In [3] the authors introduce a SBB, IC, IR mechanism that meets efficiency in expectation equal to that of [6] however requires the use of prior distributions along which valuation (reservation) prices fall. These distributions are assumed to be public knowledge.

A random sampling approach is taken by [4] and extended to a prior-free setting by [9]. The general concept is to halve the market and use each half to calculate the trading prices of the other. To achieve SBB, the authors introduce randomization in the form of a lottery price, which has a slower rate of convergence toward asymptotic efficiency as compared to [6].

An alternate IR, IC, SBB mechanism called SBBA is introduced by [10] with expected GFT equal to that of McAfee $1 - \frac{1}{k}$. In SBBA randomization is introduced in the allocation rule A for the case in which a trade reduction is necessary. This is extended to the multi-unit case in [11] and extended to the multi-lateral case in [5].

An alternate randomized extension is presented in [1] this mechanism introduces randomization in the payment rule P which comes at the cost of IC. An additional no loss constraints (NLC) is required for Strategy-proofness limiting bidders to not consider strategies that could in any way result in a negative utility (Table 73.1).

BORS Mechanism

Examining the relationship between the LFT and the least inefficient traders, b_{k+1}, s_{k+1}, allows for the revelation of the circumstances in which existing TR mechanisms guarantee efficient trades. Through this lens an undiscovered TR mechanism is presented.

$$X_\psi = \begin{cases} s_{k+1}, b_{k+1} \in [s_k, b_k] & (\boldsymbol{\Gamma} - \{b_k, s_k\}, \{b_k, s_k\}) \\ s_{k+1} \in [s_k, b_k], b_{k+1} \notin [s_k, b_k] & (\boldsymbol{\Gamma}, \{s_{k+1}, s_{k+1}\}) \\ b_{k+1} \in [s_k, b_k], s_{k+1} \notin [s_k, b_k] & (\boldsymbol{\Gamma}, \{b_{k+1}, b_{k+1}\}) \\ s_{k+1}, b_{k+1} \notin [s_k, b_k] & (\boldsymbol{\Gamma} - \{b_k, s_k\}, \{b_k, s_k\}) \end{cases} \tag{73.1}$$

The proposed mechanism, entitled BORS, achieves the same desired properties of IR, IC, BB, and has asymptotic efficiency of $1 - \frac{1}{k}$ as at most the LFT is reduced. Finally, the mechanism is deterministic and prior-free. The mechanism gains it's name by implementing either the least inefficient buyer b_{k+1} or seller s_{k+1} but not both, b or s.

Generalized TR Framework

An additional finding is detected at this stage, namely that although the McAfee TR rule is used here, the mechanism could also be implemented with an SBBA rule if SBB is favored over determinism. This key finding suggests a framework can be built based on the specifications of a market designer. The TR mechanism can be broken into two components, the initial pricing option that guarantees efficient trade, and the rule by which a trade reduction is implemented. As the initial pricing option for SBBA has the potential to result in a price equal to on of the LFT, the SBBA mechanism cannot be paired with the McAfee TR rule. However, because the McAfee and the BORS mechanism presented above do not depend on any efficient trader, these mechanisms can be paired with the SBBA TR rule. The only remaining question then becomes under what conditions a pricing mechanism may be preferable to any other.

Market Simulation

While considerable work can be done to more directly characterize the settings under which each pricing rule results in higher expected efficiency, the scope of this paper is to show that any given pricing mechanism may be preferred based on the market in which it is to be implemented. Four settings have been identified to present the performance of the three TR mechanism: McAfee, SBBA, and BORS. For any given market the range of valuations and reservations can be normalized [0, 1] (Table 73.2).

Table 73.2 Simulation settings

	Setting	Buyers distribution	Sellers distribution	Likelihood of buyer
a	No restrictions	Uniform (0, 1)	Uniform (0, 1)	0.5
b	Buyers > sellers	Triangular (0, 0.75, 1)	Uniform (0, 0.25, 1)	0.5
c	Sellers saturation	Triangular (0, 0.75, 1)	Uniform (0,0.25,1)	0.25
d	Buyers saturation	Triangular (0, 0.75, 1)	Uniform (0,0.25,1)	0.75

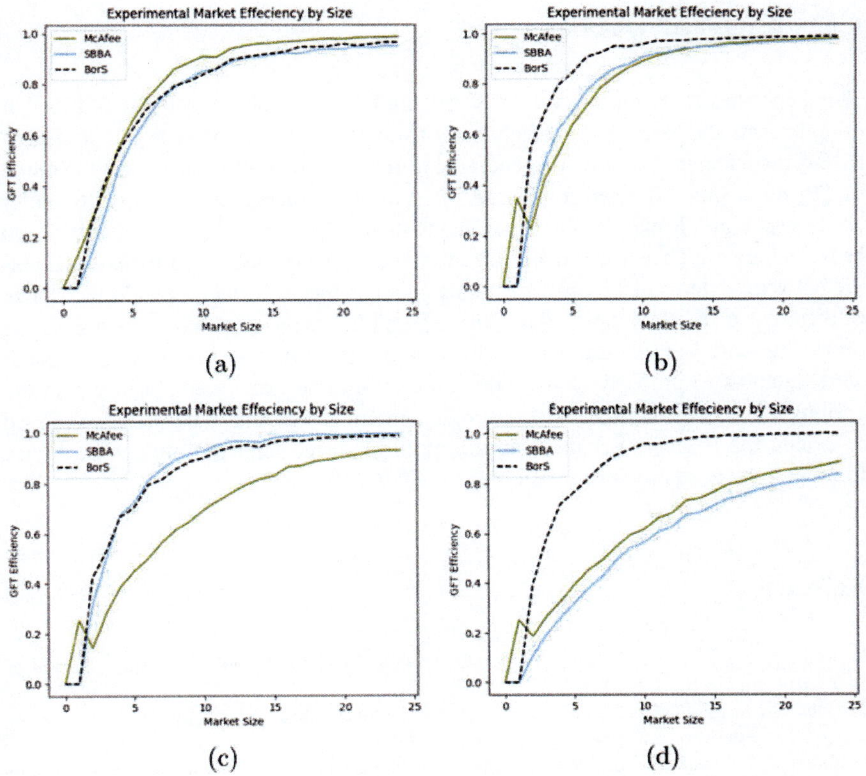

Fig. 73.1 **a** No restrictions **b** Buyers > sellers **c** Sellers saturation **d** Buyers saturation

In the first setting no restrictions are placed along this range and both buyers and sellers are drawn from a Uniform distribution with equal likelihood of a trader being assigned as either a buyer or seller. The second setting aims at making a more realistic assumption that buyers will tend to value g more than the sellers such that a potential for efficient trades exist. To showcase the scenario a simple triangular distribution is used with buyers centered at 0.75 and sellers at 0.25. Traders are equally likely to be assigned to the set of buyers or sellers. The final two scenarios aim at examining the situation in which the market is saturated by one side. Here the likelihood of being assigned to the dominant set becomes 0.75.

Overall this shows the McAfee mechanism produces consistent results as it does not favor one side of the market. As seen in Fig. 73.1a this mechanism acts as the standard. Contrarily the SBBA mechanism requires the market designer to select a mechanism that favors one side of the market. If sufficient prior knowledge of the market exists, this can work in favor of the designer as evidenced in Fig. 73.1c. Finally the proposed BORS mechanism offers similar results to SBBA in Fig. 73.1a, c, but clearly shows substantial improvement over existing mechanism in the case of buyer saturation as seen in Fig. 73.1d.

Conclusion and Outlook

This paper has revealed a novel trade reduction (TR) mechanism titled BORS that compliments the existing TR mechanism from McAfee and the recently published SBBA mechanism. Secondly, a generalization was made that incorporates the menu of TR into a single defined framework. This allows a mechanism designer to better apply a specific TR mechanism according to the market in which the mechanism is to be implemented. The framework also gives the designer the flexibility to make a trade-off between determinism and SBB. This paper is meant to highlight the existence of different markets in which the various possible baseline TR mechanisms can be preferable based on the efficiency with which they clear the market. Considerable work remains to properly characterize and outline the exact settings in which the various TR mechanisms should be implemented. Additional next steps would be to extend the BORS mechanism to the multi-unit, and later the multi-lateral case following the academic progression of the SBBA mechanism.

References

1. Anbarci, N., & Roy, J. (2018). Double auctions with no-loss constrained traders. *Journal of Theory and Decision, 84*, 1–9.
2. Babaioff, M., & Nisan, N. (2004). Concurrent auctions across the supply chain. *Journal of Artificial Intelligence Research, 21*, 595–629.
3. Babaioff, M., Nisan, N., & Pavlov, E. (2004). Mechanisms for a spatially distributed market. In *ACM Conference on Electronic Commerce (EC'04)* (pp 9–20).
4. Baliga, S., & Vohra, R. (2003). Market research and market design. In *Advances in theoretical economics* (Vol. 3). De Gruyter.
5. Gilor, D., Gonen, R., & Segal-Halevi, E. (2021). Strongly budget balanced auctions for multi-sided markets. *Journal of Artificial Intelligence*, 103548.
6. McAfee, P. R. (1992). A dominant strategy double auction. *Journal of Economic Theory, 56*, 434–450.
7. Myerson, R., & Satterthwaite, M. (1983). Efficient mechanisms for bilateral trading. *Journal of Economic Theory*, 29:265–281.
8. Roughgarden, T.: *Twenty lectures on algorithmic game theory*. Cambridge University Press (2016)
9. Segal-Halevi, E., Hassidim, A., & Aumann, Y. (2016). *MIDA: A multi item-type double-auction mechanism*. arXiv preprint arXiv:1604.06210.
10. Segal-Halevi, E., Hassidim, A., & Aumann, Y. (2016). SBBA: A strongly-budget-balanced double-auction mechanism. In *International Symposium on Algorithmic Game Theory* (pp 260–272). Springer.
11. Segal-Halevi, E., Hassidim, A., & Aumann, Y. (2018). Muda: A truthful multi-unit double-auction mechanism. In *Thirty-Second AAAI Conference on Artificial Intelligence*.

Chapter 74
Iterated Boxed Pigs Game: A Reinforcement Learning Approach

Rudy Milani, Maximilian Moll, and Stefan Pickl

Abstract This paper analyzes the iterated version of the well-known Boxed Pigs game through Reinforcement Learning. In this scenario, there are two differently sized players (pigs) that compete against each other. The core idea is about sacrificing a pay-off in order to generate some rewards. In our iterated version, these pigs play this game repeatedly using different strategies. We carry out two experiments: in the first one, we train two Q-learning agents against each other to see which equilibrium will be generated. In the second one, we pit the Reinforcement Learning agent against a fixed policy pig. The results of this experiment confirm the ability of Reinforcement Learning techniques in finding the best strategy for maximizing the return independently from the other player choices.

Keywords Simulation · Prescriptive analytics · Machine learning

Introduction

In this paper, we focus our attention on the Boxed Pigs game, which was first described in [1]. It is, in particular, well-known in the economics literature as an example of where weakness can be a strength. A general form of Boxed Pigs is described as follows: there are two pigs, a big one and a piglet, in a box. On one side, there is a lever and, on the other side, is a trough. The lever controls the food that will be in the trough. Since there is a difference in the dimensions of the animals, the game is unbalanced to one side.

An interesting interpretation of this game is given by the relations between world-leading companies in high-tech and smaller societies that are competing in the same field. The latter prefer waiting for new features discoveries from the bigger ones to copy them and have an advancement without any cost of research. However, there are

R. Milani (✉) · M. Moll · S. Pickl
Fakultät für Informatik, Universität der Bundeswehr München, Werner-Heisenberg-Weg 39, 85577 Neubiberg, Germany
e-mail: rudy.milani@unibw.de

O. Grothe et al. (eds.), *Operations Research Proceedings 2022*, Lecture Notes in Operations Research, https://doi.org/10.1007/978-3-031-24907-5_74

also other relevant applications of this abstract problem, e.g., the analogy described by McMillan [2] with the oil market (OPEC), where Saudi Arabia can be considered as the big pig and the smaller producers as the piglet. For these reasons, it is interesting to understand how the game changes if iterated over time. In the one-step game, the result is straightforward: the big pig has to press the lever and the small one will wait. In fact, there exists a dominant strategy for the piglet that consists of waiting instead of pushing the lever. This leads the great pig to press the lever or they will not receive anything.

Furthermore, this kind of problem can be also easily generalized by repeating these steps in time, and can become more complex by adding new parameters, e.g., life span and penalty parameter. However, in this research we focus only on the simplest scenario (no life span and penalty parameter). One major example of simple generalization is the iterated version of Prisoner's Dilemma [3].

An interesting approach consists of applying Reinforcement Learning techniques to this iterated game [4], since Game Theory and Reinforcement Learning are strictly correlated because of their same goal: producing an optimal policy for a given task. The major difference is in the way these strategies are obtained. While for Game Theory we need to find an equilibrium by analyzing the possible outcomes for each player action, in the case of Reinforcement Learning we derive the best strategy through data accumulated during the simulation of that particular problem. For this reason, Reinforcement Learning deals with Markov Decision Processes (MDPs). However, it is possible to compare the Game Theory results with the ones obtained from Reinforcement Learning when we study problems that can be generalized as MDPs, e.g., iterated games.

For these reasons, the purpose of this paper is to analyze the iterated version of the Boxed Pigs game through Reinforcement Learning. In particular, we want to understand if the equilibrium achieved by two Reinforcement Learning pigs, that play against each other, will be the same as in the normal scenario, and recognize the ability of these agents when they are facing fixed strategies opponents.

Theoretical Background

In the context of Reinforcement Learning (RL) we have an agent which is the learner and decision maker, which means that it has to understand how to use the information of the state to choose the best action. Everything that interacts with the agent is called environment.

Formally, at each discrete time step $t = 0, 1, 2, \ldots$ the agent receives the actual state $s_t \in \mathbb{S} \subseteq \mathbb{R}^n$ from the environment, and decides on an action $a_t \in \mathbb{A} \subseteq \mathbb{R}^m$, where $\mathbb{S}$ and $\mathbb{A}$ are the state and action spaces respectively [5]. As a consequence of the action, the agent will receive a numerical reward $r_t \in \mathbb{R}$, and the environment will update to a new state s_{t+1}. The agent has to choose actions to maximize the sum of rewards $R_t = \sum_{i=0}^{\infty} \gamma^i r_{t+i+1}$ where $\gamma \in [0, 1)$ is a discounting factor. In order to have a finite sum, it is possible to define all the rewards after a final time step T equals to zero: $r_t = 0$ for $t > T$ [5]. Fundamental for the description of the following algorithm is the *Q-function* $Q_\pi(s, a) = \mathbb{E}_\pi\left[\sum_{k=0}^{\infty} \gamma^k r_{t+k+1} \middle| s_t = s, a_t = a\right]$ evaluated under the policy π, which is defined simply as a distribution over actions given states: $\pi(a|s) = \mathbb{P}(a_t = a|s_t = s)$, where $\mathbb{P}(\cdot)$ represent the probability. The RL method we study here is *Q-learning* [6], a model-free and off-policy method. At the beginning of the learning process a Q-table ($Q(s_t, a_t) \approx Q_\pi(s_t, a_t)$), with dimensions defined by the number of the states and actions $|\mathbb{S}| \times |\mathbb{A}|$, is initialized as a matrix of zeros. The agent will always look at this matrix to determine its action using an $\epsilon - greedy$ strategy, where with probability $\epsilon \in [0, 1]$ a random action is chosen, otherwise, the action maximizing the Q-value for that state is performed. After taking an action, the Q-table is updated using the following rule:

$$Q(s_t, a_t) \leftarrow Q(s_t, a_t) + \alpha\left[r_{t+1} + \gamma \max_{a \in \mathbb{A}} Q(s_{t+1}, a) - Q(s_t, a_t)\right], \tag{74.1}$$

where $\alpha \in (0, 1]$ is the learning rate.

Related Work

To the best of our knowledge, this is the first paper investigating the iterated version of the Boxed Pigs game. However, there are plenty of works associated with iterative games, e.g., for the iterated Prisoner's Dilemma, and its solutions obtained using Reinforcement Learning techniques [4, 7]. In the particular case of the Prisoner's Dilemma, there is a Nash equilibrium given by the defection of the players which leads to a worse reward for both, rather than collaborating. In the iterated version, the usual solution is given by the cooperation until one of them decides to betray. After that action, the other player will always defect. In the context of Boxed Pigs game, we have also a Nash equilibrium which corresponds to the best reward of the small pig while there are better possibilities for the bigger pig. Consequently, the asymmetry of the game that is presented in this case is a major difference. For these reasons, we can not directly transfer the results achieved in the Prisoner's Dilemma.

Methods and Discussion

In this scenario, we consider as state the encoded previous actions taken by the two pigs, i.e., the tuples $s_t = (a^B_{t-1}, a^s_{t-1})$ with $a^B_{t-1}, a^s_{t-1} \in \{l, w\}$, where a^B_{t-1} represents the big pig action, and a^s_{t-1} the piglet move; while l indicates the action "press the lever" and w "wait". Then, the action space is given by the combination of the two possible actions between the two pigs. The asymmetry of the game between the big pig and the piglet is represented in the different rewards collected by them in the same situations. To provide the meal, at least one of the pigs must press the lever, or

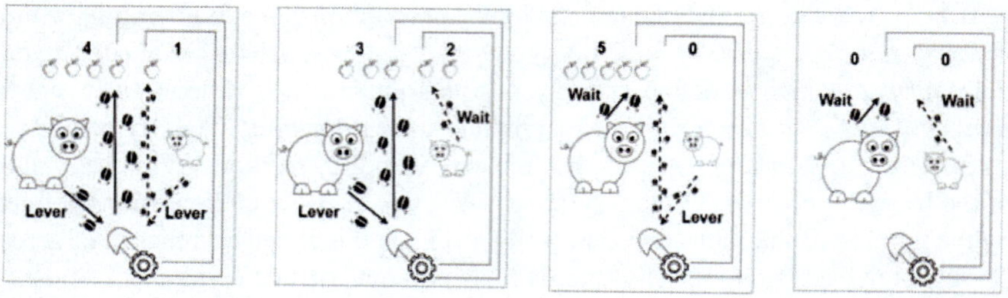

Fig. 74.1 Representation of the boxed pigs game. The piglet's arrows are dashed to show that it is slower than the big pig

nothing will happen, consequently, both rewards are $r_t^B = r_t^s = 0$, where, r_t^B and r_t^s represent the rewards at time step t of, respectively, the big pig and the piglet. If both the pigs go to press the lever, then, the bigger one will arrive first at the trough and will eat more ($r_t^B = 4$) than the small one ($r_t^s = 1$). In the case when the big pig waits ($r_t^B = 5$) and the piglet press the lever, the latter will not eat at all ($r_t^s = 0$). Vice versa, if the big pig presses the lever and the small one waits, the piglet will be able to feed itself ($r_t^s = 2$) and leave some leftovers to the first one ($r_t^B = 3$). In general, the small pig has only one possibility for eating: wait until the big one will do all the job. All the possible combinations of this game, together with the respective rewards, represented by the apples, are portrayed in Fig. 74.1. For our aims, we focus only on the finite horizon formulation, where there is a maximum number of $T = 50$ steps that can be reached.

The first experiment was based on training an agent for each pig and examining which policy is obtained. In Fig. 74.2, all the Q-tables obtained after the training phase for both the experiments are listed. In particular, Fig. 74.2a, b present the Q-tables derived after the training using the following parameters: learning rate $\alpha = 0.1$, discounting factor $\gamma = 0.95$, epsilon $\epsilon = 0.4$, decay (subtracting) of the epsilon $\Delta\epsilon = 0.001$ and 1000 episodes. In this scenario, both the RL agents come up to the Nash equilibrium that is given by the combination of pushing the lever for the pig and waiting for the piglet after a few iterations; but to obtain a visible difference between the Q-values we considered a longer training. This result is in contrast with the equilibrium obtained in the case of the Prisoner's Dilemma [7], where the solution for the iterated version consists in the cooperation of the players while the classic problem converge to the betrayal attitude. In the case of the Boxed Pigs game, the results are the same in both the versions. This is principally caused by the asymmetry of the rewards.

The second experiment consists of training only one of the pigs using Q-learning against the other that is playing with a fixed strategy. The policies that are considered are the following: always press the lever, always wait, random choice, tit-for-tat (TFT) and inverse tit-for-tat (Inv-TFT). The choice of these strategies is supported by the simplicity of the first three and the efficiency of the last two. In fact, Tit-for-tat was the winning strategy for an iterated Prisoner Dilemma tournament held in 1980

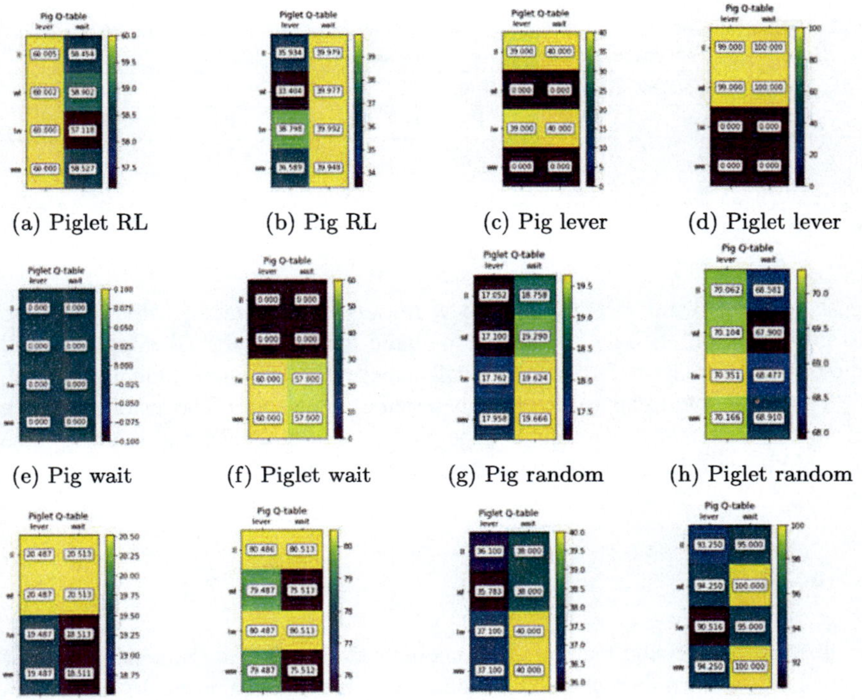

(a) Piglet RL (b) Pig RL (c) Pig lever (d) Piglet lever

(e) Pig wait (f) Piglet wait (g) Pig random (h) Piglet random

(i) Pig TFT (j) Piglet TFT (k) Pig Inv-TFT (l) Piglet Inv-TFT

Fig. 74.2 List of all the Q-tables obtained after the training against fixed policy pigs

by Robert Axelrod [8]. The idea behind this policy is to start with a cooperative choice and then copy the action that has been done by the opponent in the previous step. In the case of the inverse tit-for-tat, instead of taking the same previous action of the opponent, the opposite is chosen, e.g., if the piglet in the step before has waited, then the pig will press the lever in the next step. The parameters used in these training phases are the same as in the previous simulation.

In the following, we analyze the results obtained for this second simulation. As we can notice from Fig. 74.2c–f, when the other pig policy is deterministic, the RL pig will be almost always able to find the optimal solution for itself, i.e., which consist of respectively waiting and pressing the lever when the other one presses the lever or waits. Only in the scenario when the big pig always waits, the piglet will not be able to change its reward despite its choice. Figure 74.2g, h report another interesting aspect: when the opponent is taking random choices, then the bigger pig policy will converge to always push the lever, while the piglet will wait. This enlightens the fact that following the one-step optimal policy is the best option also for the iterated scenario against a random player, as seen in the previous experiment. For the tit-for-tat strategy, the RL pig understands that it has to alternatively do the action wait and

Table 74.1 Rounded mean of the returns obtained at the end of each training of the RL pig against all the possible combination of strategies for 10 episodes

RL agent	Opponent strategy					
	Lever	Wait	Random	Tit-for-tat	Inv-TFT	RL
Pig	250	150	150	200	247	150
Piglet	100	0	100	50	99	100

push the lever to maximize the cumulative reward-food obtained. Lastly, against the inverse tit-for-tat, the agents learn fast to do the action wait since the opponent will directly press the lever. In this way, it will achieve the maximum amount of reward.

The quality of this approach is finally tested in 10 episodes. The results obtained, reported in Table 74.1, confirm the good performance of the RL pigs to achieve the best profit against a fixed policy agent.

Conclusion

In this paper, we found that the solution obtained using RL of the iterated form of Boxed Pigs game is the same of the regular version (without any penalty parameter) and we confirm the efficiency of RL in tackling the problem of finding an optimal strategy against a fixed policy agent for the iterated version of the Boxed Pigs game.

Future works involve the analysis of the iterated version of the extended Boxed Pigs game. In this scenario, with one more pig, the application of Multi-Agent RL techniques will be necessary to find the optimal solution. Another possible experiment, that we are going to execute, concerns the addition of a penalty parameter and a minimum amount of food that each pig has to eat in order to survive, to make this problem more complex and realistic. In this way, it will be more similar to analyzing the hitting probabilities and times of an absorbing Markov chain. It is also interesting to study theoretically the iterated version in order to have a formal description of the solutions that can be found.

References

1. Rasmusen, E. (1990). Games and information. An introduction to game theory. *Theory and Decision, 29*(2).
2. McMillan, J. (1992). *Strategies and managers—How managers can use game theory to make better business decisions*.
3. Axelrod, R. (1980). Effective choice in the Prisoner's Dilemma. *Journal of Conflict Resolution, 24*(1), 3–25.
4. Sandholm, T. W., & Crites, R. H. (1996). Multiagent reinforcement learning in the iterated Prisoner's Dilemma. *Biosystems, 37*(1–2), 147–166.

5. Sutton, R. S., & Barto, A. G. (2018). *Reinforcement learning: An introduction*. MIT Press.
6. Watkins, C. J. C. H., & Dayan, P. (1992). Q-learning. *Machine Learning, 8*(3), 279–292.
7. Harper, M., et al. (2017). Reinforcement learning produces dominant strategies for the iterated Prisoner's Dilemma. *PLoS ONE, 12*(12), e0188046.
8. Axelrod, R. (1980). More effective choice in the Prisoner's Dilemma. *Journal of Conflict Resolution, 24*(3), 379–403.

Chapter 75
Monte Carlo Based Machine Learning

Sara Shashaani and Kimia Vahdat

Abstract Even though simulation is mainly used for computer models with inexact outputs, there are direct benefits in viewing results from samples of an existing dataset as replications of a stochastic simulation. We propose building Machine Learning prediction models with the Monte Carlo approach. This allows more specific accountability for the underlying distribution of the data and the impact of uncertainty in the input data in terms of bias. We opt for nonparametric input uncertainty with multi-level bootstrapping to make the framework applicable to large datasets. The cost of Monte Carlo-based model construction is controllable with optimal designs of nested bootstrapping and integrating variance reduction strategies. The benefit is substantial in providing more robustness in the predictions. Implementation in a data-driven simulation optimization problem further indicates the superiority of the proposed method compared to the state-of-the-art methods.

Keywords Bias correction · Robustness · Data-driven optimization · Simulation

Introduction

This paper provides a new perspective on data-driven modeling and its promise. Viewing machine learning (ML) as a simulation experiment entails viewing results from samples of an existing dataset as replications of a stochastic simulation. Analyzing the input-output dependency for ML models with Monte Carlo (MC) methodology benefits are many, among which are:

- seeing the model uncertainty effect on the outputs,
- distinguishing risks and errors and enabling input uncertainty (IU) analysis,
- identifying input data and model inter-dependencies,
- understanding the interaction of training set and validation and characterizing their optimal configuration,

S. Shashaani (✉) · K. Vahdat
North Carolina State University, Raleigh, NC 27695, USA
e-mail: sshasha2@ncsu.edu
URL: https://shashaani.wordpress.ncsu.edu

O. Grothe et al. (eds.), *Operations Research Proceedings 2022*, Lecture Notes in Operations Research, https://doi.org/10.1007/978-3-031-24907-5_75

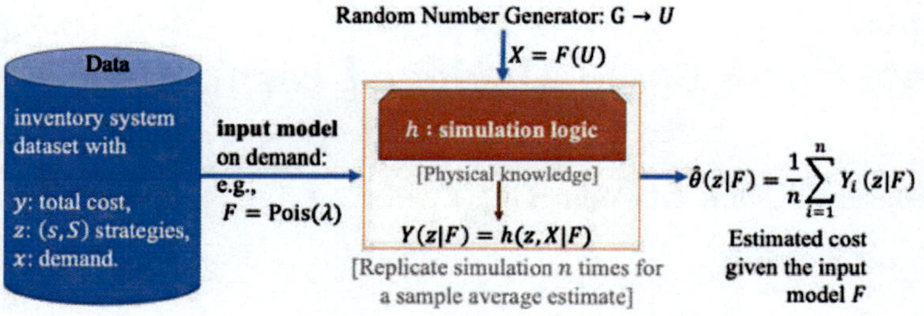

Fig. 75.1 Stochastic simulation input-output dependence for (s, S) problem

- tackling the curse-of-dimensionality with the dimension-independent MC procedures,
- leveraging variance reduction for efficiency, and
- combining external system knowledge into learning with MC and quasi-MC based methods.

To set ground notations/definitions, we start with a stochastic simulation framework on a famous stochastic problem and then revisit that as a data-driven problem.

Viewing Machine Learning as a Simulation Experiment

We introduce a unified stochastic simulation and ML framework (Figs. 75.1 and 75.2) to clarify sources of uncertainty. To illustrate this framework, consider the (s, S) inventory problem with continuous demand, whose objective is to identify the re-order and order-up-to levels that minimize the expected per-period total cost—the sum of back-order, order, and holding costs.

With stochastic simulation, a fixed logic model h is present, which requires an input model F for the random demands X, typically generated by fitting a known family of distributions to real-world collections of x. One run of the simulation, $Y(z|F)$, represents a realization of the system performance with decision $z = (s, S)$. MC then yields estimating $\theta(z) := \theta(z|F^c)$, the true expected system performance at z under the true demand distribution F^c, with sample average $\hat{\theta}(z|F)$ of $i = 1, 2, \ldots, n$ i.i.d. $Y_i(z|F)$ replications [1]. Conventional stochastic simulation assumes the logic is unbiased, i.e., $\theta(z) := \mathbb{E}[\hat{\theta}(z|F)]$. We note that $Y(z|F) = \hat{\theta}(z|F) + \epsilon_U(z|F)$, where U is the random number leading to output $Y(\cdot|F)$, and ϵ_U is a mean-zero random variable. To also include the IU error in the decomposition, we write $\hat{\theta}(z|F) = \theta(z) + \epsilon_F(z)$, where $\epsilon_F(z)$ may not be a mean-zero random variable, suggesting an existence of bias due to IU. Estimating the mean of ϵ_F relies on repeating the stochastic simulation over replicas of F [2]. With this insight, we write

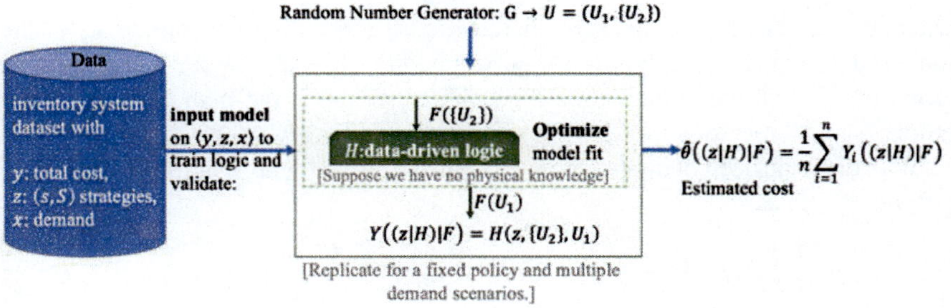

Fig. 75.2 ML input-output dependence for (s, S) problem

$$\mathrm{MSE}(\hat{\theta}(z|F)) = \; = \overbrace{\mathbb{E}_F[\mathrm{Var}(\hat{\theta}(z|F))] + \mathrm{Var}_F(\theta(z|F))}^{\mathrm{Var}(\hat{\theta}(z|F))} + \overbrace{\mathbb{E}_F[(\hat{\theta}(z|F) - \theta(z))^2]}^{\mathrm{Bias}^2(\hat{\theta}(z|F))},$$

where the variance effect is known to dominate the bias [3]. Quantifying $\mathrm{Var}(\hat{\theta}(z|F))$ and more recently parametric $\mathrm{Bias}(\hat{\theta}(z|F))$ in a low-dimensional input space has been studied [4, 5]. With high-dimensional input space, the bias effect may be significant enough. We propose two nonparametric bias estimation methods, that are easily generalizable to stochastic simulation and ML problems.

MC-Guided Input-Output Formulations of ML

Suppose, the logic of the (s, S) inventory problem was unattainable and instead we had access to historical records of the system performance (total cost) under various levels of re-order and order-up-to quantities and demands, i.e., $\{\langle(z, x), y\rangle_i\}_{i=1,2,\ldots,n}$, where $z = (s, S)$. In complicated systems, many more variables would be recorded as part of z that may or may not be important. The domain-agnostic approach ML would still include all recorded variables in the learning. With a selected family of learning algorithms (e.g., linear regression), a data-driven logic H would minimizes its outputs' discrepancy from the observed responses (total cost) over the *training set* or $\{U_2\}$ in Fig. 75.2 by choosing the optimal learning parameters (e.g., regression coefficients). Once the logic model is formed, it will be evaluated on a new point U_1 also sampled from F but with a specific order to ensure that no overlap would corrupt the process, will be generated, i.e., the model output $Y((z|H)|F)$.

Importantly, H is random and dependent on the observed data points in ML. To clarify, we assume that the learning algorithm is the perfect choice for the logic model. The uncertainty of what H truly is, will be added to the total variability. We call this error *logic uncertainty*, which is dependent on the input data uncertainty but specifically tracks the effect of a wrong logic model (logic misspecification) on the outputs. Hence a compound uncertainty encompassing the logic and input data risk needs to be estimated. H can more broadly be used to represent *model complexity*, where the choice of z would entail characteristics controlling the complexity of the model, e.g., degree of polynomials corresponding to each feature, regularization terms [6, 7], and the optimization algorithms used to fit the model among many more.

With the logic itself being random, we write out an analogue of output $Y((z|H)|F)$ decomposition as,

$$Y((z|H)|F) = \theta(z) + \epsilon_F(z) + \epsilon_H(z|F) + \epsilon_U((z|H)|F), \tag{75.1}$$

where $\epsilon_H(F)$ represents the risk associated with logic uncertainty for an input model F. By nature, the ML model is more dependent on data, more sensitive to the problem dimension, and significantly noisier than simulation models. A nested simulation scheme can quantify each uncertainty term in (75.1) [8].

Estimating Bias Due to Misspecifying the Input Distribution

Statistical variance $\epsilon_U(\cdot)$ is the most studied error term in stochastic simulation, and there are well-established methods for estimating the effect of ϵ_F on the output variance [9]. The logic uncertainty is difficult to estimate, particularly with no prior assumption. We study estimating the input induced bias and variance using nonparametric methods, where F is an empirical distribution.

In a nested simulation, the outer level varies the risk factor (in this case, F), and the inner level varies the system's best performance (fit) under the realized risk factors. For ease of exposition, we drop z since the same steps are repeated for every z. We let F^* represent a bootstrap of F and for each F^* and collect

$$\left\{\{Y_i(H_1|F^*)\}, \ldots, \{Y_i(H_r|F^*)\}\right\} \xrightarrow{\text{average}} \left\{\bar{Y}(H_1|F^*), \ldots, \bar{Y}(H_r|F^*)\right\},$$

where H_j is the logic model obtained from a replication of fitting data using F^*. This enables tracking the conditional effect of logic uncertainty. We also denote the average of all outputs on the RHS with $\bar{\bar{Y}}(F^*)$. Since we want the outputs to be i.i.d., it is important that we first sample the *test data* with $\{U_{1,1}, \ldots, U_{1,n}\}_j$ from F^*, and then take $\{U_{2,j}\}$ to build H_j. Note, this is in contrast to the usual sampling routine for ML where the training data is drawn first. Let the bootstrap distributions from which these training and test sets are drawn be F_j^{**} $j = 1, 2, \ldots, r$ and re-write $H(F_j^{**}) := H_j$ for clarity. To compute the bias within each logic, we again use one bootstrap of F_j^{**}, i.e., F_j^{***} and obtain

$$\left\{\bar{Y}(H(F_j^{**})|F^*) - \bar{Y}(H(F_j^{***})|F^*)\right\} := \{\text{bias}_j(F^*)\} \xrightarrow{\text{average}} \overline{\text{bias}}(F^*).$$

This technique is called *fast iterated bootstrapping* (FIB) [10]. However, this bias estimator has a high variance and we will need a variance reduction approach for

FIB. Defining a control variate $W_j(F^*) = \bar{Y}(H(F_j^{**})|F^*) - \bar{\bar{Y}}(F^*)$ can help reduce the variance of the logic bias estimator, since $\mathbb{E}[W(F^*)] = 0$ and $\text{Var}(W(F^*))$, $\text{Cov}(W(F^*), \text{bias}(F^*))$ can be easily estimated. We hence use the bias estimator $\{\text{bias}_j(F^*) + c \times W_j(F^*)\} \xrightarrow{\text{average}} \overline{\text{bias}'}(F^*)$, where

$$c = \frac{-\widehat{\text{Cov}}(W(F^*), \text{bias}(F^*))}{2\widehat{\text{Var}}(W(F^*))} = \frac{\frac{1}{r}\sum_{j=1}^{r} \text{bias}_j(F^*) \times W_j(F^*)}{\frac{1}{r-1}\sum_{j=1}^{r} W_j^2(F^*)}.$$

With the variance-reduced bias estimate above, we can *de-bias* our average output from input model F^* with $\bar{\bar{Y}}'(F^*) := \bar{\bar{Y}}(F^*) - \overline{\text{bias}'}(F^*)$. Recall this de-biased quantity is an estimate of $\hat{\theta}(F^*) - \mathbb{E}[\epsilon_H(F^*)]$, where $\mathbb{E}[\epsilon_H(F^*)]$ is the bias due to conditional logic uncertainty. In simulation, this bias is often neglected as the model is fixed. In ML, however, it may be the dominating source of corrupting the outputs. In Section "Numerical Example", we show how it helps with the robustness of estimation.

Unlike the bias due to logic uncertainty, the bias due to input uncertainty (the misspecification of F^c by F) is an important factor not only in ML but also in stochastic simulations, which can be tracked by allowing multiple replications of F^*. Our following nonparametric approach for estimating this bias is a contribution to both frameworks. Following the Taylor expansion and Von-Mises' path-wise differentiability [11], we develop an input bias estimator via *higher order influence functions* (HOIF). We write

$$\theta(F_\varepsilon) \approx \theta(F) + \varepsilon(F^c - F)\nabla_\varepsilon\theta(F_\varepsilon) + \frac{1}{2}(\varepsilon(F^c - F))^2\nabla^2\theta(F_\varepsilon), \tag{75.2}$$

where $F_\varepsilon = F + \varepsilon(F^c - F)$ is the perturbed input distribution. Note, bootstrap theory allows us to approximate $F^c - F$ with $F - F^*$ when enough bootstrap samples are applied. Hence the second orders approximation in (75.2) is appropriate. The first and second order derivatives can be approximated with score functions, as suggested by [12]. See [13] for the elaborated derivation of these terms. Then the estimate of $\theta(F_\varepsilon) - \theta(F)$ will be the bias that will be deducted from the final output estimate $\bar{\bar{Y}} = b^{-1}\sum_{k=1}^{b} \bar{\bar{Y}}'(F_k^*)$.

Numerical Example

Figure 75.3 compares predictions in a stochastic (s, S) inventory using the existing and proposed approaches. The nonparametric de-biased outputs for 3 scenarios correctly identify the optimal scenario (20, 50) while ignoring bias would mislead the decision-maker. The total budget used in each approach is the same.

A data-driven (s, S) inventory problem as laid out in Section "MC-Guided Input-Output Formulations of ML" reveals in Table 75.1 confidence intervals (CI) for

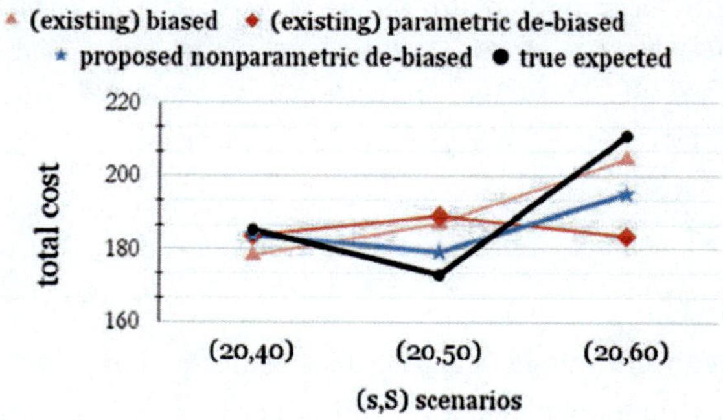

Fig. 75.3 The proposed de-bias estimator finds the correct trend

Table 75.1 ML predicts the performance of two policies in a (s, S) inventory problem with 2000 total budget

Data size	Method	z_1	z_2	90% Diff. CI
50	Biased with IU variance [9]	178.3 ± 1.9	187.1 ± 2.9	[−16.6, −0.8]
	Parametric de-biased [5]	183.7 ± 2.9	189.1 ± 10.4	[−26.9, 16.3]
	Nonparametric de-biased	183.8 ± 3.9	179.3 ± 4.6	[−8.5, 17.9]
	Optimal nonparametric de-biased	187.3 ± 5.0	170.5 ± 4.5	[4.7, 28.7]
100	Biased with IU variance [9]	181.5 ± 1.4	181.5 ± 2.7	[−7.1, 7.1]
	Parametric de-biased [5]	185.0 ± 4.2	187.1 ± 9.1	[−22.8, 18.7]
	Nonparametric de-biased	187.1 ± 3.0	170.5 ± 5.0	[4.1, 28.8]
	Optimal nonparametric de-biased	184.1 ± 5.6	171.5 ± 3.8	[1.5, 23.6]

Nonparametric de-biased estimators enable correct selection, i.e., paired residual CI above 0

$z_1 = (20, 40)$ and $z_2 = (20, 50)$, where z_2 is the optimal solution and hence ought to provide smaller predictions for the total cost, i.e., $\theta(z_1) - \theta(z_2) < 0$. A successful prediction places the paired residual 95% CI fully above 0. We experiment with different sizes of available data. When the available data is small (50), the *optimal* (i.e., budget-optimal) nonparametric de-biased estimator where the number of repeats in each layer of bootstrapping is follows a budget allocation strategy [13], is the only method correctly finding z_2 as the better solution. When the available data is large (100), either with optimal budget allocation or not, the proposed method finds the correct solution. In both cases, the parametric de-biased and biased estimators fail to find the correct solution.

Concluding Remarks

The proposed unified framework illuminates that viewing ML as a simulation can track the propagation of bias of data/logic into output. Without prediction bias, the estimates of future outcomes of a decision can mislead to a worse/riskier option. The proposed nonparametric bias estimator successfully identifies the correct trend and provides better predictions. While quantifying the dependency between ϵ_F and ϵ_H, etc., is a remaining open question, our current approach exhibits asymptotic and numerical success despite ignoring the dependency terms.

References

1. Kim, S., Pasupathy, R., & Henderson, S. G. (2015). A guide to sample average approximation. In *Handbook of simulation optimization* (pp. 207–243). Springer.
2. Barton, R. R. (2012). Tutorial: input uncertainty in output analysis. In *Proceedings of the 2012 Winter Simulation Conference (WSC)* (pp. 1–12). IEEE.
3. Lam, H. (2016). Advanced tutorial: Input uncertainty and robust analysis in stochastic simulation. In *2016 Winter Simulation Conference (WSC)* (pp. 178–192). IEEE.
4. Song, E., & Nelson, B. L. (2019). Input-output uncertainty comparisons for discrete optimization via simulation. *Operations Research, 67*(2), 562–576.
5. Morgan, L. E., Nelson, B. L., Titman, A. C., & Worthington, D. J. (2019). Detecting bias due to input modelling in computer simulation. *European Journal of Operational Research, 279*(3), 869–881.
6. Tibshirani, R. (2011). Regression shrinkage and selection via the lasso: A retrospective. *Journal of the Royal Statistical Society: Series B (Statistical Methodology), 73*(3), 273–282.
7. Hinton, G. E. (2012). A practical guide to training restricted Boltzmann machines. In *Neural networks: Tricks of the trade* (pp. 599–619). Springer.
8. Shashaani, S., & Vahdat, K. (2022). Improved feature selection with simulation optimization. *Optimization and Engineering* (pp. 1–41).
9. Song, E., & Nelson, B. L. (2017). Input model risk. In *Advances in Modeling and Simulation* (pp. 63–80). Springer.
10. Ouysse, R. (2013). A fast iterated bootstrap procedure for approximating the small-sample bias. *Communications in Statistics—Simulation and Computation, 42*(7), 1472–1494.
11. Van der Vaart, A. W. (1998). Asymptotic statistics. In *Cambridge series in statistical and probabilistic mathematics*. Cambridge University Press.
12. Lam, H., & Qian, H. (2021). Subsampling to enhance efficiency in input uncertainty quantification. *Operations Research*. https://doi.org/10.1287/opre.2021.2168.
13. Vahdat, K., & Shashaani, S. (2022). Robust prediction error estimation with Monte-Carlo methodology. arXiv preprint arXiv:2207.13612.

Part XVIII
Software Applications and Modeling Systems

Chapter 76
Xpress Mosel: Highlights from 20 Years of Software Development and New Advanced Programming Features

Susanne Heipcke and Yves Colombani

Abstract Twenty years after its first commercial release, the Xpress Mosel software keeps evolving driven by user requirements, usage patterns and technological advances. This contribution takes the reader through the major phases of its development: Mosel was initially designed as an optimization modeling language that also provided programming features. Over time the Mosel distribution has been enriched with numerous components and tools addressing a variety of purposes. The increasing use of Mosel as general-purpose programming language was recognized by turning it into a free software a few years ago. Motivation and use cases for major new programming features are discussed in detail.

Keywords Mathematical modeling · Optimization applications · Language design · Testing systems

Introduction

The first commercial publication of Xpress Mosel in 2001 introduced various new concepts for mathematical modeling: Mosel was designed as an optimization modeling language that also provided programming features, with an open, modular architecture allowing developers to add new functionality according to their needs. The first part of this paper takes the reader through the major development phases of Mosel, directing particular focus at the phase preceding the first publication during which the fundamental questions of whether, why, and how to replace an established software (namely Mosel's precursor mp-model [1], commercialized by Dash Associates since 1983) had to be addressed.

---

http://www.fico.com/xpress.

S. Heipcke (✉) · Y. Colombani
Xpress Optimization FICO, FICO House, Starley Way, Birmingham B37 7GN, UK
e-mail: SusanneHeipcke@fico.com

Y. Colombani
e-mail: YvesColombani@fico.com

O. Grothe et al. (eds.), *Operations Research Proceedings 2022*, Lecture Notes in Operations Research, https://doi.org/10.1007/978-3-031-24907-5_76

During the 20 years of Mosel's existence a shift in usage patterns can be observed, moving from largely stand-alone optimization projects to multi-user web apps that directly interact with many other systems and programs. This evolution has entailed an increasing use of Mosel as general-purpose programming language that was recognized by turning Mosel into free software a few years ago. Recent releases of Mosel further build out its programming capabilities, addressing advanced programming needs of large software development projects. The second part of this paper discusses some examples of advanced new programming functionality.

Phases of Software Development

Phase 0: Inception and Design (1997–2001, Versions 0–1.2)

Target: match functionality of existing systems; provide easy transitioning from the precursor standalone command-line tool mp-model without any disruptive changes to user habits whilst switching to an entirely new architecture that provides openings for a large variety of future developments (in-memory data exchange when embedding model execution into a host language [2], publication of the Mosel Native Interface for the implementation of user modules [3]).

- *First step:* proof of feasibility—reading and executing existing mp-model files on the new architecture (replacing an interpreted/script language written in Fortran and in parts Assembly by parsing and compilation to a virtual machine implemented in C) without any loss of efficiency or functionality (specifically the handling of sparse data structures).
- *Second step*: creating a new language in close collaboration with expert academic and industrial users (most notably the group of Laurence Wolsey at CORE, Univ. Louvain-la-Neuve, and the team of mathematicians at BASF AG, Germany, led by Anna Schreieck).
- *Third step*: development of tools and interfaces: 'mod2mos' converter, library APIs (C, Java, VB), Xpress IVE (Interactive Visual Environment) as development environment on Windows.
- *Fourth step*: finding a name—the working name 'Mmod2' of the prototype phase for 'mp-model version 2' with file extension 'mm2' (already using 'bim' ='Binary Model' for compiled files) was finally replaced by the name 'Mosel' (with the file extension 'mos') that has no specific meaning and is easy to pronounce; the naming convention 'dso' = 'Dynamic Shared Object' was adopted for Mosel modules (C libraries extending the Mosel language) after using system-specific dynamic library file extensions during the prototype phase.

Phase 1: New Directions for Modeling: Debugger and Profiler, Multi-processor Parallelism (2002–2008, Versions 1.4–3.0)

Target: make programming tools available for a modeling language; exploit newly available multi-processor machines for parallelising (decomposition) algorithms and other computational tasks on the model level.

- Programming tools and functionality: creation of debugger and profiler for the Mosel language; new programming-style data structures and functionality; introduction of the notion of packages (libraries written in Mosel).
- Multi-processor parallelism: concurrent model execution within a Mosel instance coordinated by event-based message queues; also: definition of multiple problems within a model (handled sequentially).
- Modular building block: concept of generalized file handling (I/O drivers).
- New solvers and problem formulation paradigms (Nonlinear, Constraint Programming).

Phase 2: Distributed Computing (2008–2017, Versions 3.2–4.8)

Target: anticipate emerging new usage patterns, creating technology to enable working in distributed settings including cloud architectures.

- Mosel Distributed Framework [4] and remote launcher (XPRD) provided the underlying technical framework for Xpress Insight (platform for web-based apps with scenario handling based on Mosel models, first release in 2014).
- The newly introduced remote invocation protocol is used by a new browser-based development environment Xpress Workbench (first release July 2017) and equally by automated testing systems for large Mosel projects.
- Support for cloud platforms: *aec2*, *hadoop*, and *mmhttp* components; related new features: encryption, Unicode, internationalization (message catalog selection based on system language configuration).
- Connectivity: data exchange with/invocation of other languages (Matlab, R, Java, Python).
- Possibility to execute Mosel programs in fully secure restricted mode.

Phase 3: Advanced Programming Needs (2017–now, Versions 5.0–6.2)

Target: address programming functionality needs of increasingly large software development projects including connectivity, testing systems, and expectations of developers using mainstream programming languages; provide tooling (high-level packages and low-level functionality for their implementation) for low-code development of end-user apps.

- End of 2017 Mosel was turned into a free software in recognition of its increasing use as general programming language, also opening up the matrix manipulation routines accessible from the Mosel Native Interface (NI) to provide access to alternative LP/MIP/NLP solvers [5].
- New advanced programming features for large-scale projects with multiple contributors such as dynamic packages, definition of namespaces, shared data, and inline documentation [6]; recent additions that are discussed more in detail below include union types and function pointers.
- Improved implementations of existing functionality including revised matrix handling to match ever-increasing (average and maximum) problem sizes [6].
- Creation of high-level packages for advanced programming tasks distributed as open-source (inline documentation: *moseldoc* [7], remote execution: *compsrv*, distributed computing system: *jobqueue*, testing system: *moseltest* [5])

Advanced Programming Functionality

Recent additions to Mosel comprise new maths/optimization features, notably the handling of general constraints, that is, certain nonlinear relations that are directly recognized by MIP solvers [8], and (forthcoming with Mosel 6.2) support for optimization with multiple objectives. However, the vast majority of new features relate to advanced programming functionality, a trend that may be expected to continue. We now take a closer look at some representative examples.

Union Types

A union is a container capable of holding an object of arbitrary type/structure or one of a predefined set of types that can be used, among others, for reading and storing input data of a-priori unknown type. It can also be used for the retrieval of information from another Mosel model without prior knowledge of its structure/data model, a feature that can be exploited in the implementation of generic libraries and in particular for testing systems. A union is defined by specifying the set of compatible types or using the predefined union type 'any' like the type definitions of package *json* in Fig. 76.1 that implements generic access routines for JSON files.

The code snippet in Fig. 76.2 employs the package functionality for reading a JSON file of an a priori unknown structure.

```
public declarations
  jobj=array(string) of any          !@doc.descr Type definition: JSON object
  jarr=array(range) of any           !@doc.descr Type definition: JSON array
  jnull=string                       !@doc.descr Type definition: null value (stored as a string)
 !@doc.descr Type definition: JSON value
 !@doc.info A 'jval' is either a scalar of type text, real, boolean, or jobj, jarr or jnull
  jval=text or real or boolean or jobj or jarr or jnull
end-declarations
```

Fig. 76.1 Type definitions in package 'json.mos' [7] with documentation annotations

```
uses "json"
declarations
  expl: jval                                          ! Root element (arbitrary json type)
end-declarations

if loadjson("my.json",expl)=0 and expl is jarr then   ! Read and check structure type
  writeln("Indices: ", expl.jarr.range)               ! Index numbers start with 1
  if expl.jarr(1) is jobj then                        ! Check json structure type
    with firsti=expl.jarr(1).jobj do                  ! Local definition as shorthand
      writeln("First entry: ", jsontext(firsti))      ! Display JSON-format subtree
      writeln("Fields: ", firsti.fields)              ! Display defined fields
      if isdefined(firsti("id")) then                 ! Test for field labeled 'id'
        writeln("'id' of first entry: ", firsti("id")) ! Display the (scalar) value
      end-if
    end-do
  end-if
end-if
```

Fig. 76.2 Using the functionality of package *json* in a Mosel program

Subroutine References

Subroutine references can ease the integration of user inputs (e.g. from Xpress Insight) into Mosel models, and they also provide new ways for implementing callbacks (e.g. from optimization solvers or data handling interfaces). The code snippet in Fig. 76.3 uses subroutine references to apply a function selection made by the user. This approximation problem is described in [9].

Reflection

Reflection or Reflective Programming is the ability of a process to examine, introspect, and modify its own structure and behavior [10]. It helps programmers to create generic software libraries and it can also be used for observing and modifying program execution at runtime, meaning that it is often employed for the implementation of testing systems.

The new Mosel module *mmreflect* provides functionality for retrieving and calling subroutines, retrieving and modifying array entries, scalars and sets, and it defines the array `iterator` type for enumerating arrays of unknown structure and type. The enumeration speed achieved with an iterator is equivalent to standard enumeration for the different Mosel array types (dense/dynamic/hashmap) with up to 2 dimensions, and more than one order of magnitude faster for higher numbers of dimensions (time measures averaged over 3 runs on Win10 64bit). The code snippet in Fig. 76.4

```
parameters
  FCT="sin"            ! User input: Function to approximate
end-parameters

function userfunctanh(x:linctr): nlctr
  returned:=(exp(2*x)-1)/(exp(2*x)+1)
end-function

declarations
  f: function(linctr): nlctr       ! NL constraint function
  fconst: function(real): real     ! Math (real-valued) function
end-declarations

case FCT of        ! Apply the selection made via runtime parameter 'FCT'
 "sin": do
          f:= ->sin                ! Built-in NL constraint
          fconst:= ->sin           ! Built-in math function
        end-do
 "tanh": do
          f:= ->userfunctanh       ! NL constraint definition by user
          fconst:= ->tanh          ! Built-in math function
        end-do
 ! (...)
end-case
F0:= fconst(X0); F1:= fconst(X1)  ! Evaluate function 'fconst' on numerical values
DefObjA:= abs(f(x)-((F0+S0)+SLOPE*(x-X0)))  ! Use 'f' in constraint definitions
```

Fig. 76.3 Working with subroutine references (file 'optgrid2.mos' [7])

```
uses "mmreflect"
public procedure solarray(x:any, s:any)
  declarations
    it: iterator
  end-declarations
  if x is array of mpvar and s is array of real then
    if x.array.nbdim = s.array.nbdim and
       and(i in 1..x.array.nbdim) x.array.index(i).eltype = s.array.index(i).eltype then
      inititer(it,x.array)
      reset(s.array)
      while (nextcell(it)) s.array(it).real:= x.array(it).mpvar.sol
    else
      writeln("SOLARRAY: array indices don't match")
    end-if
  else
    writeln("SOLARRAY: arguments must be arrays of type mpvar / real")
  end-if
end-procedure
```

Fig. 76.4 Using union types, reflection and array iterator (file 'solarrayanypkg.mos' [7])

shows the generic implementation of a subroutine that copies solution values from a decision variable array into another array after inspecting the provided arrays for matching structures.

Conclusion

Since its first publication Mosel has evolved driven by user requirements and to exploit technological advances, reinforcing in particular its programming functionality. An important difference to mainstream programming languages that provide extensions to support mathematical modeling, such as Python, resides in that the Mosel language has been designed to directly handle large-scale algebraic expressions for stating constraints without any need for reformulations by the developer or the use of add-on components. Furthermore, through the concept of I/O drivers it is easy to isolate a Mosel program from the systems it is embedded into for standalone development and testing purposes.

References

1. Ashford, R. W., & Daniel, R. C. (1987). LP-MODEL: XPRESS-LP's model builder. *IMA Journal of Mathematics in Management, 1*, 163–176.
2. Ciriani, T. A., Colombani, Y., & Heipcke, S. (2003). Embedding optimisation algorithms with Mosel. *4OR, 1*(2), 155–168.
3. Colombani, Y., Daniel, B., & Heipcke, S. (2004). Mosel: A modular environment for modeling and solving problems. In J. Kallrath (Ed.), *Modeling Languages in Mathematical Optimization* (pp. 211–238). Kluwer Academic Publishers, Norwell
4. Heipcke, S. (2012). Xpress-Mosel: Multi-solver, Multi-problem, Multi-model, Multi-node modeling and problem solving. In J. Kallrath (Ed.), *Algebraic Modeling Systems: Modeling and Solving Real World Optimization Problems* (pp. 81–114). Springer, Heidelberg.
5. Mosel Open Source repository. (2017). https://github.com/fico-xpress/mosel
6. Heipcke, S., & Colombani, Y. (2020). Xpress Mosel: Modeling and programming features for optimization projects. In J. S. Neufeld et al. (Eds.), *Operations Research Proceedings 2019* (pp. 677–683). Springer.
7. FICO Xpress Examples Repository. https://examples.xpress.fico.com/example.pl
8. Heipcke, S., & Colombani, Y. (2021). New MIP modeling constructs in Xpress Mosel to handle logical relations and certain nonlinear constraints. In *Proceedings of ROADEF 2021*. https://roadef2021.sciencesconf.org/350472
9. Kallrath, J. (2021). *Business optimization using mathematical programming—an introduction with case studies and solutions in various algebraic modeling languages* (2nd ed.). Springer Nature.
10. Wikipedia. https://en.wikipedia.org/wiki/Reflective_programming

Part XIX
Supply Chain Management

Chapter 77
Data-Driven Prediction of Order Lead Time in Semiconductor Supply Chain

Xin Shen, Patrick Moder, Christian Pfeiffer, Grit Walther, and Hans Ehm

Abstract This study proposes an AI-empowered order lead time prediction integrating a multidimensional real-world dataset from a semiconductor manufacturer's supply chain. Examined features capture order–, delivery–, planning–, customer–, and product– related information. We thoroughly analyze a broad spectrum of machine learning algorithms ranging from linear regression and tree-based models to neural networks and compare them with respect to prediction performance, computation time, and understandability. We find that boosting algorithms demonstrate solid predictive performance with the highest accuracy and most efficient computation time. Our results allow supply chain experts to obtain data-informed estimations of order lead times and an understanding of the predictive mechanisms.

Keywords Machine learning · Supply chain · Predictive analytics

Motivation

Despite COVID-19, the semiconductor business is booming at a breakneck pace due to a significant increase in demand for new-generation chips [1]. Downstream manufacturers expect a reliable delivery window and must communicate with semiconductor suppliers in advance to hedge against unpredictable shortages. Accurate order lead time predictions not only leverage the accuracy of semiconductor manufacturers' own supply chain planning, but also facilitate customer's procurement and production planning. However, the magnitude of order lead time is affected by multiple business activities and external factors, such as market dynamics, cycle time, and technology upgrades. The fluctuation of these parameters enhances the difficulty of order lead time approximation. Semiconductor manufacturers seek solutions that

X. Shen (✉) · C. Pfeiffer · G. Walther
RWTH Aachen University, 52062 Aachen, Germany
e-mail: xin.shen@rwth-aachen.de

P. Moder · H. Ehm
Infineon Technologies AG, Am Campeon 1 -15, 85579 Neubiberg, Germany

O. Grothe et al. (eds.), *Operations Research Proceedings 2022*, Lecture Notes in Operations Research, https://doi.org/10.1007/978-3-031-24907-5_77

produce accurate prediction results in their complex supply chain environment to increase credibility towards the customer and efficiency of their own planning operations [2].

A considerable amount of researchers and companies strive to predict order lead time with the objective of customer retention and attraction. Indirect and direct methods can be employed. The indirect method calculates the lead time based on sequencing and queueing theory, whereas the direct method utilizes historical data. Burggräf et al. systematically review literature related to direct lead time prediction using machine learning and operations research in engineer-to-order domains and identify data classes and origins that we based our data selection on. They find that including more data classes capturing different lead time-related aspects may improve prediction results and that machine learning models are a promising way to make accurate predictions [3]. Singh and Soni for instance compare the predictive performance of multiple machine learning models for order lead time in just-in-time manufacturing. The authors prove that machine learning models are able to predict lead times and propose the inclusion of larger data sets to increase accuracy [4]. Lingitz et al. apply machine learning algorithms for a semiconductor manufacturer with limited features. Still, the authors only calculate the lead time for a simple three-step sequence process without considering the whole production system [2]. To the authors' best knowledge, there is a lack of semiconductor industry-oriented prediction frameworks involving specific attributes and operational datasets. Given these shortcomings, we seek to explore the ML-supported lead time prediction on real-world datasets from the semiconductor domain. The research requirements are derived from both a theoretical perspective (prediction performance on real-world data and advanced machine learning model usage) and practical considerations (computation time and understandability).

In this paper, we precisely define order lead time in alignment with the case company, considering the order modification behavior and specific granularity. Furthermore, we apply multidimensional real-world datasets from diverse internal information systems. In addition, a broad spectrum of algorithms ranging from linear regression and tree-based machine learning models to neural networks are analyzed and compared in terms of accuracy, computation time, and feature importance. The results enable supply chain experts to obtain a holistic understanding of the predictive performance and features that drive lead time behavior.

Research Design

In the given research environment, order lead time refers to the time span from the receipt of the last modified sales order until this order is physically available in a distribution center. This definition (i) accounts for changes of order characteristics after an order gets placed (due to updated customer requirements or operational deviations at the semiconductor manufacturer) and (ii) seeks to remove uncertainties that appear after the arrival at the distribution center (due to delayed order handling on site or delivery deviations of freight forwarders). We select data and indicators for a single product line from 2018 to 2019. The selected datasets are cleaned by addressing incomplete and noisy observations. We consolidate data from multiple sources into a single table on schedule line granularity. Table 77.1 lists the attributes utilized for prediction. We use robust scaler to rescale numerical features to the same range, thus mitigating the influence of outliers [8]. Categorical features are transformed into a machine-readable format using encoding techniques: cyclical features like months are converted into sine and cosine pairs, while one-hot encoding transforms the remaining categorical features into the new binary columns [6].

To address the problem of data leakage, i.e., preventing inadvertently involving unavailable information during the prediction phase, we rearrange data based on the initial order entry date. The data is subsequently partitioned into four distinct splits with hold-out test set and expanding window on the training set [7], as illustrated in Fig. 77.1. Such partition aims to examine the data-efficient performance of individual machine learning models.

This paper presents the evaluation of nine predictive models: (1) linear regression (lr), (2) lasso regression (las), (3) ridge regression (rig), (4) random forest (rf), (5) support vector regression (svr), (6) gradient boosting (reg), (7) extreme gradient boosting (xgbr), (8) light gradient boosting machine (lgbm), and (9) artificial neural network (NN). Boosting algorithm delivers promising predictive results by ensembling all predictors, which has not been intensively investigated in the field of lead time prediction [8]. Therefore, this paper compares the performance of three boosting algorithms to close the research gap. For overviews on models and methods see [9]. Nested cross-validation combined with Bayesian Optimization is utilized to determine the optimal hyperparameters in the search space [10]. As shown in Fig. 77.1, while the inner loop serves for hyperparameter tuning, the outer loop evaluates the model generalization on hold-out testing data with optimized hyperparameters. The separate loop design aims to provide a less biased estimation of model generalization, which evaluates performance with tuned hyperparameters on a completely separate test set. Consequently, the overfitting risk is mitigated. Evaluation metrics contain mean absolute error (MAE), root mean squared error (RMSE), and coefficient of determination (R^2). Computation time in this paper indicates the total time required to execute the training and test process on a local 2.4GHz system. Shapley additive explanation (SHAP), a cooperative game-theoretical based approach, is employed to explain the individual feature contribution.

Results and Discussion

Training process with Split 4 (35260 instances) obtains the best predictive performance in comparison to other splits, as shown in Table 77.2. Continuous improvement indicates that models constantly learn from features as the training set expands. More training instances can leverage prediction accuracy by providing additional information. Overall, we find that advanced machine learning models outperform conven-

Table 77.1 Independent variables for order lead time prediction with brief description (FF: frontend, BF: backend, PO: order entry date)

Category	Variables	Description	Types
Delivery	$DW+$	Positive delivery window [days]	Numerical
	$DW-$	Negative delivery window [days]	Numerical
Customer	$Region$	Shipping-to regions	Categorical
	$Country$	Shipping-to countries	Categorical
Order	$Order\ Volume^{update}$	Updated order volume [pieces]	Numerical
	$Order\ Volume^{initial}$	Initial order volume [pieces]	Numerical
	$Order\ Cost^{update}$	Updated order cost in Euro	Numerical
	OLT_{req}^{update}	Updated requested order lead time [days]	Numerical
	$OLT_{req}^{initial}$	Initial requested order lead time [days]	Numerical
	$Change_{req}$	$\Delta(OLT_{req}^{update}, OLT_{req}^{initial})$	Numerical
	$Delta_{req}$	$\Delta(OLT_{req}^{update}, SDT^{update})$	Numerical
	$PO\ Month^{update}$	Month of latest order modification	Numerical
	$PO\ Month^{initial}$	Month of initial order entry	Numerical
	$PO\ CW^{update}$	Calendar week of latest order modification	Numerical
	$PO\ CW^{initial}$	Calendar week of initial order entry	Numerical
	$PO\ Year^{update}$	Year of latest order modification	Numerical
	$PO\ Yearinitial$	Year of initial order entry	Numerical
	$FF\ category$	Category of freeze fence	Categorical
Planning	SDT^{update}	Standard delivery time for updated order [days]	Numerical
	SDT^{update}	Standard delivery time for initial order [days]	Numerical
	$Yield_{FE}$	Cumulative planned yield at FE [%]	Numerical
	$Yield_{FE}$	Cumulative planned yield at BE [%]	Numerical
	CT_{FE}	Cumulative planned cycle time at FE [weeks]	Numerical
	CT_{BE}	Cumulative planned cycle time at BE [weeks]	Numerical
	FF_{FE}	Cumulative planned freeze fence at FE [weeks]	Numerical
	FF_{BE}	Cumulative planned freeze fence at BE [weeks]	Numerical
	FF_{total}	$FF_{total} = FF_{FE} + FF_{BE}$ [weeks]	Numerical
	HFG	Main product group	Categorical
Product	CPW	Chip per wafer [pieces]	Numerical
	$Wafer\ diameter$	Wafer size [mm]	Numerical
	$Delta_{rampup}$	$\Delta(PO^{update}, Ramp\ up\ Date)$	Numerical
	BT	Chip category	Categorical

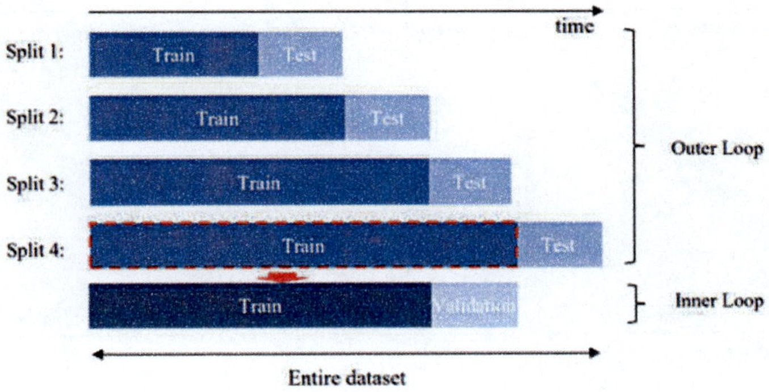

Fig. 77.1 Four-split prediction experiment with expanding training sets

Table 77.2 Prediction results on test set of Split 4, most accurate predictions and shortest computation time in bold

Model	MAE (days)	RMSE (days)	R^2	Computation time (s)
lr	16	21	0.83	0.13
las	25	28	0.67	0.20
rig	16	21	0.83	**0.11**
svr	15	25	0.82	178.04
rf	22	26	0.76	9.24
reg	12	18	0.90	2.26
xgbr	**12**	19	0.89	0.78
lgbm	**12**	**17**	**0.91**	0.23
NN	13	**17**	0.89	166.86

tional regression approaches and tree-based models yield superior prediction results. The fine-tuned feed-forward neural network achieves reasonable predictions with optimized hyperparameters, consisting of two hidden layers of 80 neuron units each and a dropout rate of 0.3. However, given the similar extensive computing time, ANN (similarly for svr) is only conditionally applicable in industrial operations. In contrast, boosting algorithms exhibit strong predictive power, while requiring low computation time with usage of stage-wise additive expansion. Tuning the hyperparameters alleviates the overfitting problem of boosting models by reducing model complexity. Among them, LightGbM with unique leaf-wise growth mechanism obtains the lowest MAE, RMSE, highest R^2 and meanwhile least computation time for Split 4.

The scatter plots in Fig. 77.2 illustrate the individual prediction and the corresponding true values, confirming the results from Table 77.2. Random forest produces the same prediction values despite the different magnitude of the independent variables. The abnormal prediction behaviour is attributed to the node split criterion based on certain input variables. This is since the tested models usually generate either overestimated ($OLT_{pred} > OLT_{real}$) and underestimated predictions ($OLT_{pred} < OLT_{real}$). Lasso regression overestimates most observations (94.07 %), while gradient boosting (62.97 %) gives the most frequent underestimated results.

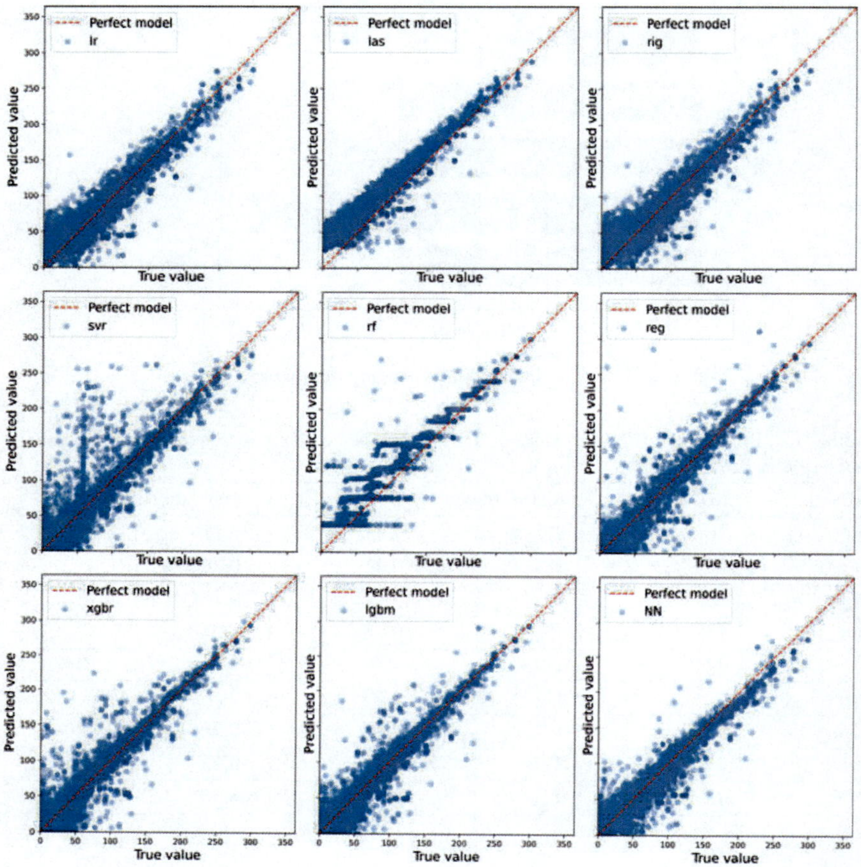

Fig. 77.2 Prediction observations on test set of Split 4

We analyze SHAP values for the best performing boosting algorithms. In addition to order information, business cycle (ramp-up date) and planning parameters (cycle time and freeze fence) are most important for prediction, which is in line with related literature and expert proposals. The ramp-up date reflects the maturity of the product and production process. An earlier ramp-up date indicates a more mature high-yield manufacturing process with less occurrence of bottlenecks. Consequently, shorter order lead time can be expected. Planned cycle time is an expert-defined average cycle time for the certain sales product. The cycle time is usually included in the lead time, which measures exclusively the time frame throughout the manufacturing process. Additionally, production freeze fence represents the frozen time frame, when the modification of production scheduling is prohibited. The longer freeze fence is usually for products with complicated manufacturing processes and excessive cycle time to secure order fulfillment.

Conclusion

Accurately predicting order lead times is critical, particularly for high-tech enterprises with complex global supply chain networks such as semiconductor manufacturers. We propose an AI-empowered prediction framework that is tested with real-world data. Notably, boosting algorithms demonstrate reliable predictive performance using the experimental design. The best performing lgbm model yields prediction errors of less than two weeks, which is a reasonable performance in practice given lead times of multiple months at the semiconductor manufacturer. Considering the potential of neural networks, it is worth investigating advanced network architectures to improve the accuracy and computation time. Asymmetrical penalty costs should be integrated into models to account for the consequences of prediction errors. This is important as overestimation of lead times in the semiconductor supply chain indicates a decrease in commercial competitiveness since competitors may offer shorter promised lead times. Yet, underestimation leads to unsatisfied customers and subsequent penalties. Furthermore, investigating other product lines and time frames is necessary. The latter may include the observation of periods longer than two years and/or during heavy demand fluctuations as occurred after the Covid pandemic outbreak. Together with analyzing additional exogeneous variables (business cycle, weather information, etc.), we may be able to further assess the model's generalizability and robustness. Finally, specific strategies and actionable items need to be defined on how to use the predictive information and underlying features for enhanced supply chain operations.

References

1. Schelthoff, K., Jacobi, C., Schlosser, E., Plohmann, D., Janus, M., & Furmans, K. (2022). Feature selection for waiting time predictions in semiconductor wafer fabs. *IEEE Transactions on Semiconductor Manufacturing*.
2. Lingitz, L., Gallina, V., Ansari, F., Gyulai, D., Pfeiffer, A., Sihn, W., & Monostori, L. (2018). Lead time prediction using machine learning algorithms: A case study by a semiconductor manufacturer. *Procedia Cirp, 72*.
3. Burggräf, P., Wagner, J., Koke, B., & Steinberg, F. (2020). Approaches for the prediction of lead times in an engineer to order environment-A systematic review. *IEEE Access* .
4. Singh, S., & Soni, U. (2019). Predicting order lead time for just in time production system using various machine learning algorithms: A case study. In: *2019 9th International Conference on Cloud Computing, Data Science and Engineering (Confluence)* (pp. 422-425). IEEE.
5. Zhang, C., Yella, J., Huang, Y., & Bom, S. (2021). Learning from failures in large scale soft sensing. *IEEE International Conference on Big Data* .
6. Mahajan, T., Singh, G., Bruns, G., Bruns, G., Mahajan, T., & Singh, G. (2021). An Experimental assessment of treatments for cyclical data. In *Proceedings of the 2021 Computer Science Conference for CSU Undergraduates, Virtual* , (vol. 6).
7. Vien, B. S., Wong, L., Kuen, T., Rose, L. F., & Chiu, W. K. (2021). A machine learning approach for anaerobic reactor performance rediction using long short-term memory recurrent neural network. *Structure Health Monitoring, 18*, 61

8. Zhang, Y., & Haghani, A. (2015). A gradient boosting method to improve travel time prediction. *Transportation. Research. Part C: Emerging Technologies*.
9. Kuhn, M., & Johnson, K. (2013). *Applied predictive modeling* (Vol. 26). Springer.
10. Parvandeh, S., Yeh, H. W., Paulus, M. P., & McKinney, B. A. (2020). Consensus features nested cross-validation. *Bioinformatics, 36*(10).

Chapter 78
Impact Analysis of Extended Payment Terms in Food Supply Chains During a Demand Shortfall

Alexander Zienau, Mahdi Alazzeh, Ole Hansen, Christina Imdahl, Marcus Wiens, and Frank Schultmann

Abstract The current pandemic has disrupted many supply chains, among them food supply chains, and their physical and financial flows. Therefore, companies with high bargaining power may extend their payment terms to downstream suppliers as a reaction to decreased financial cash flow. We model a stylized three-stage food supply chain and use simulation to analyze the effects of a demand shortfall. We then investigate the effects of payment term extensions by different companies. We find that although extending payment terms can be beneficial for a single company in the short run, it will harm the supply chain in the long-run and conclude that incentives should be put into place to motivate companies accordingly.

Keywords Supply chain management · Disaster and crisis management · Simulation

Introduction

The COVID-19 pandemic has caused many different disruptions and challenges for single businesses, but also for whole supply chains. Companies fell short on paying their liabilities as less goods were sold, or they struggled receiving the supplies as producers were closing their productions [1].

A. Zienau · M. Alazzeh · O. Hansen
4flow AG, Hallerstr. 1, 10587 Berlin, Germany

A. Zienau (✉) · F. Schultmann
Karlsruhe Institute of Technology, Institute for Industrial Production, Hertzstr. 16, 76187 Karlsruhe, Germany
e-mail: a.zienau@4flow.com

C. Imdahl
Industrial Engineering and Innovation Sciences, Eindhoven University of Technology, PO Box 513, 5600 MB Eindhoven, Netherlands

M. Wiens
Technical University Bergakademie Freiberg, Akademiestraße 6, 09596 Freiberg, Germany

O. Grothe et al. (eds.), *Operations Research Proceedings 2022*, Lecture Notes in Operations Research, https://doi.org/10.1007/978-3-031-24907-5_78

In response to disruptions during the pandemic, companies with high market power extended their own payment terms to increase their liquidity in the short-term [2]. However, this can harm long-term supply chain performance [3]. This can be especially impactful in food supply chains, where profit margins are low and few companies control a large share of the market downstream (i.e. retail and wholesale companies).

In this work, we study the relation between the physical flow of goods and the financial flow. Although there is a direct relationship between extended payment terms and increased delivery risks, the consequences of extending payment terms in a supply chain are so far scarcely researched. To our knowledge, Esenduran et al. [3] is one of the few studies which looks into the extension of payments in supply chains and the related effects. Esenduran et al. [3] analyze the impact of extended payment terms in combination with supply chain risk management measures such as quick response and backup suppliers and show how extended payment terms can harm the buyer's operations. In another study, Tsai [5] model cash flow risk in supply chains and show how decreasing a firm's cash conversion cycle (CCC) increases cash flow risks.

For a stylized multi-echelon food supply chain consisting of a supplier, a producer and a wholesaler, we develop a model in which we design physical flows to meet the downstream service level with a (s,S) order policy. Inspired by lock-downs during the COVID-19 pandemic, we analyze how a shortfall of demand in gastronomy and payment extensions of upstream companies impact supply chain performance using simulation. In this regard, we also evaluate the results of adjusted financial flows using each stage's CCC.

We find that while extending payment terms can increase the CCC of a company in the short run, it can greatly harm the supply chain, and thus the company which increased its payment terms, in the long-run. Our study enables researchers and practitioners to analyze the interplay between physical and financial flows in crises and to evaluate corporate decisions on an extension of payment. Thereby, it can also serve as decision support for public actors' crisis management that affects commercial supply chains.

Model

We set up a stylized model of a three-echelon food supply chain, consisting of a supplier (*Stage 3*), a producer (*Stage 2*), and a wholesaler (*Stage 1*), where the latter serves external aggregated demand from the gastronomy sector (*Stage 0*), see Figure 78.1. This will allow us to analyze how adjusting extended payment terms at the wholesaler or producer can affect performance and balance of the whole supply chain. Decisions within the supply chain can be made each day t. Aggregated demand from the gastronomy received by the wholesaler is assumed to be uniformly distributed: $D_{1,t} \sim \mathcal{U}(900, 1100)$.

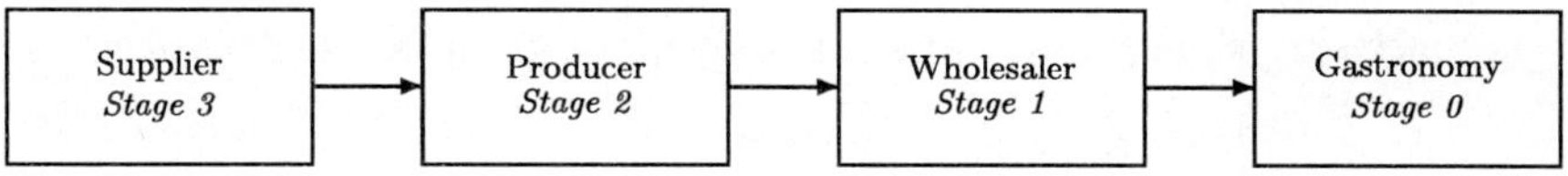

Fig. 78.1 Serial system

The wholesaler however, will only accept this demand (partially or fully) conditionally. To avoid accumulating too much backlog, the amount of demand from the gastronomy, which the wholesaler can accept per day, is capped at the sum of the current inventory position and a maximum backlog $B_1^{\max}$. Hence, the net (accepted) demand at the wholesaler is defined as:

$$\tilde{D}_{1,t} = \min(B_{1,t-1} + D_{1,t}, I_{1,t} + B_1^{\max})$$

The physical flow from the producer to the wholesaler is denoted as $S_{1,t}$ and the flow from the wholesaler to the gastronomy as $S_{0,t} = \min(\tilde{D}_{1,t}, I_{1,t})$ accordingly. Therefore, the wholesaler's on-hand inventory at the end of any day can be described as:

$$I_{1,t} = I_{1,t-1} + S_{1,t} - S_{0,t}.$$

Backlog is calculated at the end of a day as $B_{1,t} = (\tilde{D}_{1,t} - S_{0,t})^-$. The wholesaler places an order $O_{1,t}$ at the producer following a (*s*,*S*) order policy. Hence, we introduce the target inventory level S_1, the reorder level s_1 and the producer's delivery time L_2 and define orders that have been placed at the producer but not yet delivered: $O_{1,t}^o = \sum_{i=1}^{t-1} O_{1,i} - \sum_{i=1}^{t-1} S_{1,i}$.

$$O_{1,t} = \begin{cases} S_1 - (I_{1,t} + O_{1,t}^o) + L_2 \cdot \mathbb{E}(D_{1,t}) + B_{1,t} & \text{if } I_{1,t} + O_{1,t}^o \leq s_1 \\ 0 & \text{if } I_{1,t} + O_{1,t}^o > s_1 \end{cases}$$

We denote the available monetary funds at the wholesaler before production at day t as $M_{1,t}$, and the planned time until payment from the gastronomy to the wholesaler as r_1 days. Given that p_1 is per unit price the gastronomy pays to the wholesaler, the payment received by the wholesaler at the beginning of a day is then $M_{0,t}^- = S_{0,t-r_1} \cdot p_1$. Daily operations at the wholesaler also generate fixed cost C_1^{fix} and the wholesaler has to cover inventory-dependent storage costs $C_{1,t}^h = I_{1,t-1} \cdot c_1^h$.

Each day, the wholesaler pays the producer an amount $P_{2,t}$, which includes planned payments and delayed payments. These payments are subject to the per unit price p_2 the wholesaler agreed upon with the producer. It follows that $P_{2,t} = S_{1,t} \cdot p_2$ is the planned part of the payment of the wholesaler to the producer. Additionally, we define the payment extension of a liability as $\Delta_{2,t} = r_2 + k$. Here, we denote u as the interest rate for late payment and k as the earliest delay at which the wholesaler can pay his liabilities plus interest $P_{2,t} \cdot (1 + k \cdot u) \leq M_{1,t+\Delta_{2,t}}$.

In conclusion, all payments by the wholesaler for a given day are defined as

$$M_{1,t}^{-} = \sum_{n:\{n+\Delta_{2,n}=t\}} P_{2,n}.$$

Hence, the available monetary funds at the wholesaler before production starts are

$$M_{1,t} = M_{1,t-1} - C_1^{fix} + M_{0,t}^{-} - M_{1,t}^{-} - C_{1,t}^{h}.$$

Similar processes apply to the producer. The producer receives payments $M_{1,t}^{-}$ from the wholesaler each day, and has to pay his liabilities, fixed costs, inventory holding costs and production costs. The payables are $P_{3,t} = S_{2,t} \cdot p_3$ and in t the producer pays $M_{2,t}^{-} = \sum_{n:\{n+\Delta_{3,n}=t\}} P_{3,n}$. Inventory holding costs are assessed for raw material and finished good, such that $C_{2,t}^{h} = I_{2,t-1}^{raw} \cdot c_{2,raw}^{h} + I_{2,t-1} \cdot c_2^{h}$. The production costs $C_{2,t}^{p}$ are dependent on the financial ability to fund production and the available money for production is thus

$$M_{2,t} = M_{2,t-1} - M_{2,t}^{-} - C_2^{fix} + M_{1,t}^{-} - C_{2,t}^{h} - C_{2,t-1}^{p}.$$

Based on the available monetary funds the production quantity $Q_{2,t}$ is determined. The producer aims to produce a standard quantity q equivalent to his mean demand until he reached his target inventory s_2, which is set to ensure downstream availability. The producer has limited flexibility to increase or decrease his production dependent on demand realization as follows:

$$Q_t^{plan} = \begin{cases} 0.9q & \text{if } I_{2,t-1} \geq s_2 \\ q & \text{if } 0 \leq I_{2,t-1} < s_2 \\ 1.1q & \text{if } B_{2,t-1} > 0 \end{cases}$$

The realized production quantity is determined by the amount of available raw material and the available monetary funds $M_{2,t}$ because production generates costs:

$$Q_{2,t} = \min(Q_t^{plan}, M_{2,t}/c_2^{p}, I_{2,t-1}^{raw})$$

Consequently, production costs of $C_{2,t}^{p} = Q_{2,t} \cdot c_2^{p}$ emerge each day.

Following production, the producer ships the number of goods ordered by the wholesaler: $S_{1,t+L_2} = \min(O_{1,t}, I_{2,t-1} + Q_{2,t})$.

Hence, by the end of the day, the inventory level of raw and finished goods can be calculated as follows:

$$I_{2,t}^{raw} = I_{2,t-1}^{raw} - Q_{2,t} + S_{2,t} \quad \text{and} \quad I_{2,t} = I_{2,t-1} + Q_{2,t} - S_{1,t}$$

The producer orders new raw material from the supplier according to a (s,S) policy. Utilizing the outstanding orders $O_{2,t}^{o} = \sum_{i=1}^{t-1} O_{2,i} - \sum_{i=1}^{t-1} S_{2,i}$ this is

$$O_{2,t} = \begin{cases} S_2^{raw} - I_{2,t}^{raw} + O_{2,t}^o + q \cdot L_3 + B_{1,t} & if I_{2,t}^{raw} + O_{2,t}^o < s_2^{raw} \\ 0 & if I_{2,t}^{raw} + O_{2,t}^o \geq s_2^{raw} \end{cases}$$

where s_2^{raw} is set to cover a 90% service level and S_2^{raw} to cover average lead time demand.

The supplier receives the incoming payments from the producer $M_{2,t}^-$ and pays its fixed costs C_3^{fix} and the production costs from the previous day $C_{3,t-1}^p$.

$$M_{3,t} = M_{3,t-1} + M_{2,t}^- - C_i^{fix} - C_{3,t-1}^p$$

After that, the supplier aims to supply the number of units ordered by the producer. If the supplier's money is not sufficient for production, as many units of raw material are purchased as the the supplier can financially afford. c_3^p is a price per unit.

$$Q_{3,t} = \min(O_{2,t}, M_{3,t}/c_3^p)$$

Consequently, production costs of $C_{3,t}^p = Q_{3,t} \cdot c_3^p$ emerge each day. After his delivery time, the supplier delivers the goods to the producer, i.e. $S_{2,t+L_2} = \min(O_{2,t}, I_{3,t-1} + Q_{3,t})$.

Results

During normal times, the physical and financial flows in supply chain are balanced around the mean demand per day - cost and prices are in a relation that generates a margin for each stage. The price at which the supplier sells the unprocessed good is set at $p_3 = 10$. He encounters fixed costs per day of $C_3^{fix} = 5000$ and production costs of $c_3^p = 4$ per unit. This leaves him with a margin of 10%. After processing, the producer sells the good for a price of $p_2 = 20$ per unit. With this revenue, he covers fixed costs of $C_2^{fix} = 6000$, production costs of $c_2^p = 2$ per unit and inventory costs (Weighted Average Cost of Capital) for both his stock of raw and finished goods at the end of each day, at a rate of $c^h = 0.05$ times the unit price (p_3 for raw, p_2 for finished goods), per year. This equals a margin of slightly less than 10%, depending on the daily inventory costs that apply. The margins used are in line with prevalent values in the food processing industry.

The wholesaler uses his large network to sell the finished good to the gastronomy, at a price of $p_1 = 30$. He pays fixed costs of $C_1^{fix} = 8000$ and is subject to the inventory holding cost rate c^h. He is left with a margin of over 6%, which is less than in the food processing industry, but a commonly low value for food trade and retail in general, as it is a very competitive market.

The maximum backlog the wholesaler will allow is half of average daily demand at $B_{1,t} = 500$. Each payment outside a crisis is due after 30 days, for all stages, based on

Table 78.1 Scenarios and results

Scenario	Service level	Disruption				CCC		
		Sup	Prod	Wh	Breakdown	Sup	Prod	Wh
Normal	90.61	0.00	0.10	0.00	No	31.99	6.87	6.55
Crisis	84.70	0.00	0.60	1.70	No	32.20	7.88	7.46
$\Delta r_1 = 15$	85.01	0.00	0.30	1.80	No	32.20	22.78	– 3.44
$\Delta r_1 = 30$	85.04	0.00	0.60	1.30	No	32.23	40.48	– 13.57
$\Delta r_2 = 15$	42.01	180.60	121.90	113.80	311.90	41.85	– 10.41	1.75
$\Delta r_2 = 30$	34.75	233.30	207.30	199.90	200.40	49.82	– 19.46	3.75
Δr_1 and $\Delta r_2 = 15$	42.95	177.40	117.90	118.20	321.10	42.12	3.08	– 11.58
Δr_1 and $\Delta r_2 = 30$	34.22	233.10	207.90	201.30	200.20	49.69	0.58	– 22.67

average values from practice. The interest rate for late payments is set at $u = 8.12\%$ per year, in line with the number currently issued by the German Central Bank. The starting capital for the supplier amounts to over one month of fixed production cost, the values for the producer and wholesaler are set to cover four months, in line with their higher market power. Delivery times between stages are set at $L_2 = L_3 = 5$ days. As a reaction to the fluctuation in demand, the reorder points and the target inventory levels are set to $s_1 = s_2^{raw} = 5400$ and $S_1 = S_2^{raw} = 10400$ accordingly. This ensures meeting the downstream service level of 90% at the wholesaler stage.

To analyze the consequences of a negative shock in demand, as it was caused by the current pandemic during the first lockdown(s), we assume that the base value of demand (the average) is reduced to 40% of its former value. Hence, in case of a lockdown $D_{1,t}^{shock} \sim \mathcal{U}(360, 440), \forall t \subseteq T^{shock}$ applies in the model and production capacities, reorder points and target inventory levels are adjusted accordingly. The demand shock lasts from day 40 (warm-up period) to 100 in the simulation.

We evaluate the effects of this disruption on the supply chain in Table 78.1 using different indicators. The *Service level* describes the delivery reliability downstream, where the wholesaler satisfies the demand of the gastronomy. A *Disruption* at the supplier stage is captured by the total days the supplier has insufficient monetary funds to produce any raw material. Each day the producer does not have any raw material, production is disrupted at this stage. Lastly, any day the wholesaler has to turn down demand from the gastronomy because it exceeds his maximum backlog level, is also counted as a disruption. The column *Breakdown* indicates whether a supply chain member went bankrupt and at which day of the simulation this occurred. We also evaluate the financial performance of each supply chain stage using the CCC. It measures how many days it takes to convert expenses into income from customers by taking into account *Days Inventory Held*, *Days Sales Outstanding* and *Days Payables Outstanding*. The lower the CCC, the better the financial performance of a company. A negative CCC is possible for companies with high bargaining power who receive money from their customers before they pay their suppliers [4]. All results displayed in Table 78.1 are the average over 10 simulation runs.

When comparing the first two scenarios, *Normal* and *Crisis*, we find that the crisis causes minor disruptions at downstream stages and also negatively affects the downstream service level and the CCC of each stage. The decreased financial performance can lead companies with high bargaining power to extend their payment terms in order to bring their CCC down. Hence, we investigate what happens within the supply chain when the wholesaler or the producer, or both, extend their payment terms to 45 or 60 days. Δr_1 (wholesaler) and Δr_2 (producer) describe scenarios with extended payment terms of either 15 or 30 days. Note that a value of 60 days is currently the legally allowed maximum, set by Directive 2019/633/EU.

It improves the wholesaler's CCC compared to the crisis scenario, when he is the only supply chain partner that extends his payment terms (15 or 30 days). It comes at a disproportionate cost for the producer in terms of CCC though. Furthermore, the downstream service level does not improve significantly. If only the producer extends his payment terms, it also decreases the producer's CCC. However, this time the cost for the supply chain is even greater. While this intervention also lowers the CCC of the wholesaler, it greatly increases the CCC of the supplier, to the degree where disruptions occur frequently at all stages. This culminates in the supplier going bankrupt on average at day 312 or 200, depending on the degree of the payment term increase. If both the wholesaler and the producer extend their payment terms at the same time, the effects are similar to those of an extension by only the producer.

Conclusion

Although the crisis has negative effects on all stages of the supply chain, in our scenarios it is actually best if no company in the supply chain extends its payment terms and the whole supply chain bears the consequences of the crisis. This underlines the complicated situation a company might find itself in. While extending payment terms can be an attractive option in the short-run, it is harmful for the whole supply chain in the long run, and thus also for the company that extends the payment terms itself. Yet, there is evidence from practice that this was often the case [2].

An implication for managers is thus that they should consider carefully whether extending payment terms to achieve a small financial advantage is worth the potential aftermath, such as liquidity risks, along the supply chain. Typically, companies are not linked to a single upstream and a single downstream partner. Consequently, a company's liquidity risk that results from the downstream partner's payment term extension increases with a greater sales share with the downstream partner. Moreover, future research could investigate price changes as a response to extended payment terms. We also conclude that, while the Directive 2019/633/EU already limits the possible magnitude of payment term extensions, additional incentives to avoid extensions or affordable (free) loans for companies in need can be worthwhile options.

Acknowledgements This research was supported by the German Federal Ministry of Education and Research (BMBF) in the NOLAN project (grant numbers 13N14457 and 13N14459).

References

1. Barman, A., Das, R., & De, P. K. (2021). Impact of COVID-19 in food supply chain: Disruptions and recovery strategy. *Current Research in Behavioral Sciences, 2*, 100017. https://doi.org/10.1016/j.crbeha.2021.100017
2. Di Marcantonio, F., Solano-Hermosilla, G., Ciaian, P. (2022). The COVID-19 pandemic in the agri-food supply chain: Impacts and responses (EUR 31009 EN). *Publications Office of the European Union*. https://doi.org/10.2760/911133
3. Esenduran, G., Gray, J. V., & Tan, B. (2022). A dynamic analysis of supply chain risk management and extended payment terms. *Production and Operations Management, 31*(3), 1394–1417. https://doi.org/10.1111/poms.13620
4. Stewart, G. (1995). Supply chain performance benchmarking study reveals keys to supply chain excellence. *Logistics Information Management, 8*(2), 38–44. https://doi.org/10.1108/09576059510085000
5. Tsai, C. -Y. (2008). On supply chain cash flow risks. *Decision Support Systems, 44*(4), 1031–1042. https://doi.org/10.1016/j.dss.2007.12.006

Chapter 79
The Lot-Size Adaptation Approach for the Two-Level Stochastic Capacitated Lot-Sizing Problem

Markus Mickein and Knut Haase

Abstract We introduce a two-level stochastic capacitated lot-sizing problem with random demand and service level constraints under the static and static-dynamic uncertainty strategy. While the static strategy determines setup periods and lot-sizes at the beginning of the planning horizon, the static-dynamic strategy allows adjustments of lot-sizes during the planning horizon. We present a model formulation of the demand difference adaptation policy for a multi-stage production system. Thereby, lot-size adaptations depend on demand differences between the realized and expected demand. We evaluate 144 test instances with different parameter settings to quantify the economic efficiency of the uncertainty strategies for the multilevel lot-sizing problem. The computational study demonstrates that additional costs of semifinished goods and scarcity of storage capacity on upstream precesses reduce the cost saving potential of lot-size adaptations.

Keywords Stochastic programming · Production and inventory systems · Supply chain management

Introduction

Production planning is subject to demand uncertainty. Neglecting stochastic demand leads to service level violations. Therefore, Bookbinder and Tan (1988) introduce the static, dynamic, and static-dynamic uncertainty strategies for lot-sizing under random demand [1]. The static strategy determines setup periods and lot-sizes at the beginning of the planning horizon. The dynamic strategy allows adjustments of setup and lot-size decisions during the planning horizon. The static-dynamic strategy fixes setup periods for the entire planning horizon but enables subsequent adjustments of lot-sizes.

While the static strategy determines robust production schedules, the static-dynamic strategy provides flexible schedules associated with higher planning efforts.

M. Mickein (✉) · K. Haase
Institut für Verkehr, Logistik und Produktion, Universität Hamburg, Hamburg, Germany
e-mail: Markus.Mickein@uni-hamburg.de

O. Grothe et al. (eds.), *Operations Research Proceedings 2022*, Lecture Notes in Operations Research, https://doi.org/10.1007/978-3-031-24907-5_79

However, the static-dynamic strategy achieves lower costs than the static strategy for the single-level lot-sizing problem. Nevertheless, adaptable lot-sizes of finished goods require sufficient provision with semifinished goods. Therefore, it is mandatory to incorporate potential previous production stages.

Because the well-known inventory-oriented order-up-to-level policy is only suitable for uncapacitated problems without material requirements, we apply the demand-oriented demand difference adaptation policy with limited adjustments [2]. Thereby, lot-size adaptations depend on demand differences between realized and expected demand. In this paper, we present the model formulation for two-level stochastic capacitated lot-sizing problem (2L-SCLSP) of the limited demand difference adaptation policy.

This study quantifies the economic efficiency of applying the static-dynamic strategy in multilevel production systems under additional costs for semifinished goods and scarcity of storage capacity on upstream processes.

Problem Description

The considered two-level lot-sizing problem is subject to demand uncertainty. Several scenarios s within a scenario sample $\mathcal{S}$ model the stochastic demand. The production problem contains $|\mathcal{T}|$ periods t and $|\mathcal{J}|$ products j. The multilevel bill of material $\mathcal{J}_j$ defines the predecessor-successor relationship between finished and semifinished goods $\hat{\mathcal{J}}$.

We make two assumptions for the different production stages: First, the semifinished goods follow the static strategy and the finished goods follow the static or rather static-dynamic strategy. Applying the static strategy for semifinished goods smooths the fluctuation beyond the production system to avoid the bullwhip effect. Second, the demand for finished goods can be backordered under consideration of the γ service level, but the need for semifinished goods must be immediately fulfilled. Satisfying the need for semifinished goods is necessary to ensure the production capability of finished goods.

The classical lot-sizing problem determines setup periods $X_{t,j}$, lot-sizes $Q_{t,j}$, and inventories $I_{s,t,j}$. Due to the acceptance of backorders, the net inventory consists of backlog $I^-_{s,t,j}$ and physical stock $I^+_{s,t,j}$. We limit backlogs to fulfil a given γ service level. The objective function minimizes the expected holding and setup costs by the cost rate of holding h and setup v. The production capacity c_r limits production time corresponding to the time rate of production p for products $\mathcal{J}_r$ that are allocated to resource r. The production factor $f_{j,i}$ defines the required quantity of product i to produce one unit of product j.

We allow lot-size adaptations $A_{s,t,j}$ based on cumulative demand differences $\Delta_{s,t,j}$ between scenario demand $d_{s,t,j}$ and expected demand $\bar{d}_{t,j}$. Previous adjustments must be considered to avoid double counting (79.2).

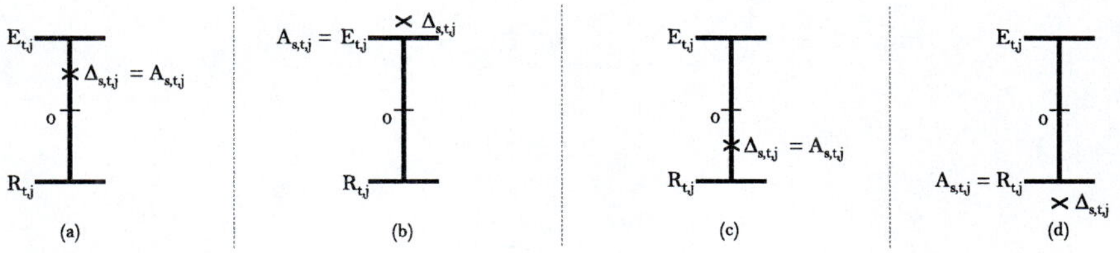

Fig. 79.1 Mechanism of the lot-size-adaptation approach

$$\Delta_{s,t,j} = \sum_{\tau<t} (d_{s,\tau,j} - \bar{d}_{\tau,j}) \qquad \forall s, t, j \tag{79.1}$$

$$A'_{s,t,j} = \Delta_{s,t,j} - \sum_{\tau<t} A_{s,\tau,j} \qquad \forall s, t, j | X_{t,j} = 1 \tag{79.2}$$

We limit the lot-size adaptation by a maximum extension $E_{t,j}$ and maximum reduction $R_{t,j}$. The binary decision variables $Y_{s,t,j}$ and $Z_{s,t,j}$ indicate adaptations by the maximum extension or reduction quantity and prevent arbitrary values. The variable $Y_{s,t,j}$ is one if the theoretical adaptation exceeds the maximum extension. Analogously, $Z_{s,t,j}$ is one if the theoretical adaptation falls below the maximum reduction.

$$A_{s,t,j} = \min\{E_{t,j}, A'_{s,t,j}\} \quad \forall s, t, j | A'_{s,t,j} > 0 \tag{79.3}$$

$$A_{s,t,j} = \max\{-R_{t,j}, A'_{s,t,j}\} \quad \forall s, t, j | A'_{s,t,j} < 0 \tag{79.4}$$

Figure 79.1 illustrates the mechanism of the lot-size adaptation approach. Note that the example shows the first production cycle without previous adjustments. Figure 79.1(a) shows the case where the lot-size adaptations correspond to positive cumulative demand differences. Figure 79.1(b) shows that the positive cumulative demand differences exceed the maximum extension quantity. Figure 79.1(c) shows the case where the lot-size adaptations correspond to negative cumulative demand differences. Figure 79.1(d) shows that the negative cumulative demand differences fall below the maximum reduction quantity.

Model Formulation

The model formulation of the 2L-SCLSP based on [3]. We adapt the original formulation by using a γ service level, considering a multilevel production system, and applying the static-dynamic uncertainty strategy.

$$\min \quad F = \frac{1}{|S|} \sum_{s} \sum_{t} \sum_{j} h \cdot I^{+}_{s,t,j} + \sum_{t} \sum_{j} v \cdot X_{t,j} \tag{79.5}$$

s.t.

$$I_{0,j} + Q_{t,j} + A_{s,t,j} - d_{s,t,j} - \sum_{i \in \mathcal{J}_j} f_{i,j}(Q_{t,i} + A_{s,t,i}) = I_{s,t,j} \quad \forall s, t = 1, j \tag{79.6}$$

$$I_{s,t-1,j} + Q_{t,j} + A_{s,t,j} - d_{s,t,j} - \sum_{i \in \mathcal{J}_j} f_{i,j}(Q_{t,i} + A_{s,t,i}) = I_{s,t,j} \quad \forall s, t > 1, j \tag{79.7}$$

$$I^{+}_{s,t,j} \geq I_{s,t,j} \quad \forall s, t, j \tag{79.8}$$

$$I^{-}_{s,t,j} \geq -I_{s,t,j} \quad \forall s, t, j \tag{79.9}$$

$$Q_{t,j} \leq m \cdot X_{t,j} \quad \forall t, j \tag{79.10}$$

$$\sum_{t} Q_{t,j} \geq \sum_{t} \bar{d}_{t,j} + \sum_{t} \sum_{i \in \mathcal{J}_j} f_{i,j} \cdot \bar{d}_{t,i} \quad j \tag{79.11}$$

$$\sum_{j \in \mathcal{J}_r} p(Q_{t,j} + E_{t,j}) \leq c_r \quad \forall t, r \tag{79.12}$$

$$\sum_{t} \sum_{s} I^{-}_{s,t,j} \leq (1-\gamma) \cdot |S| \cdot \sum_{t} \bar{d}_{t,j} \quad \forall j \tag{79.13}$$

$$X_{t,j} \in \{0, 1\} \quad \forall t, j \tag{79.14}$$

$$I^{+}_{s,t,j}, I^{-}_{s,t,j}, I_{0,j}, Q_{t,j} \geq 0 \quad \forall s, t, j \tag{79.15}$$

The objective function (79.5) minimizes the expected holding and setup costs. The inventory balance for the first period (79.6) and the remaining planning horizon (79.7) contains the net inventory, current production quantity, demand for finished goods, and need for semifinished goods. Equation (79.8) determines the physical inventory by the positive net inventory and Equation (79.9) determines the backlog by the negative net inventory. Equation (79.10) ensures the setup condition for the production. Equation (79.11) makes sure to produce the expected demand within the planning horizon to guarantee the production capability beyond the planning horizon. The capacity constraint (79.12) limits the production time by the available capacity. The service level constraint (79.13) limits backlogs by a γ service level.

Equations (79.18) to (79.25) employ the demand difference adaptation policy. Note that lot-size adaptations and backlogs of semifinished goods are fixed to zero due to the required immediately demand fulfilment and applied static uncertainty strategy.

$$A_{s,t,j} \leq \Delta_{s,t,j} - \sum_{\tau|\tau<t} A_{s,\tau,j} + m \cdot (1 - (X_{t,j} - Z_{s,t,j})) \quad \forall s, t, j \tag{79.16}$$

$$A_{s,t,j} \geq \Delta_{s,t,j} - \sum_{\tau|\tau<t} A_{s,\tau,j} - m \cdot (1 - (X_{t,j} - Y_{s,t,j})) \quad \forall s, t, j \tag{79.17}$$

$$A_{s,t,j} \leq E_{t,j} \quad \forall s, t, j \tag{79.18}$$

$$-A_{s,t,j} \leq R_{t,j} \quad \forall s, t, j \tag{79.19}$$

$$A_{s,t,j} \geq E_{t,j} - m \cdot (1 - Y_{s,t,j}) \quad \forall s, t, j \tag{79.20}$$

$$-A_{s,t,j} \geq R_{t,j} - m \cdot (1 - Z_{s,t,j}) \quad \forall s, t, j \tag{79.21}$$

$$E_{t,j} \leq m \cdot X_{t,j} \quad \forall t, j \tag{79.22}$$

$$R_{t,j} \leq Q_{t,j} \quad \forall t, j \tag{79.23}$$

$$Y_{s,t,j}, Z_{s,t,j} \in \{0, 1\} \quad \forall s, t, j \tag{79.24}$$

$$E_{t,j}, R_{t,j} \geq 0 \quad \forall s, t, j \tag{79.25}$$

Equation (79.16) and (79.17) determine the lot-size adaptation by the cumulative demand difference and previous adjustments. Equations (79.18) and (79.19) limit the lot-size adaptation by the maximum extension and reduction quantity. Equation (79.20) determines the lot-size adaptation when the cumulative demand and previous adjustments exceed the maximum extension. Analogously, Equation (79.21) determines the lot-size adaptation when the cumulative demand and previous adjustments fall below the maximum reduction. Equations (79.22) and (79.23) ensure the setup condition for adjustments and nonnegative production quantities.

Equation (79.26) considers the storage limitations on upstream processes.

$$\sum_{j \in \hat{\mathcal{J}}} (I_{0,j} + \sum_{\tau=0}^{t} Q_{j,\tau}) - \sum_{j \notin \hat{\mathcal{J}}} \sum_{\tau=0}^{t} (Q_{j,\tau} - R_{j,\tau}) \leq w \qquad \forall t \tag{79.26}$$

The storage capacity constraint (79.26) contains the initial inventory and production quantities of semifinished goods as well as the minimum need for producing finished goods. The approximated capacity represents the maximum storage capacity for semi-finished goods which is limited by the storage capacity w.

Computational Study

We evaluate 144 test instances to quantify the economic efficiency of the uncertainty strategies under different parameter settings. The test instances differ regarding the service level (SLV = {0.90, 0.98}), demand variation coefficient (VCO = {0.1, 0.3}), time between orders (TBO = {1, 5}), target capacity utilization (TCU = {0.80, 0.95}), cost ratio between semifinished and finished goods (CPL = {0.1, 0.2, 0.5}), and scarcity of storage capacity on upstream processes (SSC = {1.00, 0.50, 0.25}). We consider a divergent product structure with 4 finished and 2 semifinished goods. The expected demand is uniformly distributed over time and the scenario demand is normally distributed over the scenarios. To generate scenario samples that meet the statistical properties of the distribution more accurately, we apply a discrete sampling technique for multi-period problems [4]. We assume the production factor, production time, and holding costs to be one unit. The production and storage capacity depend on given parameters such as the demand, time between orders, production time, target capacity utilization, and assumed scarcity.

We solve the optimization problem by the scenario extension approach [4]. This approach iteratively extends the scenario sample to improve the accuracy of the uncertainty approximation. Since we apply the static uncertainty strategy for semifinished products, only the number of binary variables of the setup decision rises. However, the additional constraints for considering semifinished goods increase the planning complexity. For this reason, we extend the acceptable computation time to guarantee feasible solutions.

Table 79.1 reports the objective value of the static and static-dynamic strategy depending on additional costs for semifinished goods and scarcity of storage capacity on upstream processes. All test instances fulfil the required service level in an out-of-sample simulation. The objective value of the static strategy changes only slightly despite additional costs and capacity limitations. The reason is that the static strategy does not require storing of semifinished goods. In contrast, the static-dynamic strategy needs storing due to the adjustments of lot-sizes. Thus, additional costs and capacity limitations affect the objective value.

Additional costs for semifinished goods reduce the cost savings by the static-dynamic strategy. Lower costs (CPL = 0.1) achieves cost savings of 0.6 % and higher costs (CPL = 0.5) reduce the cost savings to 0.1 %. This decrease is caused by a 55 % reduction of lot-size adaptation quantities under higher costs.

The static-dynamic strategy achieve less cost savings under scarcity of storage capacity on upstream processes. Even a moderate capacity limitation (SSC = 0.50) reduces the cost savings to 0.3 %. That is a reduction of 30 % compared to low capacity limitation (SSC = 1.00). This lower capability for storing for storing semifinished goods reduce the lot-size adaptation quantities by 48 %.

Table 79.1 Objective value depending on the additional cost for semifinsihed goods and scarcity of the storage resource on upstream processes.

	Static\Static 2L-SCLSP			Static-dynamic\Static 2L-SCLSP		
F(SSC)	CPL = 0.1	CPL = 0.2	CPL = 0.5	CPL = 0.1	CPL = 0.2	CPL = 0.5
F(1.00)	42009	42029	42034	41763	41860	41989
F(0.50)	42029	42032	42033	41858	41886	42016
F(0.25)	42027	42032	42034	41865	41917	42009

Conclusion

This study introduces the two-level stochastic capacitated lot-sizing problem under the static and static-dynamic uncertainty strategy. The computational study demonstrates that costs for semifinished goods and scarcity of storage capacity on upstream processes reduce the economic efficiency of the static-dynamic strategy. The results indicate that the multilevel problem setting and associate constraints lower the benefits of possible adaptations. Nevertheless, other reasons must be considered when selecting the uncertainty strategy, such as the flexibility of the production system and the planning effort. Additionally, since the variation coefficient has a decisive influence on economic efficiency, further research studies should address even heavier forecast errors.

References

1. Bookbinder, J. H., & Tan, J. Y. (1988). Strategies for the probabilistic lot-sizing problem with service-level constraints. *Management Science, 34*(9), 1096-1108. https://doi.org/10.1287/mnsc.34.9.1096
2. Mickein, M., & Haase, K. (2022). A novel demand difference adaptation policy for applying the static-dynamic uncertainty stratgey for the lot-sizing problem. *Working Paper*.
3. Helber, S., Sahling, F., & Schimmelpfeng, K. (2013). Dynamic capacitated lot sizing with random demand and dynamic safety stocks. *OR Spectrum, 35*(1), 75-105. https://doi.org/10.1007/s00291-012-0283-6
4. Mickein, M., & Haase, K. (2022). Technical summary: Scenario generation and solution approach for scenario-approximated problems. Preprint. https://doi.org/10.25592/uhhfdm.10799

Author Index

O. Grothe et al. (eds.), *Operations Research Proceedings 2022*, Lecture Notes in Operations Research, https://doi.org/10.1007/978-3-031-24907-5